Springer Theses

Recognizing Outstanding Ph.D. Research

Aims and Scope

The series "Springer Theses" brings together a selection of the very best Ph.D. theses from around the world and across the physical sciences. Nominated and endorsed by two recognized specialists, each published volume has been selected for its scientific excellence and the high impact of its contents for the pertinent field of research. For greater accessibility to non-specialists, the published versions include an extended introduction, as well as a foreword by the student's supervisor explaining the special relevance of the work for the field. As a whole, the series will provide a valuable resource both for newcomers to the research fields described, and for other scientists seeking detailed background information on special questions. Finally, it provides an accredited documentation of the valuable contributions made by today's younger generation of scientists.

Theses may be nominated for publication in this series by heads of department at internationally leading universities or institutes and should fulfill all of the following criteria

- They must be written in good English.
- The topic should fall within the confines of Chemistry, Physics, Earth Sciences, Engineering and related interdisciplinary fields such as Materials, Nanoscience, Chemical Engineering, Complex Systems and Biophysics.
- The work reported in the thesis must represent a significant scientific advance.
- If the thesis includes previously published material, permission to reproduce this must be gained from the respective copyright holder (a maximum 30% of the thesis should be a verbatim reproduction from the author's previous publications).
- They must have been examined and passed during the 12 months prior to nomination.
- Each thesis should include a foreword by the supervisor outlining the significance of its content.
- The theses should have a clearly defined structure including an introduction accessible to new PhD students and scientists not expert in the relevant field.

Indexed by zbMATH.

Paul Ramond

The First Law of Mechanics in General Relativity & Isochrone Orbits in Newtonian Gravity

Doctoral Thesis accepted by
Université Paris Cité, Paris, France

Springer

Author
Dr. Paul Ramond
Paris Observatory
Paris, France

Supervisors
Prof. Jérôme Perez
Unité de Mathématiques Appliquées
ENSTA Paris - Paris Observatory
Paris, France

Dr. Alexandre Le Tiec
Laboratoire Univers et Theories
Observatoire de Paris - CNRS
Paris, France

ISSN 2190-5053 ISSN 2190-5061 (electronic)
Springer Theses
ISBN 978-3-031-17963-1 ISBN 978-3-031-17964-8 (eBook)
https://doi.org/10.1007/978-3-031-17964-8

This Springer imprint is published by the registered company Springer Nature Switzerland AG
The registered company address is: Gewerbestrasse 11, 6330 Cham, Switzerland

« Mais on devient pas Chef parce qu'on le mérite, andouille ! On devient Chef par un concours de circonstances... et on le mérite après ! Il m'a peut-être fallu dix ans pour le mériter, mon grade, moi ! Si c'est pas vingt ! Tous les jours, j'ai travaillé pour pas nager dans mon uniforme ! Il y a pas trente-six solutions, Arturus : fais semblant. Fais semblant d'être Dux, fais semblant de mériter ta place, fais semblant d'être un grand Chef de Guerre. Si tu fais bien semblant, un jour, tu verras, t'auras plus besoin. »[1]

CÆSAR IMPERATOR à ARTURUS,
Kaamelott Livre VI
(2009)

1 "But one does not become a leader because one deserves it, you fool! One becomes a leader by a lucky turn of events... and then, much later, one deserves it! I had to wait about ten years to deserve my rank, if not twenty! Everyday I worked hard to fit in my uniform! There is no easy way around it Arturus: fake it. Fake being a Dux, fake deserving your rank, fake being a great War Leader. If you fake it well, you'll see: eventually you will make it." Kaamelott, Book VI (2009).

« Ce n'est que par la science et par l'art que valent les civilisations. On s'est étonné de cette formule : la science pour la science, et pourtant cela vaut bien la vie pour la vie, si la vie n'est que misère; et même le bonheur pour le bonheur, si l'on ne croit pas que tous les plaisirs sont de même qualité, si l'on ne veut pas admettre que le but de la civilisation soit de fournir de l'alcool aux gens qui aiment à boire... Toute action doit avoir un but. Nous devons souffrir, nous devons travailler, nous devons payer notre place au spectacle, mais c'est pour voir, ou tout au moins pour que d'autres voient un jour. »[2]

HENRI POINCARÉ,
la Valeur de la Science
(1905)

[2] "It is only through science and art that civilisation is of value. Some have wondered at the formula: science for its own sake; and yet it is as good as life for its own sake, if life is only misery; and even as happiness for its own sake, if we do not believe that all pleasures are of the same quality, if we do not wish to admit that the goal of civilisation is to furnish alcohol to people who love to drink. Every act should have an aim. We must suffer, we must work, we must pay for our place at the game, but this is for seeing's sake; or at the very least that others may one day see." Poincaré, the Value of Science (1905).

To Alexandre, Jérôme, Jérôme, Mathieu, Adam, Jacques, Tanja, Luc, Thierry, Jérôme, Philippe, Jeans-Michel, Micaela, Philippe, Éric, Karim and Aurélien: I am grateful for all your invaluable help and goodwill. This PhD marks not only the end of years of support from my family and friends, but also the beginning of a new life. A warm and loving "merci" to my Grandparents up in the stars, to my Mum and Dad, to my Brother and to my Wife.

Supervisors' Foreword

Classical theoretical physics celebrated its four hundreds birthday a few years ago. As a matter of fact, in May 1618, after centuries of qualitative description, Nature's regularity was correctly and quantitatively modelled for the first time by Johan Kepler, and crystallized in a unique law for all planets of the Solar system: his celebrated third law. Then, just half a century later, Isaac Newton proposed his own laws in the Principia and gave, among a lot of other things, the first mathematical demonstration of Kepler's findings.

After these pioneering works, physics (and particularly the field of mechanics) became analytic (in the mathematical sense) through the works of Lagrange, Hamilton, Jacobi, and later sublimed by Poincaré's dynamical systems' theory. In modern textbooks, theoretical classical mechanics is very often built and developed from the examination of two important examples: the gravitational two-body problem and the harmonic oscillator[3]. These fundamental problems share a peculiar property: their period of oscillation does not depend on the angular momentum of the orbit.

In the late fifties, the French physicist and dynamical astronomer Michel Hénon suggested that this property could be relevant to characterize the gravitational potential of stellar systems like globular clusters or spherical galaxies. In a paper full of clever ideas and insights, he found a gravitational potential that he named "isochrone", for it shared the same isochrone feature as the harmonic and two-body problem. Except for the fact that Hénon's model produced a complete analytical set of the variable to be used in numerical simulations, it did not really find its place in dynamical astronomy as it did not reproduce exactly the potential and radial density of realistic gravitational clusters, at least not during their entire dynamical evolution.

More recently, with two colleagues, we presented a modern version of Hénon's theory of isochrony. We extended many mathematical aspects of his pioneering work on isochrone potentials. In particular, new potentials with the isochrone property

[3] Interestingly, the former is but an example of the latter if one uses the Binet variable, i.e., the inverse of the radial distance.

were found and classified using elements of group theory and Euclidean geometry, which revealed a strong and one-to-one relation between these peculiar potentials and parabolas in the plane. After this rebirth of an old idea, a great and unsuspected landscape appeared in the field of classical mechanics: the Land of Isochrony. In my mind, we had explored the problem for such a long time and so thoroughly that, except for one or two dark clouds in the sky, this beautiful region of theoretical physics was fully mapped. The first cloud consisted of the geometrical content of the theory. Why parabolas are so central to isochrony? The second had to do with isochrone orbits themselves and their general description in a dynamical sense.

I did not expect that, after Paul's thesis, these problems would have been resolved in such an elegant and natural way. Moreover, entirely new and unexpected regions were discovered in the Land of Isochrony, with a new and synthetic way of presenting its most fundamental results through the Hamiltonian formulation as the pinnacle of this journey.

The results derived in this thesis show in a definitive manner that the century old, intricate symmetries associated with the harmonic and Kepler potentials (e.g., to name but a few, Kepler's three laws of motion, Bertrand's theorem, Kepler's equation, the Bohlin–Levi-Civita transformations, etc.) are actually subcases of a much broader paradigm that takes place in what one could simply call Isochrone Mechanics. The most simple and fundamental potentials that are presented in every mechanics textbook turn out to be degenerate cases of the set of isochrone potentials, and similarly for their elliptic orbits. In addition, one should not forget that Paul's work did not just re-invent the theory of Isochrony, he also went a long way into the history of the concept, revealing the modernity of Galileo and Huygens' works on the matter, in an insightful and thorough historical account.

What is even more remarkable is that all this work on isochrony is just a part (perhaps two fifth?) of his thesis' content. Indeed, this manuscript contains not one, not two, but at least three pieces of work that could have been independently published as a thesis. Through carefully crafted figures, well-written phrases and elegant equations, I hope that the reader will take as much pleasure to turn these pages as I have had while sharing these three singular years with Paul.

Paris, France
April 2022

Prof. Jérôme Perez

Ever since September 14, 2015, the direct detection of gravitational waves generated by the inspiral and merger of binary systems of compact objects, composed of black holes or neutron stars, has marked the dawn of a new era in astronomy. Over the last seven years, the sensitivity of up and running ground-based detectors has been steadily increasing, and planned space-based missions such as LISA will continue to bring new data, allowing the exploration of relativistic dynamics and fundamental physics in extreme astrophysical environments. Crucially, such gravitational-wave observations are made possible thanks to highly accurate template waveforms, built by solving the Einstein field equations using a combination of numerical relativity simulations and analytical approximation methods.

Back in 2012, along with two colleagues, I uncovered a simple and mathematically elegant variational formula that captures key aspects of the gravitational dynamics of binary systems of compact objects that move along a sequence of quasi-circular orbits, the slow adiabatic inspiral being driven by gravitational-wave backreaction. This so-called 'first law of compact binary mechanics', coined by analogy with the celebrated first law of black hole mechanics, relates some kinematical properties of the individual compact bodies to dynamical properties of the binary system, such as the gravitational binding energy and the orbital angular momentum. Over the last decade, this first law of mechanics has proven instrumental in improving our understanding of the orbital motion of such gravitational-wave sources, in a regime of strong gravitational fields and relativistic speeds.

However, until recently, the domain of validity of the first law was essentially restricted to the simplest case of nonspinning compact binaries modelled as structureless point masses. While this idealized description captures the main physical effects associated with the orbital dynamics of such relativistic systems, astrophysical black holes are known to carry significant spins and neutron stars possess a nontrivial internal structure, which manifests itself, for instance, through spin-induced rotational flattening and tidally-induced quadrupolar (and higher multipolar order) deformations.

The main challenge tackled by Paul Ramond during his Ph.D. has precisely been to extend the 'monopolar' first law of compact binary mechanics to account for such spin and internal structure effects. Physically, this can be achieved by endowing each compact body with a set of multipole moments that encode the proper rotation state and the tidal deformability. The idea of such a 'multipolar gravitational skeleton' dates back to the pioneering work of Mathisson during the 1930s, and was later refined by Papapetrou and Tulczyjew during the 1950s. Whenever the typical linear size of each compact object is much smaller than the ambient curvature length scale, the lowest order multipoles are dominant, and the multipolar expansion can safely be truncated.

As a first step, Paul worked at the dipolar order, thereby accounting for the leading spin effects, and showed that his elegant, differential geometric derivation of the first law is in full agreement with a related result previously derived in the context of the canonical Hamiltonian formalism, applied to a binary system of gravitationally interacting, spinning point particles. Moreover, by exploiting the systematic and algorithmic nature of the multipolar gravitational skeleton framework, Paul then pushed his calculations at the next quadrupolar order and derived preliminary results that will most certainly prove very useful, once completed, to constrain the equation of state of neutron stars through gravitational-wave observations.

Readers from various backgrounds will find that Paul Ramond' thesis is very well-structured and remarkably well-written in a very clear and pedagogical style. All mathematical developments are completed by illuminating physical interpretations, and the work accomplished is of impressively high scientific level. It has been a great pleasure for Jérôme Perez and myself to guide Paul Ramond in these doctoral projects, and his remarkable results are a source of real joy and satisfaction.

Meudon, France
July 2022

Dr. Alexandre Le Tiec

Introduction

The material presented in this thesis summarizes results obtained for two problems of gravitational dynamical systems. The first one, presented in the first part entitled *The First Law of Mechanics in General Relativity*, aims at exploring and extending an important variational equation in the context of relativistic mechanics. A second part, entitled *Isochrone Orbits in Newtonian Gravity*, aims at solving a problem of mathematical physics motivated by astrophysical considerations, in the Newtonian (non-relativistic) context. Although both topics are rather independent, they do share one common, central feature, *symmetry*, which will be the guiding principle throughout this thesis.

The First Law of Mechanics in General Relativity

Einstein's theory of General Relativity has never been more alive. As I am writing these lines in 2021, it seems that every two years in the last decade has seen an unprecedented, astonishing breakthrough in relativistic astrophysics or cosmology. In 2013, the unprecedentedly accurate heat map of the early Universe was gathered by the Planck satellite. In 2015, the first direct detection of gravitational waves, emitted by a black hole binary, was made by LIGO's two giant interferometers. In 2017, the coalescence of two neutron stars was simultaneously observed by a hundred telescopes in the electromagnetic spectrum and three different gravitational-wave detectors. In 2019, the first direct picture of a supermassive black hole's accretion disk was released by the Event Horizon Telescope network. These revolutionary steps in the everlasting quest for knowledge would not have been possible without our modern understanding of the theory of General Relativity.

In particular, these discoveries regarding gravitational waves are the result of decades of theoretical and practical effort to build, not only the detectors, but also an effective method to extract, from their output, gravitational-wave signals. In this context, a fundamental feature has been the ability to accurately model these gravitational waves from various types of sources. A number of methods have been devised and used conjointly to solve the Einstein equation, the fundamental equation of General Relativity, in order to extract from its intrinsic nonlinearities the shape and form of gravitational-wave signals emitted by binary systems of compact objects. These objects, typically black holes and neutron stars, were for a long time among the most promising sources of gravitational waves. Since 2015, they have fulfilled this promise: around fifty binary systems of compact objects have been observed directly through the detection of gravitational waves.

Although they rely on different assumptions, the domains of applicability of all these theoretical frameworks overlap, and are becoming wider with time. It is fundamental to understand the interplay between them and to compare their outputs. To perform these comparisons, one key ingredient has been the so-called "first law of mechanics". This equation, which relates various physical quantities of interest characterizing a system of compact objects, gets its name from an analogy with the first law of thermodynamics. In General Relativity, this "first law" was first derived in the 1970s for isolated black holes. Since then, it has been generalized to numerous, more complicated systems, including compact object binaries. In spite of the different gravitational systems they describe, the various forms that the first law embodies are strikingly similar, and their simplicity, given the nonlinear intricacies of General Relativity, is rather remarkable.

The first part of this thesis is dedicated to an exploration of the First Law of Mechanics, and its generalization for binary systems of extended compact objects on circular orbits. The six chapters contain the following material:

- In Chap. 1, we present a historical summary of the century preceding the advent of gravitational-wave astronomy, from Einstein's original publications on General Relativity to the most recent gravitational-wave detections to date. This chapter contains a lot of historical references and pedagogical illustrations of key concepts of gravitational-wave physics.
- The content of Chap. 2 is a thorough discussion of the model used to describe compact objects and derive their relativistic equations of motion. In a nutshell, this model replaces extended compact objects with point particles endowed with a finite number of characteristics (mass, spin, etc.). First, in Sects. 2.1 and 2.2 we present an overview of the different approximation schemes this multipolar-point-particle model comes from. Then, in Sect. 2.3 we summarize the output results from this model that shall be used in subsequent chapters.
- In Chap. 3 we explore the notion of circular orbit as enforced by the existence of a helical Killing vector field. In particular, Sect. 3.1 contains a discussion on this circular hypothesis, Sect. 3.2 contains two key results about the behavior of multipolar particles in helically symmetric spacetimes, and Sect. 3.3 explores their geometrical consequences for spinning particles.
- Chapter 4 is dedicated to the first law of mechanics. The first two sections contain an overview of the different forms of various first laws found in the literature (Sect. 4.1), while Sect. 4.2 focuses on their applications, illustrating how these laws are used in practice. Lastly, Sect. 4.3 contains the derivation of a

central identity, the first law of mechanics for arbitrary matter fields in helically symmetric spacetimes.

- In Chap. 5, we combine all the ingredients of the previous three chapters to derive the first law of mechanics for multipolar particles at dipolar order. This derivation is detailed in Sect. 5.1, which contains the most general formulation of this law. Then, in Sects. 5.1 and 5.3, we explore some consequences of this law and provide alternative formulations of it.
- Lastly, Chap. 6 contains the integral formulation of the first law of mechanics at quadrupolar order, taking into account finite-size effects of the body described by the particle. First, in Sect. 6.1 we discuss two sources of finite-size effects, namely spin-induced and tidally-induced quadrupoles. Then, the integral law is derived in Sect. 6.2 for an arbitrary quadrupole model, applied to our spin/tidal case and shown to coincide with that found in the literature in the dipolar limit. Lastly, Sect. 6.3 presents the anticipated result of the first law at quadrupolar order.

Isochrone Orbits In Newtonian Gravity

In the late 1950s, the French astronomer and astrophysicist Michel Hénon proposed a very simple model describing a peculiar gravitational system known as a globular cluster. Although his motivations were purely practical, unknowingly, Michel Hénon provided, in his seminal paper on the topic, the seed for a very rich and beautiful problem of mathematical physics: *the (celestial) isochrone problem.* The question that one tries to answer is the following: in classical gravity, are there any spherically-symmetric, gravitational potentials that generate bounded orbits with a radial period independent of one of the constants of motion? The answer to this question is *yes*. For example, in a harmonic potential, the radial period is independent of any constant of motion. A less trivial example is the classical two-body problem, in which the radial period only depends on the conserved mechanical energy of the system, and not on its conserved angular momentum.

Remarkably, Michel Hénon found the third potential with this property. The problem then remained untouched for decades, as Hénon and the community were satisfied enough with his findings on the one hand, and because it was not obvious that Hénon's question could be turned into a mathematical physics' problem with much richer answers. Only in the late 2010s did this problem re-emerge, as it was explored by Alicia Simon-Petit, Jérôme Perez and Guillaume Duval. Hénon's isochrone problem turned out to be much richer than expected, and another family of isochrone potentials was found.

The second part of this thesis is dedicated to a thorough exploration of the isochrone problem in celestial mechanics. The four chapters contain the following material:

- In Chap. 7, we give a historical overview of the most well-known notion of isochrony in physics: the (near-)isochrony of the pendulum. In particular, we revisit Huygens' remarkable work of the isochrone, cycloid pendulum in Sect. 7.1, giving valuable insight into the solution to the celestial isochrone problem. Then, in Sect. 7.2 we provide some necessary preliminaries, as well as the astrophysical context that motivated Hénon's work, and a precise statement of the problem to be solved (Sect. 7.3).
- The content of Chap. 8 is a geometrical solution to the isochrone problem. By extending Hénon's original method, we are able to make a one-to-one correspondence between isochrone potentials and parabolae in the plane, using a characterization of parabolae due to Archimedes (Sect. 8.2). We then classify the complete set of isochrone potentials into five distinct families in Sect. 8.3.
- In Chap. 9, we use our classification of isochrone potentials to study the dynamics of test particles within them. In particular, in Sect. 9.1, we show that all isochrone potentials satisfy Kepler's third law for the period and a similar one for the apsidal angle. Then, in Sect. 9.2, we solve the equations of motion in a parametric form, using geometrical constructions with parabolae. Isochrone orbits are then depicted and classified in Sect. 9.3, allowing for a comparison of our analytical results to a direct numerical integration of the equations of motion.
- Lastly, in Chap. 10 we revisit the isochrone problem from the point of view of Hamiltonian mechanics. This analysis completely solves the problem using a well-adapted set of angle-action coordinates (Sect. 10.1). We show, in particular, that all isochrone orbits satisfy a generalization of Kepler's equation. Then, applying to the problem the powerful machinery of Birkhoff normal forms in Sect. 10.2, we derive in a self-consistent way a number of central isochrone results in Sect. 10.3, and show how they are elegantly encoded into the Birkhoff invariants of the associated isochrone Hamiltonian.

The Abel–Ruffini Theorem

On an unrelated matter, the reader will also find in Appendix D a new, geometric, group-theory-free proof of the Abel–Ruffini theorem, which states that the general solution to an algebraic equation of degree five or more cannot be written using radicals, that is, using its coefficients and arithmetic operations $+$, $-$, $\times$, $\div$ and $\sqrt{}$. This proof is based on the ideas of Vladimir Arnold, and only requires knowledge of complex numbers, whereas traditional, textbook proofs rely on advanced mathematics. It is completely self-contained and has the advantage of giving some insight as to why closed-form formulae exist for equations of degree four or less (and how they are constructed), and why they do not for degree five or more.

How to Read This Thesis

The work presented in the first part of this thesis is based on an (ongoing), three-part series of papers that, ultimately, aim at deriving the first law of mechanics at quadrupolar order for a circular binary system of compact objects. This program was subdivided into a derivation of key necessary, geometrical results (Paper I), a derivation of the master variational identity and application to dipolar particles (Paper II), and the derivation of the first law at quadrupolar order (Paper III). We chose, in order to present this work, not to include into the thesis these articles in their entirety, after an introductory paragraph. Rather, we followed a more traditional way, dispersing our work (and thus the papers' content) throughout a structured, and more logical sequence. This leads to a subdivision into six chapters, and the reading logic is depicted in the top part in Fig. 1. The color-coding in this figure is such that, after an introductory chapter (Chap. 1), the content of Paper I is in blue (Chap. 3), Paper II in red (Chaps. 2, 4 and 5) and Paper III in orange (Chap. 6). Note however that a number of results presented in the latter are still preliminary, due to ongoing calculations at quadrupolar order.

Regarding the second part of the thesis, the problem at hand was to explore and solve as thoroughly as possible the isochrone problem bequeathed by Michel Hénon. This was done in a two-part series of papers, where we first elaborated on previous works in order to complete the classification of isochrone potentials, derive a purely geometrical proof of the isochrone theorem and solve the equations of motion (Paper I). Then, we revisited the problem from a Hamiltonian point of view, providing an ultimate, self-consistent solution to Hénon's isochrone problem and revealing the profound symmetries of isochrone mechanics (Paper II). Like the first part of the thesis, all these results are dispersed to follow the logic presented on the lower part in Fig. 1. After an introductory chapter (Chap. 7), the same color-coding applies: the content of Paper I is in blue (Chaps. 8 and 9), and that of Paper II is in red (Chap. 10).

Material presented in this thesis has been published in peer-reviewed journals. For the First Law of Mechanics, Paper I has been published in *Quantum and Classical Gravity* [1], Paper II in *Physical Review D* [2] and Paper III is still being written. Regarding Isochrony, Paper I has been published in the journal *Celestial Mechanics and Dynamical Astronomy* [3], and Paper II in the *Journal of Mathematical Physics* [4]. Finally, the Paper on the Abel-Ruffini theorem has been published in *The American Mathematical Monthly* [5].

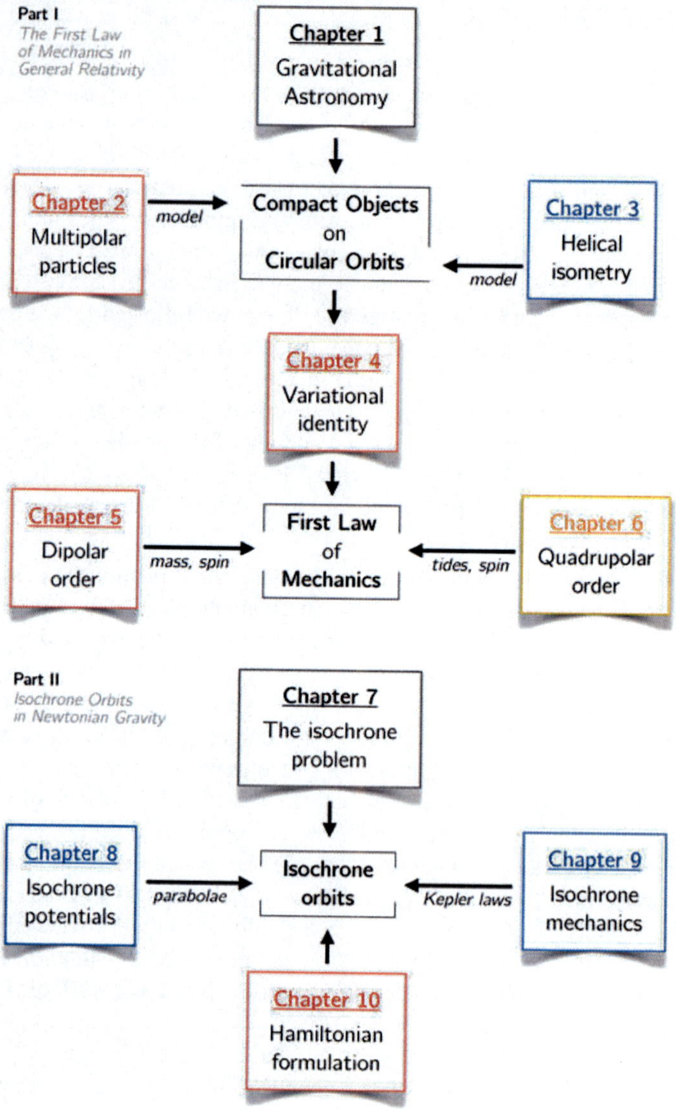

Fig. 1 General organization of the thesis

Contents

Notations

Differential Geometry

As is customary with differential geometry in GR, we specify our conventions. They follow those of Wald [6]: the metric signature is $(-, +, +, +)$, the Riemann tensor R_{abcd} satisfies $2\nabla_{[a}\nabla_{b]}\omega_c = R_{abc}{}^{d}\omega_d$ for any 1-form ω_a, and the Ricci tensor is defined by $R_{ab} \equiv R_{acb}{}^{c}$. Except in rare, mentioned cases, in the first part of this thesis we always use geometric units, such that $G = c = 1$.

Tensorial Indices

For tensor indices, we always use abstract indices using letters $(a, b, c, \dots)$ from the beginning of the Latin alphabet. Other types of indices include the following:

- Greek letters $(\alpha, \beta, \gamma, \dots)$ vary in $\{0, 1, 2, 3\}$ and denote spacetime components of tensor in a given basis,
- letters $(i, j, k, \dots)$ from the middle of the Latin alphabet denotes internal Lorenz indices, (for example, an orthonormal triad is denoted (e_i^A), with $i \in \{1, 2, 3\}$),
- capital Latin letters $(A, B, C, \dots)$ from the beginning of the Latin alphabet are mere indexing labels varying between 0 and 4,
- capital Latin letters $(K, L, M, \dots)$ from the middle of the alphabet denote multi-indices of length $k, \ell, m, \dots$,
- the Roman font subscript $\mathrm{i} \in \{1, 2\}$ labels the two components of a binary system,
- index-free, boldface symbols denote differential forms, as in $\mathbf{X} = X_{a_1 \cdots a_p}$,
- the covariant derivative ∇ is always the Levi-Civita, metric compatible one, and the classical abuse of notation $(\nabla\omega)_{ab} = \nabla_a \omega_b$ is used throughout,
- the Lie derivative along an arbitrary vector field ξ^a is denoted $\mathcal{L}_\xi$,
- the exterior derivative of a p-form $\mathbf{X}$ is denoted $\mathrm{d}\mathbf{X} = (\mathrm{d}\mathbf{X})_{a_1 \cdots a_{p+1}}$,
- subscripts that encode an information of a quantity (e.g., Ξ_{BH} or J^{abcd}_{spin}) are sometimes denoted with sans-serif font and should not be confounded with indices.

Miscellaneous

The symbol $\equiv$ is used when the quantity on the right defines the quantity on the left. We sometimes use index-free notation when there is no ambiguity. For example, the dependence of the Einstein tensor on a metric may be written $G_{ab}[g]$ instead of $G_{ab}[g_{bc}]$, or the parametrization of a curve $x = z(\lambda)$ instead of $x^\alpha = z^\alpha(\lambda)$ with $\alpha \in \{1; 4\}$. All these should be obvious from the context.

Acronyms

We gather here the most frequently used acronyms:

BH	Black Hole
BNF	Birkhoff Normal Form
EM	Electro-Magnetic
EMRI	Extreme Mass Ratio Inspiral
GKF	Generalized Killing Vector Fields
GR	General Relativity
GW	Gravitational Wave
HKV	Helical Killing Vector
IMRI	Intermediate Mass Ratio Inspiral
KAGRA	KAmioka GRAvitational wave detector
LIGO	Laser Interferometer Gravitational wave Observatory
LISA	Laser Interferometer Space Antenna
MPTD	Mathisson–Papapetrou–Tulczyjew–Dixon
NS	Neutron Star
ODE	Ordinary Differential Equation
PDE	Partial Differential Equation
PN	Post-Newtonian
SEM	Stress–Energy–Momentum
SF	Self-Force
SMBH	Super-Massive Black Hole

Symbols Used in Part I

The following table gathers symbols used frequently in Part I of the thesis, covering Chap. 1 through Chap. 6, and the dedicated conclusion, in Sect. 10.3.3. (Table 1)

Table 1 List of frequently used symbols in Part I of the thesis

Symbol	Description	Definition
Sets		
$\mathscr{E}$	Spacetime manifold	
$\mathscr{V}$	4-volume in $\mathscr{E}$	
$\mathscr{S}$	3-hypersurface in $\mathscr{E}$	
$\mathscr{U}$	2-surface in $\mathscr{E}$	
γ	Particle worldline	
x, x'	Points in $\mathscr{E}$	
z, z'	Points on γ	
Geometry		
g_{ab}	Metric tensor	
∇_a	Covariant derivative	
ε_{abcd}	Canonical volume form	
R_{abcd}	Riemann curvature tensor	
E_{ab}	(Gravito-)electric part of R_{abcd}	(6.5)
B_{ab}	(Gravito-)magnetic part of R_{abcd}	(6.5)
Particle		
τ	Proper time	
$u^a, \dot{u}^a$	4-velocity, 4-acceleration	
p^a, S^{ab}	4-momentum, spin tensor	
S^a, D^a	Spin vector, mass dipole	(2.63)
J^{abcd}	Quadrupole tensor	
T^{ab}	Stress-energy-momentum tensor	
h_{ab}	Projector orthogonal to u^a	(2.62)
δ_4	Invariant Dirac distribution	(A.17)
Multipoles		
$T^{abc_1\cdots c_\ell}$	2^ℓ-pole of T^{ab} (Ansatz)	(3.16)
$\mathscr{T}^{abc_1\cdots c_\ell}$	2^ℓ-pole of T^{ab} (normal form)	(3.17)
$\tilde{\mathscr{T}}^{abc_1\cdots c_\ell}$	Geodesic extension of $\mathscr{T}^{abc_1\cdots c_\ell}$	(3.19)
$\mathscr{L}^{abc_1\cdots c_\ell}$	Lie derivative of $\tilde{\mathscr{T}}^{abc_1\cdots c_\ell}$ along k^a	(3.19)
Isometry		
ξ^a	Generic Killing vector	
$\mathcal{L}_\xi$	Lie derivative along ξ^a	(A.20)
k^a	Helical Killing vector	(3.7)
z	Redshift parameter	(3.32)

Symbols Used in Part II

The following table gathers symbols used frequently in Part II of the thesis, covering Chap. 7 through Chap. 10, and the dedicated conclusion, in Sect. 10.3.3. (Table 2)

Table 2 List of frequently used symbols in Part II of the thesis

Symbol	Description	Definition
Classical mechanics		
(r, θ)	Polar coordinates (orbital plane)	
$\psi(r)$	Radial potential	(7.15)
ξ	(Mechanical) specific energy	(7.16)
Λ	Specific angular momentum	(7.17)
J	Radial action	(7.22)
T	Radial period	(7.20)
Θ	Apsidal angle	(7.21)
r_A, r_P	Apoastron, periastron	p. 211
Isochrone potentials		
$x = 2r^2$	Hénon variable	(8.3)
$Y(x)$	Potential in Hénon's variable	(8.3)
(a, b, c, d, e)	Parabola parameters	(8.27)
δ	Parabola discriminant	(8.27)
$(\varepsilon, \lambda, \omega, \mu, \beta)$	Potential parameters (Greek)	(8.3.2)
x_v	Abscissa of vertical tangent	(8.31)
Isochrone orbits		
E	Eccentric anomaly	(10.18)
ε	Eccentricity	(10.19b)
α	Semi-major axis	(10.22)
$\Omega = 2\pi/T$	Angular frequency	(10.17)
Hamiltonian mechanics		
H	Hamiltonian	
(J, Λ)	Action variables	p. 275
$N(J)$	Birkhoff normal form	(10.33)
$\mathfrak{b}_i$	Birkhoff invariant	(10.54)

Part I
The First Law of Mechanics in General Relativity

Chapter 1
Gravitational Astronomy

Contents

Proper separation oscillation between two freely falling bodies could be used in principle to detect the passage of a gravitational wave, as monitored by laser interferometry. However, such detectors have so low a sensitivity that they are of little experimental interest.

C. Misner, K. Thorne and A. Wheeler, Gravitation (1973)

GALILEO'S first celestial observations in 1609, using the freshly invented refracting telescope, can be rightfully called the birth of electromagnetic astronomy. His observations, among which new stars, the Moon's mountains and Jupiter's four biggest natural satellites, were published on 13 March 1610 in a short astronomical treatise,

P. Ramond, *The First Law of Mechanics in General Relativity & Isochrone Orbits in Newtonian Gravity*, Springer Theses,
https://doi.org/10.1007/978-3-031-17964-8_1

Sidereus Nuncius. Four-hundred and five years later, at 09:50:45 UTC on September the 14th, a new type of astronomy was born. The coalescence of two black holes emitted around 3 Solar masses worth of gravitational radiation and provided the first direct detection of a gravitational wave, whose existence was predicted in 1916 by Albert Einstein, as a consequence of his general theory of relativity. The work presented in the first part of this thesis is placed in this context of gravitational astronomy. The aim of this first chapter is to provide some background and context about the field. We will start, in Sect. 1.1, by a brief historical account on Einstein's general theory of relativity. Then, in Sect. 1.2, the focus will be on the theory of gravitational waves. Lastly, in Sect. 1.3 we will discuss what the direct detection of these waves has brought to science since 2015, and the exciting future that awaits.

1.1 Einsteinian Gravitation

Like every major scientific revolution, the advent of the theory of General Relativity is one of the most remarkable achievement in the history of mankind. What led Albert Einstein to the discovery/invention[1] of his theory is a fascinating story, details of which are still being discussed and debated by science historians today. I have found myself very often baffled by the ingenuity of Einstein and the serendipitous nature of his quest. Before going into details that are much more easily conveyed by equations in the following sections, I would like to start this journey with a brief, succinct summary of this remarkable piece of scientific history: the backstory of General Relativity.

1.1.1 Backstory

The advent of the theory of special relativity at the beginning of the nineteenth century marks the dawn of a new chapter in Natural Sciences, that of *modern physics* [7, 8]. On the one hand, special relativity would provide the four-dimensional spacetime structure on which *quantum mechanics* would be based, and eventually account for three of the four fundamental interactions. On the other hand, special relativity would be generalized in a far richer theory called *general relativity* (GR), providing the most accurate description of the fourth interaction, gravity. Remarkably enough, only one man can be associated to each of these three fundamental, theoretical constructions. This man is Albert Einstein. While his name resonates with that of Planck, Heisenberg and Schrodinger for the foundations of quantum mechanics, and with Poincaré, Lorentz and Minkowski for special relativity, the development of GR can be largely attributed to Einstein alone, albeit for small but crucial interactions with his friend

[1] We shall skip over the eternal debate of whether scientific facts are discovered, invented, or both, and let the reader decide for himself.

Marcel Grossmann. Traditionally, it is said that it took eight years of intense thinking and work for Einstein to construct his theory [9]. The beginning of Einstein's quest is set somewhere in the summer of 1907 while he was in Bern, and its completion on May 11 1916, when the first final, mistake free review of GR was published. Today, historians and scientists have accomplished a remarkable work in trying to understand how Einstein constructed, piece by piece, this remarkably simple and elegant theory of space, time and gravitation. These eight years of intense work can be split into three rather distinct phases [10]:

- the implications of the equivalence hypothesis in 1907,
- the use of Riemannian geometry to encode gravitation in 1912,
- the search for covariant field equations in 1915.

The following brief account on the genesis of GR is mostly based on Einstein's almost-autobiography [11], the two biographies [12, 13], the following historical references [9, 14–18] and Einstein's numerous publications. The large majority of his original scientific work, as well as English translations, can all be found on the Digital Einstein Papers web page [19], provided by the Princeton University as part of the Einstein Papers Projects [20].

Equivalence Hypothesis

Although Einstein's will to "generalise" the special theory of relativity[2] was motivated by numerous reasons, two of them seem to stand out. On the one hand, Newton's Universal Law of Gravitation was a two century old, perfectly well-tested and verified theory of gravitation.[3] However, its instantaneous, action-at-a-distance nature was clearly contradictory to the principles of SR. On the other hand, Einstein was not satisfied with the idea that, just as in Galilean relativity, a particular set of reference frames seemed to stand out of SR: the non-accelerated, inertial frames of reference. Gravitation and acceleration. Two words that resonate as two sides of the same coin today, but which were on Einstein's mind already in 1907 and would have to wait eight years to be unified geometrically in the context of GR.

The Equivalence Hypothesis (EH), nowadays called the Equivalence principle, was first introduced by Einstein himself in 1907 (see in particular the paragraph 17 in Einstein's account [21] on the principle(s) of relativity). At that time, Einstein had found[4] a job as a patent examiner at the Swiss Patent Office, in Bern. There, after lunch, he had the habit of doing what he called "semi-awaken naps", an activity consisting in letting his body at rest but otherwise virtually "physically feel" situations he would put himself in [14]. Einstein performed these "Gedankenexperi-

---

[2] Interestingly, Einstein mostly uses the phrasing *General Theory of Relativity* instead of the now common GR, thinking of the *relativity theory* as general, rather than *relativity* itself.

[3] Except for Mercury's perihelion advance, which we will get back to shortly.

[4] Thanks to the father of his friend Marcel Grossmann [16], who we will encounter again soon.

mente" (thought experiments) quite often and they were of prime importance while developing his ideas on special relativity [17].

The patent office's Gedankenexperiment consisted in reinterpreting Galileo's law of free fall in two frames of references: one free-falling with a man and his hand tools from a building's roof, and one at rest with the building. As Galileo proved two centuries before, both the man and his hand tools would fall at the same rate. Einstein then realized that the unfortunate roofer may as well consider himself at rest, for all his tools would remain at hand's reach during the fall. With this thought, Einstein realized that no local experiment could be able to distinguish between a reference frame free-falling in an homogeneous gravitational field, and one that is uniformly accelerated. This equivalence between a static, homogeneous gravitational field and a uniformly accelerated frame of reference, suggested to transpose any gravity-related fact into a geometrical-dependent one.

Metric Tensor

Until 1911, Einstein, who was now a Physics Professor at the German University of Prague, did not published a lot on the "problem of gravity" as he calls it. He was rather occupied with various topics in (what would later become) statistical physics and quantum mechanics [13]. In the mean time, people started to look into the consequences of the new theory of special relativity on Newtonian gravitation. Notable works and insight were provided by M. Abraham, G. Nordström and M. van Laue on the topic, with various attempts to construct a relativistic version of the Poisson equation (therefore opting for a scalar gravitational field). These theories were all successful in deriving covariant field equations that reduced to Newtonian gravity, but lacked the (rather important) ability to predict correctly light deflection and periastron advance (cf. historical notes in [22]).

Einstein got back to the gravitation problem in 1911. By examining the new status of non-inertial forces (Coriolis and centrifugal) in the light of the equivalence principle, and examining the role of a varying-speed of light theory to encode the gravitational field, Einstein became convinced to quit the realm of flat, affine geometry, be it Euclidean or Minkowskian. In particular, he realized sometime around August 1912 that it was necessary to consider the more general notion of differentiable manifold. Fearing that his mathematical skills would not be enough to understand this new[5] geometry, he asked for help to his good friend and mathematician Marcel Grossmann. The two accomplices ventured together into the realm of Riemannian geometry, or, in this case, Lorentzian geometry. Grossmann guided Einstein to the recent literature on differentiable manifolds and showed him how they are akin to the use of curved coordinates in affine spaces (like in SR), which Einstein knew. This part of Einstein's

[5] At the time, differential geometry was still new, following the founding works of Bernhard Riemann [23] and Felix Klein's Erlangen program [24] in Germany, as well as the great school of Italian geometers, where Luigi Bianchi [25] and then Tullio Levi-Civita [26] played a central role. I refer to Chap. 7 of the interesting book [27] for more historical details.

quest for a theory of gravity was the most intense, mentally and physically, as he said himself [11]. The reward for this intense and abstract interlude was the conviction that a relativistic theory of gravity would necessarily be geometrical, and the gravitational field of tensorial nature.

Covariant Field Equations

Einstein and Grossmann built the field equation from the Riemann curvature tensor first. They arrived at a first equation, of the form $R_{ab} = \kappa T_{ab}$, which was rapidly discarded as it was not able to recover Newton's law of gravitation in the weak-field, slow-speed limit. After some refinement, the closest thing they could get to a covariant theory with the proper Newtonian limit was crystallized in an "*Outline* of a Generalized Theory of Relativity and of a Theory of Gravitation" published in 1913 [28]. This work contains many traits that GR would inherit a few years later, among which, a mathematical introduction of tensor calculus, the use of the metric tensor to encode the gravitational field, and field equations (but not fully covariant and incorrect). Together, these three features came to be known as the "Entwurf" theory (Entwurf is German for "Outline") and would later be the basis on which Einstein would work to develop the final form of GR (notably the correct field equations). This theory was mature enough in Einstein 's mind that he wrote a synthetic review of it in 1914 [29], using for the first time the qualifier "general" (and not "generalised") for the new principles of relativity it contained.

The Entwurf theory was wrong. Yes, it had the correct Newtonian limit, but it predicted the incorrect value of Mercury's perihelion advance, and Einstein knew this from the beginning. Yet, having found nothing better, for months after the publication of the Entwurf theory he would not hesitate to take the (incorrect) consequences of his theory as features that would be "expected" from a relativistic theory of gravitation, hence the infamous "hole argument" to justify the non-covariance of the field equation [10]. In 1914, errors in the covariance proof of the Entwurf field equations were reported by others. Adding to this the fact that the special relativity spacetime in rotating coordinates was found to contradict the new field equations, Einstein abandoned the Entwurf theory and began looking for new (correct and fully covariant) field equations.

Reading the first review of 1914 [29] on his general theory of relativity and the four successive papers of November 1915[6] reveals an interesting quirk. It is almost like Einstein is explaining in each publication how the previous one was wrong and how he is trying to convince the reader that he is on the right track. His issue was to find an equation that was at once *correct* (reproducing well-known results) and *fully covariant* (not restricted to a class of coordinates). But these two problems had to be solved simultaneously, for the correct equation would readily be fully covariant. The key was to include a new term in the equation. Einstein did not account for this term

[6] Those papers were submitted on the 4th [30], 11th [31], 18th [32] and 25th [33], each Thursday of every week !.

Die allgemein kovarianten zehn Gleichungen des Gravitationsfeldes in Räumen, in denen »Materie« fehlt, erhalten wir, indem wir ansetzen

$$G_{im} = 0. \qquad (2)$$

Ist in dem betrachteten Raume »Materie« vorhanden, so tritt deren Energietensor auf der rechten Seite von (2) bzw. (3) auf. Wir setzen

$$G_{im} = -\varkappa\left(T_{im} - \frac{1}{2} g_{im} T\right), \qquad (2a)$$

wobei

$$\sum_{\varrho\sigma} g^{\varrho\sigma} T_{\varrho\sigma} = \sum_{\sigma} T_{\sigma}^{\sigma} = T \qquad (5)$$

gesetzt ist; T ist der Skalar des Energietensors der »Materie«, die rechte Seite von (2a) ein Tensor.

Fig. 1.1 The first appearance of the (correct and complete, cf. text) Einstein equation, published by Albert Einstein himself on December 2nd 1915 [33]. Equation (a) is the vacuum field equation (G_{im} stands for the Ricci tensor, not the Einstein tensor), and equation (2a) includes the SEM tensor for non-vacuum spacetimes. The minus sign in front of the Einstein constant (χ here) comes from the metric signature $(+, +, +, -)$ used by Einstein

until late 1915, even though Grossmann already hinted that it could be the solution to achieve covariance, two years earlier. In Einstein's defense, this term vanishes in many cases, including the case of electromagnetic sources. In any case, we find on the first page of the last November paper [33]:

> *I now quite recently found that one can get away without this hypothesis about the energy tensor of matter merely by inserting it into the field equations in a slightly different way,*

the hypothesis in question being, essentially, that matter behaves like electromagnetic fields. With this last improvement, the search for a relativistic theory of gravity was finally over, and GR was officially born. The final form of field equations were published on page 1 and 2, right after Einstein's comment. They are depicted in Fig. 1.1, directly extracted from Einstein's publication.

1.1.2 Einstein's Equation

After various unsuccessful attempts to find the correct equation for his general theory of relativity, the exact field equations are finally published on December 2nd of 1915 (see Fig. 1.1). Today, they are collectively known as the Einstein (field) equation, and it is customary to write it slightly differently than its initial formulation. It reads

$$G_{ab} + \Lambda g_{ab} = \kappa T_{ab} \,, \qquad (1.1)$$

where Λ and κ are two fundamental constants, G_{ab} depends explicitly on the metric tensor g_{ab} which, in turn, contains all information about the geometry of some region of spacetime $\mathscr{E}$, and T_{ab} is a tensor encoding all (non-quantum) properties of the matter and/or energy content in $\mathscr{E}$. In physical terms, we cannot do much better than Archibald Wheeler's famous words to explain what Eq. (1.1) truly means: *spacetime grips mass, telling it how to move, mass grips spacetime, telling it how to curve.*[7] Indeed, the left-hand side of (1.1) contains all the information on the curvature of spacetime, while the right-hand side encodes the physical characteristics about the mass (and energy) distribution in this region of spacetime. Equation (1.1) is one of the two fundamental postulates of GR, the second being that spacetime $\mathscr{E}$ is well-described by a four-dimensional, torsion-free, Lorentzian manifold $(\mathscr{E}, g_{ab})$, whose metric g_{ab} is the fundamental object, and main unknown of (1.1). The Einstein equation is either postulated as such or derived from an equivalent action postulate formulation.[8]

The Constants

Equation (1.1) is the most general form of Einstein's equation that satisfies all defining principles of GR. In particular, it includes two fundamental constants: κ and Λ. The constant κ is called Einstein's constant and can be expressed in terms of other fundamental constants of physics by examining the low-velocity and weak-field limit of (1.1). In SI units, it is given in terms of Newton's gravitational constant G and the vacuum speed of light c as

$$\kappa \equiv \frac{8\pi G}{c^4} \quad \Rightarrow \quad \kappa \simeq 10^{-43} \frac{\mathrm{m}^{-2}}{\mathrm{J/m}^3}, \tag{1.2}$$

elegantly expressing the fact that GR is a geometric (8π) and relativistic (c) theory of gravitation (G). The astronomically (!) small value of the coupling constant κ implies that a gigantic amount of energy density (in units of $\mathrm{J/m^3}$) is necessary to produce noticeable spacetime curvature (in units of m^{-2}). This explains at once why GR was discovered much later than Newtonian gravitation, and why, regarding applications, GR is only required for either extreme precision (e.g., GPS tracking at the nanosecond level or Mercury's tiny perihelion advance) or when looking at extreme astrophysical phenomena (e.g., black holes or the expansion of the Universe).

The other constant, Λ, is called the cosmological constant. It was manually added by Einstein in 1917 [36], in order to get a static solution of the equations obtained when applying GR to the Universe as a whole. After Edwin Hubble's demonstration that the Universe is expanding, in the 1930s Einstein discarded this constant "with a flicker of his pen" [37]. The cosmological constant the re-appeared in the nineties, as a natural explanation for the freshly observed accelerated expansion of the Universe. In any case, since the low-velocity, weak-field limit of (1.1) (i.e., Newton's Law of

---

[7] This is the actual original phrasing, found on page xi of [34].

[8] This action is called the Einstein-Hilbert action, because of its independent discovery by mathematician David Hilbert one month prior to Einstein's final publication [35].

Universal Gravitation), has been perfectly well-tested on Solar system scales without Λ, a strong upper bound exists on it, of order $10^{-42}\,\mathrm{m}^{-2}$ [38]. It is therefore safe to neglect the effect of the term Λg_{ab} in the Einstein equation while working on astrophysics, and we shall do so for the remaining of this work.

Curvature

The left-hand side of Einstein's equation (1.1) has, a priori, nothing to do with physics, It can be written purely in terms of the metric tensor g_{ab} of the underlying 4-dimensional manifold representing spacetime, through the Riemann curvature tensor R_{abcd} and its contractions $R_{ab} \equiv g^{cd} R_{acbd}$ (the Ricci tensor) and $R \equiv g^{ab} R_{ab}$ (the Ricci scalar), as

$$G_{ab} \equiv R_{ab} - \frac{1}{2} R g_{ab} \,. \tag{1.3}$$

The tensor G_{ab} is called the Einstein tensor, and is nothing but the trace-reversal of the Ricci tensor.[9] In particular, whenever $T_{ab} = 0$, the Einstein equation shows that, necessarily, $R_{ab} = 0$. Physically, vacuum spacetimes are thus described by Ricci-flat manifolds. The Einstein tensor is the object of a remarkable unicity theorem for GR, proved in its strongest form by David Lovelock in 1972.[10] Let $(\mathscr{E}, g_{ab})$ be a four-dimensional spacetime $\mathscr{E}$ with metric g_{ab}. Then [40] *any tensor that is (1) divergence-free and (2) constructed from the metric and its derivatives up to second order, is necessarily a linear combination of the Einstein tensor and the metric tensor.* As all fundamental equations in physics have at most a second-derivative dependence, it can be said that the Einstein equation is the simplest yet unique field equation of relativistic gravity on a four-dimensional spacetime.

Matter

On the right-hand side of (1.1) lies the stress-energy-momentum (SEM) tensor T_{ab}. In general, it depends on both the metric and the matter fields at play in spacetime. Physically, in some observer's frame represented by a local vector tetrad (u^a, e_i^a), with u^a that observer's four-velocity and the triad (e_i^a) spanning its local rest space, the components of T_{ab} correspond to

- the energy density $T_{00} \equiv T_{ab} u^a u^b$,
- the ith component of the momentum density vector $T_{0i} \equiv T_{ab} e_i^a u^b$,
- the ijth component of the stress tensor $T_{ij} \equiv T_{ab} e_i^a e_j^b$,

---

[9] The equivalence between Einstein's way of writing his equation in Fig. 1.1 and Eq. (1.1) comes from this remark, and from the different use of units for T_{ab} (see [39] for an excellent account on the dimensional analysis of tensors in GR).

[10] This unicity theorem is sometimes attributed to Veirmeil, Cartan and/or Weyl, who wrote on the matter fifty years before Lovelock. For the reader's interest, we provide in Appendix A.1 a brief summary and references for the contributions of these men.

In general relativity, the SEM tensor is necessarily symmetric,[11] implying that the momentum density vector $T_{0i} = T_{ab}e_i^a u^b$ coincides with the energy flux vector $T_{i0} = T_{ab}u^a e_i^b$. This extends the mass-energy equivalence $E = mc^2$ of SR [22], and was Einstein's way of thinking about the equivalence principle "relativistically" [42]. As sources, the stresses T_{ij} are typically weaker (by a factor of c^{-1}) than the momentum density T_{0i}, which is itself weaker than the energy density T_{00}. In particular, at the lowest order of approximation in c^{-1}, only the mass density $\epsilon \simeq \rho c^2$ sources the Einstein equation, which simply reduces to the Poisson equation of Newtonian physics. The explicit form of the SEM tensor depends on the type of matter/energy being described. Typical examples imply that of a perfect fluid (4.13) or the electromagnetic field (4.17). A given SEM tensor can be either postulated, or derived from a Lagrangian. We provide some examples in Chaps. 2 and 3.

Motion in GR

Unlike other classical field equations (e.g., electrodynamics), the Einstein equation is nonlinear and strongly couples each components of the metric with the matter fields. However, in most cases it does not require the addition of an equation of motion for these fields. Indeed, the Bianchi identity $\nabla_a G^{ab} = 0$, which holds irrespective of the Einstein equation and comes from pure geometry [43], readily implies with (1.1)

$$\nabla_a T^{ab} = 0\,. \tag{1.4}$$

Mathematically, since the expression of T^{ab} usually includes both the metric and the field's properties, (1.4) comes as an integrability condition of the Einstein equation. Physically, it is the manifestation of local conservation of stress-energy-momentum.[12] That the equation of motion directly follows from pure geometry is one of the most elegant features of GR. For a given metric g_{ab}, Eq. (1.4) is a PDE for the matter fields. In general, it is not sufficient to determine completely their evolution. However, for sufficiently small and/or simple objects, (1.4) completely determines the motion of the source. For example, it turns into a geodesic equation for nonspining point masses, and into Maxwell's equations for the electromagnetic field. However, for extended objects or perfect fluids, (1.4) will not be enough. One needs to add some kind of equation of state (or results from microphysics in general) to close the PDE system. The motion of extended bodies will be our main point of interest in this part of the thesis, and will be thoroughly discussed in Chap. 2.

---

[11] This comes from the assumption that the manifold representing the spacetime has no torsion. When this assumption is relaxed, one works in the Einstein-Cartan theory, a perfectly reasonable extension of GR that naturally provides some insight about the dark issues of cosmology [41].

[12] Except for the energy of the gravitational field itself, which, contrary to Newtonian gravity, becomes ill-defined in GR [6].

1.1.3 Solutions and Tests

The Einstein equation is notoriously complicated. In a given coordinate system, it is equivalent to a system of ten[13] nonlinear, coupled, second-order, partial differential equations, whose unknowns are the components of the metric tensor $g_{\alpha\beta}$, and possibly unspecified features of the matter fields. Finding exact solutions remains, to this day, an incredibly difficult exercise, and has become a whole field of research in itself [44]. Throughout the years, a remarkably small number of solutions have been found, thanks either to symmetries (e.g., black hole metrics or cosmological spacetimes) or by constructing perturbations of these exact solutions (e.g., gravitational waves or post-Newtonian spacetimes). In these cases, the symmetries or the perturbative regime allows the nonlinearity of Einstein's equation to be tamed [45, 46], but whenever a system is highly dynamical and presents strong gravitational fields (for example the merging of two black holes), numerical methods must be used. Yet, almost all of GR's experimental verification come from analytical predictions. Here, we give an overview of the main steps that made GR one of the most (if not the most) well-tested scientific theory. For more than a hundred years now, GR has passed all these tests and remains to this day the most accurate description of space, time and gravity.

Classical Tests

The classical tests of GR refer to the gravitational redshift of light, the bending of light as it passes in a gravitational field and the advance of Mercury's perihelion. These three tests were the first effects that could be experimentally verified and would favor GR over SR or Newtonian gravitation. Einstein himself discussed different applications of his theory. For example, we find in the last paragraph of his 1916 review on the foundations of GR [47]:

- *Therefore, a clock goes slowly when it is placed in the neighborhood of ponderable masses. It follows from this that the spectral lines in the light coming to us from the surfaces of big stars should appear shifted towards the red end of the spectrum.*
- *with reference to the co-ordinate system, the rays of light must appear curved [...]. A ray of light just grazing the Sun would suffer a bending of 1,7" whereas one coming by Jupiter would have a deviation of about 0,02".*
- *The Ellipse of Planetary motion suffers a slow rotation in the direction of motion [...] The calculation gives for the planet Mercury, a rotation of path of amount 43" per century, corresponding sufficiently to what has been found by astronomers.*

Like its SR analogue (Doppler redshift), gravitational-induced time dilation could, in principle, be measured through the frequency shift of light emitted by a source in a gravitational potential. Einstein's prediction of the gravitational redshift can already

[13] Although at most six of them are independent, thanks to the Bianchi identity $\nabla_a G^{ab} = 0$.

be found in his 1907 article [21] discussing the relativity principle(s). Remarkably, the sole inspection of careful light-measurements between two inertial observers in SR readily shows, that under the Equivalence hypothesis, the metric cannot be Minkowskian [48]. In particular, gravitational redshift holds for any relativistic theory of gravity and is not intrinsic to GR. The first direct experimental evidence for it did not came from "big stars" as Einstein hoped, but from the tiniest light emitters: individual atoms. In the sixties, with the advent of nuclear emission lines of radioactive isotopes, the first experiment was led by Robert Pound and Glen Rebka in 1960 [49], showing consistency with GR at the 10% level, improved five year later at 1% [50] (see [22] for subsequent experiments). Very recently, two remarkable measurements of the gravitational redshift from starlight have been made, both compatible with GR. One from the light of the star S2, which regularly visit the close, strong-field neighborhood of Sgr A* (the supermassive black hole at the center of the Milky way) [51]; and another one [52] directly from the Sun, made by observing…the Moon !

The bending of light (or light deflection, or gravitational lensing) when interpreting Newton's laws strictly, does not exist in Newtonian mechanics, as light carriers (photons) have no mass. However it was suggested long before the advent of GR that the path of photons could be influenced by the presence of a gravitational field [53, 54]. All this required somewhat of an interpretation of Newton's laws of motion and light's propagation theory. In 1911, Einstein proposed [55] a revision of his analysis of 1907 [21] regarding the propagation of light in curved spacetime, emphasizing the fact that it could very well be tested by observations. What's more, the value predicted by GR and the "Newtonian" theory were different by a factor of two, allowing one to distinguish between the two for observations with enough precision. Expeditions were led throughout the world to test this prediction during a total Solar eclipse, as this allowed the location of stars surrounding the Sun to become temporarily visible thanks to the Moon's shielding. Most of these expeditions would prove unsuccessful, due to bad weather or the advent of World War I [56]. However, the total Solar eclipse of May 29th 1919, seen from the Principe Island off the west coast of Africa, allowed British astronomers Arthur Eddington and Frank Dyson to discriminate GR from Newtonian gravity, by a reasonable level of accuracy. The results were published in 1920 [57] and, although still containing significant experimental errors, were in favor of Einstein's theory. Two years later, a similar eclipse experiment was conducted by a team lead by William Campbell and Robert Trumpler in two Australian observatories.[14] The results, depicted on the left, in Fig. 1.2, showed remarkable accuracy and validated once and for all the predicted value of GR [58]. After a century at the disposal of astronomers (see the review [59]), gravitational lensing has led to remarkable astronomical images and discoveries, a beautiful example being given on the right in Fig. 1.2. It is not a simple test of GR anymore, but one of the most important tools used on a regular basis in astrophysics [60] and cosmology [61].

---

[14] There seems to be a mistake in the original article's title: the eclipse was on 21 September, 1922 and not 1923, as can be checked on https://eclipse.gsfc.nasa.gov/. See also this beautiful article by writer and photo editor Anika Burgess for the story behind the 1922 eclipse experiment.

Fig. 1.2 *Left:* figure showing the displacement of the location of stars around the Sun during the 1922 eclipse, as extracted from the original article by Campbell and Trumpler [62]. *Right:* an example of modern day gravitational lensing images. The figure depicts distorted images of far-away galaxies, located beyond the galaxy cluster Abell 370. The lensing produces multiple copies of the same background galaxies and stretched images thereof (see Fig. 6 in Johnson and collaborators' paper [63], as well as [64] for details). The original image is a superposition of visible and IR composite images taken by the Hubble Space Telescope [65]

Perhaps the most important test GR had to pass was the explanation of an anomaly in Mercury's motion, namely the observed advance of its perihelion as compared to the theoretical value. In 1840, under the impulse of Louis Arago at the Paris Observatory, Urbain Le Verrier conducted a ten-years long, thorough analysis of the celestial motion of Mercury, both theoretically (using Newton's laws) and observationally (with an impressive review of all available observations at the time). Published in 1959 [66], Le Verrier's results showed that the value of Mercury's perihelion advance could not be completely accounted for by Newtonian mechanics, even with apparent (Earth's rotation, equinox precession) or direct (other planets pull, flattening of the Sun) biases. Le Verrier proposed, in vain, to apply the ontological solution that led him to the discovery of Uranus a few years earlier in order to explain Mercury's anomaly. The new planet, named Vulcan (Vulcain, in French), whose orbit must have been within Mercury's, was never found, although Le Verrier's fame had already convinced academic representatives to include Vulcan on celestial maps, as depicted in Fig. 1.3. The end of the story is well-known: it would be a *legislative* solution that would at once explain the correct value of Mercury's perihelion precession, and sign the birth act of GR. As Le Verrier puts it himself [67]

"L'altération des lois de la gravitation serait une dernièe ressource á laquelle il ne pourrait être permis d'avoir recours qu'après avoir épuisé l'examen des autres causes, qu'après les avoir reconnues impuissantes á produire les effets observés",

essentially saying that, to Le Verrier, changing the law of gravitation would be but the last solution to Mercury's perihelion advance. Knowing that Einstein's quest

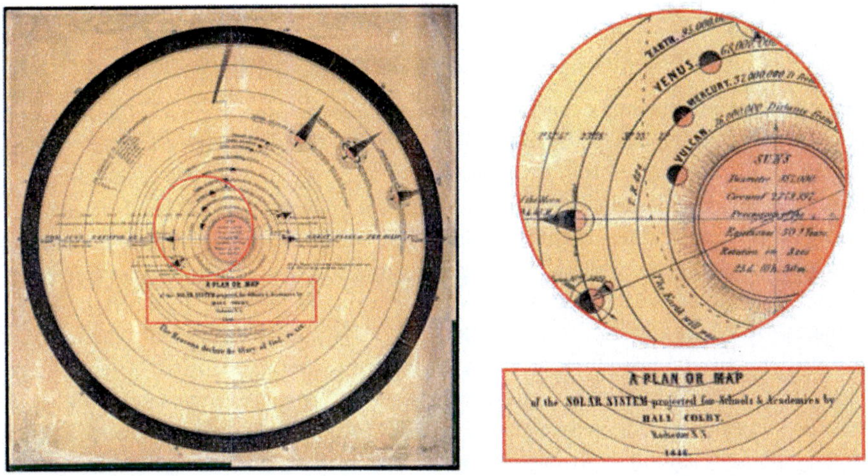

Fig. 1.3 A celestial map (lithography) presented to students and academics in American universities. It already contains the planet Vulcan, located between Mercury and the Sun. Interestingly, Neptune, of which Le Verrier actually predicted the existence, is not on the map. *Yet, this map represents an unsolved mystery for historians of physics. We encourage the reader to examine it carefully (in particular the rectangular insert) to understand why.* More information can be found on the Library of Congress website., where the original lithography stays

for the correct field equation was partly influenced by his will to find the correct precession value from his new theory [68], one cannot help but wonder how this whole story would have changed if our Solar system had not had a planet this close to the Sun...Today, the relativistic perihelion precession has been compared to GR predictions at the 10^{-5} level for some bodies of the Solar system [69].

Astrometry

The classical tests above can be classified as post-Newtonian tests, because they are but small modifications of otherwise well-known Newtonian results. There exists numerous other such slight differences between Newtonian gravity and GR, or any other competing relativistic theory of gravity. A particularly convenient way of distinguishing between all of them is the so-called Parametrized Post-Newtonian (PPN) formalism [70, 71]. In this framework, theories of gravity are expanded in the weak-field limit, and compared to one another by examining the coefficients of said expansion. The coefficients together define a set of PPN parameters that may take different values depending on the theory. As an illustration, the two most important parameters, γ and β,[15] appear in the PPN expression of the laws of light deflection (of a lightray grazing the Sun) and periastron advance (of Mercury) laws [72]:

[15] They measure how much space-curvature is produced by unit rest mass and how much "nonlinearity" there is in the superposition law for gravity, respectively [72].

$$\delta\phi = \frac{4GM_\odot}{c^2R_\odot}\frac{1+\gamma}{2} \quad \text{and} \quad \delta\omega = \frac{6\pi GM_\odot}{c^2a(1-e^2)}\frac{2+2\gamma-\beta}{3}, \tag{1.5}$$

where $M_\odot$, $R_\odot$ are the Sun's mass and radius, and a, e are Mercury's orbit semi-major axis and eccentricity. In GR, $(\gamma, \beta) = (1, 1)$, so that (1.5) coincides with the values derived by Einstein in 1916 [47]. Although the PPN parameters are constrained by many other observables, the most precise values for γ, β come from light-bending and perihelion precession measurements, thanks to two spacecrafts: one sent to observe Saturn (Cassini mission [73] for γ) and the other to observe Mercury (Messenger mission [74] for β). The values agree with GR value at the 10^{-5} level. Many other measurements have reported constraints on the PPN parameters, all of which agree with GR's predictions. Although most of them are obtained using Solar system experiments (see Chap. 8 of [75] for a recent review), measurements at the galactic and cosmological scale have been made too. Clifford Will's review [76] on GR tests remains the unavoidable reference about GR tests and experimental checks.

Black Holes and Neutron Stars

Although the idea of a "body with such gravity that not even light could escape from it" was already on the minds of eighteenth century scientists [77], black holes (BHs) only became astrophysical objects decades after the advent of GR, as the physical counterpart of peculiar mathematical solutions to the Einstein equations. The first chapter of this story starts with the first (non-flat) exact solution to the Einstein equation, derived in 1916 by Karl Schwarzschild while he was serving on the Russian front in the German army (see Jean Eisenstadt's review article on Schwarzschild's solution [78]). In his two articles [79, 80], Schwarzschild's primary goal was twofold: derive an exact solution to the full nonlinear Einstein equation (for the first time), and show that Einstein's derivation of Mercury's perihelion advance could be recovered exactly from this result. Upon knowledge of this discovery, Einstein wrote to Schwarzshild:

> *I have read your paper with the utmost interest. I had not expected that one could formulate the exact solution of the problem in such a simple way. I liked very much your mathematical treatment of the subject.*

Here, Einstein makes reference to the elegant symmetry arguments used by Schwarzshild to derive the eponymous metric: he assumes that the metric is static, spherically symmetric and Ricci-flat (vacuum solution). In particular, it can describe a black hole if there is no matter at all or the exterior of a spherically symmetric and static star. Around the same time, Johannes Droste derived independently the same result [81, 82], along with the solution of the equations of motion in the Schwarzshild spacetime in terms of the Weierstrass elliptic function $\wp$.[16] At the end of the 1930s,

[16] Since the polar equation of motion of Schwarzshild geodesics involves the Weierstrass elliptic function [82], exact and analytical formulae for the light deflection $\delta\phi$ and perihelion advance $\delta\omega$

it was known that both white dwarfs and neutron stars had a maximum mass, beyond which the gas of electron (or neutrons, respectively) would not be able to counteract the star's weight, leading to instability. These considerations culminated in 1939 with a groundbreaking paper by Robert Oppenheimer and his student Hartland Snyder [84] wherein they showed explicitly that a collapsing star (with enough mass) does lead to the formation of a black hole.

In spite of these theoretical results, experimental evidence for BHs would have to wait for the "golden age" of GR, starting with the first observations of X sources in 1962 [85], quasars a year later [86], and pulsars in 1968 [87]. While most of these sources could have been attributed to neutron stars, the observation of the X binary Cygnus X-1 constitutes the first strong BH candidate. Indeed, a measurement of its mass ruled out the possibility of it being a neutron star. Experimental identifications of BH candidates became mainstream, while theoretical works on BHs continued and showed that the perfect, theoretical spherical symmetry of black holes is not a prerequisite. In this context, the discovery by Roy Kerr [88] of the metric describing a rotating BH (see Saul Teukolsky's nice review [89]) and the singularity theorem(s) by Roger Penrose and Stephen Hawking [90–93], are of major importance. To summarize, GR predicts BHs not just because of the Schwarzshild and Kerr exact solutions, but also because these exact BHs remain stable under small perturbations [94–96] and that there exists physically realistic situations that lead to their formation (although this is still an intense field of research [97]). Of course, the direct detection of BH binaries through their gravitational wave emission, and more recently the direct imaging of a supermassive BH's surroundings [98], gave a final *YES* to this fascinating question (and key prediction) of GR: *Do black holes exist?*

Contrary to white dwarfs, the relativistic description of neutron stars is rather different from its Newtonian counterpart. This is due to their large compactness, which is 100 to 1000 times higher than white dwarfs', making their very existence a GR prediction on its own. In fact, five years after the proposed idea of their existence by Walter Baade and Fritz Zwicky [99], Robert Oppenheimer and Vladimir Volkoff provided the first explicit, relativistic calculation of a neutron star model [100]. They showed that GR predicts a universal limit to the mass of neutrons stars around $2.2M_{\odot}$. The fact that all neutron stars discovered so far have a mass below this critical value [101] is also a major success of GR.

Modern Cosmology

Modern cosmology, which entered the last decades the "precision era" [102], relies entirely on the curved spacetime description of the Universe provided by GR. The current theoretical and observational paradigm in cosmology, called the standard model of cosmology for its unparalleled success, has been verified on multiples

---

can be given in terms of the Elliptic integrals. These beautiful formulae were derived in chap IX and X of [83], and would make for an interesting academic exam or a nice illustration of the elliptic functions/integrals duality.

scales in various ways. Gravitational lensing, either strong (Einstein crosses and rings) [103–105] or weak (cosmic shear) [106, 107], allows one to probe the geometry of spacetime on cosmological distances. But among the most important probes of GR on cosmological scales is the observation and analysis of the cosmic microwave background (CMB). The most recent results from the Planck Collaboration [108] provides a direct measurement of the temperature anisotropy spectrum from the early universe, and provides information on the Universe's large scale geometry, early structure formation mechanisms and particle combination processes. Fine effects (high-multipoles modes of the spectrum), in particular the Integrated Sachs–Wolfe effect and CMB lensing, have been used to check that GR predictions are favored over alternative theories gravity (see [109, 110] as well as Sect. 4.4 of the review [111] for details.) Many other tests of GR are accessible through cosmological observations and we refer to the recent review [111] and references therein for more on this topic.

1.2 Gravitational Waves

The prediction of general relativity that will be of primary importance in the present work is the existence of gravitational waves (GW). These were first directly detected in 2015 by the LIGO observatories. The period from 2015 forth will be discussed thoroughly in Sect. 1.3. First, we will start with a short historical account in Sect. 1.2.1, split into two distinct phases: 1916–1973 and 1974–2015. Then, we will move to more theoretical aspects, discussing the mathematical derivation of GWs (Sect. 1.2.2), their different sources (Sect. 1.2.3) and the process(es) by which one can detect them (Sect. 1.3.1).

1.2.1 History

The Reviewer Strikes Back

As already mentioned, another key prediction of Einstein's GR, which he himself made as soon as 1916 [35], is the existence of ripples in the fabric of spacetime, propagating at the speed of light: gravitational waves (GWs). Prior to their direct detection in 2015, GWs went through two very distinct periods: between 1916 and 1974, they were but theoretical predictions and their existence was rather controversial. After his first paper, Einstein reconsidered GWs two years later [112] and started to question their physical existence. He showed that out of the three "types" discovered, longitudinal-longitudinal (LL), longitudinal-transverse (LT), and transverse-transverse (TT), the first two were "gauge waves" in the sense that they were not associated to spacetime curvature. The following years, major contributions were made by Arthur Eddington (again). He clarified the nature of these gauge waves and corrected Einstein's quadrupole formula [113], and discussed for the first time the implications of GW emission on a binary's orbit [114]. Despite these clarifications, GWs were hardly considered as physical objects at that time, as evidenced by the famous Physical Review controversy[17] between Einstein, Rosen and the editors of the journal [115].

The following years then saw a decrease in the physicists' interest for GWs, and for GR as a whole. Primarily because of the lack of available experiments and the consequences of the World War(s), but also because most physicists were occupied with quantum mechanics. Only in the 1950s did the interest in GWs came back, with the notable works of Bondi, Pirani and Robinson [116–118] who clarified the question of whether GWs transported energy (and thus if they could be detected). It is Pirani again who discussed [119] for the first time the effect of GWs on two point particles, which would ultimately become the most efficient way of detecting them. If GWs transport energy, a system emitting them must lose some. The final piece of the GW puzzle thus consisted in computing the backreaction of GW emission on a source's motion. Various attempts, using post-Newtonian (PN) expansions (or variations thereof) to tackle this problem were already provided in the early years [120–122] but the real stepping stone was Landau and Lifshitz's rewriting of the Einstein equation [123] in a form particularly convenient for the coupled problem of motion and radiation. Finally, after some discussions in the 1970s (again) [124] on the validity of earlier improvements [125, 126], rigorous and self-consistent derivations of a GW source's motion and its emission were obtained in the early 1980s, with notable pioneering works by Thibault Damour, Nathalie Deruelle and collaborators [127–130].

Yet, all these considerations were purely theoretical, and it is fair to say that all hesitations and controversies they underwent would have benefited from a little help by experiments and/or observation. Thankfully, a remarkable discovery came in 1974 and would, finally, provide a genuine interest for GWs from the whole community and restore hope in the quest for the (direct) detection of GWs (Fig. 1.4).

A New Hope

Neutron stars are arguably the most extreme kind of stars, with two main properties: they are strongly magnetized and have a very fast proper rotation. For some of them, the first trait makes them emit a continuous beam of radio waves along their two magnetic poles, while the second makes this continuous signal appear as pulses every time the beam crosses an observer's path. The first such star, called a *pulsar*, was discovered by Jocelyn Bell in 1967 [87], and today about three thousand of them are cataloged [131, 132], with a few tens visible in the optical wavelength [133, 134]. While isolated pulsars already provide an important test bench for GR [135], they become even more valuable when part of a binary system, allowing one to test many strong-field effects of GR dynamics.

[17] There is some irony at the core of this controversy: a strong claim in Einstein and Rosen's paper was refuted by the referee because it simply came from a bad choice of coordinates. Let this be a lesson for any student taking a course in GR: even Einstein himself fell into the trap of his beloved diffeomorphism invariance!

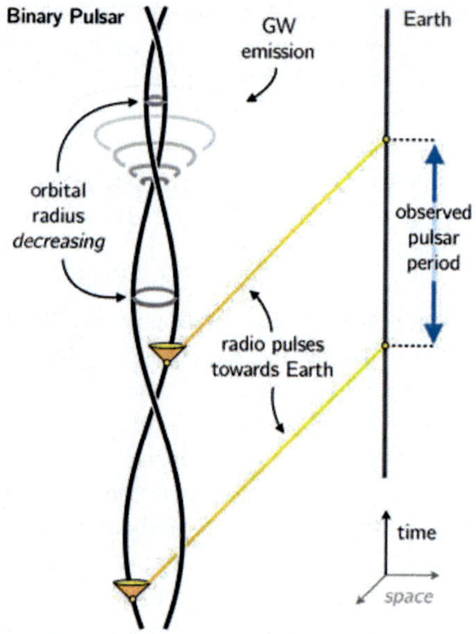

Fig. 1.4 This 2+1 spacetime diagram depicts the main geometry of the Hulse–Taylor pulsar. On the left, the double helix shows the pulsar in orbit with its companion. The arrival times are measured when the radio beam (yellow geodesic) crosses the Earth worldline (timelike geodesic, on the right). The time measured between two successive arrival times slowly decreases due to GW emission, which also makes the orbital radius decrease over time, as exaggeratedly depicted here

The Hulse–Taylor binary, technically known as PSR B1913+16,[18] is one of them. It is made of two NS, one of them being an observed pulsar, of which we detect the faint, periodic radio signal. It was detected in 1974 by Russel Hulse and his thesis advisor Joseph Taylor Jr [137], using data from the Arecibo Observatory.[19] It was the first binary pulsar ever discovered.

General relativity predicts that GW are emitted by a binary system, taking away some energy and angular momentum from it. This loss of energy and angular momentum induces a slow, secular variation of the parameters describing the Keplerian ellipse. For example, one expects an increase of the orbital frequency $\dot{\Omega}$ and decrease of the eccentricity $\dot{e}$, an (orbital-time averaged) precession of the periastron position $\delta\omega$, etc. These numbers, which vanish identically in Newtonian dynamics, are naturally called post-Keplerian parameters. In 1963, using among other things the first quadrupole formula (1.17), Philip Peters and his thesis advisor Jon Matthews provided the first expression of these post-Keplerian parameters in terms of the binary's properties. They found relations like [139, 140],[20]

---

[18] For the interested reader, this technical nomenclature encodes PSR for Pulsating Source of Radio, and 1913+16 for the equatorial coordinates (α, δ), with right-ascension $\alpha =$ 19hrs13min and declination $\delta = +16°$. See page 302 of [136] for additional details.

[19] As I am writing these lines in December 2020, parts of the Arecibo telescope have just collapsed, making additional damages to its already crumbling dish. The Arecibo telescope is responsible for a long history of discoveries, as beautifully accounted for in the review [138].

[20] The accordance between the GR prediction and the observed value allows for an *astrophysical derivation* of nontrivial series formula for the Bessel functions (see the Appendix of [139]).

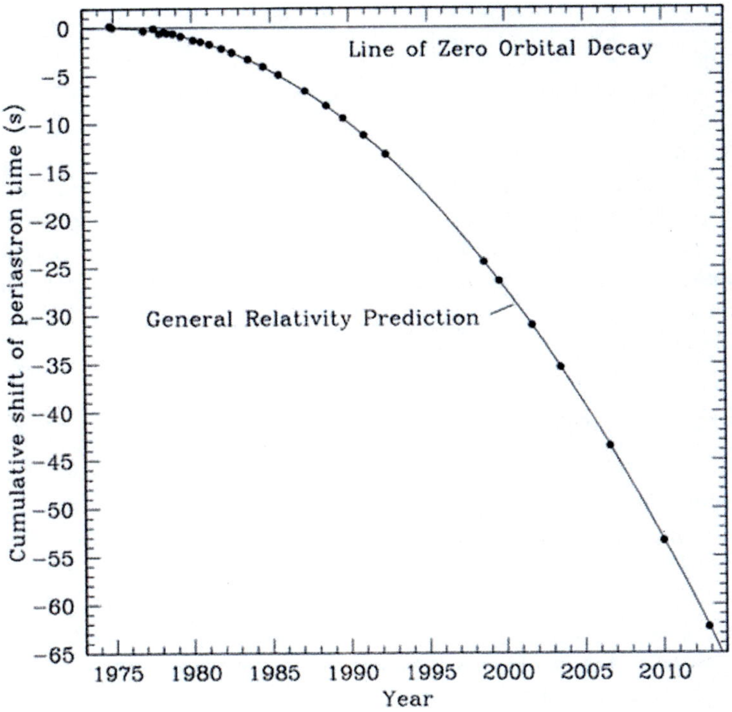

Fig. 1.5 Data points showing different measures of periastron time shift, relative to an unchanging orbit, for the Hulse–Taylor pulsar over 38 years. They all lie perfectly (error bars are too small to show) on the parabola corresponding to the GR prediction. The curvature of the parabola is given by $-\dot{\Omega}/2\Omega$, allowing one to compute the frequency shift of the orbit through the years (see [136] for details). This particular plot comes from Weisberg and Huang's most recent update on the Hulse–Taylor pulsar characteristic [143]

$$\delta\omega = \frac{3(GM)^{2/3}}{c^2}\frac{\Omega^{5/3}}{1-e^2} \tag{1.6}$$

which expresses the periastron precession parameter in terms of the total mass of the binary M, its orbital frequency Ω and eccentricity e. Thanks to the measurement of the arrival times of binary pulsars, these parameters can be estimated and compared to the prediction of GR. This provides remarkable tests, the most well-known one being depicted in Fig. 1.5.

Secular variations such as (1.6) are well-known from Newtonian, celestial mechanics for a binary perturbed by other bodies [141]. But here, the binary is isolated: these variations are solely due to a general relativistic effect: GW emission. Consequently, and provided one has a way of inferring the Keplerian and post-Keplerian parameters in the first place, the HT binary allows one to test the GR prediction. This is where the pulsar nature comes into play. Indeed, thanks to the beautiful theory of *pulsar timing*, or the art of extracting the physical properties of a pulsar from the arrival times of its radio signals. The *tour de force* of Hulse and Taylor was to find a clever way to analyze the data [142] and fit most of the HT binary parameters. In a few years, one will not be able to observe the pulsar since the beam's trajectory is expected to drift and miss the Earth due to the geodetic precession, a relativistic spin-orbit coupling within the binary.

These observations are indirect hints converging towards the existence of GWs. Yet, a direct detection, meaning an actual GW interacting with a dedicated GW detector had to wait until 2015. Before covering this in more details in Sect. 1.3, let us dive into more theoretical aspects, to understand the main physical characteristics of GWs.

1.2.2 *Linearized Gravity*

At the mathematical level, GWs are nothing but approximate solutions to the Einstein equation. They may be understood either on pure geometrical grounds as ripples in the curvature or as independent fields propagating in some background spacetime. In the latter viewpoint, they have physical features akin to EM waves: they are transverse, and propagate in vacuum at the speed of light. However, from a mathematical perspective, they share a common point with hydrodynamic waves, as they arise from the linearization of a more fundamental, nonlinear equation (Navier–Stokes in hydrodynamics, Einstein equation for GWs). The remaining of this section is dedicated to the basics of GW theory: mathematical derivation, propagation, sources, and detectors. We hide most of the technical details and refer to classical textbooks [136, 144] for further information on this subject.

Let us consider a *background* metric $\mathring{g}_{ab}$: an exact solution to Einstein's equation sourced by some arbitrary SEM tensor T_{ab}. We look for a solution to the Einstein equation in the form of a (formal) small-parameter expansion, with an Ansatz for the solution metric g_{ab} of the type

$$g_{ab} = \mathring{g}_{ab} + \epsilon h_{ab} \,, \tag{1.7}$$

where h_{ab} is the perturbation, and ϵ is a dimensionless, small parameter. Typical examples include the parameter v/c, such as in post-Newtonian theory [145], or the mass ratio as in the gravitational self force scheme (see Sect. 1.3.1). Here, we just treat ϵ as a formal, bookkeeping parameter that counts the nonlinearities in the perturbation h_{ab}. By viewing the Einstein tensor G_{ab} as an operator acting on the metric $g \equiv \mathring{g} + \epsilon h$, the Ansatz (1.7) can be used to linearize the left-hand side of the Einstein Eq. (1.1). A straightforward computation (e.g., Sect. 7.5 of [6]) provides, to linear order in ϵ

$$\mathring{\nabla}_c \mathring{\nabla}^c \bar{h}_{ab} + 2\mathring{R}_{acbd}\bar{h}^{cd} - 2\mathring{\nabla}_{(a} H_{b)} = -16\pi T_{ab} \,, \tag{1.8}$$

where $\mathring{R}_{abcd}$ is the Riemann tensor associated to the background metric $\mathring{g}_{ab}$, with which the covariant derivative $\mathring{\nabla}_c$ is compatible. Equation (1.8) is written in terms of the convenient *trace reversal* of h_{ab}, defined by $\bar{h}_{ab} \equiv h_{ab} - \frac{1}{2}\mathring{g}_{ab}\mathring{g}^{cd}h_{cd}$, as well as its "divergence" $H_a \equiv \mathring{\nabla}^b \bar{h}_{ab}$. Equation (1.8) is the linearized Einstein equation in the perturbation $\bar{h}_{ab}$, and is at the basis of many applications. It is very general and valid in any coordinate system where the splitting (1.7) holds. However, for illustration purposes, it is much more convenient to take advantage of the diffeomorphism invariance of GR to simplify this equation.

Gauges

Since the background metric $\mathring{g}_{ab}$ is fixed, we may see Eq. (1.8) as that of an independent field h_{ab} (or equivalently its trace-reversal $\bar{h}_{ab}$) evolving on the background spacetime. We would then like to simplify Eq. (1.8), for example by choosing coordinates in which $H^a = o(\epsilon)$, so that the third term on the left of (1.8) can be neglected at linear order in ϵ. A particularly convenient (and sufficient) way to do this is to perform an ϵ-infinitesimal transformation, i.e., a change of coordinates of the form $x^\alpha \mapsto x^\alpha + \epsilon\xi^\alpha$, with ξ^a an arbitrary vector field. Because the coordinates differ only at order ϵ, this induces a change in the metric g_{ab} only at the level of the perturbation. Indeed, an easy computation reveals that under this mapping, $\mathring{g}_{ab} \mapsto \mathring{g}_{ab} + o(\epsilon)$ but

$$h_{ab} \mapsto h_{ab} - \mathcal{L}_\xi \mathring{g}_{ab} + o(\epsilon)\,, \tag{1.9}$$

where $\mathcal{L}_\xi$ denotes the Lie-derivative along ξ^a. As we can see, this mapping generates an isometry ($\mathring{g}_{ab}$ is invariant) up to ϵ-corrections: the perturbation "absorbs" the ϵ-part, and leaves the background untouched. Since this is but a mere change of coordinates, it does not alter any physical observable: it is a gauge transformation. The idea is to chose this gauge by fine-tuning the arbitrary field ξ^a so that $H_a = o(\epsilon)$ in the new coordinates. To do this, simply notice that (1.9) induces

$$H_a \mapsto H_a - \mathring{\nabla}_b \mathring{\nabla}^b \xi_a + o(\epsilon)\,, \tag{1.10}$$

where metric compatibility $\mathring{\nabla}_c \mathring{g}_{ab} = 0$ was used. To summarize, if H_a was not $o(\epsilon)$ in the original coordinates, then Eq. (1.10) tells us to introduce a vector field ξ^a such that $\mathring{\nabla}_b \mathring{\nabla}^b \xi_a = H_a$ (always possible). In the new $x + \epsilon\xi$ coordinates, we have $H = o(\epsilon)$, and the third term in Eq. (1.8) can be neglected, resulting in a classical, hyperbolic wave equation

$$\mathring{\nabla}_c \mathring{\nabla}^c \bar{h}_{ab} + 2\mathring{R}_{acbd}\bar{h}^{cd} = -16\pi T_{ab}\,. \tag{1.11}$$

Although Eq. (1.11) is simpler than (1.8), it is also less general since it holds in a particular set of coordinates: those that satisfy the *gauge condition* $H^a = o(\epsilon)$, that is, at linear order in ϵ,

$$\mathring{\nabla}_b \bar{h}^{ab} = 0 \,. \tag{1.12}$$

This gauge is traditionally called the Lorenz gauge[21] by analogy with classical electrodynamics [151]. The system (1.11)–(1.12) is the linearized Einstein equation in the Lorenz gauge, and is the basis of many calculations of GW-related physics. Particularly relevant cases correspond to vacuum metrics ($T_{ab} = 0$), where $\mathring{g}_{ab}$ is either the Schwarzshild or Kerr metric. In this case, the full solution $\mathring{g}_{ab} + h_{ab}$ describes a perturbed isolated BH and h_{ab} contains the GWs emitted by the system. In the non-vacuum case ($T_{ab} \neq 0$), prototypical examples include: binary systems of compact objects, where $\mathring{g}_{ab}$ is the Minkowski metric and the matter is described as post-Newtonian source [145]; or cosmological perturbations, for which $\mathring{g}_{ab}$ is the FLRW metric and the SEM tensor is necessarily that of a perfect fluid, for symmetry reasons [53].

Before going further let us stress that while the linear system (1.11)–(1.12) is sufficient to get a grasp on how GW generation mechanisms work, and even for some applications (like the slow orbital decay of binary pulsars, see Sect. 1.2.1), it is not sufficient for the direct detection of GWs. In particular, non-linearities in h of high-order must be accounted for when building the GW templates used in a GW detector (see Sect. 1.3.1). What's more, already at the quadratic level, self-interactions of the field h come into play, with terms that can be of the same order of magnitude as linear ones, something that is often overlooked in deriving the usual quadrupole formula (see for example [152]).

Propagation

Now let us move on to solving the system (1.11)–(1.12) and consider a system of Lorenz coordinates x^α. For simplicity, we place ourselves on a flat background, so that $\mathring{R}_{abcd} = 0$ (enforcing $\mathring{g}_{ab}$ to be the Minkowski metric). Studying this free, vacuum-propagation of h_{ab} is particularly relevant for gravitational astronomy, where the GWs travel mostly in the vacuum of spacetime far away from any source. Moreover, this setup is the easiest way to distinguish from the mathematical solution h_{ab} the non-physical and physical degrees of freedom. With these assumptions made, the system (1.11)–(1.12) becomes

$$\partial_c \partial^c \bar{h}_{ab} = 0 \quad \text{and} \quad \partial_a \bar{h}^{ab} = 0 \,, \tag{1.13}$$

where we assumed Cartesian-like coordinates so that $\eta_{\alpha\beta} = \text{diag}(-1, 1, 1, 1)$, and therefore $\mathring{\nabla}_a = \partial_a$. Equations (1.13) can be simplified once more by further specifying the gauge. Indeed, notice that in the Lorenz gauge we have the liberty to add to

[21] Not to be confused with Hendrik Antoon Lorentz who gave his name to the Lorentz force, Lorentz transformations and Lorentz factor [22]. Several classical textbooks [6, 123, 136, 146, 147] make this mistake. The Lorenz gauge (or coordinates that satisfy it) is also often called "harmonic" [145, 148], and more rarely De Donder [149] or Hilbert [150].

the previous gauge vector ξ^a another arbitrary vector field ξ_o^a, such that $\partial_b\partial^b\xi_o^a = 0$. Clearly, the new coordinate system $x^\alpha + \epsilon(\xi^\alpha + \xi_o^\alpha)$ still satisfies the Lorenz gauge condition, but we may now freely chose ξ_o^a so as to remove superfluous degrees of freedom in h_{ab}. In particular, this new sub-gauge enforces 4 conditions (one for each component of ξ_o^a) which we may as well chose to be

$$\bar{h}_{0i} = 0 \quad \text{and} \quad \bar{h} \equiv \eta_{ab}\bar{h}^{ab} = 0\,. \tag{1.14}$$

In other words, we set to zero the space-time components and the trace of h_{ab}. This new system of coordinates defines a (sub-)gauge of the Lorenz one, called the transverse-traceless (TT) gauge. Noting that these conditions imply $h_{00} = 0$ (thanks to (1.12)) and $\bar{h}_{ab} = h_{ab}$ (since $\bar{h} = 0$,) we obtain as a final result a simple d'Alembert equation

$$\partial_c\partial^c h_{ij} = -16\pi T_{ij}\,. \tag{1.15}$$

This equation is valid in the TT gauge, on a Minkowskian background and when nonlinearities in h_{ab} are neglected. The point of the TT gauge is not just to simplify Eq. (1.15), as it also removes all possible non-physical degrees of freedom. Indeed, in the initial, arbitrary x coordinates, h_{ab} has 10 independent components. The Lorenz gauge (1.12) lowers this number to 6, and the TT sub-gauge to 2. The only freedom left in the TT gauge is a rigid rotation of the axes, which may be used to uncouple the two remaining degrees of freedom, as follows.

Let us consider the solution of Eq. (1.15) away from the source, so that $T_{ab} = 0$. We obtain a simple homogeneous equation $\partial_c\partial^c h_{ij} = 0$, whose solution can be written as a superposition of plane waves, say $h_{ij} = p_{ij}(k)\exp(ik_\alpha x^\alpha)$ where k^α are the components of the wave vector and p_{ij} is the mode's amplitude, or polarization tensor. Choosing coordinates (t, x, y, z) such that the vector k^a is along the z-axis, the condition (1.12) simply becomes $k^i h_{ij} = 0$: the wave is transverse to the direction of propagation. In that case, symmetry and traceless-ness of h_{ij} readily imply

$$h_{\alpha\beta} = \begin{pmatrix} 0 & 0 & 0 & 0 \\ 0 & h_+ & h_\times & 0 \\ 0 & h_\times & -h_+ & 0 \\ 0 & 0 & 0 & 0 \end{pmatrix}, \tag{1.16}$$

where $h_\times$, h_+ denote the two intrinsic polarization states of the GW, called the *plus* and *cross* polarization. The reason for these names is best explained by looking at the effect of each polarization on a ring of test masses placed in a plane orthogonal to the GW's direction. The h_+ makes the ring contract and extend (making it oval) along two perpendicular axes (say x and y), while the $h_\times$ does the same, albeit rotated by $\pi/4$, along the diagonals. Figure 1.6 depicts a spacetime diagram with these test mass rings.

The main conclusions obtained here with $\mathring{g}_{ab} = \eta_{ab}$ still hold in a general background and with matter [153, 154]. In particular, the metric perturbation h_{ab} always contains three types of degrees of freedom: (1) non-physical and pure gauge (2)

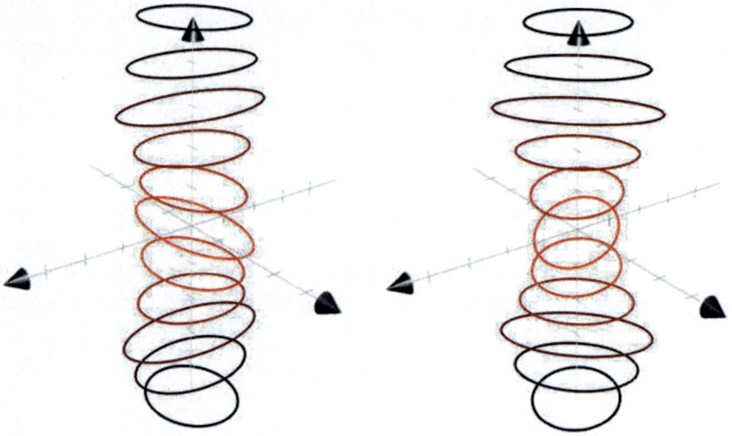

Fig. 1.6 A spacetime diagram depicting a ring of point particles. The GW propagates in the z direction (not depicted here), and the oscillations of the ring take place in the horizontal plane (transverse wave). The $+$ polarization on the left has the same effect of the $\times$ polarization, on the right, albeit rotated by $\pi/4$

physical but non-radiative, and (3) physical and radiative. The non-physical ones have been discarded with the use of the Lorenz and TT gauges. The non-radiative degrees of freedom generate the so-called gravito-magnetism effects [6, 53]. The two radiative degrees of freedom are crystallized in the h_+ and $h_\times$ polarizations of the GW. From a field-theory perspective, this implies that the field h_{ab} is associated to a spin-2 boson (the graviton) [136].

1.2.3 Sources

The solution to the wave Eq. (1.15) can be turned into an explicit formula for the components h_{ij} as a function of the TT coordinates $(t, \mathbf{x})$ and the source's physical properties. This computation is classical [53, 136, 141] and we will only sketch the general ideas, which were already in Einstein's second article on GW [112]. The first step consists in inverting the d'Alembert operator appearing in Eq. (1.15) in the usual way (e.g., by means of its retarded Green function). This gives an expression of h_{ij} terms of the TT components T_{ij} of the source's SEM tensor. From there, a rather involved but otherwise very elegant calculation allows one to simplify that expression drastically, by combining the conservation equation for the SEM tensor and a multipole expansion thereof. The leading-order result of said calculation is called Einstein's *first quadrupole formula*, and reads

$$h_{ij}(t, \mathbf{x}) = \frac{1}{|\mathbf{x}|} \frac{2G}{c^4} \ddot{Q}_{ij}(t_{\mathrm{ret}}) \,, \tag{1.17}$$

where $|\mathbf{x}|$ is the distance to the source from the origin of TT coordinates $(t, \mathbf{x})$, a dot stands for a t-derivative and the right-hand side is evaluated at the retarded time $t_{\text{ret}} \equiv t - |\mathbf{x}|/c$. This formula holds for generic, isolated sources.[22] Equation (1.17) is particularly simple as it only depends on the (second derivative of the) quadrupole moment of the source, linked to its mass density $\rho(t, \mathbf{x})$ through the classical, Newtonian expression

$$Q_{ij}(t) = \int_{\mathcal{S}} \rho(t, \mathbf{x})\Big(\mathbf{x}^i\mathbf{x}^j - \tfrac{1}{3}|\mathbf{x}|^2\delta_{ij}\Big)\mathrm{d}^3\mathbf{x}, \tag{1.18}$$

where the integral is over the source's spatial support $\mathcal{S}$. Although the leading order result (1.17) is usually not enough for the actual detection of a GW [136, 155, 156], it readily shows that GW are generally produced by time-varying *quadrupoles*. The quadrupolar nature of GWs is intrinsic to gravitation, much like the dipolar nature of EM waves is reminiscent of electrodynamics. In linearized GR, the conservation of energy and linear momentum of the source prevents monopole and dipole radiation, respectively. Again, in the full picture this is not so simple, as GW do, in fact, transport energy and linear momentum which they extract from the source [136].

GW Amplitudes

Still, from Eqs. (1.17) and (1.18), one can obtain a good estimate for a typical GW amplitude. In particular, if a non-spherical source has a rotational frequency Ω, mass M and length scale L, then $Q_{ij} \simeq AML^2$, with A a numerical coefficient measuring the asymmetry of the source. At some distance $R \equiv |\mathbf{x}|$, we thus get a typical GW amplitude of

$$h_{\mathsf{GW}} \simeq A\,\frac{L}{R}\,\frac{2GM}{Lc^2}\left(\frac{\Omega L}{c}\right)^2. \tag{1.19}$$

The first two terms show that non-spherical systems ($A \simeq 1$) that are not too far away (L/R reasonably small) will produce GW with greater amplitude. The third term we recognise as (twice) the compactness parameter $\Xi \equiv GM/Lc^2$, a first hint that we should expect compact objects to be the most promising sources of GW. In particular, systems involving black holes (BH) and neutron stars (NS) ($\Xi_{\mathsf{BH}} = 1/2$ and $\Xi_{\mathsf{NS}} \simeq 0.2$) should be excellent candidates. The last term in (1.19) shows that among these sources, those with relativistic internal velocities $v \simeq \Omega L$ produce GW with a greater amplitude. This can be either the proper rotation of an isolated compact object, or the relative orbital velocity in a binary system. Naturally, these characteristics lead us to extreme astrophysical sources which are, by definition, far away from us. Conse-

[22] It should be noted that in linearized theory, a rigorous derivation of (1.17) is impossible. Precisely because the equation of motion, which follows from the Lorenz gauge condition, implies that sources moves on a geodesic of Minkowski's spacetime. This undeniably excludes any gravitationally-driven orbit. The rigorous, self-consistent derivation includes a careful inspection of the nonlinearities [136, 145].

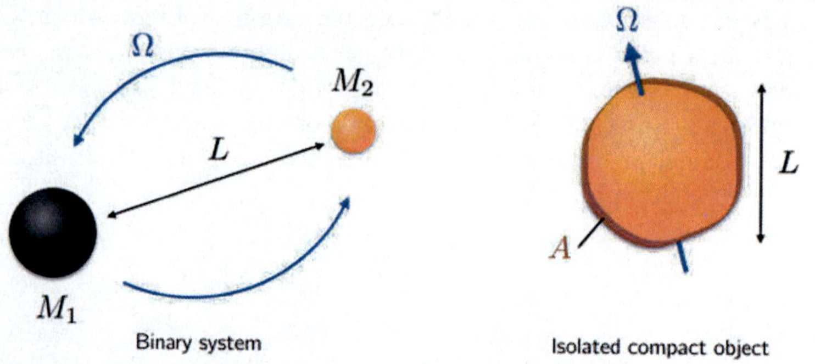

Fig. 1.7 Physical characteristics of astrophysical sources involved in the typical amplitude (1.19) and frequency (1.21) estimates, with A representing the deformation (distance to spherical symmetry)

quently, the main goal of GW detection is to focus on compact sources and build GW detectors sensitive enough to account for the $L/R \ll 1$ term (Fig. 1.7).

GW Frequencies

As we have just seen, promising sources of GW are rapidly rotating, asymmetric, isolated compact objects and orbiting binaries. Let us apply the first quadrupole formula (1.17) to a binary system of compact objects of total mass $M \equiv M_1 + M_2$ on a circular orbit of radius L and angular frequency Ω. We skip the details, referring to [136] for more on this calculation. We obtain, in the TT gauge, for the + polarization,

$$h_+(t, \mathbf{x}) = \frac{\iota_+}{|\mathbf{x}|} \frac{4G\mu}{c^2} \frac{\Omega^2 L^2}{c^2} \cos(2\Omega t_{\mathrm{ret}}) , \tag{1.20}$$

where $\mu \equiv M_1 M_2/M$ is the reduced mass, $t_{\mathrm{ret}} \equiv t - |\mathbf{x}|/c$ is the retarded time as in (1.17), and the coefficients ι_+ is a geometrical factors that depend on the relative inclination of the orbit with respect to the observer (see [141] for details). A similar equation holds for the $\times$ polarization, obtained by replacing (cos, ι_+) by (sin, $\iota_\times$). We see from Eq. (1.20) that the frequency f_{GW} of the GW is twice that of the source's orbit: $f_{\mathrm{GW}} = 2\Omega$. By combining Kepler's third law in the form $\Omega^2 L^3 = GM$ and the definition of the compactness Ξ, we get the estimate

$$f_{\mathrm{GW}} \simeq \frac{c^3 \Xi^{3/2}}{GM} \quad \Rightarrow \quad f_{\mathrm{GW}}(M) \simeq \frac{M_\odot}{M} \cdot \Xi^{3/2} \cdot 100\,\mathrm{kHz} \tag{1.21}$$

This formula already shows that more massive systems generate GW with lower frequencies. For compact binary systems, Ξ is much less than the compactness of the individual bodies ($\Xi_{\mathrm{BH}} = 1/2$ and $\Xi_{\mathrm{NS}} \simeq 0.2$) because L is the orbital separation.

This estimate allows one to compute the typical frequency-range of GW emitters of astrophysical origin, as discussed in the next section.

Coalescence of Binary Systems

Binary systems of BH and/or NS[23] are undoubtedly the most important and promising sources of GWs. The main reason behind this is that (1) they are made of the most compact objects in the Universe, and (2) they are naturally occurring systems with the crucial "varying-quadrupole" nature, due to their gravitationally bound orbit. The orbital evolution of these sources is typically split into three phases:

- the *inspiral*, during almost all of a binary's lifetime, when the orbital separation L gradually decreases due to GW emission, but remains otherwise large compared to the individual objects radii R_S;
- the *merger*, describing a highly nonlinear and relativistic region of spacetime where the two bodies are close to one another and then make contact, in the sense that the length scales verify $L \simeq R_S$;
- the *ring-down*, when the newly formed, compact object relaxes towards a steady-state, either a BH in quasi-equilibrium or a big NS close to the mass limit.

The typical waveform for these systems, depicted in Fig. 1.8, clearly shows these three different stages. Binary systems are usually divided between sub classes that depend on the masses of the compact objects: stellar-mass binary systems, extreme and intermediate mass-ratio inspirals, and supermassive BH binaries. Since general relativistic effects prevent NSs from having a mass higher than around 2.2 to $2.6M_\odot$ [97, 101], this classification primary depends on the BH masse(s) involved in the binary.

Stellar mass binary systems refers to those with one or two BHs that have masses in the range $1M_\odot - 100M_\odot$. The astrophysical origin of such binaries is still an intense field of research, in particular for the more massive objects. Stellar/collapse formation scenarii favor BH with masses in the range $5M_\odot - 50M_\odot$. The lack of evidence for BH with masses less than $5M_\odot$ is refereed to as the *mass gap*. Reasons for its existence are still debated, from physical theories to observational bias [97]. Still, their most likely progenitors are expected to be binary stars with one (or both) undergoing a gravitational collapse, forming a BH or a NS, depending on its initial mass, or perhaps through gravitational capture, in highly populated regions of galaxies [97].

On a drastically different scale, there is a general consensus today that most galaxies are host to a supermassive black hole (SMBH) with a typical mass of at least a million, up to a few billion Solar masses [161]. As a consequence, when two galaxies merge, their SMBH undergo a slow inspiral which may result into a

[23] Theoretically, more exotic types of objects, such as dark stars [157] or boson stars [158] could also be included into this class. White dwarfs on the other hand, although compact, have too large of a radius to coalesce: they never reach the last stable orbit that would provoke a merger. Yet, they do produce GWs, which could be detected as part of a GW stochastic background or as resolved individual sources [159] (see also Sect. 1.3.3).

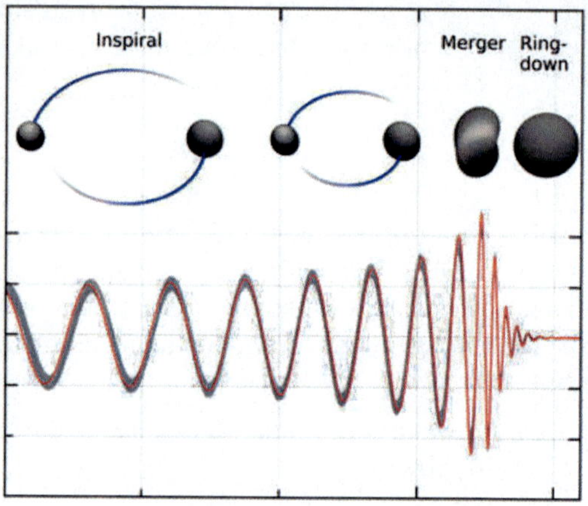

Fig. 1.8 A typical GW signal emitted by a binary system of compact objects, with its three phases: inspiral, merger and ring-down. This signal is a reconstruction of the first detection event, GW150914, extracted from the detection article by the LIGO team [160]. We clearly see that both the amplitude and the frequency rise until merger

coalescence over the billion years timescale, depending on their environment and their masses [162, 163]. These mass and time scales imply that SMBH binaries, if they exist, would be GWs emitters in frequency range the mHz and nHz. As GWs extract linear momentum from the system, an interesting byproduct of these binaries is that the post-merger SMBH may be ejected from its host galaxy. This effect, called gravitational recoil, may have major astrophysical consequences [164]. More details about SMBH binaries may be found in the recent review [97].

Lastly, the asymmetric combination of very large and low mass objects results in systems called extreme mass ratio inspirals (EMRIs). In practice, these systems are expected to exist close to the centre of galaxies where an SMBH may trap a compact star or a massive BH. They are of particular interest because the small body acts as a probe of the background BH spacetime. Consequently, the detection of GW from EMRIs are anticipated tests of GR in the strong-field regime. In particular, GWs from these systems will allow a test of the no-hair theorem, which states that BHs are uniquely determined by two numbers, their mass and spin [165–167]. The natural angle of attack for theoretically describing such systems is black hole perturbation theory, which takes advantage of the small mass ratio (between 10^{-4} and 10^7) to solve the Einstein equation in a perturbative manner. We shall come back to this formalism in Chap. 2. Contrary to stellar-mass binaries, which enter the detector only around the last orbits of the inspiral, the merger and ring-down), GW signals from EMRIs will consists in many (several hundred up to hundreds of thousand) orbital cycles, spanning anything from a few weeks to several years. The typical trajectory of the small body is very complex: it is three-dimensional and combines large spin precessions with high eccentricities, cf. Fig. 1.9.

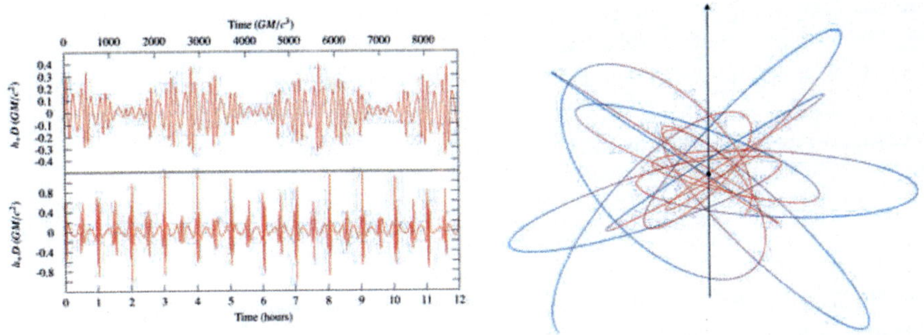

Fig. 1.9 *Left:* the "+" polarization of a typical GW signal from an eccentric, inclined EMRI obtained with a stellar-mass object orbiting a $10^6 M_\odot$ Kerr BH, with spin parameter $a = 0.9$. This data has been obtained by Drasco and Hughes in [168, 169]. The amplitude modulation on the top is due to the precession of the orbital plane, while the sharp peaks on the bottom coincide with passage at the pericenter (closest point to the central BH). *Right:* typical (Euclidean-mapped) trajectory of a test particle in the Kerr spacetime, represented by a timelike geodesic, as found in Mihaylov and Gair's article [170]

Isolated Sources

Some isolated NSs are expected to emit a steady, very clean, sine-wave-like GW signal. However, this can only be the case if they are not axisymmetric, in which case their mass quadrupole is almost constant and therefore the quadrupole formula (1.18) indicates a very low amplitude for the GW. A typical order-of-magnitude estimate gives, for such a rotating body

$$h_{\rm GW} \simeq \frac{1}{|\mathbf{x}|}\frac{4G}{c^4}\Omega^2 \delta I \,, \tag{1.22}$$

where $\delta I \equiv |I_1 - I_2|$ is the difference between the principal moments of inertia of the body orthogonal to the spin axis. As an example, a mountain of mass m located at the equator of a body of radius L generates $\Delta I = ma^2$. This estimate strongly depends on the equation of state of the NS [171], and gives $h_{\rm GW} \simeq 10^{-25}$ for typical values of a typical, kHz-rotating neutron star located in the galaxy $|\mathbf{x}| \simeq 10$kpc. Although this makes it a very low amplitude source, the near-constant periodicity of the amplitude should allow for a signal accumulation over long times, and thus a direct detection of these so-called "continuous GWs" in the future.

As we have seen, emitting GWs requires huge power outputs. As a consequence, core-collapse supernovae (CCSN) were the first natural candidates to be considered [172]. During a CCSN, a star reaches the end of its life: its newly formed, $\simeq 1.5 M_\odot$ iron core collapses on itself, creating a proto neutron star. The outer layers of the stars then start to fall inward and bounce off the core. The resulting shock, if energetic enough, breaks out from the star's envelope, leading to a supernovae explosion. During this event, GWs may be emitted at various stages, typically when the approximate axisymmetry of the star breaks. For example, during the bounce of the core [173], or alongside the (anisotropic) emission of neutrinos [174]. Although the typical GW amplitude for these CCSN depend on a considerable number of physical parameters (mass, size, spin, magnetization, environment, equation of state, etc.), a typical order-of-magnitude is [175]

$$h_{\mathsf{GW}} \simeq 10^{-24} \frac{\Omega}{1\mathrm{kHz}} \frac{20\mathrm{Mpc}}{|\mathbf{x}|}, \tag{1.23}$$

where Ω is the star's angular frequency. The GW signal from these systems is rather unique and very complex. Typically, it is made of a short burst followed by an erratic tail, and is much different than the smooth, continuous signals emitted by coalescing binary systems (compare a typical CCSN signal in Fig. 14 of [176] to Figs. 1.9 and 1.16 that depict waveforms from a coalescing binary). Consequently, the technique of matched filtering (see paragraphs below), so powerful when it comes to compact binary systems, is much less adapted for these type of systems, and alternative detection techniques must be used (excess-power searches in the time-frequency domain [177], deep learning algorithms [178], etc.). The expected rarity of these events (a few per century) is balanced by the fact that they should often present an EM counterpart, which would vastly help their detection [179]. Obtaining gravitational waveforms of CCSN is an incredibly difficult task, which relies almost entirely on numerical methods, as they are are dominated by stochastic processes. The field is an intense and active domain of research (see [180, 181] and references therein for recent, state-of-the-art simulations and computations).

Finally, let us mention other astrophysical sources of importance which are related to compact objects. Galactic binaries, which encompass white dwarf (WD) binaries or more generally binary systems with one compact object (BH, NS or WD). When entering the detectors, these systems will be far away from coalescence and their very large population (at least a few hundred millions in the Milky Way) will create an unresolved background of periodic GWs, with typical frequencies in the range 1–10 mHz [182, 183]. Accreting compact objects should also produce detectable GWs, for example, matter (say a main sequence star) tidally disrupting as it approaches a BH's horizon (see Sect. 4.3 of [136]), or accreting neutron stars [184, 185]. Elastic collisions between BHs (see Sect. 4.4 of [136]), which produce a characteristic wave form, and close-by inspiralling globular clusters may be detectable as well [186].

Cosmological Sources

Promising sources of GW are not limited to astrophysics, as we also expect to be bathing in a GW background, part of which should have been generated during the early phases of the Universe. Just like the incoherent background of astrophysical

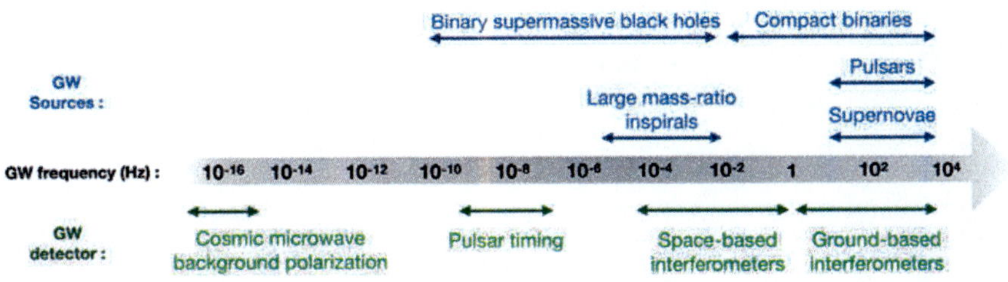

Fig. 1.10 Taxonomy of different GW sources (top, blue) with their associated, typical frequency-range (middle, black) and detection methods (bottom, green)

sources described above, GWs of cosmological origin must be described statistically, by means of an ensemble average.[24] Among the sources, we find [153]

- quantum density fluctuations amplified during the inflation period,
- various matter creation mechanisms during the reheating epoch,
- first-order phase transitions, at the epoch of electroweak symmetry breaking,
- networks of topological defects, such as cosmic strings, monopoles or textures. strings.

The stochastic background of GW generated during inflation contains a scale-invariant, irreducible component that is inflation model-independent. To this day, it has not been observed, although it should have left B-modes imprints into the CMB polarization, as well as distortions in the CMB black-body spectrum. However, this absence of detection puts stringent limits as to the frequencies of this GW background [188]. Thankfully, an increasing number of experiments are underway to try and detect these B-modes, while pulsar-timing methods (see Sect. 1.3.1) provide another promising way of putting bounds on the GW stochastic background, if not measuring it directly. For more on GW cosmology, we refer to the recent and thorough review [153].

As a summary, Fig. 1.10 recapitulates the different sources of GWs discussed above. For reviews and additional details on GW sources, see [171, 189].

1.3 A New Astronomical Era

Nowadays, it has become customary to read (part of) the following sentence "gravitational waves mark the dawn of a new era –or revolution– in astronomy". This is true both in public media, blog articles and in particular in the first introductory paragraph of scientific articles of peer-reviewed journals. Indeed, since the first detection of gravitational waves in 2015, the field of gravitational wave astronomy has had

[24] As always in cosmology, we must rely on some kind of ergodic hypothesis to argue that an ensemble average (impossible as we have only access to one Universe) coincides with a time average [136, 187].

a tremendous amount of success. In this last section, to conclude this first chapter, we will come back on the major discoveries allowed by GW observations from the last seven years (Sect. 1.3.2), present a non-exhaustive selection of applications that have been made thanks to these GW detections and what the future generation of GW detectors will bring (Sect. 1.3.3)

1.3.1 GW Detection

In this section, we examine how GWs interact with point particles and how GW templates are built, in order to understand how current ground-based GW interferometers work. We also discuss two other types of "detectors", namely the Weber bars, and pulsar-timing arrays.

GWs and Test Particles

The best way to understand how GWs are detected is to first describe how they interact with matter. The canonical way of explaining this interaction is to consider two test particles at rest with respect to one another in the absence of the GW. In a system of TT coordinates $x^\alpha = (t, x^i)$, the spacetime interval $\mathrm{d}s$ between two nearby events reads, by definition of the metric (1.7),

$$\mathrm{d}s^2 = -\mathrm{d}t^2 + (\delta_{ij} + h_{ij})\mathrm{d}x^i \mathrm{d}x^j + O(h^2)\,. \tag{1.24}$$

To observe a true physical effect of the passing of a GW, we must study physical observables, for example, a distance or time measured operationally by an observer. Suppose that an observer in the spacetime (1.24) sends a light ray in an arbitrary direction, until the ray bounces off a mirror and comes back. Between these two events (emission and reception) the observer measures, with its clock, a proper time interval $\Delta\tau$ (on the left in Fig. 1.11). Since this is one round trip of a light ray traveling at $c = 1$, the observer defines a "proper distance" L between themselves and the mirror "á la Einstein-Poincaré" [22, 190]

$$L \equiv \frac{1}{2}\Delta\tau\,. \tag{1.25}$$

To link the distance L (defined as a proper time interval) to the GW (encoded in the TT-coordinate expression of h_{ij}), one must appeal to two things. The spacetime events "emission" and "reception" are, by definition, the intersections of two distinct geodesics: a timelike one, the observer's worldline (in blue in Fig. 1.11), and a null one, the light ray's path (in yellow in Fig. 1.11). Solving the geodesic equation for the observer's worldline is a classical exercise (see e.g. [136] or [191]). It reveals that (1) the TT space coordinate of the observer remain constant, and (2) the TT time coordinate along its worldline coincides with its own proper time (all this holding at linear order in h). These two properties are rather important: the former shows that the observer does not move in the TT coordinates, even though a GW passes, and the latter implies the key relation

$$\Delta\tau = \Delta t + O(h^2) \tag{1.26}$$

between the proper time interval $\Delta\tau$ he measures and the TT time interval Δt elapsed between emission and reception. Since we expect $h = O(10^{-21})$ for a typical GW, we can safely work at linear order. Now, between these two events, the light ray travels on a null geodesic, such that $\mathrm{d}s^2 = 0$. Inserting this in (1.24), allows one to solve the resulting equation for $\mathrm{d}t$ in terms of $\mathrm{d}x^i$, and integrating along the null geodesic gives

$$\Delta t = 2\Delta x\left(1 + \frac{1}{2}h_{ij}n^i n^j\right) + O(h^2)\,, \tag{1.27}$$

where $\Delta x^2 \equiv \delta_{ij}\Delta x^i \Delta x^j$ is the coordinate spatial distance between the two events as measured in flat spacetime and n^i is the (Euclidean) unit vector joining them, see Fig. 1.11. In Eq. (1.27), the 2 comes from the light path being a round trip and the $1/2$ from expanding a square root to stay linear in h, and Δx^i is the TT coordinate spatial distance between the two events. Since Δx coincides with the proper distance L_0 between the observer and the mirror in flat spacetime (with $h_{ij} = 0$ in (1.27)), we conclude, thanks to Eqs. (1.25), (1.26) and (1.27), that the proper distance L between the observer and the mirror varies when a GW passes, according to

$$L = L_0\left(1 + \frac{1}{2}h_{ij}n^i n^j\right) + O(h^2)\,. \tag{1.28}$$

Given the typical waveform (1.20) of h_{ij}, we see from (1.28) that the distance between the test masses oscillates as the GW passes. This length variation induces a change in the optical path that a light ray follows between the test masses. These tiny length variations are measured by interferometry, the test masses being the mirrors of the interferometer. We will describe the current GW interferometers and their detection method below.

GW Templates and Matched Filtering

Reducing the noise in the GW detectors is a considerable effort, but still insufficient to detect a GW. Indeed, except for very loud (and thus rare) events, the GW signal is usually hidden in the detector's output. In order to extract the GW data, a systematic search is performed using a technique called *matched filtering*, which we now describe briefly. See otherwise [136] and references therein for details.

The first step in realizing this program is to solve the Einstein equation. More precisely, we need to know as accurately as possible the potential signals $h(t)$ that

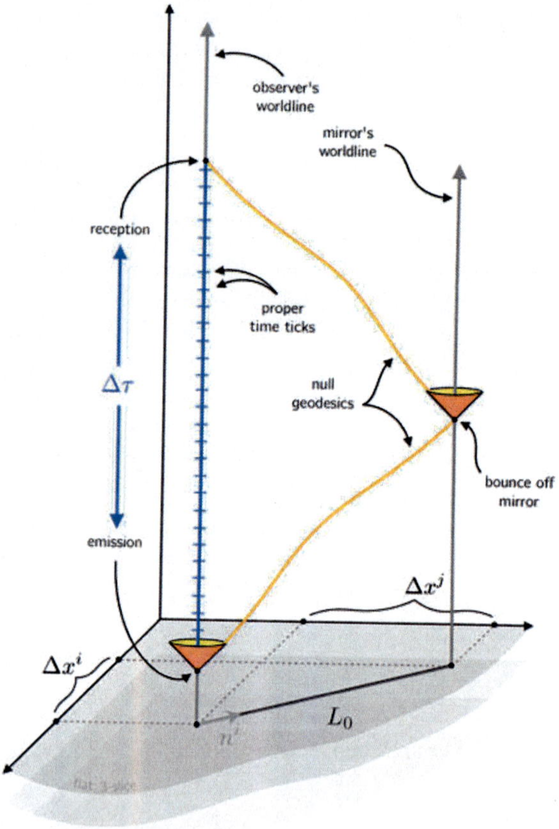

Fig. 1.11 Spacetime diagram showing the geometry of a distance measurement in presence of a GW. The distance L of (1.28) is defined as (half) the proper time interval $\Delta\tau$ measured by an observer (left), while he waits for the return of the light ray that bounces off a mirror (right). This measurement can be compared to the Euclidean distance L_0 measured in the flat spacetime when there was no GW (bottom)

may be coming from a given source of GW. Depending on the physical parameters of the source (e.g., masses, spins, orbital parameters, etc. for compact objects) a theoretical/numerical work allows theoretical physicists to build these GW templates. They are theoretical signals, solution to the Einstein equation, obtained by a combination of approximation schemes and numerical methods. In particular, focusing on compact binaries, the tools available to theoreticians are:

- post-Newtonian (PN) expansions [145, 192–195], which aim at solving the Einstein equation in an expansion with respect to the small parameter $\epsilon \equiv v^2/c^2$, where v is the typical orbital velocity of the source. This method allows one to express various quantities of interest at a given order in ϵ, for example the equations of motion of the binary, from which one can extract the binding energy E; and the metric perturbation h_{ab} far away from the source, allowing one to compute the GW energy flux $\mathcal{F}$. GW templates are then constructed adiabatically, i.e., by stitching together multiple circular orbits[25] of decreasing orbital radii (see Chap. 3 for details). In particular, if both E and $\mathcal{F}$ are known at some given PN order for a circular orbit of orbital frequency Ω, then the balance equation $\dot{E}(\Omega) = -\mathcal{F}(\Omega)$ can be integrated to find the evolution of the orbital phase, as: $\phi \equiv \int \Omega \mathrm{d}t = -\int \Omega(\mathrm{d}E/\mathrm{d}\Omega)\mathcal{F}(\Omega)^{-1}\mathrm{d}\Omega$. Since $v/c \simeq GM/Lc^2$ for self-gravitating binary systems with total mass M and separation L (thanks to the virial theorem), the PN formalism is particularly adapted for describing the inspiral phase of compact binaries, where the separation L remains large compared to the object's proper size and the orbital speed remains small $v \ll c$. For up to date, state-of-the-art PN results, see [196].

- black-hole perturbation techniques, are based on the idea that some relativistic systems are close to an isolated BH, and look for solutions to the Einstein equation as the sum of a BH metric plus a small perturbation. Two important applications are: the gravitational self-force (GSF) theory [197, 198, 198–200], and quasi-normal modes (QNM) analysis [201–203]. The former applies to EMRIs: one expands the perturbation h_{ab} in powers of the small mass ratio, and derives both the metric and the equation of motion of the small body in the BH spacetime background (see Sect. 2.2.4). This method covers the inspiral up to the plunge, as well as the ring-down of the binary, but is limited to EMRIs (mass ratio $\simeq 10^{-7} - 10^{-4}$).[26] The gravitational waveforms are then computed directly from the metric perturbation h_{ab}. In practice, the linear order perturbation is sufficient for GW detection but second order results will be necessary for parameter estimation. QNM theory aims at computing the relaxing modes of a vacuum metric representing an isolated, perturbed BH. This perturbation, encoded in the vacuum metric, can be used to derive the waveform of the ring-down (post-merger) of binary inspirals, or of BH collisions.
- numerical relativity (NR) consists in solving the Einstein equation numerically, typically by decomposing it into a 3+1 space/time system of partial differential equations. Since the simultaneous breakthrough in the early 2000s by independent groups [209–211], NR is able to simulate efficiently the formation and ring-down of a Kerr BH after the merger of two smaller ones. As of today, different NR groups release waveform banks that cover most of a binary's coalescence, including spins and eccentricities [212–214]. Although it has a priori no intrinsic physical limitations (no approximation scheme is involved), the computational cost is the limiting factor for this method, which is still complementary to other analytical approximation schemes. References for current state-of-the art results of NR may be found in Sect. 4 of [97] and references therein.

As we mentioned, these methods are particularly relevant to a portion of the whole coalescence, and/or to a subclass of sources. Consequently, to obtain com-

---

[25] PN theory has also been developed to handle eccentric orbits, in which case one also needs to rely on the balance of angular momentum in addition to the balance of energy.

[26] Although a number of recent results seem to converge on the fact that GSF theory may very-well hold for mass ratios that are not that small (for the so-called intermediate mass ratio inspirals (IMRIs), with mass ratio $\simeq 10^{-4} - 10^{-2}$), or even with mass ratio flirting with unity [204–208].

plete waveforms that span the early inspiral up to the ring-down, effective and phenomenological schemes, among which the effective-one-body (EOB) model [215, 216] and the Phenomenological (Phenom) formalism [217], are used. The EOB formalism consists in mapping the true metric of a relativistic two-body system to that of an effective metric, parametrized by the binary's physical characteristics (masses, spin, ...). The model was initially introduced by Alessandra Buonanno and Thibault Damour in [218, 219] and has since been successful at modelling binary systems with a wide range of parameters. By construction, the EOB dynamics reduce to the geodesic motion of a test particle in a Schwarzschild background in the extreme mass-ratio limit, and to the Newtonian 2-body dynamics in the weak-field/small-velocity limit. Beyond these two cases, the EOB formalism needs to be "calibrated" using a combination of results from NR, GSF and PN expansions. EOB is able to provide "factorized" waveforms for the whole coalescence of a binary system, by smoothly stitching together the inspiral-plunge and merger-ring-down waveforms. As such, the EOB framework plays a central role in the detection of GW events by current ground-based detectors [220, 221]. The Phenom model [217] is based on the simple idea that a GW signal, however complicated, may always be written in the form $h(f, \vec{p}) = A(f, \vec{p}) \exp(i\Phi(f, \vec{p}))$, with A the amplitude and Φ the phase of the waveform, that depend on the instantaneous frequency f and a set of physical and phenomenological parameters encoded in the vector $\vec{p}$. These parameters are fixed by comparing with other frameworks (PN, GSF) and/or calibrating with NR results. Both the EOB and Phenom schemes are now able to include spin precession, orbital eccentricity as well as tidal effects. Details and references may be found in Sect. 6 of [97]).

An overview and historical perspective results for all these schemes can be found in the review [222]. It should be noted that, although all these methods have strength and weaknesses, they are not limited to some well-defined number of contexts, but rather overlap in their application range. What's more, we find in the literature a number of studies that conducted comparison between all these approximation, numerical and effective schemes, essentially showing agreement between all methods whenever they can be compared [205, 223, 224] (Fig. 1.12).

Once GW templates have been constructed, they are stored into GW template banks, which are regularly updated and improved. These templates depend on intrinsic parameters that enter the source's description: masses, spins, and other geometric parameters: location in the sky, orientation with respect to the detector etc. These waveform banks are crucial for the detection of the GW itself, as they allow (1) to extract the GW from the noisy detector's output by making a systematic search (see below), and (2) estimate the source's parameters with best confidence once the GW signal is extracted.

The second step is to analyze the detector's output signal and extract the GW data. To illustrate, let us think of the detector's output as a superposition $h(t) + n(t)$, of a "true" GW signal $h(t)$ and some random noise $n(t)$. Even with all the noise reduction techniques involved in the detectors, one typically has $|h| \ll |n|$, so that the GW signal is hidden in the output. Thankfully, there exists a detection method called "matched filtering" that is tailored to the situation, taking advantage of the fact

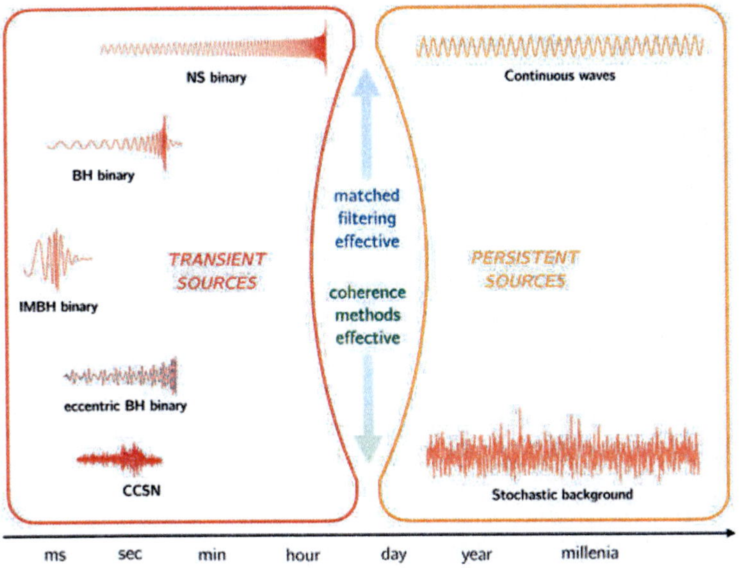

Fig. 1.12 This diagram shows the different types of GW sources. The horizontal location determines the (order of magnitude of) waveform duration in LIGO-Virgo-Kagra, with transient sources (compact binary mergers and CCSN) on the left, and persistent (i.e., continuous) sources on the right. The vertical location determines the degree of effectiveness of the two main detection techniques: matched filtering method for the top (phase-fitting with GW template, adapted for long-lasting, slowly varying waveforms) and the coherence method (looking for coherent, excess power in several detectors, most effective for erratic signals). Note that intermediate-mass BH binaries are in between the two

that (1) the GW and the noise have very different frequency characteristics (the GW has a well-defined, regularly evolving frequency whereas the noise is erratic with a wide range of frequencies), and that (2) we do know what kind of $h(t)$ to expect from typical sources, thanks to GW templates. In a word, matched filtering consists in comparing a GW template to the detector's output and compute the cross-correlation between them. The higher the correlation, the stronger our confidence in having detected a GW signal. If the same template accounts for different detectors involved in the detection, then there is no doubt that a GW has passed. Matched filtering is a very elegant detection technique which we briefly describe below, referring to [136] for details and [225] for an introduction to other approaches in GW detection.

Let us think of the detection as a linear process with some given transfer function $\mathcal{D}(\omega)$ (Fourier-domain equivalent of its impulse response). In order to know if some GW signal $h(t)$ is present in the detector's output, one constructs the signal-to-noise ratio (SNR) classically, as

$$\mathrm{SNR}[\mathcal{D}] \equiv \frac{S}{N}, \quad S \equiv \int_{\mathbb{R}} \mathcal{D}(\omega)\tilde{h}(\omega)\mathrm{d}\omega, \quad N^2 \equiv \int_{\mathbb{R}} |\mathcal{D}(\omega)|^2 S_n(\omega)\mathrm{d}\omega, \tag{1.29}$$

where S is the output power of the filtered GW signal $\mathcal{D}(\omega)\tilde{h}(\omega)$ (recall that time-convolution is equivalent to Fourier multiplication), and N^2 is the output power of the filtered noise, described by its power spectral density $S_n(\omega)$. To check whether the GW $h(t)$ is in the output, we want to make SNR[$\mathcal{D}$] maximal. To this end, we can try different transfer functions, or filters $\mathcal{D}$, but then the question becomes: which $\mathcal{D}$ will maximize the value S/N? An elegant solution to this (classical) optimization problem is obtained by writing (1.29) into a form that involves a scalar product. Indeed, if we control the typical noise entering the detector (and thus its S_n) we can define for any two complex functions $(\tilde{f}, \tilde{g})$ a positive definite scalar product by

$$(\tilde{f} \mid \tilde{g}) \equiv \int_{\mathbb{R}} \tilde{f}(\omega)^* \tilde{g}(\omega) \frac{\mathrm{d}\omega}{S_n(\omega)} \quad \Rightarrow \quad \mathrm{SNR}[\mathcal{D}] = \frac{(\mathcal{D}\, S_n \mid h)}{||\mathcal{D}\, S_n||} \tag{1.30}$$

with $||\tilde{f}||^2 \equiv (\tilde{f} \mid \tilde{f})$. The expression for the SNR on the right of (1.30) is particularly appealing, and one may even say that it already contains the solution to the maximization problem. Indeed, let us recall the Cauchy-Schwarz inequality $(\tilde{f} \mid \tilde{g}) \leq ||\tilde{f}||\, ||\tilde{g}||$, with equality if and only if $\tilde{f} \propto \tilde{g}$. Applying this to the right-hand side of (1.30), we readily conclude that the SNR is always bounded above by $||h||$. Moreover, from the equality case we find that this bound $||h||$ is the SNR's maximum, reached when $\mathcal{D}S_n \propto \tilde{h}$. In other words,

$$\mathrm{SNR_{max}} \equiv \sqrt{\int_{\mathbb{R}} \frac{|\tilde{h}(\omega)|^2}{S_n(\omega)} \mathrm{d}\omega}\,, \quad \text{obtained when} \quad \mathcal{D}(\omega) \propto \frac{\tilde{h}(\omega)}{S_n(\omega)}\,. \tag{1.31}$$

To summarize, when some GW enters the detector, the output is a noisy signal, with a low SNR. Heuristically, one can think of matched filtering as sliding thousands of stencils over the output and finding the best matching one. In practice, this amounts to trying many different filters: one explores the GW templates bank (using a standard MCMC method) to try and find the filter with the highest SNR. By construction, this template must be very close to the real GW (in the sense of the above scalar product). If $n(t)$ were a white noise, then S_n would be constant, and therefore the best filter would be the GW signal itself (recall (1.31)). In general, $n(t)$ is much richer than white noise, and the integration measure $\mathrm{d}\omega/S_n(\omega)$ acts as a weight that dictates which frequencies should be tuned up or down in the filter to maximize the SNR. An illustration of this GW extraction from a noisy signal is depicted in Fig. 1.13, with signals taken in the second detection of GWs by the LIGO interferometers [226].

Lastly, we should mention that matched filtering only works well for signals that are well-modeled, to give enough time for the matching to align the template with the true signal in order to produce a large SNR. For example, strongly modulated GW signals, such as those produced by highly eccentric orbits are currently not well-modeled to be detected via matched filtering. Less predictable signals, such as short-lasting "burst" produced by CCSN, rely on other detection methods, typically a search for coherent GW power across the detectors network. Recent readout techniques, involving for example deep-learning algorithms [178] or the so-called

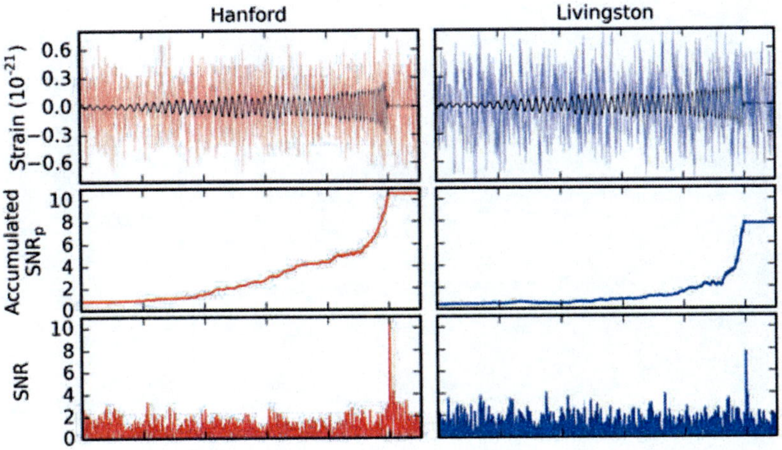

Fig. 1.13 Top row: (slightly filtered) strain data from the detectors output during the detection of the event GW151226 [226]. In black, the best-match template is highlighted, whose amplitude lies way below the noise level

balanced homodyne schemes [227] are being developed and are very promising, in particular for the new generations of detectors.

Terrestrial GW Interferometers

From Eq. (1.28), we see that the typical relative displacement $(L - L_0)/L_0$ to be measured is of the order of the GW amplitude h. Even for a rest length $L \simeq 1$ km, already $\Delta L \simeq 10^{-18}$ for a typical binary system GW, which is smaller than a thousand of an atom's radius. Measuring such a relative variation in distances is a remarkable challenge. To achieve such, light rays are sent to bounce back and forth between the test masses, which are realized by suspended mirrors. Since the distance between these mirrors changes with time, so does the optical path for the light ray, which naturally calls for laser interferometry. Today's GW detectors are therefore nothing but giant interferometers. More precisely, a single laser is split in two, thanks to a semi-reflecting lens, and injected into two perpendicular, km-long vacuum tubes. At the end of both arms, a suspended mirror reflects the laser, which is then recombined.[27] This recombination between two coherent rays creates an interference pattern whenever the optical path differs between the two: one obtains an output of typical amplitude 10^{-21}, which encodes all sources of noise as well as a potential GW signal.

[27] In practice, the light makes around 150 round trips in the vacuum cavity thanks to additional mirrors to make one Fabry-Perrot cavity in each arm, before being recombined. This effectively enhances the length of the detector's arms.

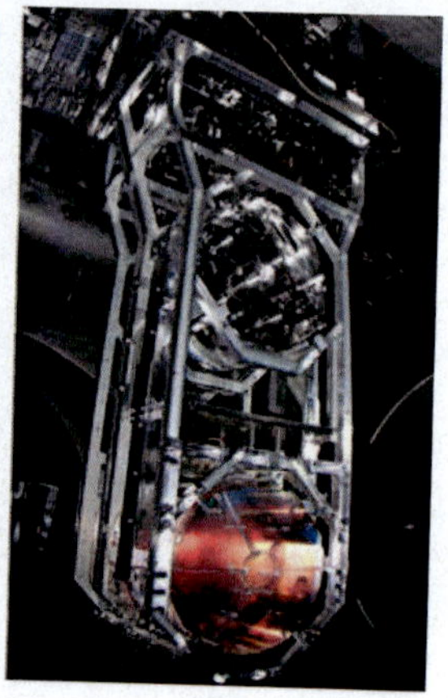

Fig. 1.14 *Left:* exterior view of the Virgo observatory, with the Alpes in the background. Visible are the two perpendicular, 3 km-long arms extending from the main building. Credit: *Nicola Baldocchi*. *Right:* the 40 kg mirror (bottom) in the LIGO detector, suspended by silica fibers to the seismic noise-reducing pendulum system (top). Credit: Map data ©2021 Google (left) and LIGO (right)

The mirrors are suspended to an ingenious system of pendulums to (1) counteract Earth gravity and make it a virtually free-falling particle, and (2) isolate against seismic noise. Speaking of noise, there lies the most challenging aspect of GW detection at the technical level: GW detectors have to be able to distinguish between a GW signal and all sorts of noise. Current detectors have a sensing bandwidth from tens to thousands of Hertz. At low frequency (between 10 and 50 Hz), the noise is dominated by scattered light and seismic activity, while quantum noise dominates high frequencies (700 Hz) noise. See [228, 229] for the different types of noise, and how they can be controlled/reduced in the LIGO [230] and Virgo [231] detectors.[28]

As of July 2021, six independent such interferometers are dedicated to the detection of GWs and are active. The three most important make the so-called LVK collaboration: the two Advanced LIGO observatories (based in the USA), the Advanced Virgo detector (based in Italy) (cf. Fig. 1.14) and the KAGRA detector (based in Japan). The other three AIGO (Australia), CLIO (Japan), and GEO600 (Germany) have a much lower sensibility than LVK but have been essential to the development and improvement of their technology (Fig. 1.15).

Resonant Mass Detectors

Although the first indirect observation of GWs dates from 1974, we find in the literature as soon as 1969 an article published by Joseph Weber [234], announcing the first direct detection of GWs using an apparatus invented by himself [235] and today

[28] See [232] for a noise hunt in the detector's output, showing how external sources such as wind, ocean waves, human activity, airplanes, thunderstorms, helicopters, etc. couple to the detector.

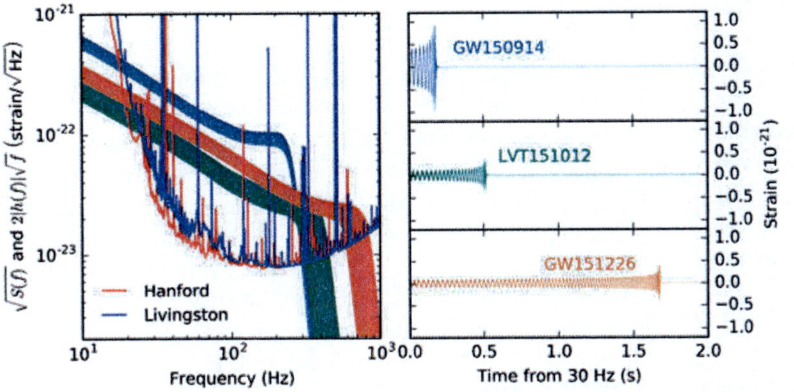

Fig. 1.15 The left figure shows the amplitude spectral density of the total strain noise of both LIGO detectors, along with the path followed by the three most significant events, during the O1 run. By construction, the SNR of each event can be read as the area between these paths and the noise curves. On the right, the (recovered) time evolution of these three signals is depicted, starting at 30 Hz when they enter. Notice the different duration of the signals and the unexpected "loudness" of the very first event, GW150914. Images extracted from the LIGO-Virgo collaboration article [233]

known as a "Weber bar". This detection technique consists in tuning the resonant frequency of a metal mass around 1kHz, a frequency associated to a number of GW sources. Typically, these bars are 1 m-long, 30 cm wide cylinders made of aluminum-alloys, and thus weight a few tons. As a GW passes, it squeezes/extends the bar slightly (recall Eq. (1.28)), and these contractions are captured by a clever read-out electro-mechanical system called a "resonant transducer" [136]. By construction, although this apparatus has a very narrow frequency range (few hundred Hertz), cooling and isolating systems allow it to reach formidable sensitivity and measure length changes of order $10^{-19} - 10^{-18}$m. This sensitivity should be sufficient to detect events that are very loud and close, typically in our galaxy, i.e., rare.

Following Weber's pioneering works, many other teams launched similar experiments around the world until the 1980s [236], making the so-called second generation of resonant-mass experiments. However, Weber's discovery was never replicated and is nowadays considered to be due to noise.[29] The third generation of resonant-mass detectors, which include the Mario Schenberg GW Antenna [238] (based in Brazil) and the MiniGRAIL Gravitational Radiation Antenna [239, 240] (based in the Netherlands), use spherical designs and are cooled down at the mK level, allowing one to reach peak sensibilities of order $10^{-21} - 10^{-20}$ [241].

[29] I recommend taking a look at Janna Levin's brilliant book *Black Hole Blues and Other Songs from Outer Space* [237]. Chapter 5 tells the story of Weber's experiment and its outcome(s).

Pulsar Timing Arrays

We have seen in earlier paragraphs that binary pulsars allowed the first (indirect) detection of GWs, in particular thanks to the remarkable stability of their period which ranges from about 1.4ms to 30ms. Today we observe hundreds of such pulsars and control the arrival time of the pulses with an accuracy of microseconds for up to several decades into the future. Consequently, any latency in the arrival times means that something happened to the pulse as it was traveling towards us. If this "something" is a passing GW, then this should be reflected in the arrival timings of several pulsars, and one expects a coherent shift in the arrival times, depending on the GW's direction, frequency and amplitude. The pioneering calculations on the effect a GW has on a pulse propagating from a pulsar to Earth date from the 1970s [242, 243]. Measuring the arrival times of various pulsars allows, through a cross-correlation analysis, to disentangle the residuals from a potential GW signal. PTAs will be able to measure GW signals from isolated SMBH and GW backgrounds, either cosmological or astrophysical (typically from unresolved SMBH binaries).

Today, three major PTAs exist: the European PTA Consortium [244], the Parkes PTA [245] (Australian) and the NanoGRAV [246] (North-American). Together with Indian PTA Project [247], they make the International PTA (IPTA) [248], while promising new PTA projects are emerging in China and South Africa (see references in [249]). Updated data releases have recently been published [250–252] and provide more constraints on GW backgrounds. As more binary pulsars are discovered and new telescopes are added to the collaboration, the range and sensitivity of PTAs will improve. In this context, the Square-Kilometer-Array telescope [253, 254], which should be operational this decade, is expected to drastically improve GW searches by PTAs.

1.3.2 Seven Years In

As of today, the LIGO and Virgo detectors are the only detectors to have directly measured GWs. Although they had been observing for more than a decade, these GW observations were made possible thanks to the "advanced" design of the detectors, which benefited from many improvements technological compared to the original ones. The nomenclature for these GW events is of the form "GWYYMMDD" with the date YYMMDD.[30] The first two *runs* (period when the detectors are online and actively looking for GWs), went from September 2015 to January 2016 (O1), and November 2016 to August 2017 (02), with Virgo joining the two LIGO detectors at the end of O2. During O1 and O2, the LIGO-Virgo collaboration observed 11 GW signals [256], 10 of which were produced by the coalescence of binary BHs, and 1

[30] The new nomenclature will also include UTC time of detection [255], in the form GWYYMMDD_HHMMSS, (Hour-Minute-Second), as several events are expected to be detected on the same day, in the near future.

from a NS merger. The first part of the third run, O3a, spanned April to September 2019 and added 39 GW events, bringing the GW catalog to a total of 50 detections [255]. Although observations stopped officially on March 2020 due to the COVID-19 pandemic, the O4 run will hopefully start in late 2022 with the KAGRA detector joining and many improvements made to LIGO-Virgo.

During the first seven years of gravitational astronomy, a remarkable diversity of GW events have been observed. Without exception, all of these detections are compatible with GR, which underwent (yet again!) a number of unprecedented tests in the radiative sector and strong field regime. In addition to validating compact binary models and raising new questions as to their formation channels, these observations allowed measurements of key GW features, such as their speed, their polarization state, multipolar structure (GW harmonics), etc. In what follows we present a non-exhaustive selection of six of the most interesting GW events detected so far. All data and companion papers are conveniently available on the dedicated page [257] on the LIGO website.

The First One: GW150914

Ninety-nine years and three months. This is the time that passed between Einstein's first publication on GWs (June 9, 1916) and their first direct detection by the LIGO interferometers (September 14, 2015) [160]. This GW event lasted 0.2 s, and encompasses the inspiral, merger, and subsequent final black hole ring-down. The two LIGO detectors observed the same signal 6.9 milliseconds apart, which corresponds to the light-time taken by the GW to travel between their respective location in Handford and Linvingston. The waveform is perfectly compatible with predictions of GR, as that emitted by a binary system of black holes of 36 and 29 Solar masses, at a luminosity distance of 410 Mpc.[31] It confirmed at once the existence of GW, of binary BHs, of BH mergers, and BHs with masses much higher than those inferred from X an gamma-ray observations. Most remarkably, this detection was registered just before the actual start of the first observing run O1. The strain from the detector shows a very loud event (SNR of 24), and the GW waveform is visible by eye (see Fig. 1.16).

The Collaborative One: GW170814

Virgo joined the LIGO detectors at the end of O2, on August 1, 2017. Thirteen days later, the three detectors conjointly observed a GW signal compatible with that emitted by a BH binary located 540 Mpc away, of 30 and 25 Solar masses [259]. The signal is again visible by eye, especially in the LIGO detectors which presented at the time a better sensitivity and noise reduction system (see Fig. 2 in [259]). The delay between the detection (8 ms for Linvingston-Handford, and 12 ms

[31] See [258] for a first principles, pedagogical derivation of these parameter estimates.

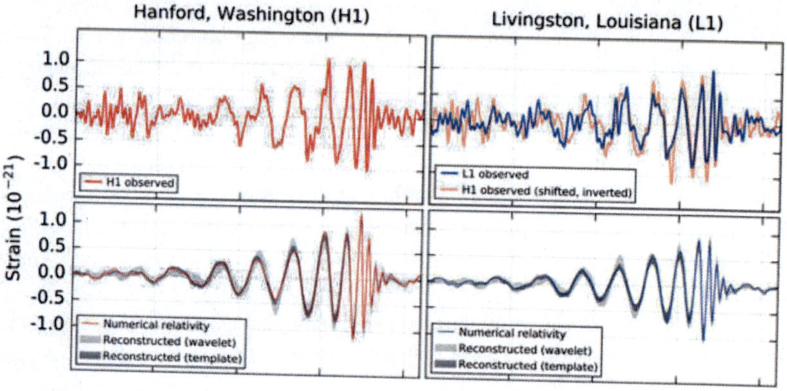

Fig. 1.16 Gravitational waveforms extracted from the output of the two LIGO detectors (Handford on the left, Linvingston on the right) for the first ever GW event, GW150914, as presented in the detection paper [160]. On the top, the direct, filtered strain. On the bottom, reconstructed waveforms are shown, obtained with the best parameters extracted from the signal

between Linvingston-Virgo) is again compatible with the speed of light, as predicted by GR the three-detectors network allow for a much more precise localization of the source on the sky. The most important feature of this detection is the ability to measure the polarization of the GW. Technically, the polarization could not be measured accurately by the LIGO detectors only because of their very small relative orientation. Thanks to Virgo, it was possible to show that the measurement strongly favors GR and its two-polarization prediction (Fig. 1.17).

The Bright One: GW170817

Three days after the first three-detectors detection, we witnessed one of the most incredible scientific events in history: the joint detection of GWs from a binary NS merger along with its EM counterpart. On August 17, 2018 LIGO-Virgo detect[32] a very loud signal (SNR of 32.4). Lasting for more than 100 s, it was identified through matched filtering with that of a binary NS merger [260]. Thanks to triangulation, the location of the source was confined in a 31 deg^2 region on the sky, at a distance of 40 Mpc. The signal was compatible with a binary NS merger with masses constrained between 0.86 and 2.26 Solar masses. But then, around 1.7s after GW170817, the Fermi telescope observed a gamma-ray burst (RB170817A) in the same direction. Much like an Earth-wide treasure hunt,[33] the updated, 28 deg^2 localization map was

[32] In fact, Virgo, although fully operational, did not see anything of this event. This indicates that the source was in the detector's blind direction (i.e. the GW produced the same displacement in both arms of the detector) which helped to find its location in the sky.

[33] This analogy works pretty well since it has been reported that an exceedingly large amount of gold (between 3 and 13 Earth masses !) was produced during this event [261].

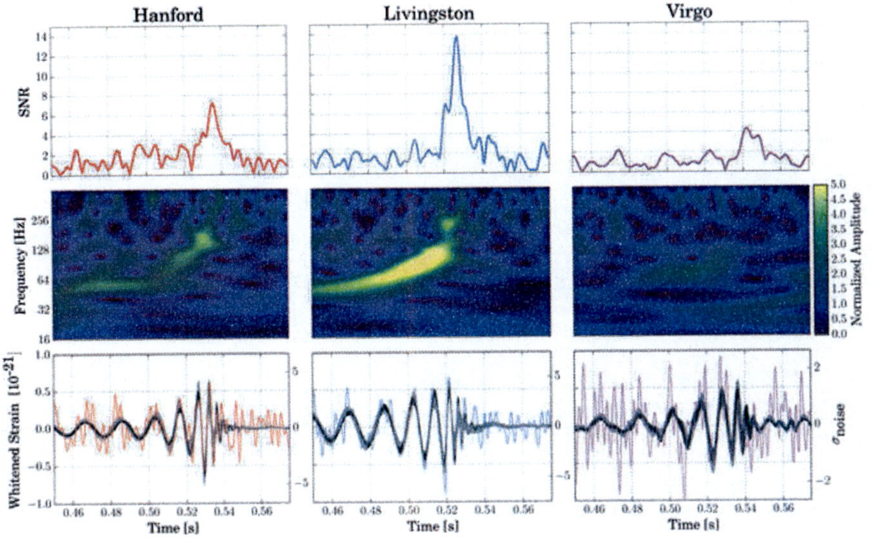

Fig. 1.17 This figure shows the first three-detector detection, GW170814, with different characteristics of the signal as a function of time: SNR at the top, power in time-frequency domain in the middle, and the reconstructed waveforms on the bottom. From LIGO-Virgo collaboration's detection paper [259]

issued publicly and a wide observation campaign began. Less than 11 h after the merger, an optical transient was detected in the NGC 4993 galaxy by the SWOPE telescope. During the following days, more than a hundred telescopes across the seven continents observed the event, and data was collected in the whole EM spectrum [262]. This event is, to date, the most fruitful multi-messenger observation and has had a tremendous impact in astronomy, cosmology and virtually all branches related to astrophysics (Fig. 1.18).

The Massive One: GW190521

With the GW190521 event, the first observation of an intermediate mass BH was made. These objects have a mass above the 100 Solar mass threshold, a number not favored by current stellar collapse models, making it particularly interesting. The signal was emitted by two BHs of 85 and 66 that merged into one of 142 Solar masses [264, 265]. The signal detected from this coalescence is unlike those described above, as can be seen in Fig. 1.19. Indeed, these massive systems merge at lower orbital frequencies, which means they are very close to merger when they enter the detector's bandwidth, giving only a few cycles. These events are thus less sensitive to matched filtering methods (that work best for long-duration signals) [229]. The 85 Solar mass BH belongs to the so-called "second mass gap" produced by pair-instability supernovae processes. Explaining the formation channels that

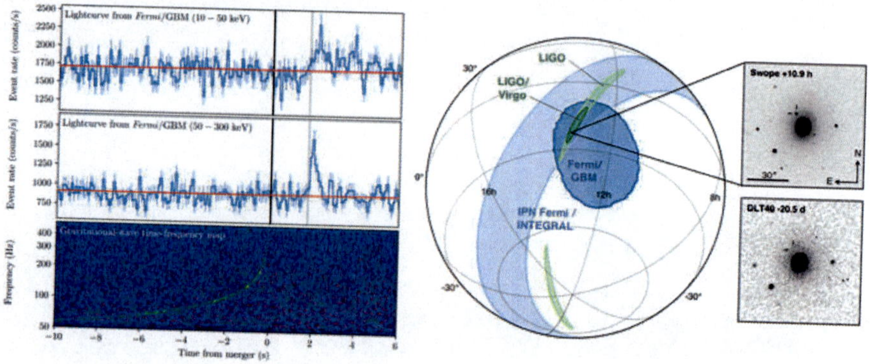

Fig. 1.18 This figure shows key features of the NS merger event GW170817. In the left panel, as a time series, the lightcurves (number of gamma photons per unit of time) of the Fermi telescope (top and middle). The black vertical line corresponds to the merging of the NSBH binary, as indicated by the tip of the GW frequency curve (bottom). We see that the peak of gamma photons arrives less than 2 s after merger. On the right panel, the colored patches show the localization of the source from different instruments. The inset magnifies the location of the host galaxy NGC 4993 before (bottom) and after (top) the event, in visible light. From LIGO-Virgo collaboration's GW detection paper [263] and multi-detection paper [262]

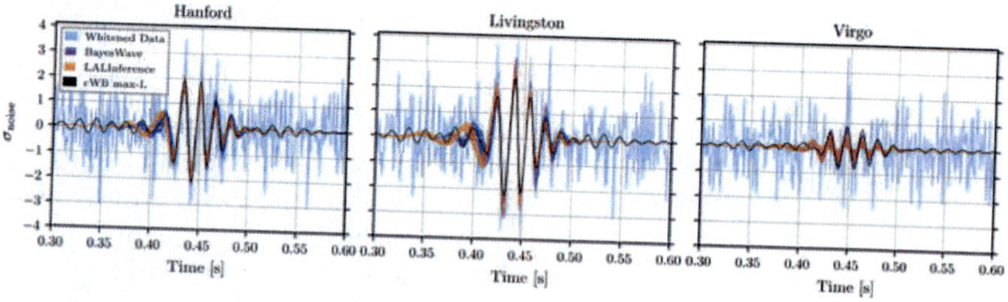

Fig. 1.19 This figure shows the three time-domain strains from the LIGO-Virgo detectors during the event GW190521, with reconstructions of the signal superimposed over the data (light-blue). Notice the small number of cycles in this waveform, due to the larges masses of the components. From the original detection paper [264]

lead to such BH populations remains a challenge today, and GW astronomy provides a way of estimating these merger rates and thus discriminate between formation models [97].

The Strange One: GW190814

On August 14, 2019, GWs from a compact binary system were observed by the LIGO-Virgo detectors [266]. The system is composed of a 23 Solar masses BH of with a smaller object, between 2.50 and 2.67 Solar masses. This object is rather unexpected: on the one hand standard NS models prevent them from being that massive [101], on the other such a small BH has never been observed before with EM observations. The nature of this secondary object remains somewhat of a mystery to this day, as most known formation mechanisms do not account for compact objects within this so-called "first mass gap", between massive NSs $2.2M_\odot$ and low-mass BHs ($\simeq 5M_\odot$).[34] Another important feature of this event is its mass ratio of 0.11, making it the most asymmetric binary observed to date. This property allowed to resolve, with very high confidence, the signature of the subdominant mode of the GW multipolar structure. For this event, the subdominant $(\ell, m) = (3, 3)$ mode was clearly visible alongside the dominant, quadrupolar mode $(\ell, m) = (2, 2)$, at 1.5 times its frequency[35] (see [267] for explanations). This higher-multipole measurement was expected from GR and was also made in the earlier, second most asymmetric event GW190412 [268].

The Awaited One: GW200105

During the first two-and-a-half runs, no GWs emitted by a NS-BH merger were detected. The most probable candidates were either a marginal event that could not be distinguished from a detector noise artifact, or the "strange" event GW190814 with a secondary mass above the NS mass limit. It is only in the data of the O3b run that not one but two such events were seen and confirmed for the first time [269]. The first, GW200105 was detected by one of the two LIGO detectors and VIRGO, while the second, GW200115, was detected by all three. The signal is compatible with sources that have a primary object of mass above the NS mass limit, and a secondary object with mass below it, as can be seen in Fig. 1.20. The effect of tidal disruption between the bodies near merger, if there were some, was not found in the imprint of the GW signal, and no EM counterparts was identified either. Even though these observations (tidal effects and EM) are not expected from such far-away systems, it is not possible to rule out that this event was a binary BH merger with a component of very low mass. Either way, this event is the first of its kind to be a confirmed, and many more NS-BH mergers are expected to be observed during the next runs.

1.3.3 Applications and Future

The field of GW astronomy has just started and yet, many upcoming projects and experiments are already being built and/or planned. Thankfully, the tremendous impact these last seven years have had on astrophysics, cosmology and fundamental physics helped raise at least as many questions as it answered. In this last section, we provide a selection of major applications and results that have been possible thanks to the first seven years of GW astronomy, before taking a glimpse at its very promising future with the next generation of space- and ground-based detectors.

---

[34] The second mass gap comes from the scarcity of observed BHs above fifty Solar masses [97].

[35] Coincidentally, this event is also the *fifth* in this selected list....

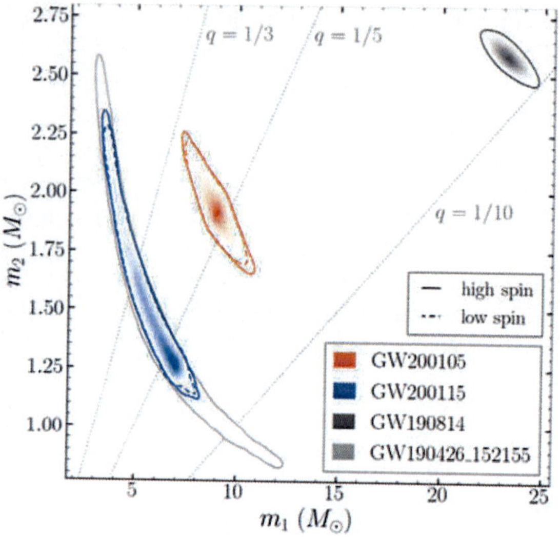

Fig. 1.20 This diagram shows the posterior distribution for the component masses of the first two confirmed NS-BH events (red and blue). In grey, two probable, but not confirmed events from the O3a run, as presented in the detection paper [269]

Astrophysics with GW

As all currently detected GW have been emitted by astrophysical sources, numerous applications from GW astronomy deals with astrophysics. If we only count the GW events with a false alarm rate less than 1/year, then the catalog contains 44 BH binaries, 2 NS binaries and 1 system made of a BH and light object (GW190814). A Bayesian analysis of the parameters of these systems, including the component masses, the mass ratio, the effective spin and luminosity distance, allowed to put constraints on population models and binary system formation mechanisms [270].

The first binary NS merger (GW170817) allowed to show evidence for the creation of hypermassive neutron stars, which was the most probable outcome of this merger [271]. The simultaneous detection of the GW and EM signals from the binary NS mergers have directly confirmed the long lasting hypothesis that binary system coalescence is the source of high-energy transients, in particular short gamma-ray bursts [263, 272] and kilonovae [273, 274]. It also provided strong evidence that NS mergers are an important formation mechanism of heavy elements, constraining nucleosynthesis models [275], and provided a new and unique way of understanding the formation and structure of relativistic jets [276, 277].

Fundamental Physics with GW

The joint observation of EM and GW emission from the binary NS merger also allowed for test of GR. For example, the speed of GWs, which was found compatible with the speed of light with a 10^{-16} precision, allowed to rule out a number of modified gravity theories [278]. Additionally, it was possible to test the Equivalence Principle using the Shapiro delay, again finding agreement with its GR value at the 10^{-6} level [263]. Similar tests enabling a discrimination between GR and competing theories will be possible with LISA [279] and third generation detectors [280].

A number of constraints have already been put on microphysics models describing the state of matter in NS's core. This is possible thanks to a measurement of the deformability coefficients (Love numbers) of the neutron stars in the event GW170817 [260, 281]. In particular, the constraints on the equation of state from this event showed that they were neither very stiff nor very soft [282, 283]. Similarly, numerical estimates on high-order PN parameters have been obtained, constraining two-body dynamics of a compact-object binaries in the strong-field, nonlinear regime of GR, thanks to the early detections GW150914 and GW151226 [233].

Cosmology with GW

Although the ΛCDM model has brought a remarkable understanding of the Universe's history and composition, its very name hides two major unresolved mysteries. First, while the observed acceleration of the expansion of the Universe can be accounted for by introducing a nonzero cosmological constant (Λ, in ΛCDM), there are good reasons to look for alternative explanations, gathered under the name "dark energy". Second, the ΛCDM model can only fit observations if about 25% of the Universe's content is assumed to be a pressure-less matter fluid that interacts only gravitationally, thus called Cold Dark Matter (CDM in ΛCDM). These two problems represent a major challenge for the community and are the object of much research today. GW observations have already helped shedding light on these dark issues. As an example, the first GW detections have been reanalyzed under the assumptions that they were emitted by primordial BHs, which could enter (a fraction of) the dark matter component [284, 285]. Regarding dark energy, constraints on its equation of state are already possible from GW observations and will be improved during the next observational runs [286].

More recently, the first multi-messenger observation of a binary NS merger through both GW and EM emission [260] allowed to measure the local expansion rate of the Universe through a measurement of the Hubble constant H_0 in a new, independent way. Although not competitive yet due to several uncertainties, the result shows a difference between the value of H_0 from CMB data [108] and that of supernovae of type Ia [287]. Consequently, one will have to wait to witness a possible resolution of the so-called Hubble tension [288, 289].

LISA

The Laser Interferometer Space Antenna (LISA) is a spaced-based GW interferometer operated by the European Space Agency and planned to launch in the mid-2030s [159]. Planned to launch in the mid-2030s, it consists in three spacecrafts, arranged at the corners of a virtual equilateral triangle of sides about 2.5 million km (the "arms" of the detector), cf. Fig. 1.21. Free falling masses inside these spacecraft

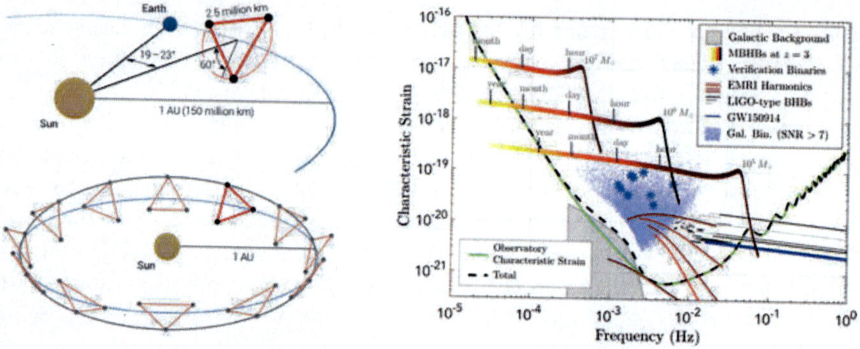

Fig. 1.21 The left panel depicts the geometry of LISA's orbit. Each spacecraft follows its own heliocentric, elliptic orbit, inclined with respect to the ecliptic, in order to maintain an equilateral triangle configuration. The right panel shows the characteristic strain of various astrophysical and cosmological sources expected to be detected by LISA. Some of them, e.g., massive binary black holes (MBBHs) should be observed for months to years. These figures can be found in the LISA project proposal [159]

will act as test masses and six lasers (one in each direction per spacecraft pair) will allow to measure by interferometry distance variations between them. As one of the most ambitious scientific projects ever designed, it will require an unprecedented level of precision, calibration and data management. To check whether the planned technology (interferometer optics, drag-free test masses and reliability of various components) will meet the LISA's requirements, the LISA Pathfinder mission was launched in 2015. It consisted in a single spacecraft located at the L1 Lagrange point, within which a 38 cm interferometer sensed the distance between two free-falling masses. The mission was successful as the results obtained were better than expected (noise levels were more than tenfold below the target value) [290]. LISA's dimensions are such that it will be able to detect GWs with frequencies in the range of microhertz up to hundreds of millihertz. LISA should thus be able to target various GW sources, among which

- early BH binaries of (cosmological) redshift up to $\simeq$20 [291], shedding some light on the very first BH populations
- intermediate-mass and SMBH binary systems, in the $10^2 - 10^7$ Solar masses range, allowing one to constrain evolutionary models for BHs from early ages until the star formation era [292],
- the early inspiral of stellar-mass BH binaries which could then be detected by terrestrial GW interferometers years later [293],
- galactic binaries and help understand the local population of compact object binaries [294, 295],
- EMRIs, which, although the computation of waveform represent a challenge, will allow to probe the neighborhood of massive BHs [296]
- unresolved GW sources, in particular the cosmological background [297] and that of unresolved galactic binaries [298].

LISA will allow routine multi-messenger astronomy [299], GR tests in nonlinear, strong regimes yet unexplored and open the door to GW cosmology. A more complete discussion of GW sources as well as the countless physical applications it will allow may be found in the LISA white paper [300] and the recent reviews [97, 301]. Other space-based interferometers have been proposed, such as the Japanese DECi-hertz Interferometer Gravitational wave Observatory (DECIGO) [302], a 1000 km equilateral triangle interferometer. A proof-of-concept mission, B-DECIGO, should launch in the 2020s [303].

Other Projects

Current terrestrial GW interferometers (Advanced LIGO and Advanced Virgo) will continue to be updated with technological improvements and detection methods updates. The new designs, called A+ and AdV+, will improve sensitivity by a factor two thanks to better mirrors (lower thermal noise, optical coatings), better lasers (frequency-dependent optical squeezing) and better statistics and GW detection methods. In the meantime, the Japanese KAGRA detector [304] will join the LIGO-Virgo collaboration as soon as O4, and LIGO-India should come online in the 2020s as well [305]. The third generation of terrestrial detectors will still be based on laser interferometry but will present arms of much longer length. The two promising projects are the Einstein Telescope (ET) based in Europe and the Cosmic Explorer (CE) based in the United States. ET will be built underground and have the geometry of a 10 km equilateral triangle [306], while CE will keep the two-arms configuration with lengths up to 40 km [307]. Other projects will also aim at filling in the yet unexplored deci-Hz range [308].

Side Note

Before concluding this chapter, we would like to add a side note, in view of the previous considerations on the future of GW astronomy (it applies to the future of science in general). It is clear that future missions and experiments, although of primary importance for enlarging the field of GW astronomy, will be very demanding and expensive on various levels, in particular regarding financial investment and energy requirement. In regard to GW astronomy, the numbers that are currently associated to GW data analysis will be increased by several orders of magnitude for the next-gen detectors, requiring a significant increase in energy consumption, e.g., to power the observatories, cool down supercomputers facilities, etc. At a time when (hopefully) climate change and the energy crisis will be on the table for every major decision, the field of GW astronomy will have to face a new challenge, that of being compatible with (at least) carbon-neutral trajectories. This is a responsibility that will have to be taken by the community, and must be discussed today.

Chapter 2
Multipolar Particles

Contents

> *But where is the simple fact ? The astronomer has found it because the distances of the stars are immense, so great that each of them appears only as a point, and because a point is simpler than a body which has shape and qualities.*
>
> H. Poincaré,
> Science et Méthode (1908)

As we saw in Chap. 1, binary systems of compact objects are among the most promising sources of GWs. These include black holes (BHs) and/or neutron stars (NSs). A critical issue in constructing GW templates for the detection is to (1) model these compact objects at a satisfying level of accuracy and (2) understand their orbital and internal dynamics. In this second chapter, the goal is to cover a family of models, which we will collectively refer to as "gravitational skeletons", that aim at tackling these issues. Although BHs and NSs are rather different objects, with the former being

P. Ramond, *The First Law of Mechanics in General Relativity & Isochrone Orbits in Newtonian Gravity*, Springer Theses,
https://doi.org/10.1007/978-3-031-17964-8_2

a vacuum region of spacetime and the latter a compact ball of magnetized fluid with extreme thermodynamics, both can be described in the skeleton approach, provided that the description is limited to dipolar order in the skeleton formalism. Such a dipolar (or pole-dipole) description consists in representing the compact object as a point particle equipped with a mass, and (possibly) a spin. In this chapter, we will start by illustrating the skeleton formalism at the Newtonian level (Sect. 2.1), before giving an overview of different, independent methods that derive its relativistic equivalent (Sect. 2.2). A summary of the important results and a number of extensions necessary for the next chapters will then be presented (Sect. 2.3).

2.1 Newtonian Skeleton

One of the first and most important things we learn while studying classical gravity is to compute the gravitational potential $\psi(\mathbf{x})$ generated at some point $\mathbf{x} \in \mathbb{R}^3$ by a body with support $\mathscr{B} \subset \mathbb{R}^3$, with mass density $\rho(\mathbf{y})$, where $\mathbf{y} \in \mathscr{B}$. In Newtonian gravity, the link between these two quantities is provided by the Poisson equation

$$\Delta\psi(\mathbf{x}) = 4\pi\rho(\mathbf{x}) \quad \Leftrightarrow \quad \psi(\mathbf{x}) = -\int_{\mathscr{B}} \frac{\rho(\mathbf{y})}{|\mathbf{x}-\mathbf{y}|}\mathrm{d}\mathbf{y}\,. \tag{2.1}$$

For a generic source $\rho(\mathbf{x})$, the integral on the right-hand side cannot be computed in closed form. However, it is possible for a spherically-symmetric body or a point mass m described by $\rho(\mathbf{y}) = m\delta_3(\mathbf{y})$, where δ_3 is the usual, 3-dimensional Dirac distribution. In fact, the result coincides for both cases and reads

$$\psi(\mathbf{x}) = -\frac{m}{|\mathbf{x}|}\,, \quad \text{where} \quad m \equiv \int_{\mathscr{B}} \rho(\mathbf{y})\mathrm{d}\mathbf{y}\,. \tag{2.2}$$

This remarkable result of Newtonian gravitation, sometimes called Newton's theorem[1] shows that the potential at the exterior of a spherically-symmetric body is independent of the mass distribution within it, and only depends on the macroscopic, total mass within $\mathscr{B}$. That gravity only cares about macroscopic features is present in GR too, and is sometimes referred to as the "effacement principle". We will come back to this in the next paragraphs.

In many cases, spherical symmetry for an astrophysical body is but an idealization, and one needs to go beyond this assumption. Indeed, the proper rotation of this body induces centrifugal forces that, in return, deform it (see Sect. 2.3 in [141]). Equilibrium configurations of such rotating bodies were already studied by Newton himself and shown to take ellipsoidal shapes. The extent of this deformation primarily depends on the rigidity of the body (spin deformability, in this case). Spherical symmetry can also be broken in the case where the body belongs to a binary system,

---

[1] The interested reader can find in [309] other types of non-spherical sources with this property.

where it is deformed under the tidal forces generated by its companion (see Sect. 2.5 in [141]). Once again, these tidal forces deform the body and create a bulge which may be locked within the direction of the companion or lagging behind that axis, depending on the viscosity and rigidity of the body.

To go beyond spherical symmetry in Newtonian theory, the general idea is to rewrite the integral in Eq. (2.1) at some point $\mathbf{x}$ in the exterior of the body (where $|\mathbf{x}| \geq |\mathbf{y}|$). For example, one may use the Legendre polynomials or the usual spherical harmonics decomposition. Integrating this expansion term by term in (2.1) gives the classical multipolar expansion of the gravitational potential

$$\psi(\mathbf{x}) = -\frac{m}{|\mathbf{x}|} - \frac{m_i \hat{\mathbf{x}}^i}{|\mathbf{x}|^2} - \frac{3}{2}\frac{m_{ij}\hat{\mathbf{x}}^i\hat{\mathbf{x}}^j}{|\mathbf{x}|^3} - \frac{5}{2}\frac{m_{ijk}\hat{\mathbf{x}}^i\hat{\mathbf{x}}^j\hat{\mathbf{x}}^k}{|\mathbf{x}|^4} + \cdots\,, \tag{2.3}$$

where $\mathbf{x}^i$ are the coordinate components of the unit vector $\hat{\mathbf{x}} \equiv \mathbf{x}/|\mathbf{x}|$, and the Euclidean tensors $(m_i, m_{ij}, m_{ijk}, \ldots)$ are the (traceless, symmetric) mass moments of the source. The vector $m^j = \int_{\mathscr{B}} \rho(\mathbf{y})\mathbf{y}^i \mathrm{d}\mathbf{y}$ is the mass dipole, and can be set to zero when the origin of the coordinates is at the center of mass. The higher order tensors are the mass quadrupole, mass octupole, etc. They are all given explicitly as integrals of the mass density over the source, as

$$m^{i_1\ldots i_\ell} \equiv \int_{\mathscr{B}} \rho(\mathbf{y})\, \mathbf{y}^{\langle i_1\ldots i_\ell\rangle} \mathrm{d}\mathbf{y}\,, \tag{2.4}$$

where $\langle\cdots\rangle$ denotes the usual symmetric-trace-free (STF) operation on these indices (see for instance [141], Eq. (1.153)). Since $|\mathbf{y}| \leq R$ in Eq. (2.4), with R a measure of the (maximal) extension of the body, each multipole can be bounded as $m^{i_1\ldots i_\ell} \leq m R^\ell$. Consequently, each term in the expansion (2.3) is smaller than the previous one by a factor of $R/|\mathbf{x}|$. Therefore, outside the source, one expects that the main features of the body will be well-described by the first few terms of this expansion.

With the expansion (2.3) of the potential $\psi(\mathbf{x})$, the Newtonian skeletonization can now be introduced. Consider truncating the expansion (2.3) at quadrupolar order (for the sake of simplicity). This defines the quadrupolar potential

$$\psi_{\mathrm{q}}(\mathbf{x}) \equiv -\frac{m}{|\mathbf{x}|} - \frac{m_i \hat{\mathbf{x}}^i}{|\mathbf{x}|^2} - \frac{3}{2}\frac{m_{ij}\hat{\mathbf{x}}^i\hat{\mathbf{x}}^j}{|\mathbf{x}|^3}\,. \tag{2.5}$$

Of course, ψ_{q} is not the potential created by the initial body $\mathscr{B}$ of mass density $\rho(\mathbf{x})$, since its fine, post-quadrupolar details encoded in higher multipoles have been neglected. Rather, it corresponds to another, effective body, whose mass density $\rho_{\mathrm{q}}(\mathbf{x})$ satisfies the Poisson equation $\Delta\psi_{\mathrm{q}} = 4\pi\rho_{\mathrm{q}}$, with

$$\rho_{\mathrm{q}}(\mathbf{x}) = m\delta_3(\mathbf{x}) + m^i\partial_i\delta_3(\mathbf{x}) + m^{ij}\partial_i\partial_j\delta_3(\mathbf{x}) \tag{2.6}$$

which easily follows from the distributional identity $\Delta(\mathbf{x}^{-1}) = -4\pi\delta_3(\mathbf{x})$. Mathematically, ρ_q is a singular mass density, supported at the origin, whereas the original body had a non-singular compact support $\mathscr{B}$. But physically, it generates exactly the same gravitational field, so long as the octupole, hexadecapole, etc of $\mathscr{B}$ are neglected. This generalises Newton's theorem, which stated that a spherically symmetric body is equivalent to a point mass distribution, from a purely gravitational point of view. By linearity of the Poisson equation, this result can be generalised to any multipolar order, an ℓ-th order multipole being sourced by the ℓ-th spatial derivative of the Dirac distribution δ_3. Note that the initial body may have a time-dependent density, in which case the multipole moments will be time-dependent as well.

To summarise, a compact object $\mathscr{B}$ of continuous mass density $\rho(t, \mathbf{x})$ in Newtonian gravity can be skeletonized, i.e., equivalently described at any multipolar order by a singular mass density

$$\rho_{\mathsf{skel}}(t, \mathbf{x}) = m(t)\delta_3(\mathbf{x}) + \sum_{\ell\geq 1} m^L(t)\partial_L\delta_3(\mathbf{x})\,, \tag{2.7}$$

where the $m^L \equiv m^{i_1 \ldots i_\ell}$ are its STF mass moments and $\partial_L \equiv \partial_{i_1} \ldots \partial_{i_\ell}$. Because it replaces the original body $\mathscr{B}$ of three-dimensional compact support by a singular density ρ_{skel} equipped with a finite set of moments, this effective body is called a multipolar *particle*.

2.2 Relativistic Skeletons

The Newtonian skeletonization presented in the last section works well thanks to the linearity of the Poisson equation. Its GR-equivalent, the Einstein equation, is nonlinear, making it unclear how to generalize this procedure. However, it has long been known and shown in independent ways that one can still make sense of the Einstein equation when the extended compact object is replaced by a point particle, despite its nonlinearity, in some perturbative contexts. In this section, we start with a brief review of four independent "skeletonization" methods, that have been proposed in the literature through the years. Although we try to be as precise as possible, the point of this section is not to review these methods, as (1) this is done very well elsewhere (cf references in the text below), and (2) we are only interested in the end result, namely the skeleton SEM tensor and the evolution equations. Rather, the aim is to emphasize that several (independent) routes can reach the same result, but also that all are complementary in their assumptions and range of applicability.

In the present section, the emphasis will be on the monopole-dipole model (from Sect. 2.2.1 up to 2.2.4). The extension to higher multipoles is briefly described at the end (Sect. 2.2.5), before Sect. 2.3 provides a complete summary of the quadrupolar-point-particle model that will be used in the following chapters.

2.2.1 Lagrangian Approach

The Lagrangian approach is based on the classical action-variation method, already used in Newtonian mechanics and in the analysis of geodesics in curved spacetime. The latter is a good starting point and is presented first, before extending the method massive, spinning point particles. A particularly well-written account is that provided by Jan Steinhoff in [310], in which all details can be found.

Test Particle

As we saw in Chap. 1, Einstein's main insight into the construction of GR was the equivalence principle, i.e., the fact that one can always remove a gravitational field locally by going into an accelerated frame of reference. The direct consequence is that a test particle in flat spacetime follows a trajectory that maximizes its proper time . According to the equivalence principle, the same should be true for a test body embedded in a gravitational field, i.e., in GR.

Let x^α be a coordinate system covering the spacetime $(\mathscr{E}, g_{ab})$, and γ be the worldline of a test particle in it. We parametrize γ by four equations ($x^\alpha = z^\alpha(\lambda)$, $\lambda \in \mathbb{R}$), where z^α are four functions of an arbitrary parameter $\lambda \in \mathbb{R}$ varying along γ. The *action* S of this test particle is well-known [6, 22, 311] and a function of γ (or equivalently of z^α). It reads[2]

$$S(z^\alpha) \equiv \int_\gamma \left(-g_{\alpha\beta}(z)\frac{\mathrm{d}z^\alpha}{\mathrm{d}\lambda}\frac{\mathrm{d}z^\beta}{\mathrm{d}\lambda}\right)^{1/2}\mathrm{d}\lambda\,. \tag{2.8}$$

This action is a functional of the worldline γ through the metric $g_{\alpha\beta}$ evaluated at $z^\alpha \in \gamma$, and the λ-derivative thereof. Extremizing S with respect to the worldline is easily done à la Euler-Lagrange [22]. The result is classical: the curve γ that maximizes (2.8) satisfies

$$\frac{\mathrm{d}^2 z^\alpha}{\mathrm{d}\lambda^2} + \Gamma^\alpha_{\beta\gamma}\frac{\mathrm{d}z^\beta}{\mathrm{d}\lambda}\frac{\mathrm{d}z^\gamma}{\mathrm{d}\lambda} = K\frac{\mathrm{d}z^\alpha}{\mathrm{d}\lambda}\,, \tag{2.9}$$

where $\Gamma^\alpha_{\beta\gamma}$ are the Christoffel symbols evaluated at $z^\alpha(\lambda)$, given in terms of the metric coefficients $g_{\alpha\beta}$ and K, the non-affinity "coefficient", is actually a function of λ. Equation (2.9) describes a geodesic in the spacetime $(\mathscr{E}, g_{ab})$. It takes the simplest form when λ is an affine function of τ, say $\lambda = a\tau + b$ with $(a, b) \in \mathbb{R}^\star \times \mathbb{R}$ two constants. In this case, λ is called an *affine parameter*, and the quantities $\mathrm{d}z^\alpha/\mathrm{d}\lambda$ are (up to a constant factor) the components of the four-velocity u^a of the particle, the unique vector tangent to γ and normalized as $g_{ab}u^a u^b = -1$. In this case, the geodesic equation (2.9) takes the covariant form

---

[2] Strictly speaking the integral is over $\lambda \in \mathbb{R}$, but it is customary to write $\int_\gamma$. There is a one-to-one correspondence between $\mathbb{R} \ni \tau$ and $\gamma \ni z^\alpha$ through the parametrization $x^\alpha = z^\alpha(\lambda)$.

$$\dot{u}^a \equiv u^b \nabla_b u^a = 0\,, \tag{2.10}$$

with ∇_a the metric-compatible covariant derivative. For the sake of simplicity, in Chap. 3 and onward we shall always consider a proper time parametrization of the worldline and work with the unit, tangent vector u^a, i.e., the four-velocity of the particle. For the time being, it is more convenient to work with a generic affine parameter $\lambda = a\tau + b$ and then choose $\lambda = \tau$ at the end of the calculation.

Massive, Spinning Particle

A *test particle*, by definition, does not act as a source of the gravitational field. Mathematically, this is ensured by demanding that its Lagrangian L (such that $S = \int_\gamma L\,\mathrm{d}\lambda$ in (2.8)) has no functional dependence on the metric g_{ab}. In this way, when we apply the principle of least action to the Einstein-Hilbert action (vacuum GR) + test particle action (2.8), only the former gives a nonzero contribution, and the field equations for g_{ab} are the GR vacuum equations. There is a one-way decoupling: the metric is independent of the particle, but the particle depends on the metric through the geodesic equation.

The next level of approximation for a small object, beyond the test-particle limit, is to turn on the two-way coupling between the particle and gravity. This is done by adding a functional dependence on g_{ab} in the Lagrangian (integrand of the action (2.8)). Through the Einstein equation, the metric g_{ab} itself becomes dependent on the particle's property and motion. Computing the field generated by a point particle comes with a number of difficulties (essentially because of the singular nature of the particle) that may be solved using a variety of methods depending on the approximation scheme (see, e.g., [312, 313] in post-newtonian expansions and [314, 315] in the gravitational self-force formalism). For our purposes, we will not need to solve the Einstein equation in order to get crucial information of the behavior of the point particle. Everything will be obtained from first principles, just as in the test-particle case. The method discussed here uses Lagrangian mechanics and was pioneered in GR by Bailey and Israel [316], extending Hanson and Regge's work on the special relativistic spherical top [317].

Let us consider a massive, spinning point particle, represented as a timelike worldline γ. At the kinematic level, the translational degrees of freedom of the particle are encoded in the parametric equation of γ, taken, as before, to be $x^\alpha = z^\alpha(\lambda)$ where z^α are four functions of $\lambda \in \mathbb{R}$, an affine parameter along γ. The rotational degrees of freedom are introduced via an orthonormal tetrad (ϵ^a_B), free to rotate along γ. Note that ϵ^a_B is not a rank-2 tensor: a is an abstract index (meaning that ϵ^a_B is a vector), whereas B is a mere label that tells which vector of the tetrad it is. The action S of the particle is then postulated as

$$S(z^\alpha, \epsilon^a_B) \equiv \int_\gamma L[v^a(\lambda), \Omega^a{}_b(\lambda), g_{ab}(\lambda)]\mathrm{d}\lambda\,, \tag{2.11}$$

where L, the Lagrangian of the particle, now depends on the dynamical, covariant quantities conjugated to z^α and ϵ^a_B, namely the tangent vector v^a to γ and the rotation tensor $\Omega^a{}_b$, respectively. These are defined in the usual way through the rate of change of z^α and ϵ^a_B along γ, by

$$\dot{z}^\alpha \equiv v^\alpha \quad \text{and} \quad \dot{\epsilon}^a_A \equiv \Omega^a{}_b \epsilon^b_A \,, \tag{2.12}$$

with a dot denoting the covariant derivative along γ, as in Eq. (2.10). The next step to make some progress is to assume that the action $S(z, \epsilon)$ be invariant under an arbitrary change of affine-parameter. In particular, linear changes such as $\lambda \mapsto k\lambda$ with $k \in \mathbb{R}$ should leave it invariant. Given the definitions (2.12), S is invariant if and only if $kL(v/k, \Omega/k, g) = L(v, \Omega, g)$. This indicates that L is homogeneous of degree one with respect to (v, Ω). Thanks to the Euler theorem[3] for homogeneous functions, this implies

$$L = u^a p_a + \frac{1}{2}\Omega^a{}_b S_a{}^b \,, \quad \text{where} \quad p_a \equiv \frac{\partial L}{\partial u^a} \,, \quad S_a{}^b \equiv 2\frac{\partial L}{\partial \Omega^a{}_b} \,, \tag{2.13}$$

where we finally used the proper time as a parameter (hence $v^a = u^a$), now that the calculation is over. In Eq. (2.13), the spin tensor $S_a{}^b$ and the momentum form p_a are functions of $(u^a, \Omega^a{}_b)$ and the factor 2 for the spin is purely conventional. It is rather remarkable that the sole "scalar nature" requirement for the action is sufficient to explicitly express it in terms of the dynamical properties of the particle. However, one should not conclude from (2.13) that L is linear in the spin, since it still depends on the metric through p^a and $S_a{}^b$. We also see here that, fundamentally, the four-momentum is a 1-form (as in SR [22]) and the spin is a $(1, 1)$-tensor. This will be important later on for the derivation of the first law of mechanics at dipolar order.

With the expression of the Lagrangian (2.13), we can now compute in the classical way the first-order variation of the action (2.11) to obtain the Euler-Lagrange equations for the particle. Leaving the metric aside, since there are two types of degrees of freedom (translational and rotational), there are two distinct ways of varying S. First, with respect to the tetrad ϵ^a_A, keeping the worldline γ fixed. Second, with respect to γ, keeping the tetrad fixed.[4] After some computations which we do not detail here (found for instance in [310, 318, 319]), maximization of the action provides two evolution equations

$$\dot{p}_a = \frac{1}{2} R_{bcda} S^{bc} u^d \,, \tag{2.14a}$$

$$\dot{S}^{ab} = 2p^{[a}u^{b]} \,, \tag{2.14b}$$

[3] In a nutshell, if $kL(v/k, \Omega/k, g) = L(v, \Omega, g)$ for any $k \in \mathbb{R}$, differentiating with respect to k and then setting $k = 1$ readily gives $L(v, \Omega, g) = u\partial_v L + \Omega\partial_\Omega L$.

[4] Since the tetrad is defined along the worldline, it cannot remain fixed, strictly speaking, but is rather parallel-transported along the family of worldlines defined for the variation.

with an overdot denoting the covariant derivative $u^a\nabla_a$ along γ. A few comments are in order. First, we see that these two evolution equations express the derivative of (p^a, S^{ab}) along γ. Compared to the test-particle case, there is no equation for the four-velocity (such as (2.10)). However, it is possible to derive, from Eq. (2.14) such an equation, albeit now $u^b\nabla_b u^a \neq 0$. This will be discussed in Sect. 2.3.2. Second, although some insight on the physical interpretation of p^a and S^{ab} can be obtained by inspecting their definition (2.13) as conjugate quantities, we can also study the spinless limit of equations (2.14). Setting $S^{ab} = 0$ in (2.14b) implies that $p^{[a}u^{b]} = 0$ and thus that $p^a = \mathcal{m} u^a$ for some $\mathcal{m}$. But (2.14a) implies, with $S^{ab} = 0$, that p^a is parallel transported along γ. Combining these two results readily gives $\dot{\mathcal{m}} = 0$ and $\dot{u}^a = 0$, which means that a spinless particle travels on a geodesic γ and that $\mathcal{m}$, with dimension of a mass, is constant along γ. Notice that this is identical to the case of a test-particle. Naturally, the consequence is that p^a corresponds to the linear momentum of the particle, whereas S^{ab} contains the spinning degrees of freedom. The presence of spin prevents p^a from being parallel transported.

Even though the Lagrangian (2.13) is a function defined on a worldline, one can construct an action S by integrating it over a 4-region of spacetime thanks to the four-dimensional, covariant version of the Dirac distribution (defined properly in Appendix A.2). In particular, we have

$$S = \int_{\mathscr{E}} L[u, \Omega, g]\, \delta_4(x, z(\tau))\, \mathrm{d}V\,, \tag{2.15}$$

where x is a point of $\mathscr{E}$ and $\mathrm{d}V \equiv \sqrt{-g}\,\mathrm{d}^4x$ is the covariant volume element on $\mathscr{E}$. One can then derive the SEM tensor for the point particle in the usual way, i.e., by varying the action (2.15) with respect to the metric g_{ab}. Instead of the metric, in practice it is easier to do this variation by introducing a background tetrad e^a_A, such that $g_{ab} = \eta_{AB} e^A_a e^B_b$, and varying with respect to it. The result then follows from the definition of the SEM as the functional derivative of the action with respect to the background tetrad. The calculation makes use of the evolution equations (2.14) and can be found in Sect. II of [320] (see also Sect. 7 in [194]). The result is

$$T^{ab}(x) = \int_\gamma u^{(a}p^{b)}\, \delta_4(x, z(\tau))\, \mathrm{d}\tau + \nabla_c \int_\gamma u^{(a}S^{b)c}\, \delta_4(x, z(\tau))\, \mathrm{d}\tau\,. \tag{2.16}$$

This SEM tensor is distributional and non-zero only along the worldline of the particle. Thanks to the distribution bi-scalar δ_4, it is both covariant (independent of the coordinates x^α) and independent of the worldline parametrization $z^\alpha(\tau)$.

This monopole-dipole formalism does not account for the physical features of an extended body. To account for this, one needs to go back to the Lagrangian (or action) (2.11) and add other degrees of freedom. For example, if we allow the Lagrangian to depend on the Riemann curvature tensor and its successive covariant derivatives $L[u, \Omega, g, R, \nabla R, \ldots]$, then the higher multipole moments are defined as $J \equiv \partial_R L$ (quadrupole), $O \equiv \partial_{\nabla R} L$ (octupole), etc. However, imposing the scalarity of the Lagrangian just as before still gives the same *expression*, namely

$$L[u, \Omega, g, R, \nabla R, \ldots] = u^a p_a + \frac{1}{2}\Omega_{ab}S^{ab}, \tag{2.17}$$

the dependence on $(g, R, \nabla R, \cdots)$ being hidden in (p_a, S^{ab}). This formalism has been used in particular at octupolar order to compute cubic-in-spin effects in the dynamics and energy flux of compact binaries in [320]. There, it was also extended to any order in the Riemann derivatives (see Appendix A there). It should be noted that the dependence on the Riemann tensor and its derivatives can be equivalently encoded into the so-called "metric extensions", which are natural tensorial quantities obtained from symmetrized derivatives of the metric. See [321], or Appendix 2 of [322] for details on these.

Lastly, let us also mention, for the sake of completeness, the effective-field-theory (EFT) framework, which is closely related to the Lagrangian scheme presented here. This framework, initiated in [323], is inspired by ideas coming from quantum field theory-type calculations and has proven to be very powerful for applications in GR. In particular, this formalism provides one of the ways to construct explicit quadrupolar models, as will be touched upon in Sect. 6. For more on EFT, we refer to [324–326] and references therein, as well as [310] and the recent review [195] by Michèle Levi.

2.2.2 MPTD Formalism

The Lagrangian method of Bailey and Israel presented in the last section is general and can be applied to GR as well as other metric theories of gravity. It takes as an assumption the fact that an extended body can be represented by a single worldline, along which multipoles, defined geometrically, evolve and characterize the extended body. In particular, it does not show that an extended body's motion can be equivalently described by that of a point particle equipped with multipoles. This was done prior to Bailey and Israel's method, during earlier attempts to study the dynamics of extended bodies in a way closer to the Newtonian skeletonization presented in Sect. 2.1. This task was tackled by a number of authors, among which four names traditionally stand out: Myron Mathisson, Ioannis Papapetrou, Włodzimierz Tulczyjew and William Dixon, in chronological order of appearance in the literature. A thorough account on the history behind these men's work is provided by Dixon himself in [327]. In a nutshell, the MPTD formalism refers to a theory aiming at describing the motion of extended bodies in GR, that is based on early attempts by Mathisson, Papapetrou and Tulczyjew, and was then completed, extended and synthesized by Dixon, in the late seventies. More precisely

- Mathisson published his pioneering work [328] on extended bodies in GR up to (almost) quadrupolar order, and discussed the problem of center-of-mass [329]. He also made important mathematical contributions [330] to clarify earlier points. He presented several related works in flat spacetime [331].[5]

[5] Mathisson died prematurely in 1940 from tuberculosis, aged 43. In his early days, he corresponded in French with Einstein, who showed great interest in his methods [332]. His works on GR and

- Papapetrou took on the problem of center-of-mass definition in GR [335] and then proposed the first general derivation [336, 337] of the evolution equations based on an integral definition of the multipoles, in the sense that it held for arbitrary center-of-mass conditions. These methods usually lacked general covariance and only went up to dipolar order.
- Tulczyjew reformulated Mathisson's original approach in 1959 [338], proposing a fully covariant and general derivation of the evolution equations, and then refined Papapetrou's integral approach in 1962 by using an adapted set of coordinates [339].
- Dixon started working on the topic in 1964 [340] using the integral approach of Papapetrou, introducing two-point tensors. Then in [341] he explored the problem in flat spacetime to finally find the missing parts of the big picture, which he completed and clarified up until 1974 [322, 342, 343].

Dixon's recent synthetic review [327] on the topic is complete, detailed and clear. Since we only want to focus on the equations of motion of the body, from now called *evolution equations*, and on the construction of the SEM tensor, wewill only cover these derivations and chose one that is most adapted to our purpose, based on Tulczyjew's ideas. The reason for this is twofold. First, although Tulczyjew's method relies on more assumptions than Dixon's work does (making it somewhat less self-consistent), it does produce the same results regarding the evolution equations and the SEM tensors, which is all we need in this work. Second, Tulczyjew's method present the advantage of being extendable to an other purpose, which we will encounter in the following chapters for the derivation of crucial results. We will therefore leave all the details of the MPTD formalism aside except for the derivation of the SEM tensor of a multipolar, extended body, and the evolution equations of its multipoles, in the spirit of Tulczyjew's.

Tulczyjew's Reduction Process

The method of Tulczyjew extend ideas found in both Mathisson and Papapetrou's approaches. We have chosen to present it here in details because the general process behind the derivation of the evolution equations will be used in Chap. 3 for another important matter. His method is a four-step algorithmic procedure which can be summarized as follows

$$\begin{cases} \text{i. Ansatz for SEM tensor,} \\ \text{ii. SEM tensor conservation,} \end{cases} \quad \Rightarrow \quad \begin{cases} \text{iii. Reduced SEM tensor,} \\ \text{iv. Evolution equations.} \end{cases}$$

The tools hidden in the implication "⇒", in order to go from the *assumptions* (i, ii) to the *results* (iii, iv), are rather elementary. They involve an orthogonal decompo-

wave diffraction impressed Paul Dirac and David Hilbert so much that they published Mathisson's works on these problems posthumously in 1942, in [333] and [334], respectively.

sition with respect to the four-velocity and two theorems, due to Tulczyjew, that generalize the fundamental lemma of the calculus of variations.[6]

We will first present Tulczyjew's method at monopolar order in detail, so as to get the general idea, and then sketch the dipolar derivation. Although more involved, the dipolar and quadrupolar cases are algorithmically straightforward and follow the same steps as what will be done below. The detail of the dipolar case can be found in Appendix A.4.3, and the quadrupolar case in [344].

Step i. The Ansatz for the SEM tensor of a monopolar particle is constructed as a tensor distribution that vanishes outside of some wordline γ representing the trajectory of the body. To make it independent of any coordinate choice, it is helpful to introduce (again) the two-point tensor $\delta_4(x, y)$, called the 4-dimensional Dirac distribution, which generalizes the Euclidean (or *coordinate*)-delta distribution. Its rigorous definition is found in Appendix A.2. It allows us to write a covariant Ansatz of the form

$$T^{ab}(x) = \int_\gamma \mathcal{T}^{ab}(y(\tau))\, \delta_4(x, y(\tau))\, \mathrm{d}\tau\, . \tag{2.18}$$

Notice in Eq. (2.18) that the tensor $\mathcal{T}^{ab}(y)$ is arbitrary at this stage and defined only along γ, whereas T^{ab} is defined at any point $x \in \mathscr{E}$. In particular, $T^{ab}(x) \equiv 0$ whenever $x \notin \gamma$. Since $\mathcal{T}^{ab}$ is defined along γ, we can perform a time+space decomposition of it with respect to the four-velocity u^a, which is also defined along γ. Thanks to its symmetry, this "orthogonal" decomposition takes the form

$$\mathcal{T}^{ab} = \mathcal{m} u^a u^b + 2\mathcal{m}^{(a} u^{b)} + \mathcal{m}^{ab}\, , \tag{2.19}$$

where $(\mathcal{m}, \mathcal{m}^a, \mathcal{m}^{ab})$ are the time-time, time-space and space-space projections of $\mathcal{T}^{ab}$, respectively. By definition, $\mathcal{m}^a$ and $\mathcal{m}^{ab}$ are both orthogonal to u_a, and $\mathcal{m}^{ab}$ is symmetric. Therefore, $\mathcal{m}$, $\mathcal{m}^a$ and $\mathcal{m}^{ab}$ have 1, 3 and 6 independent components, respectively, and encode the same information as T^{ab} (or equivalently $\mathcal{T}^{ab}$).

Step ii. Now let us insert the 3+1 decomposition (2.19) into the Ansatz (2.18) and apply the covariant derivative ∇_a. Rearranging the result gives

$$\nabla_a T^{ab} = \int_\gamma u^a \nabla_a (\mathcal{m} u^b + \mathcal{m}^b)\, \delta_4 \mathrm{d}\tau + \nabla_a \int_\gamma \left(\mathcal{m}^a u^b + \mathcal{m}^{ab}\right) \delta_4 \mathrm{d}\tau\, , \tag{2.20}$$

where, to get the first term, a key distributional identity (given and proved in Appendix A.4.1) was used. Notice that in (2.20), the second integrand is orthogonal to the four-velocity, with respect to the index a present in ∇_a. With this property (orthogonality with respect to derivative index), the right-hand side of (2.20) is said to be into *normal form*. This normal form always exists, whatever the multipolar order, as showed initially by Tulczyjew [338]. We will refer to this as Tulczyjew's first theorem, and in Appendix A.3.1 we provide a proof of an extension of the orig-

---

[6] This is the lemma that says (details aside) "if a function $f : \mathbb{R} \to \mathbb{R}$ is such that $\int_\mathbb{R} f(x) g(x) \mathrm{d}x = 0$ for all functions $g : \mathbb{R} \to \mathbb{R}$, then f vanishes identically".

inal theorem, in the case of a binary system of particles (Tulczyjew's theorem only deals with a single particle).

Now let us impose the SEM conservation $\nabla_a T^{ab} = 0$ to Eq. (2.20). This implies that the *sum of integrals* on the right-hand side of (2.20) must vanish. In general, this does not imply the vanishing of each individual integral. But because it is in normal form, it is the case here. This result is called *Tulczyjew's second theorem*. Details (and an extension to binary systems) about this theorem can be found in Appendix A.3. In any case, the sum (2.20) vanishes thanks to SEM conservation, and since it is in normal form, Tulczyjew's theorem imply that both integrands vanish. This gives the two following equations

$$u^a \nabla_a (m u^b + m^b) = 0 \,, \tag{2.21a}$$

$$m^a u^b + m^{ab} = 0 \,. \tag{2.21b}$$

Step iii. Let us start by examining the consequences of equation (2.21b). Contracting it with u_b readily gives $m^a = 0$, which implies, from the very same equation, that $m^{ab} = 0$ as well. Consequently, wee see that only *one* component of T^{ab} (and thus of T^{ab}) "survives" the reduction process: the time-time component $m \equiv T^{ab} u_a u_b$. In particular, the *reduced* SEM tensor of a monopolar particle reads

$$T^{ab} = \int_\gamma m u^a u^b \, \delta_4 \, \mathrm{d}\tau \,, \tag{2.22}$$

where m is an arbitrary scalar field defined along γ, that depends on τ, a priori. It is naturally interpreted as the mass of the body that the particle represents. The evolution equation will therefore constrain this remaining degree of freedom.

Step iv. Finally, now that we have seen that $m^a = 0 = m^{ab}$, let us examine equation (2.21a). Using the Leibniz rule, and contracting it with u_b readily gives two evolution equations

$$u^a \nabla_a m = 0 \quad \text{and} \quad u^a \nabla_a \sqcap^{\llcorner} = 0 \,. \tag{2.23}$$

In other words, both the mass m and the four-velocity u^a of the particle are parallel-transported along γ. For the former, it means that the only degree of freedom of the particle (its mass) is in fact *constant*. For the latter, since u^a is tangent to the worldline γ, its parallel transport means that γ is a *geodesic* of the underlying spacetime. These are called the *evolution equations*, as they evolve of the multipoles along the worldline. If we construct the linear momentum vector $p^a \equiv m u^a$, then (2.23) can be written equivalently as

$$u^a \nabla_a p^b = 0 \quad \text{and} \quad 0 = 2 p^{[a} u^{b]} \,, \tag{2.24}$$

which coincides with the dipolar result (2.14) derived earlier with $S^{ab} = 0$, i.e., for a monopolar (hence spinless) particle.

Tulcyjew's Reduction at Dipolar Order

Without giving the details (which can be found in Appendix A.4) let us give an overview of the calculation involved in the Tulczyjew reduction process at *dipolar* order, using the same steps as above.

Step i. The dipolar Ansatz now involves two terms, say

$$T^{ab} = \int_\gamma \mathcal{T}^{ab}\,\delta_4\,\mathrm{d}\tau + \nabla_c \int_\gamma \mathcal{T}^{abc}\,\delta_4\,\mathrm{d}\tau\,, \tag{2.25}$$

where $\mathcal{T}^{ab}$, $\mathcal{T}^{abc}$ are arbitrary. We perform an orthogonal decomposition. For $\mathcal{T}^{ab}$ Eq. (2.19) still holds, while for $\mathcal{T}^{abc}$, it is sufficient to take

$$\mathcal{T}^{ab} = m u^a u^b + 2 m^{(a} u^{b)} + m^{ab} \tag{2.26a}$$

$$\mathcal{T}^{abc} = u^a u^b n^c + 2u^{(a} n^{b)c} + n^{abc} + o^{ab} u^c\,, \tag{2.26b}$$

where all the $\Updownarrow$'s and \'s are orthogonal to the four-velocity, and both n^{abc} and o^{ab} are symmetric in (ab).

Step ii. We take the covariant derivative ∇_b of (2.25) and bring the result into normal form. After a rather lengthy calculation detailed in Appendix A.4, we arrive at

$$\nabla_b T^{ab} = \int_\gamma \mathcal{X}^a \delta_4 \mathrm{d}\tau + \nabla_b \int_\gamma \mathcal{X}^{ab} \delta_4 \mathrm{d}\tau + \nabla_b \nabla_c \int_\gamma \mathcal{X}^{abc} \delta_4 \mathrm{d}\tau\,, \tag{2.27}$$

where the $\mathcal{X}$'s are in normal form: $\mathcal{X}^{ab} u_b = 0 = \mathcal{X}^{abc} u_b$ and $\mathcal{X}^{abc}$ is symmetric in (bc). They are given by explicit formulae in terms of u^a, R_{abcd}, the m's, n's and o^{ab} as well as their $u^a \nabla_a$-derivative. The full expressions, which are not relevant here, will be derived and used in Chap. 3, Sect. 3.2.1.

Step iii. The SEM conservation $\nabla_b T^{ab} = 0$ and Tulczyjew's second theorem imply that all $\mathcal{X}$'s in Eq. (2.27) vanish. This gives three equations. One of them readily implies $n^{(ab)} = 0 = n^{abc}$, which simplifies the other two. The remaining system of two equations naturally calls for the introduction of two particular combinations of the m's, n's and o^{ab}. They are

$$p^a \equiv m u^a + m^a + (o^{ab} + u^a n^b + n^{ab})\dot{u}_b \tag{2.28a}$$

$$S^{ab} \equiv 2u^{[a} n^{b]} + 2n^{ab}\,, \tag{2.28b}$$

Inserting them back into the Ansatz (2.25) then gives the reduced form of the SEM tensor at dipolar order (at monopolar order, we had obtained (2.22)). The result is

$$T^{ab}(x) = \int_\gamma u^{(a} p^{b)}\,\delta_4\,\mathrm{d}\tau + \nabla_c \int_\gamma u^{(a} S^{b)c}\,\delta_4\,\mathrm{d}\tau\,. \tag{2.29}$$

We see that this is the sum of two terms, the first one bringing four components via the vector p^a and the second one six, since S^{ab} is antisymmetric.

Step iv. From the definitions (2.28), we can easily compute the derivatives of (p^a, S^{ab}) along γ. The result looks rather familiar:

$$u^e \nabla_e p^a = \frac{1}{2} R_{bcd}{}^a S^{bc} u^d \quad \text{and} \quad u^c \nabla_c S^{ab} = 2 p^{[a} u^{b]} \,. \tag{2.30}$$

We recognize, as expected, the evolution equations for the dipolar particle (2.14) derived with the Lagrangian method in the last section. The right-hand side of these equations acts as a force vector and torque tensor that prevent p^a and S^{ab} from being parallel-transported.

At quadrupolar order, the four steps are the same, but the calculation is much more involved [344]. In particular, at the quadrupolar level the Ansatz fot $\mathcal{T}^{ab}$ contains a term with two covariant derivatives, in addition to the dipolar Ansatz (2.25). Consequently, when applying SEM conservation this term will pick up an addition covariant derivative and the reduction process will involve the computation of the normal form for a term like $\nabla_b \nabla_c \nabla_d \int_\gamma \mathcal{T}^{abcd}$. Although this is a lengthy and tedious calculation, it did not discourage the authors of [344], who derived the reduced form of the quadrupolar SEM tensor (given below in Eqs. (2.58)–(2.59)) and the evolution equations at quadrupolar order (Eq. (2.57)).

To summarize, Tulczyjew's method takes as an input a distributional Ansatz for the SEM tensor, directly inspired from the Newtonian skeletonized mass density (2.7), which it generalizes as

$$T^{ab}_{\text{skel}}(x) = \sum_{\ell=0}^{\infty} \nabla_{c_1 \cdots c_\ell} \int_\gamma \mathcal{T}^{abc_1 \cdots c_\ell}(y)\, \delta_4(x, y)\, \mathrm{d}\tau \,. \tag{2.31}$$

Then, by SEM conservation and following an algorithmic procedure (involving the so-called normal form) it removes all the redundant degrees of liberty of the $\mathcal{T}$'s. This allows us to rewrite the SEM in a reduced form, in term of reduced moments $(p^a, S^{ab}, \ldots)$, along with evolution equations for these. The general picture behind Tulczyjew's reduction process is depicted in Fig. 2.1. This approach is to be contrasted with Dixon's [327] (which was not presented at all here) that essentially *defines* these moments beforehand (as integrals over T^{ab}) and derives their evolution equation from the conservation of T^{ab}. A summary of the differences between these approaches is provided in Sect. 2.2.5.

2.2.3 *Generalized Killing Fields*

The realm of Newtonian mechanics is built on Universal time and Euclidean 3-space. As a consequence, it enjoys various geometrical results linked to the isometries

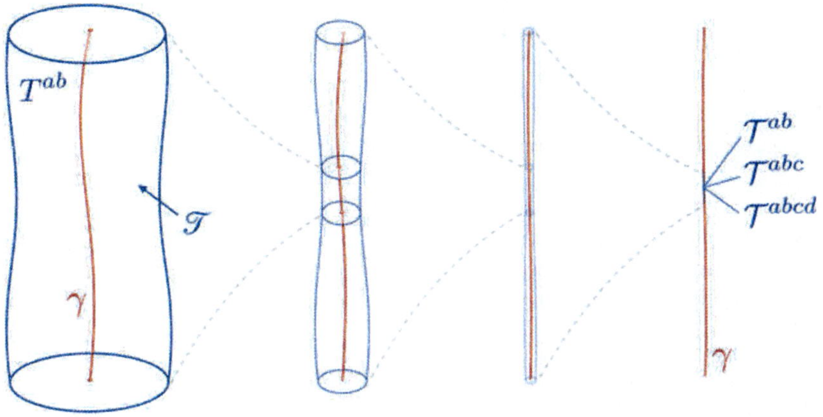

Fig. 2.1 In the multipolar gravitational skeleton model, the smooth SEM tensor T^{ab} of an extended compact body enclosed in a worldtube $\mathscr{T}$ (left) is replaced by the distributional SEM tensor (2.31) of a particle endowed with a collection of multipoles ($\mathcal{T}^{ab}$, $\mathcal{T}^{abc}$, $\mathcal{T}^{abcd}$, ...), all defined along a worldline $\gamma \subset \mathscr{T}$ (right), as if the extended body was "observed from far away"

of Euclidean space. In particular, Euclidean 3D space is a maximally symmetric manifold: it contains the maximum number of symmetries a 3D space can have. These isometries are generated by six independent Killing vector fields, three associated to rotations and three to translations. For any isolated system, the conservation of its total linear and angular momentum can be derived from the invariance by translations and rotations, respectively. Behind these conservation laws lies a general procedure for finding equations of motion, as encapsulated in the theory of Noether invariants [345].

In a nutshell, the generalized Killing fields (GKF) framework developed by Abraham Harte consists in extending this Euclidean duality between isometries and evolution laws to relativistic mechanics. There are slight differences between the two, notably because Euclidean spaces are maximally symmetric whereas GR spacetimes are usually not. Harte's program is a remarkably simple and elegant way of deriving the laws of motion close to Dixon's work [327], with one considerable advantage over it: it naturally accounts for the self-field of the object. Harte has provided several pedagogical reviews on his methods, see in particular [346, 347], which we strongly encourage the reader to give a look, even the non-relativistic sections. Let us now sketch the general ideas in the Newtonian case and then explain what changes need to be made for the GR extension.

Euclidean Mechanics

In Newtonian mechanics, more particularly in astrophysics, one is almost always interested in the bulk motion of objects. The equations of motion for a continuous distribution of matter of compact support $\mathscr{B}$, mass density $\rho(\mathbf{x})$ and velocity (3-vector) $\mathbf{v}$ follow from the local conservation of mass and linear momentum (see, e.g., [348])

$$\frac{\partial \rho}{\partial t} + \nabla \cdot (\rho \mathbf{v}) = 0 \quad \text{and} \quad \frac{\partial (\rho \mathbf{v})}{\partial t} + \nabla \cdot (\rho\, \mathbf{v} \otimes \mathbf{v} + \boldsymbol{\sigma}) = -\rho \nabla \phi\,, \tag{2.32}$$

where $\otimes$ denotes the tensorial product, the Cauchy stress tensor $\boldsymbol{\sigma}$ encodes information on the contact, small-range forces (e.g., thermodynamic pressure), and ϕ is a potential from which derives the long-ranged forces (e.g., the gravitational potential). Solving these equations to get the motion of a star or a planet is unnecessarily complicated. Rather, one usually integrates out the fine details by defining a total mass $m \equiv \int \rho \mathrm{d}^3\mathbf{x}$, and total momenta. Classically, there are two kinds of momenta, linear and angular

$$\mathbf{p} \equiv \int_{\mathscr{B}} \rho\, \mathbf{v}\, \mathrm{d}^3\mathbf{x} \quad \text{and} \quad \mathbf{S} \equiv \int_{\mathscr{B}} \rho\, (\mathbf{x} - \mathbf{z}) \times \mathbf{v}\, \mathrm{d}^3\mathbf{x}\,, \tag{2.33}$$

where $\times$ denote the Euclidean cross product, and $\mathbf{z}$ is an arbitrary point with respect to which angular momentum is defined (usually, the center of mass simplifies the equations, but let us stay general for now). Equations of motion for these quantities are easily obtained by taking the derivative of their definitions (2.33), and combining the result with the conservation equations (2.32). We directly obtain $\mathrm{d}m/\mathrm{d}t = 0$, so that the total mass is conserved, and for the momenta

$$\frac{\mathrm{d}\mathbf{p}}{\mathrm{d}t} = -\int_{\mathscr{B}} \rho\, \nabla \phi\, \mathrm{d}^3\mathbf{x}\,, \tag{2.34a}$$

$$\frac{\mathrm{d}\mathbf{S}}{\mathrm{d}t} = -\int_{\mathscr{B}} \rho\, (\mathbf{x} - \mathbf{z}) \times \nabla \phi\, \mathrm{d}^3\mathbf{x} - \frac{\mathrm{d}\mathbf{z}}{\mathrm{d}t} \times \mathbf{p}\,, \tag{2.34b}$$

The first of these equations is nothing but (the integral formulation of) Newton's second law $\mathrm{d}\mathbf{p}/\mathrm{d}t = \mathbf{F}$, expressing the rate of change of the momentum in terms of the total force $\mathbf{F} \equiv \int_{\mathscr{B}} \mathbf{f} \mathrm{d}^3\mathbf{x}$, as a volume-integral of the force densities $\mathbf{f} \equiv -\rho \nabla \phi$. The second equation expresses the total torque that drives the angular momentum of the extended body, with the spin component (first term, integrated) and the orbital component (second term).

Isometries and Motion

The previous results are well-known and taught in any graduate course on classical mechanics. There is, however, an alternative approach to motivate the definitions (2.33) of the momenta and their evolution equations (2.34), based on the isometries of Euclidean space. This method takes advantage of the highly-symmetric nature of 3D Euclidean space, where Newtonian mechanics take place. In particular, it is convenient to introduce a functional, called the *generalized momentum*, that encapsulates both forms of momentum. It is defined by

$$\mathcal{P}_\xi(t) \equiv \int_{\mathscr{B}} \rho(\mathbf{x}, t) v_a(\mathbf{x}, t) \xi^a(\mathbf{x}) \mathrm{d}^3\mathbf{x} , \tag{2.35}$$

where ξ^a is a Killing vector field of Euclidean space. At a given time t, $\mathcal{P}$ takes as an input a Killing field ξ^a and outputs a real number. For instance, if $\xi^a \equiv (\partial_x)^a$, is the Killing vector associated to translations in the x-direction, then $\mathcal{P}_\xi$ represents the x-component of the body's linear momentum. The main property of the quantity $\mathcal{P}_\xi$ is that it is a linear map on the space of Killing field ξ^a. But Killing fields satisfy a remarkable property: specifying the value of $(\xi_a, \nabla_a \xi_b)$ at *any* one point $\mathbf{x}_o$ in space suffices to know $\xi_a(\mathbf{x})$ at *every* point $\mathbf{x}$. This central result follows from the system of equations satisfied by the pair (ξ_a, Ξ_{ab}), where $\Xi_{ab} \equiv \nabla_a \xi_b$, namely

$$v^a \nabla_a \xi_b = v^a \Xi_{ab} \quad \text{and} \quad v^a \nabla_a \Xi_{bc} = -R_{bcad} v^a \xi^d , \tag{2.36}$$

where v^a is any tangent vector to a curve joining $\mathbf{x}_o$ to $\mathbf{x}$. If the metric g_{ab} is given, then so is the Riemann tensor R_{abcd} that appears in Eq. (2.36). Therefore, (2.36) is a well-posed, linear differential system and any initial data $(\xi_a(\mathbf{x}_o), \Xi_{ab}(\mathbf{x}_o))$ is sufficient to compute the value of $(\xi_a(\mathbf{x}), \Xi_{ab}(\mathbf{x}))$ at any $\mathbf{x}$. In particular, there always exists two two-point tensors $A^{\mathrm{ab}}(\mathbf{x}, \mathbf{x}_o)$ and $B^{a\mathrm{bc}}(\mathbf{x}, \mathbf{x}_o)$ such that

$$\xi^a(\mathbf{x}) = A^{a\mathrm{b}}(\mathbf{x}, \mathbf{x}_o) \xi_{\mathrm{b}}(\mathbf{x}_o) + B^{a\mathrm{bc}}(\mathbf{x}, \mathbf{x}_o) \Xi_{\mathrm{bc}}(\mathbf{x}_o) . \tag{2.37}$$

These two-point tensors simply propagate the information from $\mathbf{x}_o$ to $\mathbf{x}$. They are akin to the flow of vector fields associated to the solution of a differential system. Inserting the general decomposition of Killing fields (2.37) into the definition (2.35) of the generalized momentum then gives

$$\mathcal{P}_\xi = p^a \xi_a + \frac{1}{2} S^{ab} \Xi_{ab} , \tag{2.38}$$

where the two tensors (p^a, S^{ab}) are defined as

$$p^b \equiv \int_{\mathscr{B}} \rho v_a A^{ab} \mathrm{d}^3\mathbf{x} \quad \text{and} \quad S^{bc} \equiv 2 \int_{\mathscr{B}} \rho v_a B^{abc} \mathrm{d}^3\mathbf{x} . \tag{2.39}$$

In the 3D Euclidean setting, the two-point tensors (or propagators) A^{ab} and B^{abc} simply read [347] $A^{ab}(\mathbf{x}, \mathbf{x}_o) \equiv \delta^{ab}$ and $B^{abc}(\mathbf{x}, \mathbf{x}_o) \equiv (\mathbf{x} - \mathbf{x}_o)^{[b} \delta^{c]a}$. If we insert these expressions back in (2.39), we recover the original definitions (2.33), with the spin vector $\mathbf{S}$ there being equivalent to the antisymmetric spin tensor S^{ab} here by the usual Hodge duality (in 3D spaces).

In this context, the linear and angular momenta (p^a, S^{ab}) are defined uniquely as the coefficients appearing in the generalized momentum $\mathcal{P}_\xi$, when writing it as the linear combination of ξ_a and $\nabla_a \xi_b$. In order to get the evolution equations for p^a and S^{ab}, it suffices to take the derivative of (2.38), combine it with the evolution equations for the Killing fields (2.36) and match the coefficients in front of ξ_a and Ξ_{ab}. After some more algebra, one then obtains two equations, whose general form reads

$$\frac{\mathrm{d}p^a}{\mathrm{d}t} = \frac{1}{2} R_{bcd}{}^{a} S^{bc} v^d + F^a \quad \text{and} \quad \frac{\mathrm{d}S^{ab}}{\mathrm{d}t} = 2p^{[a} v^{b]} + M^{ab}, \tag{2.40}$$

where v^a is a vector generating time translations, and F^a, M^{ab} are the force vector and (antisymmetric) torque tensor, that can be expressed as integrals over the body and reduce to their expression (2.34) when explicitly computed. This derivation shows that the evolution equations for the momenta p^a, S^{ab} are dual to those of the Killing quantities ξ_a, Ξ_{ab}. In particular, nothing here is relativistic, and the coupling between spin and curvature (appearing in $\dot{p}^a$) is already present at the "Newtonian" level. This so-called "Papapetrou-term" is a consequence of the geometry of the problem and not the relativistic nature of the theory, a fact only revealed by the GKF approach. In fact, this is one of the main feature of the GKF method, which does not appear to be widely known: it reveals the strong connection between the laws of motion and the geometry of the underlying space(time). Treating Newtonian mechanics this way provides valuable insight on the strong link between Euclidean isometries and the laws of motion. As an example, the invariance of the Laplacian's Green function $G(\mathbf{x}, \mathbf{y}) = -|\mathbf{x} - \mathbf{y}|^{-1}$ with respect to Euclidean isometries is, in fact, equivalent to Newton's third law [346].

Lorentzian Generalization

The previous calculations may be seen, at first glance, as nothing but a complicated way of motivating the (otherwise natural) momenta definitions (2.33) and computing the (otherwise straightforward) evolution equations (2.34). The point of using the generalized momentum linear form $\mathcal{P}_\xi$ and the two-point tensors A^{ab}, B^{abc} (both of which admit trivially simple expressions in Euclidean spaces) is that their definition is, in fact, independent of the underlying manifold. Consequently, they can be extended to any Lorentzian spacetime. In the case of GR, a number of adjustments are necessary, but not fundamentally complicated. In particular

- the Newtonian time $t \in \mathbb{R}$ is replaced by a $3+1$ foliation of spacetime, with a parameter $s \in \mathbb{R}$ indexing the leaves, playing the role of time.
- the Killing fields of Euclidean space are replaced by Generalized Killing Fields (whence the name, GKF). The space of GKFs has the same dimension as that of Killing fields, but contrary to them, GKF exist for any regular manifold. Whereas Killing fields are defined by the Killing equation $\mathcal{L}_k g_{ab} = 0$ that holds on the whole spacetime, GKFs are constructed from the 3+1 foliation and a worldline

γ, and verify $\mathcal{L}_k g_{ab} = 0$ only on γ. Otherwise, they enjoy the same properties as Killing fields do.

- the generalized momentum is a functional of GKFs and depends on the leaf index s, through $P_\xi(s) \equiv \int_{\mathcal{B}} T^{ab}\xi_b \mathrm{d}S_b$, where $\mathrm{d}S_b$ is the 3-volume element on the spacelike leaf, $\mathcal{B}$ is the intersection between T^{ab}'s support and that leaf.

Except for these natural modification, the calculations remain the same in Euclidean space and in curved spacetime (details of which can be found in Sect. 4 of [347]). One advantage of the GKF results over the MPTD scheme (as synthesized by Dixon [322, 342, 349]), is that it extends Dixon's result in the case where the self-field of the object is non-negligible. Although discussing this would be outside the scope of this section, one of the strengths of the GKF approach is the ability to capture self-field effects (i.e., influence of the body's gravitational field on its own motion) in a non-perturbative framework. More precisely, where the MPTD scheme provides an excellent basis for understanding the motion of the body, the GKF approach allows one to generalize the definitions of momenta (and center of mass) so that the multipole expansions (in the SEM tensor or the evolution equations) still hold (and behave nicely) when the object's self-field is taken into account. As a consequence, although the GKF equations derived in [346, 347] may look the same as that obtained in the MPTD scheme (or in the Lagrangian formalism, for that matter), they usually apply to a wider range of cases.

As a final note, Harte himself stresses [350] that the most important aspects of the GKF method are not the results nor the fact that it applies to more general cases, but rather that it shows (1) that linear and angular momentum are actually two pieces of a more general object, which can be manipulated at once (much like the electric and magnetic fields), (2) that the evolution equations for the dipole particle (no force or torque) are dual to the Killing evolution equations, and thus more a feature of curved-geometry than relativistic motion,[7] and (3) that the dynamics (forces and torque) only start at the quadrupolar level, and can simply be seen as a measure of the deviation from (the local, Poincaré) symmetry. This short account on the GKF method skips over many other important and interesting features, in particular the discussion about self-field and renormalization of the point-particle limit for extended bodies. These are discussed in depth in [346, 347]. Fortunately, the effects of the body on its own dynamics are well captured by the self-force formalism, which we present next.

2.2.4 *Gravitational Self-force*

In the previous sections, we have seen how a compact object can be explicitly modeled as a (possibly spinning) point particle. We did not need to give the explicit form for the metric, nor to specify if it was a solution to the Einstein equation. Rather, the

[7] For example, the curvature-spin "coupling" is present in "Newtonian mechanics" on Riemannian 3-spaces, cf Sect. 2.2.7 of [347]. It is the flatness of Euclidean 3-space that makes this term absent in Newtonian mechanics.

laws of motion for the multipoles of the particle readily come from the sole energy conservation equation. In all methods, the form of the SEM tensor of multipolar particle is found to be (2.29) (at dipolar order), which universally describes spinning, massive objects. As a consequence, these methods rely on a crucial assumption: the metric tensor is implicitly defined initially. One could ask: what is the metric in these computations ? Is it the metric solution to the Einstein equation when the multipolar SEM tensor is taken as the source ? Is it a vacuum metric ? A second question could be: how do these results apply to the case of a black hole, which is made of pure vacuum and therefore has no intrinsic SEM tensor? Indeed, the Dixon and Harte's formulae for the multipoles involve integrals over the SEM tensor of the source, while the Tulczyjew algorithm makes the assumption of a multipolar expansion of the SEM tensor of the source in the first place.

One way of answering these questions is to explore the problem under the light of gravitational self-force theory (GSF). This alternative path, which is a direct application of perturbation theory to GR, in the is under intense investigation as one of the most promising way of building GW templates for the detection of EMRIs by the space-based interferometer LISA. The GSF method does lead to evolution equations and a SEM tensor that look the same as those obtained with the previous methods. On the one hand it only holds at a some given order in the perturbation, but on the other hand it sheds some light as to the nature of the metric appearing in the equations. For our purposes, the most important aspect of GSF is that it is perfectly well defined for objects with a vanishing SEM tensor (such as BHs). In the following paragraphs we provide a brief summary of the assumptions behind GSF theory and how they are used to derive the evolution equations and the multipolar SEM tensor.

Overview of GSF

The main goal of GSF is to describe a compact binary system made of a small compact object of mass m orbiting a much larger compact object of mass M. For definiteness, in this paragraph we will take the big object to be a supermassive Kerr black hole (BH) of mass M, and, to contrast, the small body will be a neutron star (NS). The mass ratio of these two provides a small-parameter $\epsilon \equiv m/M \ll 1$, which, in turn, naturally calls for a perturbative expansion in powers of ϵ of the problem. In the absence of the NS ($\epsilon = 0$), the metric describing the system is that of an isolated BH, say $\mathring{g}_{ab}$. Consequently, far away from the NS, the exact metric g_{ab} of the full system (BH+NS) can be written as

$$g_{ab}(\epsilon) \equiv \mathring{g}_{ab} + h_{ab}(\epsilon) \,, \quad \text{with} \quad h_{ab} \equiv \sum_{n\geq 1} \epsilon^n h_{ab}^{(n)} \,, \tag{2.41}$$

where h_{ab} is rightfully called the *perturbation*. The perturbation encodes the small difference between an isolated BH and a BH surrounded by the NS. It is expanded formally in powers of ϵ, with coefficients[8] $h^{(n)}_{ab}$. Decomposition (2.41) is called the *outer expansion* of g_{ab} as it only holds at some reasonable distance from the NS. Indeed, close to the NS the full metric g_{ab} does not resemble $\mathring{g}_{ab}$ but rather the metric of an isolated NS, say $\overset{*}{g}_{ab}$. Consequently, then there must exist a second decomposition of the full system g_{ab}, called the *inner expansion*, such that

$$g_{ab}(\tilde{r},\epsilon) \equiv \overset{*}{g}_{ab}(\tilde{r}) + H_{ab}(\tilde{r},\epsilon)\,, \quad \text{with} \quad H_{ab} \equiv \sum_{n\geq 1} \epsilon^n H^{(n)}_{ab}(\tilde{r})\,, \tag{2.42}$$

where H_{ab} is the perturbation, r is a measure of the distance to the NS and $\tilde{r} \equiv r/\epsilon$, such that (2.42) is valid for $\epsilon \to 0$ at fixed $\tilde{r}$, whereas (2.41) holds for $\epsilon \to 0$ at fixed r. For the reader familiar with basic perturbation theory, the vicinity of the NS generates a boundary layer [352] in the background solution, where the outer expansion (2.41) breaks down. The existence of an outer (2.42) and inner (2.41) expansions, as well as the Einstein equation (1.1), are the only necessary assumptions behind GSF theory, making it very general and powerful. In particular, if one assumes that (1) the two expansions coincide over some intermediate, *buffer* region (after all they are two expansions of the same metric g_{ab}) and that (2) the exact metric g_{ab} solves the vacuum Einstein equation outside the NS, then one readily obtains the general structure of the perturbation h_{ab}, from which everything else can be obtained.

First, and irrespective of the Einstein equation, the small object's metric $\overset{*}{g}_{ab}$ is shown to be $\tilde{r}$-asymptotically flat (i.e., $\overset{*}{g}_{ab} \to \eta_{ab}$ as $\tilde{r} \to +\infty$ for fixed ϵ). This is but a consequence of the matching condition, which encodes the dependence of $\overset{*}{g}_{ab}$ for large $\tilde{r}$. This asymptotic flatness can be used to properly define a set of multipole moments for the NS, for example by applying the very general theory of Geroch-Hansen moments [353, 354]. As a consequence, the coefficients $H^{(n)}_{ab}$ of the inner expansion can be expressed entirely in terms of these multipole moments. Then, thanks to the matching condition again, the $h^{(n)}_{ab}$ can be expressed in terms of these multipoles via $H^{(n)}_{ab}$.

Then, one may solve the Einstein equation iteratively in powers of ϵ, outside the small body. The main feature of the obtained solution (which necessitates a number of computational tricks, gauge choices and control of the boundary conditions) is that the perturbation h_{ab} can always be written a sum of two pieces

$$h_{ab} = h^R_{ab} + h^S_{ab}\,. \tag{2.43}$$

The so-called *regular piece*, h^R_{ab}, encodes the information on the background metric $\mathring{g}_{ab}$, for instance the SMBH's mass, spin, and tidal environment, and is independent of the small body's characteristics. It is both regular at the location the small body, and a vacuum solution to the Einstein equation. On the contrary, the singular piece,

[8] One may use *singular* ($h^{(n)}_{ab}$ depends also on ϵ) or *regular* (formal ϵ Taylor-expansion) expansions depending on whether the representative worldline of the body is fixed or ϵ-dependent. See [351] and references therein for a discussion about these different approaches.

h^S_{ab}, diverges when approaching the small body. It is not a vacuum solution, and explicitly depends on the body's multipole moments. Broadly speaking, the general form of the singular piece is, at quadratic order in ϵ,

$$h^S_{ab} = \epsilon \frac{2m}{r}\delta_{ab} + \epsilon^2 \frac{4S^c n^d}{r^2} u_{(a}\varepsilon_{b)cd} + O(\epsilon^3) \tag{2.44}$$

where r is a measure of the distance from the NS to the field evaluation point and n^a is the unit vector that joins them. Equation (2.44) contains information on the small body: m its mass, S^{ab} a spin tensor defined with respect to some worldline that represents the body, and u^a is that worldline's unit tangent vector. These multipoles m, S^{ab} are defined from the object's own spacetime (as if it were isolated). With the metric perturbation (2.44), we can now understand how the equation of motion and the SEM tensor of the NS are derived.

Evolution Equations

In GSF theory, the worldline that will ultimately represent the small body is introduced implicitly through the use of Fermi-Walker (FW) coordinates, denoted (t, x^i) (see also section III.B of [351]). These coordinates generalize, to a whole curve, the locally inertial normal coordinates that are traditionally built around a point, such that the metric near this point looks flat (locally) (see Eric Poisson's relativity's toolkit [355] for details). We will not discuss further the exact definition and existence condition of this worldline further, but refer to the discussion in Sect. III.B of [351] for details. To illustrate, let us look at the metric coefficients written in FW coordinates. Putting $r^2 \equiv \delta_{ij}x^i x^j$, they read [355]

$$g_{tt} = -1 - 2a_i x^i - (R_{titj} + a_i a_j)x^i x^j + O(r^3)\,, \tag{2.45a}$$
$$g_{it} = -\tfrac{2}{3}R_{tiaj}x^i x^j + O(r^3)\,, \tag{2.45b}$$
$$g_{ij} = \delta_{ij} - \tfrac{2}{3}R_{aibj}x^i x^j + O(r^3)\,, \tag{2.45c}$$

and thus include the FW components $a^\alpha = (0, a^i)$ of the four-acceleration vector $u^b\nabla_b u^a$, with u^a the unit tangent to the worldline, as well as the local spacetime curvature through the components of the Riemann tensor. Thanks to the FW coordinate system, the acceleration of the worldline is thus encoded in the metric. The Einstein equation, when solved for the metric outside the small body, gives an expression of the acceleration $u^b\nabla_b u^a$ in terms of the body's multipole moments via the singular piece h^S_{ab} of the metric (2.44) and the regular piece h^R_{ab}. These equations have been obtained at linear order[9] in ϵ for an arbitrary compact object. In this case, the acceleration along γ reads

[9] Most second-order self-force calculations are currently ongoing, and the equations of motion have, for the moment, only been obtained for spinless, spherically symmetric objects [356, 357].

$$u^b \nabla_b u^a = \epsilon \cdot \left(\frac{1}{2m} R_{bcd}{}^a u^b S^{cd} - \frac{1}{2} g^{ab} (2h^{R,1}_{bc;d} - h^{R,1}_{cd;b}) u^c u^d \right) + O(\epsilon^2) , \quad (2.46)$$

with $h^{R,1}_{ab;c} \equiv \nabla_c h^R_{ab}$ denoting the covariant derivative compatible with the background metric g_{ab} of the first order, regular piece of the perturbation. Notice that there is no zero-th order term in Eq. (2.46): this is consistent with the test-particle limit, wherein the small body indeed follows a geodesic of the background spacetime. The right-hand side of Eq. (2.46) contains two terms, of which

- the second is a first-order term proportional to the spin tensor S^{ab} of the small body (uniquely defined from the spin vector S^a of Eq. (2.44) by setting the mass dipole moment to zero, cf. Sect. 2.3.2 below). This is the term that we found in all previous calculations for dipolar particles;
- the last term is new, and involves the first-order, regular piece of the metric perturbation h^R_{ab}. If the small body was a test particle, then this term would vanish identically. Therefore, it accounts for the small body's own influence on its trajectory, hence its name: the self-force (per unit mass, to be precise).

Equation (2.46) is the fundamental and central result of GSF theory, and is known as the MiSaTaQuWa equation (named after the authors of [358, 359] who derived it for the first time). In addition to the equation of motion (2.46), one obtains, still from the Einstein equation, conservation laws for the body's multipoles, namely

$$u^a \nabla_a m = O(\epsilon^2) \quad \text{and} \quad u^c \nabla_c S^{ab} = O(\epsilon^3) . \quad (2.47)$$

Consequently, in GSF theory, the spin is parallel-transported along the (accelerated) worldline of the particle.[10] One remarkable aspect of equations (2.46) and (2.47) is that they can be combined into a form similar to the evolution equations derived using the other methods (Lagrangian, MPTD and GKF). Indeed, if we work at linear order in spin, such that $p^a = m u^a$, the conservation of mass (2.47) combines nicely with (2.46) so that

$$u^e \tilde{\nabla}_e p^a \equiv \frac{1}{2} \tilde{R}_{bcd}{}^a u^b S^{cd} + O(\epsilon^2) , \quad (2.48)$$

where, this time, $\tilde{\nabla}$ is the covariant derivative compatible with the *effective* metric $\tilde{g}_{ab} \equiv \mathring{g}_{ab} + h^R_{ab}$. This form gives an answer to the question asked earlier: what is the metric that enters the evolution equations? The answer from GSF is that it is neither the exact metric g_{ab}, nor the background metric $\mathring{g}_{ab}$. Rather, it is a vacuum metric $\tilde{g}_{ab}$, defined as the background + the regular piece of the perturbation. We see here a generalization of the equivalence principle, which traditionally refers to test objects: here the particle is accelerated in the real, physical spacetime g_{ab}, but can also be interpreted as following a geodesic in an effective, vacuum geometry

[10] This does not contradict the dipolar equation of precession derived earlier in other schemes. Indeed, to compare these results, one would need (at least) to (1) compare both result in the same spacetime, (i.e, expressed the equations with the same metric), and (2) expand the dipolar precession equation in terms of a mass ratio, which would then lead to $\dot{S}^{ab} = O(\epsilon^3)$.

$\tilde{g}_{ab}$. The presentation here follows the so-called self-consistent method initiated by Adam Pound [351]. Other routes are possible to derive the self-force equations of motion, for example the approach of Sam Gralla and Robert Wald that also provide compatible first [360] and second order [361] results.

SEM Tensor

The results for the equation of motion and the evolution equations rely on a subtle, yet central, assumption behind GSF. The metric perturbation, as obtained from solving the Einstein equation, is technically valid only outside of the small body, in vacuum. In terms of the distance r to the worldline, these results hold for r larger than $O(\epsilon)$ distances. However, extending these results all the way down to the worldline (i.e., $r \to 0$) does make mathematical sense and does not alter the physics in the outside region [351, 362]. This, in a sense, generalizes the Newton theorem: computing the field at the exterior of a spherical body or that of a point mass yields the same result.

Let us now explain the general idea behind the derivation of the SEM tensor at linear order in ϵ, first. The Einstein tensor[11] G^{ab} associated with the full metric $g_{ab} = \mathring{g}_{ab} + h_{ab}$ can be linearized as,

$$G^{ab}[\mathring{g} + \epsilon h^1] = 0 + \epsilon \cdot G^{ab}_{(1)}[h^{S,1}] + O(\epsilon^2) \tag{2.49}$$

with $G^{ab}_{(1)}$ the linearized Einstein tensor (that looks like a d'Alembert operator in a Lorenz gauge, recall Sect. 1.2.2). The vanishing zeroth-order term and the fact that only h^S contributes to $G^{ab}_{(1)}$ come from $\mathring{g}_{ab}$ and h^R_{ab} being vacuum solutions, respectively. One may now explicitly compute equation (2.49) in terms of the body's multipoles thanks to the expression of h^S given in (2.44). Since everything is taken to hold all the way to the worldline, the result takes a distributional form and reads [200, 360, 363]

$$G^{ab}_{(1)}[g] = 8\pi m \int_\gamma u^a u^b \, \delta_4(x, z(\tau)) \, \mathrm{d}\tau + O(\epsilon^2), \tag{2.50}$$

where u^a is the four-velocity along the worldline γ, τ the proper time and where the Dirac covariant distribution δ_4, all built from the background metric $\mathring{g}_{ab}$. If we interpret equation (2.50) as the linearized Einstein equation, then we recognize on the right-hand side $8\pi T^{ab}$, where T^{ab} is the SEM tensor of a point particle of mass m, moving in the background spacetime $\mathring{g}_{ab}$, just like we encountered in the previous sections. However, we should stress that, here, this result is not an assumption and does not derive from a point-particle Ansatz of the body: it is *derived* solely from the matching condition and the Einstein equation. This result is central: it shows that

[11] We choose the upstairs indices to match references cited in the text. Care must be taken while moving indices in perturbation theory because of the different metrics at play and their ϵ-dependence.

at the first order in ϵ, a small body (even a black hole) acts as a point particle, in the sense that it generates the same gravitational field (encoded in h^S_{ab}) as one.

At second order, one can proceed in (almost) the same way, namely, use the left-hand side of the Einstein equation to define the SEM tensor of the body. In particular, once we know the perturbation h_{ab} outside the body and extend it all the way to the worldline, we can take the sourced Einstein equation $G_{ab} = 8\pi T_{ab}$, compute the Einstein tensor at second order, and then *define* the SEM tensor as $(8\pi)^{-1} G_{ab}$, since this equation is to hold everywhere. This is in the same spirit as what we did in the Newtonian case in Sect. 2.1, where the skeletonized mass density (2.6) was *defined* as the quantity that generates the truncated potential (2.5) through the Poisson equation. In particular, at second order the SEM tensor is now computed through the second-order expansion of the Einstein tensor

$$G^{ab}[g] = 0 + \epsilon \cdot G^{ab}_{(1)}[h^{S,1}] + \epsilon^2 \cdot \left(G^{ab}_{(1)}[h^{S,2}] + G^{ab}_{(2)}[h^1, h^1] \right) + O(\epsilon^3) , \quad (2.51)$$

with now the second order piece $G^{ab}_{(2)}$ coming from the quadratic term in the expansion of $G_{ab}[g + \epsilon h]$ that is quadratic in h (see Appendix A in [362] for details). Due to the assumption that the Einstein equation holds all the way to the worldline, the term $G^{(2)}_{ab}$ in (2.51) involves strongly singular terms that prevent the left-hand side of (2.51) (and thus T_{ab}) to be a well-defined distribution, unlike the first order calculation. As a consequence, the second-order calculation requires some extra care: either with a method involving some kind of regularization procedure (e.g., a puncture scheme, cf. [364]), or with an exact computation but in a well-chosen gauge (which removes, by construction, the strongest singularities), as was done recently by Sam Upton and Adam Pound [362]. In both cases, and quite remarkably, the result (2.49) still holds at second order, in the sense that the first and second order terms in (2.51), computed thanks to formulae like (2.44), all combine nicely so that [200, 365]

$$(8\pi)^{-1} G^{ab}[g] = m \int_\gamma \tilde{u}^a \tilde{u}^b \, \tilde{\delta}_4 \, \mathrm{d}\tilde{\tau} + \tilde{\nabla}_c \int_\gamma \tilde{u}^{(a} \tilde{S}^{b)c} \, \tilde{\delta}_4 \, \mathrm{d}\tilde{\tau} + O(\epsilon^3) , \quad (2.52)$$

where now all *tilde* quantities are built from the effective metric $\tilde{g}_{ab} \equiv \mathring{g}_{ab} + h^R_{ab}$. Here appears the spin contribution (which finds its origin in the singular piece $h^{S,2}_{ab}$ in (2.44)). One again, we observe the SEM tensor of a spinning point particle, moving, not on the background metric, but on the effective metric $\mathring{g} + h^R$. We emphasize one last time that this is the result of an explicit calculation, namely the Einstein tensor at second-order in ϵ^2. As predicted by Detweiler in [366], the point particle model does hold at nonlinear order in GSF theory. A summary of the important results are depicted in Fig. 2.2.

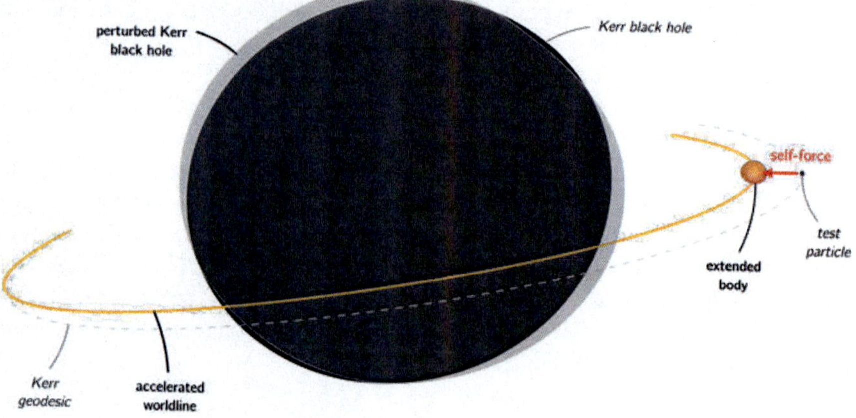

Fig. 2.2 This picture depicts the three main features of GSF. The real system consists in an extended NS (yellow ball on the right) orbiting a perturbed Kerr BH (deformations in grey) on an accelerated worldline (in yellow). This worldline is accelerated with respect to the Kerr geodesic (dotted grey curve) that would be followed by a test particle around an exact Kerr BH (cylindrically symmetric, in black). This acceleration is interpreted as resulting from a *gravitational self-force* (red arrow, on the right)

2.2.5 *Discussion*

The point of the previous sections was to present, without going into too much details, how different multipolar schemes (Lagrangian, MPTD, GKF and GSF) are used to derive (1) the SEM tensor of an extended compact object modelled as a point particle endowed with multipoles, and (2) the evolution equations for these multipoles. Only the main results will be used in the next sections, and no more reference will be made to these particular schemes. Before moving to the next section where a summary is provided and where we build on these results, we take some time to (1) discuss how these schemes differ in their assumptions and what they each bring into the picture of relativistic dynamics, and (2) give some information on how they generalize to higher multipoles, as we only provided a pole-dipole discussion up to now. These two points are discussed below, in that order.

Comparisons

Although we believe that a more thorough comparison between the multipolar schemes would be of valuable interest for the sake of clarity and for understanding precisely how their applications range overlap, we must leave this for future work. Here, we only summarize the assumptions and main results of these schemes, and pin point how some of them answer questions that others raise, emphasizing their complementarity.

The Lagrangian method relies on two main assumptions. First, that a metric tensor is given: all calculations are made relative to this metric and the dynamics take place in the spacetime described by it. Second, it assumes a point-particle model for the extended body, by the introduction of a worldline, and needs an Ansatz for the Lagrangian. However, very little assumptions about the Lagrangian are required to derive its explicit form in terms of the multipoles, which are defined as conjugated variables of the kinematic degrees of freedom. In particular, the four-momentum, spin tensor, and higher-order multipoles arise as the generalized momenta of the worldline's four-velocity, rotation coefficients, and the (successive derivatives of) the Riemann tensor. Nevertheless, this method presents the advantage of deriving very straightforwardly the evolution equations for the multipoles by a simple "action minimization" procedure. Also, the very definition of the multipoles renders their algebraic symmetries obvious (the quadrupole tensor has the same symmetries as the Riemann tensor since it is its conjugated momentum). Finally, the explicit form of the Lagrangian opens up a very effective Hamiltonian treatment of the dynamics of point particles in GR [367], that even extends in the non-conservative regime. References of interest for the Lagrangian method are Bailey and Israel's pioneering works [316] and Sylvain Marsat's paper [320] with a complete derivation at all multipolar orders. See also [194] for a recent review on the Hamiltonian formulation of binary mechanics.

The MPTD scheme was historically the first to derive the evolution equations for the multipoles, which nowadays are sometimes called the MPTD equations. The only assumption behind this framework is the conservation of the SEM tensor representing the body. From there, two routes are possible. First, in Dixon's point of view, one can define linear and angular momentum as integrals over the body's SEM tensor. These definitions, when chosen correctly (here lies Dixon's main input) lead to a set of evolution equations that are well-posed. They involve a force vector and a torque tensor that can be expressed as a sum of higher order multipoles (starting with the quadrupole), also defined as integrals. In the spirit of Tulczyjew, one can also take an alternative route. Instead of defining the momenta, start with an Ansatz for the SEM tensor, which is directly inspired by its Newtonian equivalent, albeit being covariant and adapted to a 3+1 spacetime. Then, by the sole application of SEM conservation, one derives at once the evolution equations of the non-redundant degrees of freedom of the multipoles, defined as projections of the Ansatz. Although Dixon's way is advantageous as it allows one to define momenta by integrals over the SEM tensor, and discuss the worldline unicity, it cannot account for black holes, which by definition do not have a SEM tensor over which to integrate. On the contrary, Tulczyjew's scheme already assumes a point-particle model, and can thus be applied to any kind of compact object, a priori. Finally, we note that our main interest in Tulczyjew's reduction process is to apply it for the derivation of useful conservation laws, as done in Chap. 3. References of interest for the MPTD scheme are Dixon's most recent and complete account of his work [327], and for Tulczyjew's reduction process, see Andrzej Trautman's lecture notes [311], as well as [1, 344].

The method of GKF developed by Abraham Harte provides a direct extension of Dixon's scheme. For that reason, this scheme only applies to matter sources, with a non-zero SEM tensor, excluding the case of black holes. However, it offers two considerable advantages. First, it is written in an interesting geometrical way, providing insight as to what belongs to kinematics (the pole-dipole evolution equations are pure geometry, in this sense) and what belongs to the dynamics (the effect of all higher-order multipoles starting with the quadrupole). The definition of the momenta are combined into a single entity that is defined by duality with Killing fields (generalized, if need be). The method applies to a wide range of theories of gravity, including Newtonian, GR and higher-dimensional extensions thereof (as explained in [347]). The second advantage of the GKF framework is to shed some light on the influence of a body on its own motion (self-field) in a non-perturbative setting. In particular it readily shows which metric actually enters the evolution equations and the SEM tensor, a question left unanswered in the Lagrangian and MPTD schemes. References for the GFK method are Abraham Harte's own reviews [346, 347] and his recent article [368] containing a summary and applications at quadrupolar order.

The GSF theory presents a number of advantages over the above schemes. The overall appeal of the GSF method is that it does not rely on any limiting assumption, except for the Einstein equation (which can hardly be discarded). The small object's point-particle behavior is actually a result of the theory, as is its effective SEM tensor, which actually applies for any type of compact objects, neutron stars and black holes. Yet, the most important advantage of GSF is practical: the GSF framework includes a whole algorithmic procedure destined to compute the actual waveform emitted by EMRIs [200]. This is necessary, as GSF is the most promising tool for producing templates and analyzing the data of the space-based interferometer LISA. These computations represent an ongoing challenge, with the most recent milestone being the explicit computation of the gravitational binding energy between a small object orbiting a large Schwarzschild black hole, at second-order in the mass ratio [369].

Beyond the Dipole

The evolution equations (2.14) obtained in the previous section were derived at dipolar order. This means that the body is described by two "multipoles", its momentum p^a and spin S^{ab}. Each approach (Lagrangian, MPTD, SF or GKF) can, in principle, be extended to any order $n \geq 2$ in the multipolar expansion. These extensions are present at all multipole orders in [320] for the Lagrangian method, in [322] for the MPTD method and in [346] for the GKF method. In GSF theory, the evolution equations have been derived at first order for spinning objects [351, 358, 360] and at second order for nonspinning, spherical objects [356, 357]. Quadrupole and higher-order multipoles only appear at third order or more in the mass ratio, but the general result that the small body orbits on a geodesic of an effective metric is known to hold even in the full nonlinear setting, as the GKF method readily shows [346]. Along with evolution equations, the corresponding SEM tensor at dipolar order is easily derived in each method. It should be noted, however, that, although algorithmically straightforward, the computations become cumbersome for the SEM tensor at higher than dipolar order; see [344] for the quadrupolar order, and [320] for the SEM tensor at octupolar order in the Lagrangian formalism. Finally, for the SEM tensor in GSF theory, we refer Sect. 5.5 of [200] and references therein.

The general structure of the evolution equations is known at all multipolar orders. Much like (2.14), they only contain two equations (one for p^a, one for S^{ab}) but with additional terms on their right-hand sides. In particular, at any order $n \geq 2$ in the multipolar expansion, the evolution equations (2.14) are to be replaced by

$$\dot{p}^a = \frac{1}{2} R_{bcd}{}^a S^{bc} u^d + \sum_{k=2}^{n} F^a_{(k)} \quad \text{and} \quad \dot{S}^{ab} = 2 p^{[a} u^{b]} + \sum_{k=2}^{n} M^{ab}_{(k)} , \tag{2.53}$$

where the $(F^a_{(k)}, M^{ab}_{(k)})$ denote contributions from the higher-order multipoles that appear at each order. As they drive the evolution of the momentum and the spin, they are known as the force and moment tensors, respectively. As an illustration, their expression for $n = 2$ are [320, 322, 344]

$$F^a_{(2)} = -\frac{1}{6} J^{bcde} \nabla^a R_{bcde} \quad \text{and} \quad M^{ab}_{(2)} = \frac{4}{3} R_{edc}{}^{[a} J^{b]cde} , \tag{2.54}$$

where the 4-indices tensor J^{abcd} is called the quadrupolar tensor. It possesses the same algebraic symmetries as the Riemann tensor R_{abcd} (recall the discussion around Eq. (2.17)). Therefore it has only 20 independent components. We shall come back in details to the quadrupolar case in the next section.

At the next, octupolar order ($n = 3$), a 5-indices tensor O^{abcde} appears. The evolution equations (2.53) now contain the quadrupolar contributions (2.54) as well as octupolar terms, given by [320]

$$F^a_{(3)} = -\frac{1}{12} O^{bcdef} \nabla^a \nabla_b R_{cdef} , \tag{2.55a}$$

$$M^{ab}_{(3)} = -\frac{2}{3} \nabla_c R_{def}{}^{[a} O^{b]cdef} + \frac{1}{6} \nabla^{[a} R_{cdef} O^{b]cdef} , \tag{2.55b}$$

with the octupole O^{abcde} possessing 40 independent components, thanks to its many algebraic symmetries. More generally, at any order $n \geq 2$ an additional, $(2+n)$-indices tensor called the 2^n-polar tensor, appears in the expressions of $(F^a_{(n)}, M^{ab}_{(n)})$. It possesses many algebraic symmetries and exactly $(n+2)(3n-1)$ independent components [320, 322]. In this work, we will limit ourselves to quadrupolar order and therefore only need Eq. (2.54).

Regarding the multipolar SEM tensor of a point particle, at quadrupolar order a new contribution T^{ab}_{q} (see Eq. (2.59) below) must be added to the pole-dipole SEM tensor T^{ab} given in Eq. (2.29). Its expression involves the quadrupole tensor J^{abcd} that appears in the quadrupolar force $F^a_{(2)}$ and torque $M^{ab}_{(2)}$ (2.54). It has been derived using Tulczyjew's reduction process in [344], with the Lagrangian method (see e.g., [320]) and can be computed with Dixon's or the GKF method by inverting the integral expressions of the multipoles.[12] These quadrupolar contributions are crucial to encode tidal deformations in a binary and/or quadratic-in-spin effects [370]. The SEM tensor at octupolar order includes a new term T^{ab}_{oct} that only appeared (to my knowledge) in [320] using the Lagrangian formalism. The full expression involves the octupolar tensor O^{abcde} that appears in the octupolar evolution equations, with the octupolar force and torque given in (2.55). This contribution, which must be added to the pole-dipole-quadrupole SEM tensor, reads, schematically

$$T^{ab}_{\text{oct}} = \int_\gamma (\nabla R)\, O\, \delta_4 \mathrm{d}\tau + \nabla \int_\gamma R\, O\, \delta_4 \mathrm{d}\tau + \nabla\nabla\nabla \int_\gamma O\, \delta_4\, \mathrm{d}\tau\,, \tag{2.56}$$

where R represents the Riemann tensor. The exact expression can be found in Eq. (2.25) of [320]. The octupolar piece is necessary to account for the cubic-in-spin contributions in the dynamics of spinning binaries, as shown in that same paper. We note that given the length of Tulczyjew's reduction process already at quadrupolar order (given in [344]), the same calculation at octupolar order is possible by hand but would be extremely tedious. However, it is interesting to note that this reduction process could be implemented systematically at all orders in a algorithmic way, for example with the help of a symbolic algebra system, as it shares strong similarities with combinatorics-type calculations. We leave this for future work.

2.3 Model Summary

In the previous section we saw how the description of an extended compact object in GR can be approached. These methods are rather different in their assumptions and definitions. Yet, up to minor details, they all provide the same results regarding the evolution equations and the SEM tensor of a multipolar particle. In this section, we provide a summary of our model, used in the following chapters to derive the first law of mechanics. In particular, we also define and study the conservation properties of derived quantities, like a spin vector, a mass dipole, and various other scalars.

2.3.1 *Quadrupolar Particles*

In the gravitational skeleton formalism, a compact object is represented by a point particle moving along a worldline γ. In a coordinate system x^α, the worldline γ is described by four parametric equations of the form $x^\alpha = y^\alpha(\tau)$, where $\tau \in \mathbb{R}$ denotes the proper time, and y^α are four functions $\mathbb{R} \to \mathbb{R}$ of τ.

---

[12] Although we have not been able to find an example of this particular derivation in the literature, it may (possibly [350]) be found in Ruprecht Schattner's PhD thesis.

Evolution Equations

At quadrupolar order, all the physical information about the compact object is encoded into four tensors defined along γ: the four-velocity u^a (tangent to γ), the four-momentum p^a (not parallel to u^a in general), the spin tensor S^{ab} (antisymmetric) and the quadrupole tensor J^{abcd} (same algebraic symmetries as the Riemann tensor). These multipoles evolve along γ and are coupled to one another and to the geometry, as encoded in the following evolution equations

$$\dot{p}^a = \frac{1}{2} R_{bcd}{}^{a} S^{bc} u^d - \frac{1}{6} J^{bcde} \nabla^a R_{bcde} , \tag{2.57a}$$

$$\dot{S}^{ab} = 2 p^{[a} u^{b]} + \frac{4}{3} R_{edc}{}^{[a} J^{b]cde} . \tag{2.57b}$$

Because they evolve the momentum p^a and the spin S^{ab}, Eq. (2.57a) and (2.57b) are called the equation of motion and equation of precession, respectively. The system (2.57) is made of two coupled, first order, ordinary differential equations for the components of (p, S), given (u, J) and the geometry. Indeed, as tensor fields defined along γ only, they can always be viewed as functions of the sole proper time τ.

Notice that there is no evolution equation for the quadrupole J^{abcd}. An explanation can be drawn from SEM tensor conservation (1.4), which, in general, is not sufficient to constrain the matter fields so much as to fully determine their evolution [371]. Indeed, a quadrupolar particle in GR suffers the same treatment as the perfect fluid with density and pressure (ϱ, P) in Newtonian gravity. For the latter, the Navier-Stokes equations are not sufficient to determine the dynamics, and must be completed by an equation of state relating ϱ and P. Similarly, in GR, for the system (2.57) to be well-posed, an equation expressing the quadrupole J^{abcd} in terms of the other multipoles (p^a, S^{ab}), the four-velocity u^a and/or the geometry $(g_{ab}, R_{abcd}, \ldots)$ is required. These expressions as well as more details will be given in the Chap. 6, when we work at quadrupolar order. For the moment, we do not require explicit formulae and will keep our result valid for a generic quadrupole J^{abcd}. Additionally, even with the quadrupole expressions, the system (2.57) is still not well-posed and requires another equation (called a *supplementary spin condition*), which is touched upon in the following paragraphs.

SEM Tensor

A multipolar particle curves spacetime around it by sourcing the Einstein equation via its SEM tensor. The latter is supported on the worldline $\gamma : x = z(\tau)$ of the particle. At quadrupolar order, it is the sum of three terms, each corresponding to a successive multipolar order and depending on the associated multipole (p, S, J). At dipolar order, we have

$$T^{ab} = \int_\gamma u^{(a} p^{b)} \, \delta_4(x, z) \, \mathrm{d}\tau + \nabla_c \int_\gamma u^{(a} S^{b)c} \, \delta_4(x, z) \, \mathrm{d}\tau \,, \tag{2.58}$$

where the first and second term in (2.58) form the classical monopole-dipole model of a compact object, discussed in Sect. 2.2. At quadrupolar order, one must add to the right-hand side of Eq. (2.58) the following term

$$T^{ab}_{\mathsf{quad}}(x) = \frac{1}{3} \int_\gamma R_{cde}{}^{(a} J^{b)cde} \, \delta_4(x, z) \, \mathrm{d}\tau - \frac{2}{3} \nabla_{cd} \int_\gamma J^{c(ab)d} \, \delta_4(x, z) \, \mathrm{d}\tau \,, \tag{2.59}$$

where we introduced the shorthand notation $\nabla_{cd} = \nabla_c \nabla_d$ here, used in the remaining of the thesis. We see in Eq. (2.59) that the quadrupolar contribution contains a term with two derivatives, as expected, and a term deprived of derivatives. In fact, the latter is simply the anti-symmetric, second-order covariant derivative contribution of the quadrupolar sector. It has simply been recast in terms of the Riemann tensor, thanks to its definition as the commutator of double covariant derivatives. Consequently, the right-most term on the right of (2.59) is actually the symmetrized second-derivative, and one could write $\nabla_{(cd)}$ in front of it. It turns out that, thanks to its symmetries, the quadrupole verifies $J^{c(ab)d} = J^{a(cd)b}$, so that that this symmetrization is unnecessary. Finally, we stress that in Eqs. (2.58)–(2.59), the four-velocity u^a and the three multipoles (p, S, J) are functions of the worldline only (or $z(\tau)$, or simply τ, equivalently) and it is the Dirac distribution $\delta_4 \equiv \delta_4(x, z(\tau))$ that makes T^{ab} a well-defined, distributional tensor field on $\mathscr{E}$. In particular, for any $x \notin \gamma$ we have $T^{ab}(x) = 0$.

Scalars

Next, from the variables u^a, p^a and S^{ab}, we introduce three positive scalar fields along γ: the rest mass m, the dynamical mass μ and the spin amplitude S as

$$m \equiv -p^a u_a \,, \tag{2.60a}$$

$$\mu^2 \equiv -p^a p_a \,, \tag{2.60b}$$

$$S^2 \equiv \tfrac{1}{2} S^{ab} S_{ab} \,. \tag{2.60c}$$

The coefficient $1/2$ for the spin amplitude is purely conventional. In general, none of these scalar fields is conserved along γ, and the masses m and μ need not coincide [344]. However, as we shall prove later, in the case of a binary system of quadrupolar particles moving on a circular orbit, m, μ and S are all conserved, for each particle.

Momentum-Velocity Relation

Contracting equation (2.57b) with u_b and using the mass m defined in (2.60a) readily implies the *momentum-velocity* relation

$$p^a = mu^a - \dot{S}^{ab}u_b + \frac{4}{3}R_{edc}{}^{[a}J^{b]cde}u_b\,. \tag{2.61}$$

This equation is exact. For a dipolar particle ($J^{abcd} = 0$), it shows that the 4-momentum $p^a = p^a_{\text{kin}} + p^a_{\text{hid}}$ is the sum of the timelike kinematic momentum $p^a_{\text{kin}} \equiv mu^a$ and of the spacelike hidden momentum (see e.g. Ref. [372]) $p^a_{\text{hid}} \equiv h^a{}_b p^b$, such that $u_a p^a_{\text{hid}} = 0$, where we introduced the projector orthogonal to u^a,

$$h^a{}_b \equiv \delta^a{}_b + u^a u_b\,. \tag{2.62}$$

The projector (2.62) should not be confused with the metric perturbation h_{ab} used earlier, of which no further mention should be made in the remaining chapters.

2.3.2 *Dipolar Particles*

In this last section we introduce the notion of supplementary spin condition (SSC), which allows (1) to obtain a well-posed system of evolution equations and (2) to expand key relations in powers of the spin.
Note: from now until the end of the present chapter, equations result from a dipolar calculation, where the quadrupole J^{abcd} has been set to zero anywhere it would otherwise appear from the evolution equations (2.57). Their generalization at any multipolar order is provided in Appendix A.4.4. We will need the following equations in Chap. 5 and their quadrupolar corrections in Chap. 6.

Spin Supplementary Condition

Consider for now an extended body, whose support is a worldtube $\mathscr{T}$, as well as a reference worldline $\gamma_\circ \subset \mathscr{T}$, with unit timelike tangent v^a. As shown by Dixon [342] or Harte [346], the body's momentum p^a and spin S^{ab} can be expressed as surface integrals over the SEM distribution, the later depending also on the choice of reference worldline $\gamma_\circ$ [327, 349]. The six degrees of freedom encoded in the spin tensor S^{ab} can equivalently be encoded in two spacelike vectors S^a and D^a, both orthogonal to v^a, according to

$$S^{ab} = \varepsilon^{abcd}v_c S_d + 2D^{[a}v^{b]} \quad \Longleftrightarrow \quad \begin{cases} S^a \equiv -\frac{1}{2}\varepsilon^{abcd}v_b S_{cd}\,, \\ D^a \equiv -S^{ab}v_b\,. \end{cases} \tag{2.63}$$

Physically, the vector D^a can be interpreted as the body's mass dipole moment, as measured by an observer with 4-velocity v^a, i.e., with respect to the reference worldline $\gamma_\circ$, while S^a can be interpreted as the body's spin with respect to that wordline [373].

We now go back to the quadrupolar particle described by (2.57)–(2.58). In a given basis, the equations of evolution (2.57) are equivalent to a system of 10 ordinary differential equations for 13 unknowns, namely the $4+6+3=13$ independent components of $p^\alpha(\tau)$, $S^{\alpha\beta}(\tau)$ and $u^\alpha(\tau)$. (Having specified a physical model for the quadrupole J^{abcd}, its components $J^{\alpha\beta\gamma\delta}(\tau)$ are known functions of $u^\alpha(\tau)$, $p^\alpha(\tau)$ and $S^{\alpha\beta}(\tau)$.) Since we did not specify the worldline γ representing the body, it is not surprising that such under-determinacy should occur in the gravitational skeleton model. To obtain a well-posed problem, three additional constraints on the spin tensor, known as a spin supplementary condition (SSC), thus have to be imposed, equivalent to the choice of the reference wordline $\gamma_\circ$ with tangent v^a for the actual extended body. These constraints take the form $S^{ab} f_b = 0$, where f^a is a timelike vector. In this work, we shall adopt the so-called Frenkel-Mathisson-Pirani SSC (or Pirani SSC for short), defined by setting

$$S^{ab} u_b = 0\,, \tag{2.64}$$

which states that the mass dipole D^a with respect to the wordline γ vanishes, or equivalently that the spin vector S^a in Eq. (2.63) is orthogonal to the 4-velocity u^a. This natural choice of SSC is primarily motivated by the fact that the first law of compact binary mechanics to be derived in Chap. 5 will take its simplest form if (2.64) holds. Moreover, as we shall prove in the next chapter, for circular orbits the 4-velocity of each particle is necessarily tangent to the generator k^a of the helical Killing symmetry. For such orbits, the SSC (2.64) will thus be equivalent to the geometrically-motivated, algebraic constraint

$$S^{ab} k_b = 0\,. \tag{2.65}$$

Other choices of SSC are of course possible, such as the Tulczujew-Dixon SSC $S^{ab} p_b = 0$. Our results are, for the most part, written in a covariant manner, and independent of a choice of SSC. The latter will mainly be used to expand the results in powers of the spin, as a mean to compare our results to the literature. We refer to the references [373–376] for a review of the various SSC, their physical interpretation and well-posedness.

Conservation of Rest Mass and Spin Amplitude

Together with the equations of evolution (2.57), the Pirani SSC (2.64) implies exact conservation laws *at dipolar order*.

First, contracting the equation of precession (2.57b) with $S^{ab} = S^{[ab]}$ and using (2.64) shows that the particle's spin amplitude (2.60c) is conserved at dipolar order:

$$\dot{S} = 0\,. \tag{2.66}$$

Moreover, the particle's rest mass m (2.60a) is conserved along γ as well, up to quadrupolar terms. To prove it, first take the derivative of (2.60a) to get

$$\dot{m} = -(p^a \dot{u}_a + \dot{p}^a u_a)\,. \tag{2.67}$$

To compute the first term, we contract (2.61) with u_a and find that the first two terms on the right-hand side of the resulting equation vanish, thanks to $\dot{u}^a u_a = 0$ and $\dot{S}^{ab} u_a \dot{u}_b = 0$, respectively (the latter follows from a Leibniz rule, the antisymmetry of the spin tensor and the SSC). Consequently, we find $p^a \dot{u}_a = O(J)$. For the second term in (2.67), we simply contract the equation of motion (2.57a) with u_a, and thanks to the symmetries of the Riemann tensor, the first term on the right vanishes and we also find $\dot{p}^a u_a = O(J)$. Explicitly, the result is

$$\dot{m} = 0\,. \tag{2.68}$$

Similar conservation laws hold for other choices of SSC. For instance, using the Tulczujew-Dixon SSC $S^{ab} p_b = 0$, one can establish the conservation of the spin amplitude S and of the dynamical mass μ defined in (2.60b). We emphasize that if the SSC $S^{ab} u_b = 0$ is imposed, the conservation laws (2.66) and (2.68) are *exact* at dipolar order, and not merely perturbatively valid in a Taylor series expansion in the spin. At quadrupolar order, the terms $O(J)$ can be computed explicitly by following the calculation steps provided in the above text.

Definition of a Spin Vector

Having imposed the SSC (2.64), the decomposition (2.63) implies that the three remaining degrees of freedom of the spin tensor S^{ab} can be encoded in a spin vector S^a—or equivalently in a spin 1-form $S_a \equiv g_{ab} S^b$—obeying

$$S_a \equiv -\frac{1}{2} \varepsilon_{abcd} u^b S^{cd} \quad \Longleftrightarrow \quad S_{ab} = \varepsilon_{abcd} u^c S^d\,. \tag{2.69}$$

By construction, the spin vector S^a is spacelike and orthogonal to u^a, while its norm coincides with the conserved norm (2.60c) of the spin tensor, namely

$$S^a u_a = 0\,, \tag{2.70a}$$

$$S^a S_a = S^2\,. \tag{2.70b}$$

The first result derives from the definition (2.69) and the antisymmetry of ε_{abcd}, while (2.70b) derives from Eq. (2.69) and (2.60c) together with the SSC (2.64). Moreover, the rate of change of the spin vector is easily computed from the definition (2.69) as

$$\dot{S}_a = -\frac{1}{2}\varepsilon_{abcd}\dot{u}^b S^{cd} \,, \tag{2.71}$$

where we used $\dot{\varepsilon}_{abcd} = 0$ by metric compatibility, the equation of spin precession (2.57b), and the antisymmetry of ε_{abcd}. This equation of evolution can be further simplified as follows. By substituting in (2.71) the expression (2.69) for S^{cd} in terms of u^a and S^a, and using the orthogonality $\dot{u}^a u_a = 0$, we readily obtain

$$\dot{S}^a = u^a \dot{u}^b S_b \,. \tag{2.72}$$

The spin vector S^a is found to obey the Fermi-Walker transport law at dipolar order (when $O(J) = 0$). This will be responsible for the Thomas precession that we shall encounter later in Sect. 5.2.3. Finally, we notice that by substituting the expression (2.69) of the tensor S^{ab} into the momentum-velocity relationship (2.67), the 4-momentum can alternatively be written as

$$p^a = mu^a - \varepsilon^a{}_{bcd} u^b \dot{u}^c S^d \,, \tag{2.73}$$

which readily implies $p^a S_a = 0$. At dipolar order, the 4-momentum $p^a = p^a_{\text{kin}} + p^a_{\text{hid}}$ is the sum of the timelike kinematic momentum $p^a_{\text{kin}} \equiv mu^a$, with m constant, and of the spacelike hidden momentum $p^a_{\text{hid}} \equiv -\varepsilon^a{}_{bcd} u^b \dot{u}^c S^d$, which is orthogonal to u^a, $\dot{u}^a$ and S^a. By using the condition of metric compatibility, which implies $\dot{\varepsilon}_{abcd} = 0$, as well as the equation of spin precession (2.72), the rate of change of the hidden momentum is given by $\dot{p}^a_{\text{hid}} \equiv -\varepsilon^a{}_{bcd} u^b \ddot{u}^c S^d$, and thus belongs to the two-space orthogonal to u^a and S^a. Lastly, we note that the Pirani SSC allows to express u^a entirely in terms of m, p^a and S^{ab}, as showed[13] in Sect. II.C of [376], in the form

$$mu^a = p^a + \frac{1}{S^2} S^{ab} S_{bc} p^c \,. \tag{2.74}$$

which is compatible with Eq. (2.73) (cf. Eqs. (25) and (26) in [375]). Now, it is clear from Eq. (2.74) that the tensor $S^{ab} S_{bc}$ acts as some kind of a projector. A direct calculation using (2.69) indeed reveals that

$$S^{ab} S_{bc} = -(\delta^a{}_c + u^a u_c - s^a s_c) S^2 \,, \tag{2.75}$$

with $s^a \equiv S^a / S$. This is the rank-2 tensor that projects onto the subspace orthogonal to both u_a and S_a (up to multiplication by the constant $-S^2$).

[13] To our knowledge, it was not written down prior to the publication of [376]. As noted by the authors, the existence of this "momentum-velocity" relation for the Pirani SSC is rather intriguing, given the numerous claims in the literature that such relation did not exist.

Acceleration, Momentum and Velocity

The SSC (2.64) can also be used to derive some approximate equations of evolution and algebraic relations, in the sense that they hold true only up to some order in the spin tensor S^{ab}. We shall denote by $O(S^n)$ any term that involves n spin tensors (or spin vectors). We start by introducing the gravito-magnetic part of the Riemann tensor, $B_{ab} \equiv \star R_{acbd} u^c u^d$, with $\star R_{abcd} \equiv \frac{1}{2} \varepsilon_{ab}{}^{ef} R_{efcd}$ the self-dual of the Riemann tensor. Using the definition (2.69) of the spin vector, the equation of motion (2.57a) can be written in the more compact form $\dot{p}^a = B^{ab} S_b$. Substituting this formula into the left-hand side of the derivative of equation (2.67), and using the conservation (2.68) of the rest mass m, yields

$$\dot{u}^a = \frac{1}{m} B^{ab} S_b - \frac{1}{m} S^{ab} \ddot{u}_b \,, \tag{2.76}$$

where we further used $\dot{S}^{ab} \dot{u}_b = 0$ because of equation (2.57b) and the orthogonality of $\dot{u}_a$ to both u^a and p^a. Equation (2.76) is exact, and readily shows that $\dot{u}^a = O(S)$, so that the motion of a dipolar particle cannot be geodesic. The rightmost term in (2.76) is therefore at least of $O(S^2)$. Actually, the later involves $\ddot{u}_b$, and can therefore be expanded in powers of the spin at any order by recursively taking the derivative of equation (2.76) and substituting it back into its own right-hand side. Doing this once, while using (2.72) and the orthogonality $B^{ab} u_b = 0$, we can isolate the quadratic-in-spin contribution and obtain the following spin expansion:

$$\dot{u}^a = \frac{1}{m} B^{ab} S_b - \frac{1}{m^2} S^{ab} \dot{B}_{bc} S^c + O(S^3) \,. \tag{2.77}$$

Equations (2.76) and (2.77) have a number of interesting consequences. First, by noticing that $S^{ab} S_a = 0$ by Eq. (2.69), we contract (2.76) with S_a and insert the result in the equation of precession (2.72). We then obtain the *exact* spin evolution equation

$$\dot{S}^a = \frac{1}{m} u^a B_{bc} S^b S^c \,, \tag{2.78}$$

which shows that $\dot{S}^a = O(S^2)$. Hence the driving torque that prevents S^a from being parallel-transported along γ is quadratic in spin. Second, we substitute Eq. (2.77) in the right-hand side of equation (2.67) to obtain the following expansion for the momentum:

$$p^a = m u^a + \frac{1}{m} S^{ab} B_{bc} S^c + O(S^3) \,. \tag{2.79}$$

From this equation we see that $p^a = m u^a + O(S^2)$. Therefore the 4-momentum of a dipolar particle can only be aligned with its 4-velocity up to linear order in the spin. Third, taking the norm of equation (2.67) provides a simple relation between the dynamical mass μ and the rest mass m. Using the SSC (2.64), we readily obtain

$$\mu^2 = m^2 + \dot{u}^a S_{ab} S^{bc} \dot{u}_c \, . \tag{2.80}$$

Equation (2.77) then implies $\mu = m + O(S^4)$, such that the two notions of mass coincide up to quartic-in-spin corrections. Since $\dot{m} = 0$ the dynamical mass satisfies $\dot{\mu} = O(S^4)$. Notice also that if the spin amplitude S is small enough, then $\mu^2 > 0$ and p^a is timelike.

Finally, it should be kept in mind that the contributions of $O(S^2)$ in Eqs. (2.77)–(2.79) should be interpreted with care at dipolar order, because additional terms of $O(S^2)$ would contribute to those same equations if we were to include the spin-induced quadrupole in our gravitational skeleton model of spinning compact objects. This is particularly relevant for the derivation of the quadrupolar first law discussed in Chap. 6. As a matter of fact, it is possible to obtain an exact relation for $\dot{u}^a$ in terms of nothing but p^a, S^a and curvature tensors, as showed in [376] (see Sect. II.B there). This takes the form

$$m\dot{u}^a = B^{ss} S^a - \frac{1}{S^2} S^{ab} p_b \, , \tag{2.81}$$

where $B^{ss} \equiv B_{ab} S^a S^b / S^2$. It is worth noting that this formula splits $\dot{u}^a$ into two simple terms: the first one is colinear to S^a and the second is orthogonal to S^a, since $S^{ab} S_a = 0$ by construction.

Chapter 3
Helical Isometry

Contents

The author seeks a level of rigor that is (i) high enough to give him confidence of the results derived, but also (ii) low enough to permit the treatment of real astrophysical systems in the real, non-asymptotically flat universe.

K. Thorne,
Rev. Mod. Phys. 52(2) (1980)

EXACTLY circular orbits, which in a 2+1 spacetime diagram depict perfect helices, do not exist in nature. Yet, one of the central assumptions behind the work presented in this thesis relies on an exact isometry that describes compact binary systems moving along circular orbits. In this chapter, the first section is dedicated to various definitions, motivations and classical results about circular orbits of compact binary systems and their mathematical modeling using a helical Killing vector. Then, in the following two sections we will show how this helical isometry, in the presence of multipolar particles, implies two central geometrical results. First, in Sect. 3.2.1

P. Ramond, *The First Law of Mechanics in General Relativity & Isochrone Orbits in Newtonian Gravity*, Springer Theses,
https://doi.org/10.1007/978-3-031-17964-8_3

we show that each particle must follow an helical trajectory in a helical spacetime, and in Sect. 3.2.2 that all its multipoles (momentum, spin, quadrupole) are Lie-dragged along this helical trajectory. These results are derived covariantly and are self-consistent, i.e. they do not require any additional assumption other than the helical isometry. Finally, several useful identities are derived in Sect. 3.3.

3.1 Helically Isometric Spacetimes

In this work, we will deal exclusively with binary systems of point particles on a circular orbit. In this first section we shall explain why this assumption, which clearly contradicts the very notion of *inspiral*, is a good model for compact binary systems (Sect. 3.1.1), and then show how this is enforced at the mathematical level, with the introduction of a helical Killing vector field. As the circular approximation constitutes one of the main assumptions of our work, we discuss its limitations, along with those arising from of the use of point particles, in Sect. 3.1.3.

3.1.1 *Why Circular Orbits?*

A traveling GW emitted by a binary system carries energy and angular momentum which it extracted from its source. At the leading PN order, this process is encoded in the fluxes of energy $\mathcal{P}$ and angular momentum $\mathcal{T}^i$, which are given by (see Sect. 12.3.2 of [141], or Sect. 3.3 of [136])

$$\mathcal{P} = \frac{1}{5} \dddot{I}^{ij} \dddot{I}^{ij} \quad \text{and} \quad \mathcal{T}^i = \frac{2}{5} \epsilon^i{}_{jk} \ddot{I}^{j\ell} \dddot{I}^{k\ell} , \tag{3.1}$$

in terms of the time derivatives of $I^{ij}(t)$, the mass quadrupole of the source. The fluxes (3.1), in turn, drive a slow evolution of the energy E and angular momentum L^i of the source itself, which would otherwise be constant at the Newtonian level. The relation between the fluxes and the binary's energy and angular momentum are known as "balance laws", and, when averaged over one orbit, take the natural form $\mathrm{d}E/\mathrm{d}t = -\mathcal{P}$ and $\mathrm{d}L^i/\mathrm{d}t = -\mathcal{T}^i$ [141, 145]. When applied to a binary system of masses m_1, m_2 on a Keplerian ellipse of semi-major axis a and eccentricity e, these balance laws are shown to induce a slow, secular variation of the orbital parameters e and a themselves. To derive these evolutions, at the Newtonian level we may simply use the classical formulae for the energy, $E \propto -1/a$, and for the angular momentum, $L \propto (a(1-e^2))^{1/2}$. Combining these with balance laws and Eq. (3.1) readily give [136]

$$\frac{\mathrm{d}a}{\mathrm{d}t} = -\frac{64}{5}\frac{m_1 m_2}{a^3}\frac{1}{(1-e^2)^{7/2}}\left(1+\frac{73}{24}e^2+\frac{37}{96}e^4\right), \tag{3.2a}$$

$$\frac{\mathrm{d}e}{\mathrm{d}t} = -\frac{304}{15}\frac{m_1 m_2}{a^4}\frac{e}{(1-e^2)^{5/2}}\left(1+\frac{121}{304}e^2\right). \tag{3.2b}$$

Readily from these equations, we see that both a and e will decrease with time. Consequently, the two bodies will get closer to each other, while orbiting on an ever-circularizing orbit. Clearly, from the (Newtonian) Kepler's third law $\Omega^2 = (m_1 + m_2)a^{-3}$, a decrease in a implies that the orbital frequency Ω of the binary will increase with time.[1] However, these two processes occur at different rates. The main reason is because of the eccentricity e in the numerator of (3.2b). This equation makes the eccentricity rapidly evolve towards the circular limit ($e = 0$) since for as long as the orbit is eccentric, $\dot{e} < 0$. On the other hand, when $e \simeq 0$, then $\dot{e} \simeq 0$ as well, and the orbit remains circular. In the light of dynamical systems, the system (3.2) exhibits a phase space (e, a) with a rapid horizontal direction and a slow vertical one for most of the eccentricity's values. The two extreme regimes, where $e \simeq 0$ and $e \simeq 1$, show a rapid vertical direction and slow horizontal one (see Fig. 3.1). To summarize, a (moderately massive) binary on a highly eccentric orbit looses its eccentricity so rapidly that when it is essentially circular, the orbital separation is still orders of magnitude larger than the body's typical sizes. Consequently, the binary is still far away from merger, and by the time it starts plunging, its eccentricity has become completely negligible.

To put these results in perspective, let us apply them to a real binary system. Remarkably, using Eq. (3.2) to compute $\mathrm{d}a/\mathrm{d}e \equiv \dot{a}/\dot{e}$ gives an ordinary differential equation that can be solved analytically as [136]

$$a(e) = C\frac{e^{12/19}}{1-e^2}\left(1+\frac{121}{304}e^2\right)^{870/2299} \tag{3.3}$$

with C an integration constant that is fixed e.g., by a measurement of the orbital parameters of the binary. Figure 3.1 depicts in a (e, a)-diagram the relation (3.3) for two known pulsars, the Hulse-Taylor (first pulsar in a binary ever discovered) and the double-pulsar (first and only binary made of two pulsars [377]). We see that, indeed, in the $e \in [0.1, 0.9]$ region, both e and a lose an order of magnitude along the evolution of (3.3).

It should be noticed, to be more precise, that this circularization process by the loss of angular momentum through the emission of GWs applies to relatively low-mass binary systems. Usually, these systems will be circularized for a long time when the GW enters the detector's bandwidth, around 10 Hz. But notice, due to the $m_1 m_2$ factor in Eq. (3.2), that this process is far less efficient for very massive systems, in particular SMBH binaries and EMRIs. For these systems, when the GW will enter the detector's bandwidth, the binary will not have lived long enough for the orbit to

[1] This can readily be seen on the observed waveforms of the GW events (see figures in Sect. 1.3.2 of Chap. 1), since the GW frequency is (at leading order) twice the orbital frequency.

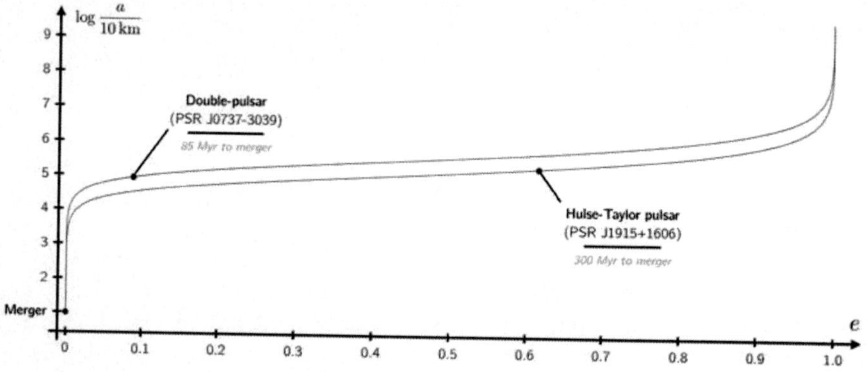

Fig. 3.1 This diagram shows the evolution of the typical orbital separation (semi-major axis a) of a binary in terms of its eccentricity e. In the region $e \in [0.1, 0.9]$, both e and a decrease by a factor of ten within the same time span. Therefore, a compact binary loses its eccentricity during its early inspiral, and has $e \simeq 0.01$ when reaching the end of the inspiral (when a is a few times the radii of the body, here 10 km for a NS). Two real examples are plotted: the only double-pulsar known to this day (PSR J0737-3039) and the Hulse-Taylor pulsar, with their corresponding time to merger

be circular. It is always a competition between the time taken to circularize the orbit (a time which is longer for massive systems) and the frequency at which the GW enters the detector's bandwidth.

Nevertheless, let us now examine a binary system on a circular orbit of radius R and angular frequency Ω. The leading-order, PN result for the secular evolution of Ω and R reads (we restore the factors of c)

$$\frac{\mathrm{d}R}{\mathrm{d}t} = -\frac{64}{5}\frac{G^3m^3\nu}{R^3c^5} + O(c^{-7}) \quad \text{and} \quad \frac{\mathrm{d}\Omega}{\mathrm{d}t} = \frac{96}{5}\frac{Gm\nu}{R^3}\gamma^{5/2} + O(c^{-7}), \qquad (3.4)$$

where $\nu \equiv m_1m_2/m^2$ is the symmetric mass ratio. These equations are central because they allow to show that two time scales are present in the problem: (1) the orbital time scale $T_{\text{orb}} \equiv 2\pi/\Omega$ and (2) the radiation-reaction time $T_{\text{rea}} \equiv \Omega/\dot{\Omega}$, which governs the secular evolution of the orbital parameters, in particular Ω. Using the (Newtonian) Kepler third law in the form $\Omega^2R^3 = Gm$ and the viriel theorem $v^2 = Gm/r$, Eq. (3.4) gives an estimate

$$\frac{T_{\text{orb}}}{T_{\text{rea}}} \propto \nu\left(\frac{v}{c}\right)^5. \qquad (3.5)$$

Consequently, during most of the inspiralling phase ($v/c \ll 1$) and/or for systems with large mass ratios ($\nu \ll 1$), we have $T_{\text{orb}} \ll T_{\text{rea}}$ and the binary makes a large number of orbits before the orbital radius decreases significantly as a reaction to

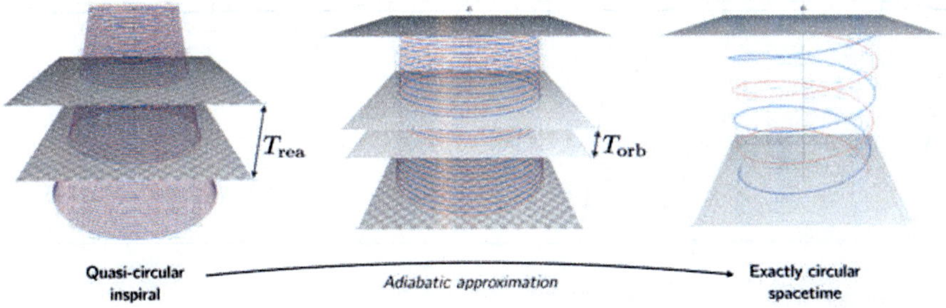

Fig. 3.2 Three 2+1 spacetime diagrams that summarize the idea of the adiabatic approximation for a quasi-circular inspiral. On the left, the inspiral spans several $T_{\rm rea}$, over which the orbital radius decreases significantly. In the middle, the zoom shows that $T_{\rm rea}$ still spans many orbital periods $T_{\rm orb}$. However, over a timescale $T_{\rm orb}$ (right), the orbital radius is constant, and the orbit can be considered circular

the GW emission.[2] The existence of two distinct time scales justifies the adiabatic approximation of an inspiral, i.e., the fact that it is well-approximated by a sequence of circular orbits. This adiabatic approximation is quite powerful, as it allows the construction of GW templates for the whole inspiral phase just by stitching together a series of circular-orbits, whose orbital parameters evolve over a radiation reaction timescale $T_{\rm rea}$. As depicted in Fig. 3.2.

To summarize, by the time the GW emitted by a binary system possesses a high enough frequency to enter the detector's sensitive bandwidth (around a a few tens of Hz for current detectors, their orbit is expected to show a negligible eccentricity.

3.1.2 *Helical Killing Vectors*

Definition

When a spacetime $(\mathscr{E}, g_{ab})$ possesses a (continuous) isometry, there exists trajectories along which the metric g_{ab} is preserved. The generator of these trajectories is called a Killing vector field (here after Killing field or Killing vector). If such a Killing vector exists, it satisfies, by definition

$$\mathcal{L}_\xi g_{ab} = 0\,, \tag{3.6}$$

with $\mathcal{L}_\xi$ denoting the Lie-derivative with respect to the vector field ξ^a. Equation (3.6) translates mathematically the fact that, along the integral curves of ξ^a, the metric is conserved. Building on the works of Bonazolla, Gourgoulhon and collaborators [379,

[2] Even though the estimate relies on the assumption of circular orbits and thus is invalid for EMRIs which are typically eccentric, the existence of two timescales (orbital and radiation-reaction), and thus the validity of an adiabatic approximation, still holds for most binaries, including EMRIs [378].

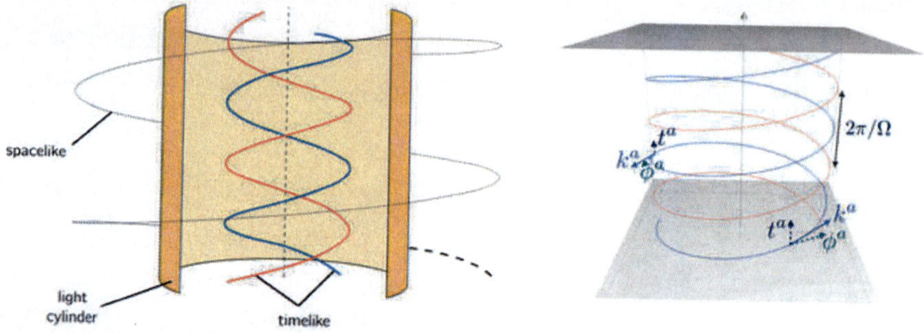

Fig. 3.3 Light cylinder and $k = t + \Omega\phi$

380], we shall define a helical Killing vector (HKV), always denoted k^a, as a Killing field admitting a decomposition

$$k^a \equiv t^a + \Omega\,\phi^a\,, \tag{3.7}$$

where t^a is an arbitrary timelike vector field, ϕ^a an arbitrary spacelike vector field with closed integral curves of parameter length 2π, and Ω is a positive constant that will be interpreted as the circular-orbit angular velocity of the binary. The decomposition is illustrated on Fig. 3.3, which justifies the *helical* in HKV. As Killing fields are defined up to a normalization factor, we choose to normalize the HKV (3.7) so that $k^a t_a \to -1$ at spatial infinity. The first occurence of the introduction of an HKV seems to be due to Steven Detweiler in [381]. This definition is equivalent to a more axiomatic one, constructed from the flow of periodic vector fields, as in [382]. A spacetime that is axisymmetric and static admits a whole family of HKVs, by setting $t^a = (\partial_t)^a$ and $\phi^a = (\partial_\phi)^a$, where t and ϕ are the coordinates that the spacetime's metric is independent of. However, we emphasize that in our context, neither t^a nor ϕ^a are Killing fields. The null hypersurface over which $k^a k_a = 0$ is known as the light cylinder. Heuristically, it corresponds to the set of points where an observer would have a circular motion around the helical axis of symmetry ($\phi^a = 0$) with a "velocity" equal to the vacuum speed of light. Excluding any black hole region, the helical Killing field (3.7) is timelike everywhere inside the light cylinder and spacelike everywhere outside of it, as depicted on Fig. 3.3. In this work, we are only interested in the region inside the light cylinder, where the integral curves of the HKV (the helices) are all timelike.

Properties of Killing Fields

Independently of its helical nature, the HKV k^a satisfies a number of identities because it is a Killing field in the first place. Some of them will be of utmost importance for the derivation of the first law of mechanics later on. Let us give some of these properties, keeping the notation ξ^a to emphasis that these are not restricted to the HKV k^a but to any Killing field. We leave their mathematical proof in Appendix A.2 for the most part.

First of all, the definition (3.6) is independent of the existence of a connection on the spacetime. However, the Levi-Civita connection can help bring (3.6) in the equivalent, and sometimes more convenient, form

$$\nabla_{(a}\xi_{b)} = 0\,. \tag{3.8}$$

Taking the trace of (3.8) shows that a Killing field k^a is divergenceless: $\nabla_a\xi^a = 0$. Killing's equation (3.8) also implies the identity $\frac{1}{2}\nabla_a(\xi^b\xi_b) = \xi^b\nabla_a\xi_b = -\xi^b\nabla_b\xi_a$ for the "acceleration" of a Killing field along its integral curves. Contracting it once more with ξ^a shows that the norm squared $\xi^b\xi_b$ of a Killing field is conserved along its integral curves $\xi^a\nabla_a(\xi^b\xi_b) = 0$.

Very important for our work is the so-called Kostant formula, which relates the second covariant derivative of ξ^a to ξ^a itself, as

$$\nabla_a\nabla_b\xi^c = R_{dab}{}^c\xi^d\,. \tag{3.9}$$

The proof combines the definition of the Riemann tensor, its algebraic symmetries and the Killing equation (3.8)

Then, from the Kostant formula, one can establish the commutation between the Lie-derivative $\mathcal{L}_\xi$ and the metric-compatible,[3] covariant derivative ∇. Indeed, for any tensor field χ

$$\nabla_a(\mathcal{L}_\xi\chi) = \mathcal{L}_\xi(\nabla_a\chi)\,. \tag{3.10}$$

A covariant proof of this result (i.e., not requiring the introduction of a coordinate system) is provided in Appendix A.2.

A number of geometrical quantities are Lie-dragged along a Killing field, as a direct consequence of the Lie-dragging of the metric (3.6). Among them, the canonical volume form ε_{abcd} and the Riemann tensor R_{abcd} associated to g_{ab} are both Lie-dragged along ξ^a

$$\mathcal{L}_\xi\varepsilon_{abcd} = 0\,, \tag{3.11a}$$

$$\mathcal{L}_\xi R_{abc}{}^d = 0\,. \tag{3.11b}$$

We provide covariant proofs for all these formulae in Appendix A.2. Surprisingly, as far as we know, some of these important results and their proof scarcely appear in classical textbooks, Ref. [383] being a notable exception.

Another important result, which is only of interest when dealing with point particles, is the Lie-dragging of the covariant, four-dimensional Dirac distribution $\delta_4(x, y)$. Heuristically, as it is built from nothing but the metric and two events in spacetime, we expect it to be Lie-dragged. A proof of this statement is provided

---

[3] Note that $\mathcal{L}_\xi$ need not commute with a generic connection, cf. Appendix A.2.2.

in Appendix A.2, where it is also rigorously defined as a (distributional) bi-tensor. Skipping over the details of the proof, we will need later on the identity

$$\mathcal{L}_\xi \delta_4(x, y) = 0 \,, \tag{3.12}$$

for any two points x and y in spacetime. Equation (3.12) implies, in particular, the Lie-dragging of the quantity $\delta_4(x, z(\tau))$, when $x^\alpha = z^\alpha(\tau)$ is a parametrization of some particle's worldline.

Finally, and perhaps most importantly, we argue that the SEM tensor T^{ab} of the matter present in spacetime must also be invariant along the integral curves of any Killing field. The proof of this result is a simple consequence of the Einstein equation $G_{ab} = 8\pi T_{ab}$. Indeed, since the metric and the Riemann tensor are Lie-dragged (Eqs. (3.6)–(3.11b)), so is the Einstein tensor G_{ab} built from them, and it follows that

$$\mathcal{L}_\xi T^{ab} = 0 \,. \tag{3.13}$$

We will still need to justify this statement in the case where T_{ab} is singular, in particular in the context of point particles. This discussion is relegated to Sect. 3.1.3 just below, where we also summarize the assumptions and limits of our work.

3.1.3 Limits of the Model

Before going into the next section where we examine the consequences of an helical isometry for multipolar particles, we discuss various assumptions that have been made so far, and which constitute the limits of our results.

Helical Spacetimes

As we discussed in Sect. 3.1.1, the validity of an exactly circular orbit holds for (1) systems that have been circularized and (2) over time scales small compared to the radiation reaction one, as crystallized in Eq. (3.5). For longer timescales, the emission of GWs draws energy from the system, implying a decrease of the orbital radius and ultimately breaking the helical symmetry. In a sense, in order to maintain the binary on a fixed circular orbit, the energy radiated in GWs would need to be compensated by an equal amount of incoming radiation. This heuristic argument of incoming GWs can actually be made mathematically precise and, as a consequence, it has long been known that in GR helically symmetric spacetimes cannot be asymptotically flat [381, 384, 385]. Again from an heuristic point of view: far away from the source, the resulting (emitted and incoming) system of standing GWs ends up dominating the energy content of the spacetime, so that the falloff conditions necessary to ensure asymptotic flatness cannot be satisfied. As emphasized in Refs. [206, 382, 386–388], however, asymptotic flatness can be recovered if, loosely speaking, the gravitational radiation can be "turned off." This can be achieved, in particular, in two instances at least. First, using the Isenberg, Wilson and Mathews approximation to GR, also known as the conformal flatness condition (CFC) approximation [389–391]. There, spacetime is foliated into conformally flat spacelike slices that satisfy a truncated set of field equations. Second, in the context of approximation methods such as PN theory [145] where one can distinguish between radiative and conservative effects, and GSF theory [199, 200] where a local wavezone can be defined from an approximate asymptotic region [382, 392].

Point Particles

Whereas the use of point particles is perfectly justified in Newtonian gravity or classical electrodynamics, their use in GR is highly questionable, because of the nonlinearity of the Einstein equation. On the mathematical side, Geroch and Traschen [393] (see also [394]) have proven that, precisely because of this nonlinearity, a point particle source will usually yield solutions that are mathematically ill-posed in any suitable class of functions. Indeed, from a more physical perspective, we know that the closest thing to a point particle in GR is a black hole. However, in the context of approximation methods (e.g. PN theory [145] or GSF [200]), the Einstein equation *can* be coupled to a distributional source such as (2.18) in a meaningful manner. In the PN framework, this is true at least at the fourth PN order [395, 396] provided that a regularization scheme (e.g. dimensional regularization [397, 398]) is used to handle the divergent self-field of each particle. Additionally, as mentioned in Sect. 2.2.4, it was shown recently that even at second order in perturbation theory, the Einstein equation is well-defined with a point-particle source, at least in (so-called) highly regular gauges [362, 366].

As most of our results will be used in the context of approximation methods such as PN or GSF, in this work we will always implicitly assume that such a regularization scheme is being used, or that a highly regular gauge is chosen, while evaluating various tensor fields along the wordline of a multipolar point particle. All tensorial equations, including Einstein's, should thus be interpreted in a regularized sense. This justifies that the helical symmetry $\mathcal{L}_k g_{ab}$ implies the Lie-dragging of the SEM tensor, as discussed above Eq. (3.13). Additionally, we mention that any "physically reasonable" regularization method yields $T^{ab}|_\gamma = 0$, so that $R^{ab}|_\gamma = 0$ by the Einstein equation. Hence the spacetimes that we shall be considering in this work, in addition of being helically symmetric, are also Ricci-flat in a regularized sense. This assumption has been also used for co-rotating binary black hole spacetimes, such as those considered in [206, 380, 382, 399].

3.2 Consequences for Multipolar Particles

3.2.1 Helical Trajectories

Detweiler's Parameter

In a seminal paper, Steven Detweiler [400] performed the first comparison of the predictions from the perturbative GSF framework and PN theory, for a binary system of nonspinning compact objects moving along a closed circular orbit. This comparison relied on the calculation of the quantity $U \equiv \mathrm{d}t/\mathrm{d}\tau$, i.e. on the time component of the 4-velocity of the smaller body, modelled as a structureless point particle, as a function of the angular velocity Ω of the binary's circular orbit. This comparison was later extended and refined in Refs. [401–405]. In Detweiler's paper, the quantity U (or rather its inverse $z \equiv \mathrm{d}\tau/\mathrm{d}t$) was coined the "redshift" parameter, for reasons that will become clear at the end of the section (cf. around Eq. (3.31)).

Soon after, the discovery of the first law of compact binary mechanics [382, 387, 406] showed that the kinematical redshift parameter U also has a *dynamical* relevance, as it can be related in a simple manner to some of the global properties of the binary system, such as its binding energy and orbital angular momentum; see also [407]. Using the first law of binary mechanics, the redshift parameter has been used to compare the predictions of GSF theory to the results from numerical relativity simulations of black hole binaries moving on quasi-circular orbits, showing remarkably good agreement, even for comparable-mass binary systems [206, 408, 409].

Using the BH perturbation theory developed in [410, 411], the GSF contribution to the redshift parameter has been computed analytically, up to very high PN orders [412–417], and numerically with extremely high accuracy [418, 419]. Combined with the first law of binary mechanics, those results have been used to complete the calculation of the 4PN dynamics of arbitrary-mass-ratio binary systems of nonspinning compact objects [420], as well as calibrate one of the potentials entering the EOB Hamiltonian that controls the conservative part of the binary's orbital dynamics [421–423] (some of these applications are detailed in Chap. 4, Sect. 4.2.3).

For compact binary systems moving along eccentric orbits, a generalization of the redshift parameter has been introduced in the context of GSF theory [424], and then used to perform a comparison with the predictions from PN theory [425]. This generalized redshift $\langle U \rangle$ appears naturally in the first law of mechanics for eccentric-orbit compact binaries [426, 427], and it has been used to calibrate the remaining potentials that enter the conservative part of the EOB Hamiltonian [428, 429].

Given the large amount of work, reviewed above, that relies on the redshift parameter U, it is arguably important to give this physical quantity a simple, geometrical and coordinate-invariant meaning. To the best of our knowledge, this question has only been addressed in the context of the perturbative GSF framework, where the redshift parameter U has been shown to be gauge-invariant under gauge transformations generated by a helically symmetric gauge vector [400, 430, 431].

Among the main results established in this thesis is the fact that the existence of the (helical) Killing field enforces the particles's worldlines to be (helical) Killing trajectories. In other words, the four-velocity u^a of a particle in a helically isometric spacetime, satisfies

$$k^a|_\gamma = z u^a \,, \tag{3.14}$$

where z is a constant along γ. By using a spherical-type coordinate system (t, r, θ, ϕ) adapted to the helical isometry, such that $k^a = (\partial_t)^a + \Omega\,(\partial_\phi)^a$ in a neighborhood of γ, the coordinate components of the 4-velocity u^a then simply read $u^\alpha = z^{-1}(1, 0, 0, \Omega)$. In particular,

$$z = \frac{\mathrm{d}\tau}{\mathrm{d}t} = \frac{1}{U} \,, \tag{3.15}$$

which shows that, while using adapted coordinates, Detweiler's redshift parameter is simply the inverse of the geometrically-defined constant that appears in the helical constraint (3.14). The formulas (3.14) and (3.15) have been written down in various papers, e.g. in [205, 415, 431–433], but have never been proven in the more general context considered here, namely without any restriction on the coordinate system, for an *arbitrary*-mass-ratio binary system of *spinning* compact bodies *with internal structure*.

Lie-Dragging Constraints on the Multipoles

To derive the central result (3.14) from the Lie-dragging of the SEM tensor, it is more convenient to use the unreduced form of a quadrupolar SEM tensor, in the sense of Tulczyjew's reduction process (described in Chap. 2, Sect. 2.2.2). Therefore, instead of the reduced and ready-to-apply form (2.58)–(2.59), our starting point is the multipolar SEM tensor of a binary system of quadrupolar particles in the form

$$T^{ab} = \sum_{\mathrm{i}} \left\{ \int_{\gamma_\mathrm{i}} T_\mathrm{i}^{ab}\, \delta_4^\mathrm{i}\, \mathrm{d}\tau_\mathrm{i} + \nabla_c \int_{\gamma_\mathrm{i}} T_\mathrm{i}^{abc}\, \delta_4^\mathrm{i}\, \mathrm{d}\tau_\mathrm{i} + \nabla_{cd} \int_{\gamma_\mathrm{i}} T_\mathrm{i}^{abcd}\, \delta_4^\mathrm{i}\, \mathrm{d}\tau_\mathrm{i} \right\} , \tag{3.16}$$

where we used the shorthand $\delta_4^\mathrm{i} \equiv \delta_4(x, y_\mathrm{i})$ and the special equal sign with Σ_i atop it for clarity, denoting a sum over the particle on the right-hand side. More precisely, we shall use the *normal form* associated with Eq. (3.16), which is obtained by performing an orthogonal decomposition of the multipoles T_i^{ab}, T_i^{abc} and T_i^{abcd} with respect to the 4-velocity u_i^a. Using the formula (A.41), the SEM tensor (3.16) can be written in the alternative form

$$T^{ab} = \sum_{\mathrm{i}} \left\{ \int_{\gamma_\mathrm{i}} \mathscr{T}_\mathrm{i}^{ab}\, \delta_4^\mathrm{i}\, \mathrm{d}\tau_\mathrm{i} + \nabla_c \int_{\gamma_\mathrm{i}} \mathscr{T}_\mathrm{i}^{abc}\, \delta_4^\mathrm{i}\, \mathrm{d}\tau_\mathrm{i} + \nabla_{cd} \int_{\gamma_\mathrm{i}} \mathscr{T}_\mathrm{i}^{abcd}\, \delta_4^\mathrm{i}\, \mathrm{d}\tau_\mathrm{i} \right\} , \tag{3.17}$$

where the dipole and quadrupole moments now obey the constraints $\mathcal{T}_{\rm i}^{abc} u_c^{\rm i} = 0$, $\mathcal{T}_{\rm i}^{ab[cd]} = 0$ and $\mathcal{T}_{\rm i}^{abcd} u_d^{\rm i} = 0$. The fact that such a normal form always exists and is unique is one of the two theorems of Tuclzyjew, which are reviewed in Appendices A.3.1 and A.3.2. As shown in Appendix A.4, the multipole moments appearing in Eq. (3.17) are explicitly given by

$$\mathcal{T}_{\rm i}^{ab} \equiv T_{\rm i}^{ab} - \left(T_{\rm i}^{abu} - (T_{\rm i}^{abuu})^{\cdot} + 2T_{\rm i}^{ab(cu)} \dot{u}_c^{\rm i}\right)^{\cdot} + R_{cde}{}^{(a}\mathcal{U}_{\rm i}^{b)edc}\,, \tag{3.18a}$$

$$\mathcal{T}_{\rm i}^{abc} \equiv T_{\rm i}^{ab\hat{c}} - 2\left(T_{\rm i}^{ab(du)}\right)^{\cdot} h^c_{{\rm i}\,d} - T_{\rm i}^{abuu} \dot{u}_{\rm i}^c\,, \tag{3.18b}$$

$$\mathcal{T}_{\rm i}^{abcd} \equiv T_{\rm i}^{ab(\hat{c}\hat{d})}\,, \tag{3.18c}$$

where, for each particle, the upper index u denotes a contraction with u^a, for example $T_{\rm i}^{abu} \equiv T_{\rm i}^{abc} u_c$, and the hat above an index denotes a contraction with the orthogonal projector (2.62), e.g. $T_{\rm i}^{ab\hat{c}} \equiv T_{\rm i}^{abd} h^c{}_d$. We also used $\mathcal{U}_{\rm i}^{bedc} \equiv 2T_{\rm i}^{beu\hat{d}} u_{\rm i}^c - T_{\rm i}^{be\hat{c}\hat{d}}$, to fit the first equation on a single line.

Heuristically, one expects that the Lie-dragging along k^a of T^{ab}, Eq. (3.13) above with $\xi^a = k^a$, implies some differential relationships obeyed by the multipoles $\mathcal{T}_{\rm i}^{ab}$, $\mathcal{T}_{\rm i}^{abc}$ and $\mathcal{T}_{\rm i}^{abcd}$ that appear in Eq. (3.17). However, these multipoles are merely defined along $\gamma_{\rm i}$. To define them as tensor fields off these worldlines, we introduce some smooth extensions $\tilde{\mathcal{T}}_{\rm i}^{ab}$, $\tilde{\mathcal{T}}_{\rm i}^{abc}$ and $\tilde{\mathcal{T}}_{\rm i}^{abcd}$. Such an extension can be chosen freely. Here, it is defined by parallel propagation along spacelike geodesics perpendicular to $\gamma_{\rm i}$. Therefore, for each particle, in a normal neighborhood of a given point $y \in \gamma$, we define the extensions

$$\tilde{\mathcal{T}}^{ab}(x) \equiv g^a{}_{a'}(x,y)\, g^b{}_{b'}(x,y)\, \mathcal{T}^{a'b'}(y)\,, \tag{3.19a}$$

$$\tilde{\mathcal{T}}^{abc}(x) \equiv g^a{}_{a'}(x,y)\, g^b{}_{b'}(x,y)\, g^c{}_{c'}(x,y)\, \mathcal{T}^{a'b'c'}(y)\,, \tag{3.19b}$$

$$\tilde{\mathcal{T}}^{abcd}(x) \equiv g^a{}_{a'}(x,y)\, g^b{}_{b'}(x,y)\, g^c{}_{c'}(x,y)\, g^d{}_{d'}(x,y)\, \mathcal{T}^{a'b'c'd'}(y)\,, \tag{3.19c}$$

where the bitensor $g^a{}_{a'}(x,y)$ is the parallel propagator from y to x (see Appendix A.2.3). As shall be proven in the next Sect. 3.2.1, the final results will not depend upon this particular choice of extension. Owing to the presence of the invariant Dirac functional $\delta_4(x, y_{\rm i})$ in each integral in Eq. (3.17), we may replace the multipoles by their smooth extensions (3.19) there. Taking the Lie derivative along k^a on both sides and using Eq. (3.13), as well as the property (3.12) and the commutation (3.10) of the Lie and covariant derivatives, we readily obtain

$$0 = \sum_{\rm i} \left\{ \int_{\gamma_{\rm i}} \mathcal{L}_{\rm i}^{ab}\, \delta_4^{\rm i}\, {\rm d}\tau_{\rm i} + \nabla_c \int_{\gamma_{\rm i}} \mathcal{L}_{\rm i}^{abc}\, \delta_4^{\rm i}\, {\rm d}\tau_{\rm i} + \nabla_{cd} \int_{\gamma_{\rm i}} \mathcal{L}_{\rm i}^{abcd}\, \delta_4^{\rm i}\, {\rm d}\tau_{\rm i} \right\}, \tag{3.20}$$

where we introduced the notation $\mathcal{L}_{\rm i}^{\cdots} \equiv \mathcal{L}_k \tilde{\mathcal{T}}_{\rm i}^{\cdots}$ for the Lie derivatives of the smoothly extended multipoles. The multipolar sums in Eq. (3.20) are not in normal form: the multipoles $\mathcal{L}_{\rm i}^{abc}$ and $\mathcal{L}_{\rm i}^{abcd}$ have the right algebraic symmetries, but they need not be orthogonal to $u_c^{\rm i}$. However, thanks to the Theorem 1 in Appendix A.3.1, this normal form exists and is unique, and as shown in Appendix A.4 it reads

$$0 = \sum_{\mathrm{i}} \bigg\{ \int_{\gamma_{\mathrm{i}}} \Big(\mathscr{L}_{\mathrm{i}}^{ab} - \big(\mathscr{L}_{\mathrm{i}}^{abu} + \mathscr{L}_{\mathrm{i}}^{abcu} \dot{u}_{c}^{\mathrm{i}} - (\mathscr{L}_{\mathrm{i}}^{abcu})^{\cdot} u_{c}^{\mathrm{i}} \big)^{\cdot} + 2 R_{cde}{}^{(a} \mathscr{L}_{\mathrm{i}}^{b)eu\hat{d}} u_{\mathrm{i}}^{c} \Big) \delta_4^{\mathrm{i}} \, \mathrm{d}\tau_{\mathrm{i}}$$
$$+ \nabla_c \int_{\gamma_{\mathrm{i}}} \big(\mathscr{L}_{\mathrm{i}}^{ab\hat{c}} - 2 (\mathscr{L}_{\mathrm{i}}^{abdu})^{\cdot} h^{c}_{\mathrm{i}\,d} - \mathscr{L}_{\mathrm{i}}^{abuu} \dot{u}_{\mathrm{i}}^{c} \big) \delta_4^{\mathrm{i}} \, \mathrm{d}\tau_{\mathrm{i}} + \nabla_{cd} \int_{\gamma_{\mathrm{i}}} \mathscr{L}_{\mathrm{i}}^{ab\hat{c}\hat{d}} \, \delta_4^{\mathrm{i}} \, \mathrm{d}\tau_{\mathrm{i}} \bigg\} . \tag{3.21a}$$

Those multipolar sums are now in normal form, so that, according to Tulczyjew's second theorem (2 in Appendix A.3.2), each integrand must be identically equal to zero along γ_{i}. This implies the following constraints, valid for any $\mathrm{i} \in \{1; 2\}$

$$\mathscr{L}_{\mathrm{i}}^{ab} = \big(\mathscr{L}_{\mathrm{i}}^{abu} + \mathscr{L}_{\mathrm{i}}^{abcu} \dot{u}_{c}^{\mathrm{i}} - (\mathscr{L}_{\mathrm{i}}^{abcu})^{\cdot} u_{c}^{\mathrm{i}} \big)^{\cdot} - 2 R_{cde}{}^{(a} \mathscr{L}_{\mathrm{i}}^{b)eu\hat{d}} u_{\mathrm{i}}^{c} , \tag{3.22a}$$

$$\mathscr{L}_{\mathrm{i}}^{abc} = - \mathscr{L}_{\mathrm{i}}^{abu} u_{\mathrm{i}}^{c} + 2 (\mathscr{L}_{\mathrm{i}}^{abdu})^{\cdot} h^{c}_{\mathrm{i}\,d} + \mathscr{L}_{\mathrm{i}}^{abuu} \dot{u}_{\mathrm{i}}^{c} , \tag{3.22b}$$

$$\mathscr{L}_{\mathrm{i}}^{abcd} = - \mathscr{L}_{\mathrm{i}}^{abuu} u_{\mathrm{i}}^{c} u_{\mathrm{i}}^{d} + 2 \mathscr{L}_{\mathrm{i}}^{abu(\hat{d}} u_{\mathrm{i}}^{c)} . \tag{3.22c}$$

These constraints will be central to prove, in the next subsection, that quadrupolar particles do follow helical Killing trajectories.

Quadrupolar Particles Follow Killing Trajectories

Let f_{ab} denote a tensor field with compact support $\mathscr{V} \subset \mathscr{E}$ that is smooth on the interior $\mathscr{V}^{\circ}$ of $\mathscr{V}$. The Lie dragging (3.13) of the distributional SEM tensor (3.17) implies that $\int_{\mathscr{V}} f_{ab} \mathcal{L}_k T^{ab} \, \mathrm{d}V = 0$, where $\mathrm{d}V$ is the invariant 4-volume element. Therefore, by using the Leibniz rule on the Lie derivative, we readily obtain

$$\int_{\mathscr{V}} \mathcal{L}_k (T^{ab} f_{ab}) \, \mathrm{d}V = \int_{\mathscr{V}} T^{ab} \mathcal{L}_k f_{ab} \, \mathrm{d}V . \tag{3.23}$$

The integral appearing in the left-hand side of Eq. (3.23) is easily shown to vanish, as follows. By using the definition of the Lie derivative of a scalar field, together with $\nabla_c k^c = 0$, and applying Stokes' theorem together with $f_{ab} = 0$ on the boundary $\partial\mathscr{V}$, we have

$$\int_{\mathscr{V}} \mathcal{L}_k (T^{ab} f_{ab}) \, \mathrm{d}V = \int_{\mathscr{V}} \nabla_c (k^c T^{ab} f_{ab}) \, \mathrm{d}V = \oint_{\partial\mathscr{V}} T^{ab} f_{ab} \, k^c \mathrm{d}\Sigma_c = 0 , \tag{3.24}$$

where $\mathrm{d}\Sigma_c$ is the surface element orthogonal to $\partial\mathscr{V}$. Next, we substitute the expression (3.17) of the binary's quadrupolar SEM tensor, in normal form, into the integral that appears in the right-hand side of Eq. (3.23). After commuting the integrals over $\mathscr{V}$ and γ_{i}, integrating by parts, using Stokes' theorem and the compact-supported nature of the tensor f_{ab}, as well as the defining property (A.15) of the invariant Dirac distribution, we obtain

$$\int_{\mathscr{V}} T^{ab} \mathcal{L}_k f_{ab} \, \mathrm{d}V = \sum_{\mathrm{i}} \int_{\gamma_\mathrm{i}} \left(\mathscr{T}_\mathrm{i}^{ab} \mathcal{L}_k f_{ab} - \mathscr{T}_\mathrm{i}^{abc} \nabla_c \mathcal{L}_k f_{ab} + \mathscr{T}_\mathrm{i}^{abcd} \nabla_{cd} \mathcal{L}_k f_{ab} \right) \mathrm{d}\tau_\mathrm{i} \, . \tag{3.25}$$

On the one hand, from the result (3.10) we may commute the covariant and Lie derivatives in the second and third terms in the right-hand side of (3.25). On the other hand, we notice that $\mathscr{T}_\mathrm{i}^{ab} = \tilde{\mathscr{T}}_\mathrm{i}^{ab}$ along γ_i, and similarly for $\mathscr{T}_\mathrm{i}^{abc}$ and $\mathscr{T}_\mathrm{i}^{abcd}$, so that the multipoles can be replaced by their smooth extensions (3.19) off γ_i. Combined with Eqs. (3.23) and (3.24), the formula (3.25) then implies

$$\sum_{\mathrm{i}} \int_{\gamma_\mathrm{i}} \left(\tilde{\mathscr{T}}_\mathrm{i}^{ab} \mathcal{L}_k f_{ab} - \tilde{\mathscr{T}}_\mathrm{i}^{abc} \mathcal{L}_k \nabla_c f_{ab} + \tilde{\mathscr{T}}_\mathrm{i}^{abcd} \mathcal{L}_k \nabla_{cd} f_{ab} \right) \mathrm{d}\tau_\mathrm{i} = 0 \, . \tag{3.26}$$

Applying the Leibniz rule to the Lie derivatives in the integrand and recalling the notation $\mathscr{L} \equiv \mathcal{L}_k \tilde{\mathscr{T}}$, the formula (3.26) now implies

$$\begin{aligned} & \sum_{\mathrm{i}} \int_{\gamma_\mathrm{i}} \mathcal{L}_k \left(\tilde{\mathscr{T}}_\mathrm{i}^{ab} f_{ab} - \tilde{\mathscr{T}}_\mathrm{i}^{abc} \nabla_c f_{ab} + \tilde{\mathscr{T}}_\mathrm{i}^{abcd} \nabla_{cd} f_{ab} \right) \mathrm{d}\tau_\mathrm{i} \\ = & \sum_{\mathrm{i}} \int_{\gamma_\mathrm{i}} \left(\mathscr{L}_\mathrm{i}^{ab} f_{ab} - \mathscr{L}_\mathrm{i}^{abc} \nabla_c f_{ab} + \mathscr{L}_\mathrm{i}^{abcd} \nabla_{cd} f_{ab} \right) \mathrm{d}\tau_\mathrm{i} \\ = & \sum_{\mathrm{i}} \int_{\gamma_\mathrm{i}} \left(\mathscr{L}_\mathrm{i}^{abu} f_{ab} - (\mathscr{L}_\mathrm{i}^{abcd} f_{ab})^{\cdot} u_c^\mathrm{i} u_d^\mathrm{i} - 2 \mathscr{L}_\mathrm{i}^{abuc} \nabla_c f_{ab} \right)^{\cdot} \mathrm{d}\tau_\mathrm{i} = 0 \, , \end{aligned} \tag{3.27}$$

where we have used the constraints (3.22) in the second equality and the fact that f_{ab} has a compact support in the last equality. Equation (3.27) must hold for any f_{ab} with compact support $\mathscr{V}$ and smooth on $\mathscr{V}^\circ$. In particular, it must hold for all tensor fields f_{ab} whose support excludes either γ_1 or γ_2, such that both proper time integrals in (3.27) must identically vanish. Therefore, for all $\mathrm{i} \in \{1, 2\}$, we have established that

$$\int_{\gamma_\mathrm{i}} \mathcal{L}_k f_\mathrm{i} \, \mathrm{d}\tau_\mathrm{i} = 0 \, , \quad \text{where} \quad f_\mathrm{i} \equiv \tilde{\mathscr{T}}_\mathrm{i}^{ab} f_{ab} - \tilde{\mathscr{T}}_\mathrm{i}^{abc} \nabla_c f_{ab} + \tilde{\mathscr{T}}_\mathrm{i}^{abcd} \nabla_{cd} f_{ab} \, . \tag{3.28}$$

Clearly, having $k^a \propto u_\mathrm{i}^a$ along γ_i is a *sufficient* condition for Eq. (3.28) to hold for any f_{ab}. Indeed, $k^a \propto u_\mathrm{i}^a$ implies $\mathcal{L}_k f_\mathrm{i} = k^a \nabla_a f_\mathrm{i} \propto \dot{f}_\mathrm{i}$, and the integral of $\dot{f}_\mathrm{i}(\tau_\mathrm{i})$ over γ_i vanishes because f_i has compact support. We now argue that $k^a \propto u_\mathrm{i}^a$ along γ_i is also a *necessary* condition for Eq. (3.28) to hold for all f_{ab}.

We summarize here the idea behind the proof and relegate the details to Appendix A.3.3. First, we perform an orthogonal decomposition of k^a with respect to the tangent 4-velocity u_i^a to γ_i, according to $k^a|_{\gamma_\mathrm{i}} = z_\mathrm{i} u_\mathrm{i}^a + w_\mathrm{i}^a$, where $z_\mathrm{i} \equiv -k^a u_a^\mathrm{i}$ and $w_\mathrm{i}^a \equiv h^a_{\mathrm{i}\,b} k^b$. With these notations, the integrand in Eq. (3.28) becomes $\mathcal{L}_k f_\mathrm{i} = z_\mathrm{i} \dot{f}_\mathrm{i} + w_\mathrm{i}^a \nabla_a f_\mathrm{i}$. Second, we let $\mathbb{F}$ denote the set of scalar fields f_i given by Eq. (3.28), with f_{ab} of compact support $\mathscr{V}$ and smooth on $\mathscr{V}^\circ$. We now consider the following proposition:

$$\forall f_\text{i} \in \mathbb{F} \, , \quad \int_{\gamma_\text{i}} (z_\text{i} \dot{f}_\text{i} + w_\text{i}^a \nabla_a f_\text{i}) \, \text{d}\tau_\text{i} = 0 \quad \Longrightarrow \quad \forall y \in \gamma_\text{i} \, , \quad \begin{cases} \dot{z}_\text{i}(y) = 0 \, , \\ w_\text{i}^a(y) = 0 \, . \end{cases} \tag{3.29}$$

Proposition (3.29) is most easily proved by contraposition. More precisely, one assumes that $\dot{z}_\text{i} \neq 0$ or $w_\text{i}^a \neq 0$ at some point along γ_i and shows that, consequently, there exists an $f_\text{i} \in \mathbb{F}$ such that the integral on the left-hand side is nonzero; see Appendix A.3.3 for details. Since this result holds for any 4-volume $\mathscr{V}$ chosen initially, we conclude that $\dot{z}_\text{i} = 0$ and $w_\text{i}^a = 0$ at any point along γ_i. Consequently, the expansion of k^a along γ_i simply reads $k^a|_{\gamma_\text{i}} = z_\text{i} u_\text{i}^a$, with z_i constant along γ_i.

The "Redshift" Parameter

In the last paragraphs we have proven that if the SEM tensor T^{ab} describes a pair of quadrupolar particles moving along a circular orbit, then its Lie-dragging $\mathcal{L}_k T^{ab} = 0$ along the helical Killing field k^a implies that for any particle $\text{i} \in \{1, 2\}$ of the system, there exists a constant scalar field z_i defined on γ_i such that

$$\forall y \in \gamma_\text{i} \, , \quad k^a(y) = z_\text{i} u_\text{i}^a(y) \, . \tag{3.30}$$

In other words, k^a is tangent to the worldlines of both particles, or equivalently γ_1 and γ_2 are integral curves of the helical Killing field k^a. Moreover, since u_i^a and $k^a|_{\gamma_\text{i}}$ are both timelike and future-directed, the constant z_i is strictly positive along γ_i. That the helical Killing field (3.7) should be colinear to the particles' 4-velocities makes perfect physical sense, because the support of the helically symmetric SEM tensor (3.16) is restricted to the worldlines γ_1 and γ_2.

Following Detweiler's seminal work [400], the scalar field z_i has been coined the "redshift" parameter/variable, e.g. in Refs. [206, 401, 409, 422, 433]. The reason for this title can be understood from the following thought experiment. Consider a binary system on a circular orbit. One of its components, $\mathcal{B}$, emits a photon of (emission) energy $E_\mathcal{B}$. This photon travels on a null geodesic, with a four-momentum p^a parallel to is received by an observer $\mathcal{S}$ (named Steve), located far away from the binary, near its symmetry axis. On the one hand, the photon's energy as measured by Steve is $E_\mathcal{S} = k^a p_a$, since its four-velocity coincides with the HKV at its location. Indeed, far away from the binary, we can write $k^a = (\partial_t)^a + \Omega \, (\partial_\phi)^a$ where t and ϕ are part of the inertial spherical-type coordinate, and $(\partial_\phi)^a = 0$ along the symmetry axis. On the other hand, the energy of the photon at emission is simply $u^a p_a$, since u^a is the object's four-velocity. Now, since the photon travels on a null geodesic, and k^a is a HKV, the quantity $p^a k_a$ is conserved.[4] Consequently, it follows from Eq. (3.30) that the ratio of the observed and emitted energy is

[4] Indeed, one has $p^b \nabla_b (p^a k_a) = k_a p^b \nabla_b p^a + p^a p^b \nabla_b k_a$. The first term vanishes thanks to the geodesic equation, and the second one because of the Killing equation.

$$\frac{E_{\mathcal{S}}}{E_{\mathcal{B}}} = \frac{k^a p_a}{u^a p_a} = \frac{k^a p_a}{z^{-1}(k^a p_a)} = z\,, \quad \text{whence} \quad z = \frac{\lambda_{\mathcal{B}}}{\lambda_{\mathcal{S}}}\,, \tag{3.31}$$

where $\lambda = h/E$ is the photon's wavelength and h Planck's constant. In most cases, the emission energy of the photon as measured close to the binary is larger than that measured at infinity. In particular, we see that z is a (coordinate-independent) measure of the redshift of a photon, emitted from the particle, when the photon is observed on the symmetry axis at large distances.

Contracting (3.30) with u^{i}_a and taking the norm of (3.30) yields two simple expressions for the redshift:

$$z_{\mathrm{i}} = -u^a_{\mathrm{i}} k_a \quad \text{and} \quad z_{\mathrm{i}} = |k|_{\mathrm{i}} \equiv (-k^a k_a)^{1/2}|_{\gamma_{\mathrm{i}}}\,. \tag{3.32}$$

In particular, the redshift coincides with the norm of k^a along γ_{i}. Since the norm of a Killing field is necessarily conserved along its integral curves, the redshift z_{i} must be conserved along γ_{i}. Indeed, the constraint (3.30) and Killing's equation (3.8) imply

$$z_{\mathrm{i}}\dot{z}_{\mathrm{i}} = -\tfrac{1}{2}\, u^a_{\mathrm{i}} \nabla_a (k^b k_b) = -u^a_{\mathrm{i}} k^b \nabla_a k_b = -z_{\mathrm{i}} u^a_{\mathrm{i}} u^b_{\mathrm{i}} \nabla_{(a} k_{b)} = 0\,. \tag{3.33}$$

This is consistent with the result (3.29). The conserved redshift (3.32) is not to be confused with the Killing energy of the quadrupolar particle, which will be defined in Sect. 3.3.2 below.

We stress that (3.30) holds irrespective of a choice of SSC for the spins S^{ab}_{i} of the particles, and irrespective of a particular physical model for the quadrupoles J^{abcd}_{i}. Moreover, while we have established this result at the quadrupolar order, we expect it to hold at *any* order in the multipolar expansion (2.18). In particular, at the monopolar order it is a classical result that the solutions to the equations of motion (2.57a) for nonspinning massive particles are timelike geodesics. Equation (3.30) thus implies that the integral curves of the helical Killing vector along γ_{i} must be timelike geodesics in this case.

Finally, we note that the constraint (3.30) implies that for any scalar field f defined along γ_{i}, the Lie derivative along k^a simply reduces, up to a factor of the constant redshift (3.32), to the ordinary derivative with respect to proper time τ_{i} along γ_{i}, namely

$$\mathcal{L}_k f|_{\gamma_{\mathrm{i}}} = z_{\mathrm{i}} \dot{f} = z_{\mathrm{i}} \frac{\mathrm{d}f}{\mathrm{d}\tau_{\mathrm{i}}}\,. \tag{3.34}$$

Introducing a spherical-type coordinate system (t, r, θ, ϕ) adapted to the helical Killing symmetry, such that $k^a = (\partial_t)^a + \Omega\, (\partial_\phi)^a$ holds everywhere (or at least in a neighborhood of γ_{i}), the coordinate components of the 4-velocity u^a_{i} simply read $u^\alpha_{\mathrm{i}} = z_{\mathrm{i}}^{-1}(1, 0, 0, \Omega)$. In particular, $z_{\mathrm{i}} = \mathrm{d}\tau_{\mathrm{i}}/\mathrm{d}t$ such that Eq. (3.34) reduces to

$$\mathcal{L}_k f|_{\gamma_{\mathrm{i}}} = \frac{\mathrm{d}f}{\mathrm{d}t}\,. \tag{3.35}$$

3.2.2 *Lie Dragging of the Multipoles*

In the last section, we showed that the existence of a Killing vector k^a in a spacetime containing multipolar particles implied that their worldline must be Killing trajectories, as crystalised in Eq. (3.30). In this section, we go further. In particular, we show that (1) the generic ($\mathcal{T}$) and reduced ($\mathscr{T}$) multipoles of the particle obey differential equations, and (2) that these relations imply the Lie-dragging of the the momentum, spin and quadrupole tensors (p^a, S^{ab}, J^{abcd}). We work at quadrupolar order, but expect the results to hold at any multipolar order.

Thanks to the colinearity (3.30) of the helical Killing field (3.7) and the tangent 4-velocity to the worldline γ_i, the Lie derivative along k^a of any tensor field defined solely along γ_i is well defined. In particular, $\mathcal{L}_k u_\mathrm{i}^a$, $\mathcal{L}_k p_\mathrm{i}^a$, $\mathcal{L}_k S_\mathrm{i}^{ab}$ and $\mathcal{L}_k J_\mathrm{i}^{abcd}$ are well-defined tensor fields along γ_i. In this section, we shall omit the subscript $\mathrm{i} \in \{1, 2\}$ whenever an equation applies for both particles. We shall establish that the 4-velocity u^a, momentum p^a, spin S^{ab} and quadrupole J^{abcd} of each particle are Lie-dragged along the helical Killing field k^a, as expected given the Lie-dragging (3.13) of the SEM tensor of the binary system.

Lie-Dragging of Velocity and Related Identities

Taking the covariant derivative of Eq. (3.30) along γ readily gives $\dot{k}^a = z\dot{u}^a$, because the redshift z is constant. Using Eq. (3.30), this equation can be rewritten as $u^b \nabla_b k^a = k^b \nabla_b u^a$, which is equivalent to

$$\mathcal{L}_k u^a = 0\,. \tag{3.36}$$

Therefore, the 4-velocity u^a is Lie-dragged along $k^a|_\gamma = zu^a$. Together with the Lie-dragging $\mathcal{L}_k g_{ab} = 0$ of the metric, the formula (3.36) implies that the projector $h_{ab} = g_{ab} + u_a u_b$ is also Lie-dragged along k^a, namely

$$\mathcal{L}_k h_{ab} = 0\,. \tag{3.37}$$

Moreover, for any tensor field χ we may combine Eq. (3.36) with the commutation (3.10) of the Lie and covariant derivatives, together with the Leibniz rule, to establish that

$$\mathcal{L}_k(\dot{\chi}) = \left(\mathcal{L}_k \chi\right)^{\dot{}}\,, \tag{3.38}$$

i.e., the Lie derivative along k^a commutes with the covariant derivative along γ. This general result will prove useful in Sect. 3.3.2 below.

Lie-Dragging of Momentum, Spin and Quadrupole

The next goal is to establish that, for each particle, the momentum p^a, spin S^{ab} and quadrupole J^{abcd} are Lie-dragged as well. The idea is to start with the general expression of the quadrupolar SEM tensor for a binary of particles, in the form $T^{ab} = \sum_{\rm i} T^{ab}_{\rm i}$, where (recall Chap. 2, Eqs. (2.58)–(2.59))

$$T^{ab} = \sum_{\rm i} \left\{ \int_{\gamma_{\rm i}} u_{\rm i}^{(a} p_{\rm i}^{b)} \delta_4^{\rm i} \mathrm{d}\tau_{\rm i} + \nabla_c \int_{\gamma_{\rm i}} u_{\rm i}^{(a} S_{\rm i}^{b)c} \delta_4^{\rm i} \mathrm{d}\tau_{\rm i} \right. \tag{3.39a}$$

$$\left. + \frac{1}{3} \int_{\gamma_{\rm i}} R^{\rm i}_{cde}{}^{(a} J_{\rm i}^{b)cde} \delta_4^{\rm i} \mathrm{d}\tau_{\rm i} - \frac{2}{3} \nabla_{cd} \int_{\gamma_{\rm i}} J_{\rm i}^{c(ab)d} \delta_4^{\rm i} \mathrm{d}\tau_{\rm i} \right\}, \tag{3.39b}$$

and then take the Lie-derivative along k^a of it. When we do so, the left-hand side vanishes because $\mathcal{L}_k T^{ab} = 0$, by assumption (3.13). On the right-hand side, $\mathcal{L}_k$ only hits (p^a, S^{ab}, J^{abcd}) in the integrals because ∇ and $\mathcal{L}_k$ commute (3.10) and u^a, δ_4 and R_{abcd} are all Lie-dragged along k^a, as established in Eqs. (3.36), (3.12) and (3.11b), respectively. Consequently, the Lie-dragging of (3.39) directly reads

$$0 = \sum_{\rm i} \left\{ \int_{\gamma_{\rm i}} u_{\rm i}^{(a} \mathscr{p}_{\rm i}^{b)} \delta_4^{\rm i} \mathrm{d}\tau_{\rm i} + \nabla_c \int_{\gamma_{\rm i}} u_{\rm i}^{(a} \mathscr{S}_{\rm i}^{b)c} \delta_4^{\rm i} \mathrm{d}\tau_{\rm i} \right. \tag{3.40a}$$

$$\left. + \frac{1}{3} \int_{\gamma_{\rm i}} R^{\rm i}_{cde}{}^{(a} \mathscr{J}_{\rm i}^{b)cde} \delta_4^{\rm i} \mathrm{d}\tau_{\rm i} - \frac{2}{3} \nabla_{cd} \int_{\gamma_{\rm i}} \mathscr{J}_{\rm i}^{c(ab)d} \delta_4^{\rm i} \mathrm{d}\tau_{\rm i} \right\}, \tag{3.40b}$$

where we introduced the following notations for the Lie derivatives of the multipoles:

$$\mathscr{p}^a \equiv \mathcal{L}_k p^a , \quad \mathscr{S}^{ab} \equiv \mathcal{L}_k S^{ab} \quad \text{and} \quad \mathscr{J}^{abcd} \equiv \mathcal{L}_k J^{abcd} . \tag{3.41}$$

Equation (3.40) does look like a normal form in the sense of Tulczyjew, but is not so in general. Indeed, although the integrands have the right symmetries required for the normal form (which boils down to $\mathscr{J}^{c(ab)d} = \mathscr{J}^{d(ab)c}$ here), they need not be orthogonal to the four-velocity. Consequently, we first need to bring Eq. (3.40) into its normal form, just like the SEM tensor (3.16) was brought into its normal form (3.17)–(3.18). Again, this is done by (1) expanding each integrand with respect to u^a and (2) using extensively the "magic" formula (A.41). Once this is done, we apply Tulczyjew's second theorem (see Appendix A.3.2), and obtain the following constraints on the Lie-dragged multipoles $\mathscr{p}^a$, $\mathscr{S}^{ab}$ and $\mathscr{J}^{abcd}$,

$$u^{(a} \mathscr{p}^{b)} = \left(u^{(a} \mathscr{S}^{b)u}\right)^{\cdot} - \tfrac{1}{3} R_{cde}{}^{(a} \mathscr{J}^{b)cde} + \tfrac{2}{3}\left(\mathscr{J}^{u(ab)u}\right)^{\cdot\cdot} + \tfrac{4}{3} R_{cde}{}^{(a} \mathscr{J}^{b)(\hat{d}u)e} u^c , \tag{3.42a}$$

$$u^{(a} \mathscr{S}^{b)\hat{c}} = -\tfrac{4}{3}\left(\mathscr{J}^{d(ab)u}\right)^{\cdot} h^c{}_d - \tfrac{2}{3} \mathscr{J}^{u(ab)u} \dot{u}^c , \tag{3.42b}$$

$$\mathscr{J}^{\hat{c}(ab)\hat{d}} = 0 . \tag{3.42c}$$

We dropped the i index since these equations hold for each particle individually, owing to our extension of Tulczyjew's second theorem. They are the consequences of the condition $\mathcal{L}_k T^{ab} = 0$ on (p^a, S^{ab}, J^{abcd}) when T^{ab} describes a quadrupolar particle, and are but ordinary differential equations for $(\not{p}^a, \mathscr{S}^{ab}, \mathscr{J}^{abcd})$. Let us now establish that the only solution to these equation is the one where $\mathscr{J}^{abcd}$, $\mathscr{S}^{ab}$ and $\not{p}^a$ vanish identically, in that precise order.

We start by showing that (3.42c) implies $\mathscr{J}^{abcd} = 0$. To this end, we perform an orthogonal decomposition of $\mathscr{J}^{abcd}$ with respect to u^a, with help of the orthogonal projector h_{ab}. Thanks to the algebraic symmetries of $\mathscr{J}^{abcd}$ (namely those of R_{abcd}), this decomposition simply reads (details about this decomposition are given in Chap. 6, Sect. 6.1)

$$\mathscr{J}^{abcd} = \hat{\mathscr{J}}^{abcd} + 2u^{[a}\mathscr{J}^{b]cd} + 2u^{[d}\mathscr{J}^{c]ba} - 4u^{[a}\mathscr{J}^{b][c}u^{d]}\,, \tag{3.43}$$

where the tensors $\hat{\mathscr{J}}^{abcd} \equiv \mathscr{J}^{\hat{a}\hat{b}\hat{c}\hat{d}}$, $\mathscr{J}^{abc} \equiv \mathscr{J}^{\hat{a}u\hat{b}\hat{c}}$ and $\mathscr{J}^{ab} \equiv \mathscr{J}^{\hat{a}u\hat{b}u}$ are all orthogonal to u^a, by definition. We then symmetrize (3.43) with respect to the indices b and c, and contract with the projector $h^e{}_a h^f{}_d$, so that Eq. (3.42c) implies

$$\hat{\mathscr{J}}^{c(ab)d} + 2\mathscr{J}^{(cd)(a}u^{b)} - u^a u^b \mathscr{J}^{cd} = 0\,. \tag{3.44}$$

Contracting (3.44) with $u_a u_b$ and $u_a h^e{}_b$ gives $\mathscr{J}^{ab} = 0$ and $\mathscr{J}^{abc} = 0$, respectively. Substituting these equations back into (3.44) yields the third constraint $\hat{\mathscr{J}}^{a(bc)d} = 0$. Finally, substituting these three constraints into the decomposition (3.43) gives

$$\mathscr{J}^{abcd} = \hat{\mathscr{J}}^{a[bc]d}\,, \tag{3.45}$$

which tautologically means that $\mathscr{J}^{a(bc)d} = 0$. Combined with the other algebraic symmetries of $\mathscr{J}^{abcd}$ this readily implies that $\mathscr{J}^{abcd}$ vanishes identically.[5] We have thus proven that the quadrupole of each particle is Lie-dragged along its worldline, i.e.,

$$\mathscr{J}^{abcd} \equiv \mathcal{L}_k J^{abcd} = 0\,. \tag{3.46}$$

Given Eq. (3.46), the system (3.42) simplifies drastically, as it reduces to that for a dipolar particle, namely

$$u^{(a}\not{p}^{b)} = \left(u^{(a}\mathscr{S}^{b)u}\right)^{\cdot}, \tag{3.47a}$$

$$u^{(a}\mathscr{S}^{b)\hat{c}} = 0\,. \tag{3.47b}$$

Contracting Eq. (3.47b) with $h^d{}_a u_b$ and $u_a u_b$ implies $\mathscr{S}^{ab}h^c{}_a h^d{}_b = 0$ and $\mathscr{S}^{ab}h^c{}_a u_b = 0$, respectively. Since $\mathscr{S}^{ab}u_a u_b = 0$ by the antisymmetry of $\mathscr{S}^{ab}$, we conclude that all the contributions to the orthogonal decomposition of $\mathscr{S}^{ab}$ with

---

[5] This can be shown by applying alternatively the antisymmetry in the middle two indices and the one of the last two indices. After fives steps, the end result is $\mathscr{J}^{abcd} = -\mathscr{J}^{abcd}$.

respect to u^a vanish. Consequently, we have shown that the spin tensor of each particle is Lie-dragged along its worldline, i.e.,

$$\mathscr{S}^{ab} \equiv \mathcal{L}_k S^{ab} = 0\,. \tag{3.48}$$

Finally, we may substitute Eq. (3.48) into (3.47a) and contract with $h^c{}_a u_b$ and $u_a u_b$ to obtain $\not{p}^a h^b{}_a = 0$ and $\not{p}^a u_a = 0$. We thus conclude that the 4-momentum of each particle is Lie-dragged along its worldline, i.e.,

$$\not{p}^a \equiv \mathcal{L}_k p^a = 0\,. \tag{3.49}$$

The physical models for a spin-induced or tidally-induced quadrupole that will be used in Chap. 6 are consistent with the Lie-dragging of u^a, p^a, S^{ab} and J^{abcd}. We naturally expect that the results (3.46), (3.48) and (3.49) extend to an arbitrary multipolar order in the gravitational skeleton formalism.

To end this section, we mention that the spin vector S^a and the mass dipole D^a of the particles, defined in Eqs. (2.63), are Lie-dragged too

$$\mathcal{L}_k S^a = 0 \quad \text{and} \quad \mathcal{L}_k D^a = 0\,, \tag{3.50}$$

since they are defined from S^{ab}, u^a and the canonical volume form ε_{abcd}, which are Lie-dragged. Interestingly, the formula (3.30) and some of its consequences—namely the Lie dragging (3.48) and (3.49) of the 4-momentum and spin tensor—were previously established in [434], in a mathematical physics journal. Their derivation is done at dipolar order and is based on Dixon's integral definition of the momenta [349]. Consequently, it is unclear how their result extends to black holes, as pointed out in Chap. 2, Sect. 2.2.5. In the context of the first law of mechanics, which is our main goal here, it is important that we remain self-consistent throughout the calculations and that we control the definition of the multipoles used in the final form of the first law. Consequently, it is more practical to use Tuclzyjew's route for our purpose. What's more, despite these slight differences, the consistency of our results with those of [434] (in the regions where our applications range overlap) illustrates once more that the two formalisms are complementary and consistent with each other.

Algebraic Constraints on the Multipoles

Finally, we discuss an interesting consequence of the Lie-dragging (3.49) and (3.48) of the momentum p^a and spin S^{ab}, in light of the helical constraint (3.30). Combining Eqs. (3.30) and (3.8), the formulas (3.49) and (3.48) can be rewritten as

$$z\dot{p}_a = p^c \nabla_c k_a = -p^c \nabla_a k_c\,, \tag{3.51a}$$

$$z\dot{S}_{ab} = 2S^c{}_{[a} \nabla_{b]} k_c = 2\nabla_c k_{[a} S^c{}_{b]}\,. \tag{3.51b}$$

By combining those Lie-dragging equations with the equations of evolution (2.57), while using the helical constraint (3.30), we obtain the following relations that must be satisfied by the momentum, spin and quadrupole of each particle:

$$p^c \nabla_c k_a = \frac{1}{2} R_{bcda} S^{bc} k^d - \frac{1}{6} z J^{bcde} \nabla_a R_{bcde} \,, \tag{3.52a}$$

$$S^c{}_{[a} \nabla_{b]} k_c = p_{[a} k_{b]} + \frac{2}{3} z R_{edc[a} J_{b]}{}^{cde} \,. \tag{3.52b}$$

Assuming that the spacetime geometry is known, so that $(k^a, \nabla_a k_b, R_{abcd}, \nabla_a R_{bcde})$ are known, and given a physical model for the quadrupole J^{abcd}, the relations (3.52) are a set of ten *algebraic* equations for the ten unknowns p^α and $S^{\alpha\beta}$. Interestingly, by recalling the Kostant formula (3.9) and the expression (3.32) for the redshift z, the formula (3.52b) appears schematically (getting rid of all tensorial indices and numerical prefactors) as a multipolar identity of the form

$$p\, k + S\, \nabla k + J\, \nabla\nabla k = 0 \,, \tag{3.53}$$

while (3.52a) has a similar multipolar structure, with an additional overall covariant derivative. It would be interesting to see if this pattern extends at higher multipolar orders, and to assess whether it carries or not any deeper meaning. Naturally, these equations are closely related to similar formulas established in the context of Dixon's and Harte's formalisms for extended fluid bodies, in presence of an isometry [349, 435].

3.3 Properties of Dipolar Particles

Along with the identities presented in Sect. 3.1.2, other results that are specific to an HKV in the context of multipolar particles will be of importance for the derivation of the first law later on. In this subsection we consider a binary system of *dipolar* particles moving along a circular orbit, and explore some of the consequences of the Lie-dragging along the helical Killing vector (3.7) of the 4-velocity u^a, the 4-acceleration $\dot{u}^a$ and the spin vector S^a, as was established in the previous section. We first introduce the notion of twist and vorticity and show that it is aligned with the spin vector, and the discuss some conserved quantities.

3.3.1 Twist, Vorticity and Spin Vector

An important notion associated with the helical Killing vector field (3.7) is the *twist*

$$\varpi^a \equiv -\frac{1}{2}\varepsilon^{abcd} k_b \nabla_c k_d \,. \tag{3.54}$$

The twist (3.54) is orthogonal to k^a and Lie-dragged along k^a, as a direct consequence of the Lie-dragging of ε^{abcd}, k_b and $\nabla_c k_d$ (cf. Chap. 3). It will be shown that the twist (3.54) does not vanish everywhere. As a consequence of the Frobenius theorem, k^a cannot be hypersurface-orthogonal [355]. Moreover, it can be shown that the twist (3.54) obeys the identity [6]

$$\nabla_{[a}\varpi_{b]} = \frac{1}{2}\varepsilon_{abcd} k^c R^{de} k_e \,. \tag{3.55}$$

For a Ricci-flat, helically symmetric spacetime we thus have $\nabla_{[a}\varpi_{b]} = 0$, so that, locally, the 1-form ϖ_a is exact, i.e. there exists a scalar field ϖ such that

$$\varpi_a = \nabla_a \varpi \,. \tag{3.56}$$

The scalar twist ϖ is then Lie-dragged along the helical Killing field k^a, as $\mathcal{L}_k \varpi = k^a \varpi_a = 0$.

From the twist, we can now endow each particle with a *vorticity* V^a, namely the restriction to the worldline γ of the twist (3.54) associated with the helical Killing field (3.7), defined as

$$V^a \equiv -\frac{1}{2}\,\varepsilon^{abcd} u_b \nabla_c k_d \,. \tag{3.57}$$

Indeed, the helical constraint (3.30) implies $\varpi^a|_\gamma = z V^a$. The vorticity V^a is orthogonal to u^a and is Lie-dragged along $k^a|_\gamma \propto u^a$. The definition (3.57) of the vorticity, which involves the Noether 2-form $\nabla_a k_b$, should be compared to the definition (2.69) of the spin vector, which involves the antisymmetric spin tensor S_{ab}. The duality between $(V^a, \nabla_a k_b)$ and (S^a, S_{ab}) is made even clearer when expressing $\nabla_a k_b$ in terms of V^a. Indeed, contracting Eq. (3.57) with $\varepsilon_{abcd} u^b$, using Eq. (3.30) and re-arranging the result readily gives

$$\nabla_a k_b|_\gamma = \varepsilon_{abcd} u^c V^d + 2 k_{[a} \dot{u}_{b]} \,, \tag{3.58}$$

which is to be compared to Eq. (2.63). We will use this formula in Sect. 5.3 below to simplify the first law of compact binary mechanics.

We now consider the rate of change of the vorticity (3.57) along γ. Using the condition of metric compatibility, which implies $\dot{\varepsilon}^{abcd} = 0$, together with the conservation of $\nabla_c k_d$ along γ (shown below in Eq. (3.78)), we simply have $2\dot{V}^a = -\varepsilon^{abcd} \dot{u}_b \nabla_c k_d$. Substituting the decomposition (3.58) into this formula while using $\dot{u}^a u_a = 0$ then yields

$$\dot{V}^a = u^a \dot{u}^b V_b \,, \tag{3.59}$$

showing that V^a obeys the Fermi-Walker transport law, like the spin vector S^a. Contracting (3.59) with V_a then implies $\dot{V}^a V_a = 0$, so that the norm of the vorticity is conserved along γ.

The vorticity and the spin vector share one more property: provided that the SSC (2.64) is satisfied, they have the same spacelike direction. In order to establish this property, consider on the one hand the spacelike vector

$$W^a \equiv \varepsilon^a{}_{bcd} u^b S^c V^d \,, \tag{3.60}$$

and on the other hand the Lie-dragging of the spin vector S^a (cf. Eq. (3.50)). The latter implies $z\dot{S}^a = S_b \nabla^b k^a$, an equation which can be simplified with the help of Eq. (3.58), to find an alternative expression for W^a as

$$W^a = z(\dot{S}^a - u^a \dot{u}^b S_b) \,, \tag{3.61}$$

which vanishes as a consequence of the equation of spin precession (2.72).[6] Since S^a and V^a are both orthogonal to u^a, the equation $W^a = 0$ holds if, and only if, S^a and V^a are aligned. Let s^a denote their common unit spacelike direction, such that $s^a s_a = 1$. Then we obtained the important result

$$V^a = V s^a \,, \tag{3.62a}$$

$$S^a = S\, s^a \,, \tag{3.62b}$$

where $V \equiv (V^a V_a)^{1/2}$ is the norm of the vorticity and S that of the spin vector, as defined in Eq. (2.70b). The norms V and S are both constant along γ, while s^a is Lie-dragged along γ. The colinearity (3.62) will prove useful in Sect. 5.1 to write the first law of binary mechanics in its simplest form, in terms of scalar quantities.

Around Eq. (3.56), we established that for our class of Ricci-flat spacetimes, the twist (3.54) associated with the helical Killing field (3.7) must derive from a scalar potential ϖ. Therefore, the vorticity (3.57) is necessarily proportional to the gradient of the scalar twist ϖ, evaluated along γ. Combining Eq. (3.56) with the orthogonality $V^a u_a = 0$ implies the conservation of the scalar twist ϖ along γ, namely

$$\dot{\varpi} = u^a \varpi_a = z u^a V_a = 0 \,. \tag{3.63}$$

For a spinning particle in a binary system on a circular orbit, we may use the Lie-dragging of the 4-acceleration $\dot{u}^a$ along k^a, namely $z\ddot{u}_b = \dot{u}^c \nabla_c k_b$ (see Paper I), to express Eq. (2.76) in the implicit form

---

[6] If the SSC (2.64) is not imposed, then the relationship $S^{[a} V^{b]} = 0$ can be generalized to $S^{[a} V^{b]} = z D^{[a} \dot{u}^{b]}$, as can easily be shown by generalizing the equation of spin precession (2.72) to a nonzero mass dipole D^a.

$$m\dot{u}^a = (P^{-1})^a{}_b B^{bc} S_c \,, \quad \text{where} \quad P^a{}_b = \delta^a{}_b + \frac{1}{zm} S^{ac} \nabla_b k_c \,. \tag{3.64}$$

The term linear in the spin tensor in the operator $P^a{}_b$ can be rewritten in terms of the spin vector S^a and the vorticity V^a by substituting the expressions (2.69) and (3.58) for S^{ab} and $\nabla_a k_b$. Using the colinearity (3.62) and introducing the projector $h^a{}_b \equiv \delta^a{}_b + u^a u_b$ orthogonal to the 4-velocity, we readily find

$$S^{ac} \nabla_b k_c = p^a_{\text{hid}} k_b + VS \, (h^a{}_b - s^a s_b) \,, \tag{3.65}$$

where we recall that $p^a_{\text{hid}} = -\varepsilon^a{}_{bcd} u^b \dot{u}^c S^d$ is the spacelike "hidden momentum" appearing in the momentum-velocity relationship (2.73), and $h^a{}_b - s^a s_b$ is the projector in the spacelike plane orthogonal to the common axial direction s^a of S^a and V^a. If e^a_1 and e^a_2 denote two spacelike unit vectors spanning that plane, such that (u^a, e^a_1, e^a_2, s^a) is an orthonormal tetrad, then the operator $P^a{}_b$ in Eq. (3.64) reads

$$P^a{}_b = \delta^a{}_b + \frac{1}{m} p^a_{\text{hid}} u_b + \frac{VS}{zm} \left(e^a_1 e_{1b} + e^a_2 e_{2b} \right) \tag{3.66}$$

Remarkably, the inverse $(P^{-1})^a{}_b$ of the operator (3.66) can be written in closed form by assuming an Ansatz of the form $(P^{-1})^a{}_b = \delta^a{}_b + \alpha \, p^a_{\text{hid}} u_b + \beta \, (e^a_1 e_{1b} + e^a_2 e_{2b})$, with α, β two constants to be solved for. Thanks to the defining identity $(P^{-1})^a{}_b P^b{}_c = \delta^a{}_c$, the orthonormality of the basis (e^a_1, e^a_2) and of $p^a_{\text{hid}} u_a = 0 = p^a_{\text{hid}} s_a$, one obtains

$$(P^{-1})^a{}_b = \delta^a{}_b - \frac{1}{1+a\omega} \frac{1}{m} p^a_{\text{hid}} u_b - \frac{a\omega}{1+a\omega} \left(e^a_1 e_{1b} + e^a_2 e_{2b} \right) . \tag{3.67}$$

Here, we introduced the Kerr parameter $a \equiv S/m$ of the spinning particle and we anticipated on the formula $V = z\omega$, with ω the invariant spin precession frequency to be defined in Sect. 5.2 below; see e.g. Eqs. (5.33) and (5.43). Substituting Eq. (3.67) back into the formula (3.64) for the 4-acceleration while using $u_b B^{bc} = 0$ then gives

$$\dot{u}^a = \frac{a}{1+a\omega} \left(B^{ab} s_b + a\omega s^a B^{bc} s_b s_c \right) . \tag{3.68}$$

This simple formula shows that the 4-acceleration is entirely sourced by the coupling of the magnetic-type tidal field with the spin vector, just like the rate of change $\dot{p}^a = B^{ab} S_b$ of the 4-momentum. Equations (2.70a) and (3.68) yield $\dot{S}^a u_a = -\dot{u}^a S_a = -\frac{1}{m} B^{ab} S_a S_b$, in agreement with the spin precession equation (2.78).

3.3.2 Conserved Quantities

In this final section, we explore the various conserved quantities associated with the isometry generated by the helical Killing field (3.7). In particular, given the Lie-dragging along k^a of the 4-velocity, 4-momentum, spin and quadrupole tensor of each particle established in Sect. 3.2.2, the result (3.34) implies that any scalar field that is constructed out of the particle's variables (u^a, p^a, S^{ab}, J^{abcd}) and the spacetime geometry (g_{ab}, k^a, $R_{abcd}, \dots$) will be conserved along γ.

Killing Energy

For a *generic* Killing vector field ξ^a, i.e., for a Killing vector field that does not necessarily satisfy the helical constraint (3.30), the Killing energy of a particle with momentum p^a and spin S^{ab} is defined as[7]

$$E_\xi \equiv p^a \xi_a + \frac{1}{2} S^{ab} \nabla_a \xi_b \,. \tag{3.69}$$

At the dipolar order, this linear combination of p^a and S^{ab} is easily seen to be conserved by substituting the equations of evolution (2.57) with $J^{abcd} = 0$ into the expression for $\dot{E}_\xi$ and by using Killing's equation (3.8) and the Kostant formula (3.9). Remarkably, the conservation of the Killing energy (3.69) holds beyond the dipolar order. Indeed, it can be shown that the scalar (3.69) is a constant of motion for an *arbitrary* extended body endowed with an infinite set of multipole moments[8] [310, 327, 347, 371].

In general, however, neither the monopolar contribution, nor the dipolar contribution to the Killing energy (3.69), say

$$E_\xi^{(p)} \equiv p^a \xi_a \quad \text{and} \quad E_\xi^{(S)} \equiv \frac{1}{2} S^{ab} \nabla_a \xi_b \,, \tag{3.70}$$

will be separately conserved. For example, if t^a and ϕ^a denote the usual Killing vector fields associated with the stationary and axisymmetry of the Kerr geometry, then a test spinning particle orbiting a spinning black hole has a conserved energy $-E_t$ and a conserved angular momentum E_ϕ, but the monopolar and dipolar contributions $\{E_t^{(p)}, E_t^{(S)}\}$ and $\{E_\phi^{(p)}, E_\phi^{(S)}\}$ to E_t and E_ϕ are not separately conserved.

However, in our case the helical nature of the Killing field k^a implies the constraint (3.30), from which we readily derive the exact conservation laws

---

[7] This constant of motion was recently used in [365] to compute a GW flux balance law in an EMRI with a spinning secondary.

[8] To make contact with the GFK formalism touched upon in Chap. 2, Sect. 2.2.3, the energy E_ξ is precisely the numerical value of the generalized four-momentum $\mathcal{P}_\xi$ with the true Killing vector field ξ^a, and not a mere generalized one, cf. Sect. 3.5 in [347].

$$z\dot{E}_k^{(p)} = \mathcal{L}_k(p^a k_a) = (\mathcal{L}_k p^a)k_a + p^a \mathcal{L}_k k_a = 0\,, \tag{3.71a}$$

$$2z\dot{E}_k^{(S)} = \mathcal{L}_k(S^{ab}\nabla_a k_b) = (\mathcal{L}_k S^{ab})\nabla_a k_b + S^{ab}\mathcal{L}_k \nabla_a k_b = 0\,, \tag{3.71b}$$

where we used Eqs. (3.48)–(3.49) and the identity $\mathcal{L}_k \nabla_a k_b = \nabla_a \mathcal{L}_k k_b = 0$ (see Appendix A.2). So, in our physical setup, the monopolar and dipolar contributions to the Killing energy E_k are *separately* conserved, irrespective of a particular choice of SSC. In particular, by combining Eq. (3.30) with the definition (2.60) of the particle's rest mass, the monopolar contribution to the Killing energy is easily seen to coincide with the red shifted rest mass:

$$E_k^{(p)} = p^a k_a = -mz\,. \tag{3.72}$$

This expression is indeed consistent with the conservation (3.33) and (3.73) of z and m. The separate conservation of $E_k^{(p)}$ and $E_k^{(S)}$ is a consequence of the constraint (3.30) on the helical Killing field, which must be satisfied here because both particles act as a source of spacetime curvature, contrary to the case of a spinning test particle in the Kerr black hole geometry, for which there exists no relationship (for a generic orbit) between the velocity u^a of the test particle and the Killing fields $t^a|_\gamma$ and $\phi^a|_\gamma$ along the particle's wordline γ.

Other Geometrically Conserved Quantities

Equations (3.36), (3.48) and (3.49) readily imply the Lie-dragging along k^a of the scalar norms (2.60). Combined with the identity (3.34) we conclude that the rest mass $m = -p^a u_a$, the dynamical mass $\mu^2 = -p^a p_a$ and the spin amplitude $S^2 = \frac{1}{2} S^{ab} S_{ab}$ are all conserved along γ, irrespective of a choice of SSC, i.e.

$$\dot{m} = 0\,, \quad \dot{\mu} = 0 \quad \text{and} \quad \dot{S} = 0\,. \tag{3.73}$$

Moreover, as shown below Eq. (3.8), Killing's equation implies the identity $k^c \nabla_c k_a = -\frac{1}{2}\nabla_a (k^c k_c)$. When evaluated along γ, this yields $z^2 \dot{u}_a = \frac{1}{2}\nabla_a |k|^2|_\gamma = z\nabla_a |k|$, where we used Eqs. (3.30)–(3.33), and the fact that k^a is necessarily timelike in a neighborhood of γ. The 4-acceleration can thus be expressed in terms of the gradient of the norm of the helical Killing field as

$$\dot{u}_a = \nabla_a \ln |k|\,. \tag{3.74}$$

Contracting with u^a we find that $|k|_\gamma = z$ is conserved along γ, as established earlier. Interestingly, for a monopolar particle, the geodesic equation $\dot{u}_a = 0$ and (3.74) imply $\nabla_a \ln |k| = 0$, so that $|k|$ is conserved not only along γ, but also along the spacelike directions orthogonal to γ.

Moreover, by applying the general result (3.38) to the particular case of the 4-velocity u^a of a given particle, while making use of the Lie-dragging (3.36) of u^a, we readily obtain the Lie-dragging along k^a of the 4-acceleration:

$$\mathcal{L}_k \dot{u}^a = 0\,. \tag{3.75}$$

More precisely, the general result (3.38) should be applied to an extension $\tilde{u}^a$ of u^a in a neighborhood of γ, as in (3.19). This result can alternatively be derived by taking the Lie derivative along k^a of the expression (3.74) of the 4-acceleration, as $\mathcal{L}_k \nabla_a \ln|k| = \nabla_a \mathcal{L}_k \ln|k| = 0$. The formula (3.75) is equivalent to

$$z\ddot{u}_a = \dot{u}^c \nabla_c k_a\,, \tag{3.76}$$

which contracted with the 4-acceleration implies $\dot{u}^a \ddot{u}_a = 0$, thanks to Killing's equation (3.8). Thus, the norm of the acceleration is conserved, in addition to that of the velocity. Moreover, contracting (3.76) with u^a and using Killing's equation along with the helical constraint (3.30) with z constant implies $u^a \ddot{u}_a = -\dot{u}^c \dot{u}_c$, so that

$$\dot{u}^a \dot{u}_a = -\ddot{u}^a u_a = \text{const.} \tag{3.77}$$

The same argument holds for the rates of change of any Lie-dragged quantity. In particular, $\dot{p}^a \dot{p}_a$, $\dot{S}^a \dot{S}_a$, $\dot{D}^a \dot{D}_a$, $\dot{S}^{ab} \dot{S}_{ab}$ and $\dot{J}^{abcd} \dot{J}_{abcd}$ are all constant along γ.

Finally, using the Kostant formula (3.9) together with the constraint (3.30), we can easily show that $\nabla_a k_b$ is conserved along γ, according to

$$(\nabla_a k_b)^{\cdot} \equiv u^c \nabla_c \nabla_a k_b = -u^c R_{abcd} k^d = -z R_{abcd} u^c u^d = 0\,, \tag{3.78}$$

the last equality following from the antisymmetry of the Riemann tensor with respect to its last two indices. As will be shown in Chap. 5, the conserved norm $|\nabla k|_\gamma$ of the conserved 2-form $\nabla_a k_b|_\gamma$ is very closely related to the precession frequency of the spin vector S^a that was introduced in Sect. 2.3.2, and which has been used extensively to compare PN and GSF calculations [436–441], and to calibrate EOB models [442].

Chapter 4
First Laws of Mechanics

Contents

Yesterday I received from Miss Noether a very interesting paper on the generation of invariants. I am impressed by the fact that these things can be understood from so general a point of view. It would have done the Old Guard of Göttingen no harm to be sent back to school under Miss Noether. She really seems to know her trade!

A. Einstein,
Letter to D. Hilbert (1918)

P. Ramond, *The First Law of Mechanics in General Relativity & Isochrone Orbits in Newtonian Gravity*, Springer Theses,
https://doi.org/10.1007/978-3-031-17964-8_4

SINCE the pioneering works on isolated black hole thermodynamics in the 1970s, the so-called *first law of mechanics* has gradually been extended to a variety of binary systems, such as black hole binaries, magnetized neutron stars or spinning point particles. They have also been generalized to account for the wide variety of relativistic orbits: quasi-circular, eccentric, and black hole geodesics. The aim of this chapter is to present these first laws, explain what they are about, and understand why they have become a central tool in the theory of relativistic mechanics and gravitational waves. To achieve this goal, we begin in Sect. 4.1 with an overview of the different kinds of "first laws" that can be found in the literature. Then, in Sect. 4.2, we review a selection of the many applications they have been used for. We shall see, in particular, that they provide a powerful tool to compare and benchmark the output of the different approximation schemes presented in Chap. 2. Lastly, in Sect. 4.3, we derive a first, general version of the first law of mechanics, which will be the basis of the calculations presented in Chap. 5.

4.1 Zoology of First Laws

The first occurrence of a "first law of mechanics" in GR dates back to the seventies with the pioneering papers on BH thermodynamics, by several physicists, namely Jacob Bekenstein, Stephen Hawking, James Bardeen and Brandon Carter, to name but the principal protagonists. Their results showed that several scalar, physical observables of an isolated Kerr BH could all be related by a differential relation that resembles the first law of thermodynamics. Since then, such differential (and algebraic) relations have been extended to a plethora of sources and come in many forms, all under the name "first law of mechanics". Broadly speaking, these extensions, can be categorized into three families, represented here chronologically (along with their principal protagonists)

- 1990's: Generalized BH thermodynamics (Wald, Iyer et al.),
- 2000's: Extension to BH and NS binary systems (Friedmann, Uryū, et al.),
- 2010's: Extension to point particle models (Le Tiec, Blanchet et al.).

The first family contains extensions of the pioneering works on BH thermodynamics, to a broader class of spacetimes and theories of gravity. One of them will be discussed separately in Sect. 4.3, where we will extend its validity for our own purposes. The second and third families deal with first laws for binary systems, with practical applications in relativistic celestial mechanics and GW astronomy, just as our work. Sections 4.1.2–4.1.4 fill in the details on their context and derivation. However, before considering binary systems, we cover in Sect. 4.1.1 the basics of their *common ancestor*, the laws of BH thermodynamics.

4.1.1 *Black Hole Thermodynamics*

As we covered briefly in Chap. 1, one of the most remarkable result of GR is the fact that each and every isolated black hole in the Universe is described by solely two numbers, its mass M and spin S. Indeed, they are all described by the Kerr metric, a two-parameter family of exact, vacuum, stationary solutions to Einstein's equations. Let us represent this metric by giving its spacetime element $\mathrm{d}s$ in the Boyer-Lindquist coordinates $(t, r, \theta, \phi) \in \mathbb{R}^2 \times \mathbb{S}^2$ (see [82] for a review a detailed account on the Kerr metric):

$$\mathrm{d}s^2 = -\left(1 - \frac{2Mr}{\Sigma}\right)\mathrm{d}t^2 - \frac{4Mr}{\Sigma}a\sin^2\theta\,\mathrm{d}t\mathrm{d}\phi + \frac{\Sigma}{\Delta}\mathrm{d}r^2 + \Sigma\,\mathrm{d}\theta^2 + \left(r^2 + a^2 + \frac{2Mr}{\Sigma}a^2\sin^2\theta\right)\sin^2\theta\,\mathrm{d}\phi^2\,, \tag{4.1}$$

where $\Delta(r) \equiv r^2 - 2Mr + a^2$ and $\Sigma(r,\theta) \equiv r^2 + a^2\cos^2\theta$. The metric (4.1) is parametrized by the BH's mass $M > 0$ and its spin S, or equivalently the (dimensionless) spin parameter $a \equiv S/M \geq 0$. The parameter a can take any value in $[0; M]$. In the limit $a \to 0$, Eq. (4.1) reduces to the Schwarzschild metric in the Schwazshild-Droste coordinates, while $a = M$ defines the so-called *extremal* Kerr BH. One advantage of the Boyer-Lindquist coordinates is the explicit independence of the metric components with respect to the coordinates t and ϕ. Consequently, the vector fields $(\partial_t)^a$, and $(\partial_\phi)^a$ are Killing. These encode the fact that the Kerr spacetime is stationary and axisymmetric, respectively. The components of the Kerr metric (4.1) diverge when $\Sigma = 0$. This corresponds to the 2-dimension locus of events $\mathscr{R}$ where $(r,\theta) = (0, \pi/2)$. It is called the ring singularity of the Kerr BH and is a true, curvature singularity of the Kerr spacetime. The metric (4.1) also diverges when $\Delta = 0$. Since Δ is a quadratic-in-r polynomial, this singularity defines two 3-dimensional hypersurfaces of constant BL radius. Contrary to $\mathscr{R}$, these two regions correspond to a mere coordinate singularity, and the Kerr *event horizon*, denoted $\mathscr{H}$, corresponds to the outer one, namely

$$\mathscr{H} : r = \hbar\,, \quad \text{where} \quad \hbar \equiv M + \sqrt{M^2 - a^2}\,. \tag{4.2}$$

The other surface $\mathscr{H}_{\text{in}}$, located on the 3-hypersurface $r = M - \sqrt{M^2 - a^2}$, is called the *inner horizon* of the Kerr BH. The laws of thermodynamics involve three fundamental, geometrical quantities associated to the event horizon $\mathscr{H}$.

Angular Velocity

The event horizon $\mathscr{H}$ is also a *Killing horizon*, in the sense that there exists a null Killing vector field χ^a that is normal to $\mathscr{H}$. This Killing field must be a linear combinations of $(\partial_t)^a$ and $(\partial_\phi)^a$, the two generators of Kerr isometries. Up to a constant normalizing factor, it can thus be written as $(\partial_\phi)^a + \Omega_{\mathscr{H}} (\partial_\phi)^a$ where $\Omega_{\mathscr{H}}$ is a constant. Requiring that this Killing vector be null on $\mathscr{H}$ provides an expression of $\Omega_{\mathscr{H}}$ in terms of the parameters (M, a). Summarizing, it is found that on $\mathscr{H}$

$$\chi^a = (\partial_t)^a + \Omega_{\mathscr{H}} (\partial_\phi)^a , \quad \text{where} \quad \Omega_{\mathscr{H}} \equiv \frac{a}{2M\hbar} , \tag{4.3}$$

with $\hbar$ given in terms of (M, a) by Eq. (4.2). The quantity $\Omega_{\mathscr{H}}$ is called the *angular velocity* of the Kerr BH, as it coincides with its apparent rotation rate as measured by an inertial, static observer located at spatial infinity.

Surface Gravity

The Killing field χ^a coincides, on $\mathscr{H}$, with the null geodesic generator of the Kerr spacetime. Accordingly, when evaluated on the event horizon, χ^a satisfies a (pre-)geodesic equation, in the form

$$\chi^b \nabla_b \chi^a |_{\mathscr{H}} = \kappa \chi^a , \quad \text{where} \quad \kappa \equiv \frac{\hbar - M}{2M\hbar} , \tag{4.4}$$

and where κ is the non-affinity coefficient. Now, considering the outer region where χ^a is timelike. The parameter κ can be given a physical interpretation, as follows. Consider an observer $\mathscr{O}$ located outside of $\mathscr{H}$, whose worldline's coincides with an integral curve of χ^a. The four-velocity of that observer is then $u^a = \chi^{-1} \chi^a$, where $\chi^2 \equiv -\chi^a \chi_a$ is the (squared) norm of χ^a. Using Killing's equation, the four-acceleration of this observer can be shown to satisfy $u^a \nabla_b u^a = \frac{1}{\chi^2} \chi^b \nabla_b \chi^a$. This relation is valid outside of the event horizon $\mathscr{H}$, where as Eq. (4.4) holds on $\mathscr{H}$. By taking the norm of these two equations, and the limit $\mathscr{O} \to \mathscr{H}$ where the observer approaches the event horizon, we find that the norm $\Gamma \equiv \sqrt{\dot{u}_a \dot{u}^a}$ of the observer's four-acceleration satisfies

$$\kappa = \lim_{\mathscr{O} \to \mathscr{H}} \chi \Gamma . \tag{4.5}$$

Consequently, κ is the value of the (normalized) acceleration of the observer, $\chi\Gamma$, as it approaches the black hole. Note, however, that this result only motivates the name "surface gravity" and does not correspond to the "gravity felt" by the observer *per se*. In particular, since χ^a is null on $\mathscr{H}$, $\chi \to 0$ as $\mathscr{O} \to \mathscr{H}$. Therefore Eq. (4.5) requires $\Gamma \to \infty$ in this limit: the "physical surface gravity" becomes infinite as one approaches the BH's horizon.[1]

[1] It is possible to give another justification of the term surface gravity, by identifying κ as the force exerted by an operator at infinity necessary to keep a unit mass at rest close to $\mathscr{H}$, see [355].

Surface Area

As mentioned previously, the event horizon of the Kerr BH is also a Killing horizon. As such, it is also a *totally geodesic null hypersurface*, or more simply, a *non-expanding* horizon (see Sect. 1.3.5 [82] for details). This feature allows for the introduction of a well-defined notion of area A, for a Kerr BH. This area is defined as that of any cross section $\mathscr{U}$ of the event horizon $\mathscr{H}$ (see Sect. 2.3.4 for a proper definition). In terms of the 3+1 Kerr coordinates $(\tilde{t}, r, \theta, \tilde{\phi})$, which are not singular on $\mathscr{H}$, such a cross section is given by $\tilde{t} = \text{cst}$ and $r = \hbar$, which induces a 2-metric q_{ab} on $\mathscr{U}$ given by

$$q_{\alpha\beta}\mathrm{d}x^\alpha \mathrm{d}x^\beta = (\hbar^2 + a^2\cos^2\theta)\,\mathrm{d}\theta^2 + \frac{(\hbar^2 + a^2)^2}{\hbar^2 + a^2\cos^2\theta}\sin^2\theta\,\mathrm{d}\tilde{\phi}^2. \tag{4.6}$$

The area of this (closed-)surface is then simply computed by integrating the covariant surface element definition $\sqrt{q}\,\mathrm{d}\theta\,\mathrm{d}\tilde{\phi}$ over $\mathscr{U}$, with $q \equiv \det(q_{ab})$ the determinant of the induced metric on $\mathscr{U}$. Since $q = (\hbar^2 + a^2)^2\sin^2\theta$, as can be read off Eq. (4.6) this integration is straightforward, and the area of the Kerr BH is found to be

$$A = 8\pi M \hbar\,. \tag{4.7}$$

We note that this area is defined for an isolated, Kerr BH. In the case where the BH presents disconnected components, such as a binary BH, the cross section $\mathscr{U}$ becomes time-dependent. In particular, it presents two disconnected components that "touch" at the time of merger, as is nicely represented in the figures of [443].

First Law

Recalling that $S = Ma$ and viewing the area A as a function of the two black hole parameters (M, S), it is straightforward to compute the partial derivatives $(\partial_M A, \partial_S A)$ from Eqs. (4.7) and (4.2). One can then compute first-order variation δA in terms of δS and δM. By combining this with the definitions (4.3) of angular velocity and (4.5) of surface gravity, the following remarkable result emerges

$$\delta M - \Omega_{\mathscr{H}}\,\delta S = \frac{\kappa}{8\pi}\delta A\,. \tag{4.8}$$

This computation was first presented in the pioneering paper by Bardeen, Carter and Hawking [444]. Surely, it looks like the classical first law of thermodynamics $\delta U + P\delta V = T\delta S$. This formal analogy was extended to a rigorous, thermodynamical interpretation by Hawking [445], who managed to perform a semi-classical calculation, i.e., account for quantum effects in a curved spacetime. In particular, he showed that BHs not only absorb but also radiate energy at a finite temperature given by $\hbar\kappa/2\pi$, where $\hbar$ is the reduced Planck constant. It then follows from the first law (4.8) that the BH entropy is simply given by $A/4\hbar$, proportional to the BH's area, a result proposed by Bekenstein [446] shortly before Hawking's paper. This groundbreaking discovery was, at once, the first milestone in the ongoing search for a self-consistent theory of quantum gravity, and the first "first law of mechanics" to be derived in the context of compact objects. The reader is invited to browse Robert Wald's review [447] for more on the thermodynamics of BHs.

Other Laws

The analogy between the geometric quantities of a Kerr BH and a thermodynamical processes does not stop at the first law (4.8). Already from Eq. (4.5), we see that the surface gravity of the Kerr BH is constant. In particular, it is defined everywhere on $\mathscr{H}$, and is therefore a constant scalar field there, in spite of the Kerr non-spherical symmetry. In fact, this fundamental result

$$\kappa = \text{cst.}\,, \tag{4.9}$$

has been shown to hold for any Killing horizon. In view of Hawking's result that κ relates to the BH temperature, this result has been coined the "zeroth law" of black hole mechanics.[2] Interestingly, this law holds irrespective of the Einstein equation, and thus is not restricted to GR (in fact it holds in virtually all metric theories of gravity [448]).

The second law of thermodynamics, stating that the entropy of an isolated system undergoing cannot decrease, has also been shown to hold for an arbitrary BH spacetime, where it takes the form

$$\delta A \geq 0\,, \tag{4.10}$$

where δA is the change in total area of the BH. The fact that it is the total area is important for the case of a BH spacetime with two disconnected components, i.e., two BHs. Remarkably, this law can be explicitly verified thanks to GW signals detected from BH mergers. For example, by analyzing the data from the first GW event GW150914, the authors of [449] showed that the second law of BH mechanics (4.10) was verified at a confidence level of at least 95%.

[2] "Mechanics" because it holds for any BH spacetime, even for those with disconnected components, for example binary BHs.

4.1.2 Compact Objects Binaries

The first law of BH thermodynamics (4.8) presented above deals with an isolated Kerr BH. In the advent of GW astronomy, it was quite natural to try and extend its validity to compact objects *binaries*, including BHs and NSs.

Two Black Holes

In [382], the authors have explored various extensions of the first law of mechanics. In particular, they showed that the case a BH binary system on a circular orbit also satisfies such a first law. To model the circular orbit, they use the HKV k^a defined in Chap. 3. Although their result is obtained after a considerable calculation unrelated to the simple, algebraic derivation of (4.8), the first law for a BH binary turns out to be very simple: it is the sum of the contributions (4.8) of two isolated BHs. In particular, the Noether charge Q associated to the helical isometry satisfies

$$\delta Q = \frac{\kappa_1}{8\pi}\delta A_1 + \frac{\kappa_2}{8\pi}\delta A_2\,, \tag{4.11}$$

where κ_i and A_i is the horizon surface gravity and area of each BH of the binary $\mathrm{i} \in \{1, 2\}$. To derive this law for a binary, the authors first show that the zeroth law holds also for this system, extending the isolated-BH case (4.9). In particular, the HKV $k^a = t^a + \Omega\phi^a$ that defines the helical isometry is shown to be tangent to the null generators of the horizon[3] of the black hole. The surface gravity κ, defined as

$$k^a \nabla_a k^b|_{\mathscr{H}} = \kappa k^b\,, \tag{4.12}$$

is then constant on $\mathscr{H}$. The area A, of the BH is defined in the same way as an isolated Kerr BH, as the area of an arbitrary hypersurface $\mathscr{S}$ transverse to k^a.

BH and NS

A very natural extension of the binary first law (4.11) is to account for neutron stars as well as black holes, in order to describe any kind of compact binary system. This was first achieved in [382] by John Friedmann, Kōji Uryū and Masaru Shibata. There, a binary system of BH and/or NS is represented as a helically symmetric spacetime, as defined in Sect. 3.1, with vacuum regions surrounded by a Killing horizon for the BHs, and balls of perfect fluid for the NS. The SEM tensor of the latter reads

$$T^{ab} = \varepsilon u^a u^b + p h^{ab} \tag{4.13}$$

[3] If there are two BHs, then this horizon $\mathscr{H}$ has two disconnected components $(\mathscr{H}_1, \mathscr{H}_2)$ and k^a, when evaluated at one component $\mathscr{H}_\mathrm{i}$ is tangent to its respective generators.

where ε and p are the perfect fluid's energy density and pressure, and $h^{ab} \equiv g^{ab} + u^a u^b$ projects orthogonally to the four-velocity field u^a. As a perfect fluid, the NS satisfies the usual first law of thermodynamics,

$$\Delta\varepsilon = h\Delta\rho + T\rho\Delta s \tag{4.14}$$

which relates the Lagrangian variations Δ of the energy density ε, the mass density ρ and entropy s per unit baryon mass. Equation (4.14) also involves the temperature T and the specific enthalpy $h \equiv (\varepsilon + p)/\rho$ of the fluid. An explicit calculation then shows that the Noether charge Q associated to the helical isometry satisfies

$$\delta Q = \frac{\kappa}{8\pi}\delta A + \int_{\mathscr{S}} \left[\bar{T}\Delta(s\mathrm{d}M_B) + \bar{\mu}\Delta(\mathrm{d}M_B) + v^\alpha \Delta(hu_\alpha \mathrm{d}M_B) \right]. \tag{4.15}$$

In equation (4.15), the first term on the RHS is the contribution from the BH, as in (4.8) for an isolated Kerr BH. The contribution from the NS involves various thermodynamics quantities associated to a perfect fluid. In particular, the (redshifted) temperature $\bar{T} \equiv T/u^t$ and chemical potential $\bar{\mu} \equiv \mu/u^t$, where u^t is the time-component of the velocity field and m_B is the baryon average mass. Similarly, $\mathrm{d}M_B \equiv \rho u^\alpha \mathrm{d}S_\alpha$ is the baryon mass of a fluid element. Finally, the Lagrangian variation Δ and Eulerian one δ are linked by $\Delta = \delta + \mathcal{L}_\xi$, where ξ^a is a displacement field between any two fluid elements. The first law (4.15) is the most general one. In practice (i.e., in numerical simulations), one restricts to perfect fluids with certain hydrodynamic configurations. One that simplifies the first law greatly is that of an isentropic fluid. It implies additional (Lagrangian) conservation laws for the mass current, the entropy and the vorticity

$$\Delta(\rho u^a \sqrt{-g}) = 0\,, \quad \Delta s = 0 \;\text{ and }\; \Delta\omega_{ab} = 0\,, \tag{4.16}$$

with $\omega_{ab} \equiv 2\nabla_{[a}(hu_{b]})$ (the fluid's vorticity 2-form). Each conservation equation in (4.16) implies the vanishing of the corresponding term[4] in the integral of (4.15). Consequently, for such isentropic, perfect fluids, the first law is identical to that of an isolated black hole (4.8) (up to the interpretation of the mass and angular momentum involved in Q). The first law (4.15), which relates the helical Noether charge variations to changes in the thermodynamic and hydrodynamic equilibrium of matter and in the area of the horizon, has been used in the numerical simulation of binary inspirals involving neutrons stars [450–452] and construction of (numerical) initial data of such systems [379, 453–455] (see also the series of papers [456–458]).

---

[4] In the case of a nonisentropic, but otherwise irrotational flow (vanishing vorticity) only the third term vanishes. Other fluid configurations are discussed in [382].

Magnetized NS

As emphasized in Chap. 1 of this thesis, one of the main features of neutron stars is their intense magnetization. To aim at a more realistic model for NS/BH binaries using a sequence of equilibria, it is therefore important to understand how this magnetization enters the first law. This was done in [459] using a similar approach to the above law, except that an additional term was added to the SEM tensor of the fluid to account for an electromagnetic field. In particular, the SEM tensor now reads $T^{ab}_{\rm pf} + T^{ab}_{\rm em}$, where $T^{ab}_{\rm pf}$ is given in (4.13) and

$$T^{ab}_{\rm em} \equiv \frac{1}{4\pi}\left(F^{ac}F^{b}{}_{c} - \frac{1}{4}g^{ab}F_{cd}F^{cd}\right) \tag{4.17}$$

is the SEM tensor of an EM field, with the Faraday tensor $F_{ab} \equiv 2\nabla_{[a}A_{b]}$ defined in terms of the EM potential 1-form A_b (see for example [22] for details on the relativistic EM field). The spacetime is still assumed to be helical, and the Noether charge Q associated to the HKV satisfies

$$\delta Q = \frac{\kappa}{8\pi}\delta A + \Phi\delta Q_E + I_{\rm pf} - \int_{\mathscr{S}} \left[\, A_b k^b \Delta(j^\alpha \mathrm{d}S_\alpha) + 2 j^{[\alpha}k^{\beta]}\Delta(A_\beta \mathrm{d}S_\alpha)\,\right], \tag{4.18}$$

where now the black hole's contribution is not limited to the surface gravity term $\propto \delta A$ but also includes an EM part, with Q_E and Φ standing for the electric charge and scalar potential on the horizon. This term appears because $T^{ab}_{\rm em}$ sources the Einstein equation and therefore the BH is not Kerr but rather Kerr-Newman. Similarly, for the NS two terms are present: $I_{\rm pf}$, which stands for the integral coming from the $T^{ab}_{\rm pf}$-part of the source, is identical to that of (4.15), whereas the integral term in (4.18) involves the electric current j^a and comes from $T^{ab}_{\rm em}$-part of the total SEM tensor. The NS being now modeled by magnetohydrodynamics (MHD), other types of assumptions can be made as to the physical properties of its fluid flow. In particular, the ideal MHD assumption, $F_{ab}u^a = 0$, readily implies a conservation law for the magnetic flux, which, in turn, implies that the second term in the integral (4.18) vanishes. Similarly, the first term in that integral vanishes under the assumption of conserved circulation of the magnetic flow. Other types of equilibrium configurations lead to a simplification of the general first law (4.18) and are discussed thoroughly in [459]. These MHD results (and the formalism around it) allows for a self-consistent inclusion of magnetic fields in the construction of quasi-equilibrium sequences for binary system inspirals [460, 461].

4.1.3 Point Particles Binaries

In the previous section, we saw how a first law of mechanics for binary systems can be derived for extended compact objects, including BHs, and Ns as perfect fluids. But as we have seen in Chap. 3, real compact objects can also be modeled accurately by extracting from them a small number of multipoles, evolving along a single worldline. The post-Newtonian formalism has shown to be a useful method of obtaining these alternative first laws. Here we discuss two such laws, one for point masses at the 3PN level, and an extended version which accounts for the nonlocal-in-time part of the dynamics.

Circular Orbits

In 2012, it was shown by Alexandre Le Tiec, Luc Blanchet, and Bernard Whiting in [387] that a first law of binary mechanics can be obtained for such a system of pointlike objects. This discovery could almost be said "serendipitous" as they were able to guess the form of the first law from a careful examination of the PN expansions of the physical quantities that enter it. In particular, they considered the 3PN expressions[5] for the binding energy E, angular momentum J and redshifts z_{i}. For circular orbits, these 3PN expressions look like

$$E = -\frac{m_1 m_2 x}{2m}\left(1 + \sum_{k=1}^{3} E_k(m_{\mathrm{i}}) x^k + o(x^3)\right), \tag{4.19a}$$

$$J = \frac{G m_1 m_2}{\sqrt{x}}\left(1 + \sum_{k=1}^{3} J_k(m_{\mathrm{i}}) x^k + o(x^3)\right), \tag{4.19b}$$

$$z_{\mathrm{i}} = 1 + \sum_{k=1}^{4} \zeta_k(m_{\mathrm{i}}) x^k + o(x^4), \tag{4.19c}$$

and z_2 is obtained from z_1 by exchanging m_1 with m_2 and $m \equiv m_1 + m_2$. Equation (4.19) are PN expansions in terms of the 1PN parameter $x \equiv (m\Omega)^{2/3}$, which is well-adapted to circular orbits. The coefficients (E_k), (J_k) and (ζ_k) are rational functions of (m_1, m_2) (see Eqs. (2.35)–(2.37) in [387] for their exact expressions). By inspecting Eqs. (4.19), the authors remark that three PDE can be written relating the quantities $(M \equiv E + m, J, z_1, z_2)$, seen as functions of (Ω, m_1, m_2). These are

$$\frac{\partial M}{\partial \Omega} - \Omega \frac{\partial J}{\partial \Omega} = 0 \quad \text{and} \quad \frac{\partial M}{\partial m_{\mathrm{i}}} - \Omega \frac{\partial J}{\partial m_{\mathrm{i}}} = z_{\mathrm{i}}\,, \tag{4.20}$$

for $\mathrm{i} \in \{1; 2\}$. These PDE's can all be grouped in one single differential expression, by writing the differentials δM, δJ, δm_{i} and reading their coefficients from (4.20). This provided the *first law of point-particle binary mechanics*:

[5] Their work also includes of logarithmic contributions of 4PN and 5PN orders, which we omit here for clarity, but which are also checked to verify the first law in their paper.

$$\delta M - \Omega\,\delta J = \sum_{\rm i} z_{\rm i}\delta m_{\rm i}\,. \tag{4.21}$$

This law is obtained by solving the Einstein equation iteratively in powers of c^{-1}, as they are sourced by two point masses. In particular, the SEM tensor of the Einstein equation is precisely the monopolar particle (2.22). Several algebraic relations can be derived from the differential expression (4.21), and these are discussed in the paper. Moreover, the comparison between the variation δQ of the Noether charge associated to the helical isometry and the quantity $\delta M - \Omega\delta J$ appearing on the left-hand side of (4.21) is discussed, reaching the same conclusions as those presented in Chap. 3. The PN first laws also have an important application: as they are expected to hold at all PN order, they can be used to constrain (and sometimes explicitly compute) PN coefficients that would otherwise be very calculation-heavy. We shall discuss this further in Sect. 4.2.

Eccentric Orbits

The first laws presented above, be it for extended compact objects or their point-particle approximation, are restricted to circular orbits. In order to extend the first law to generic bound orbits (eccentric, for short), the natural framework has turned out to be the canonical, ADM formalism of GR. Applied to binary system of point masses (m_1, m_2), it has been possible to derive up to 3PN order[6] an autonomous Hamiltonian $H(\mathbf{x}, \mathbf{p}; m_{\rm i})$ that encodes the 3PN dynamics of the system. The phase space coordinates $(\mathbf{x}, \mathbf{p})$ are the canonical position and momenta of the particles (3-vectors), defined in some fixed gauge, and they obey Hamilton's equations

$$\frac{\mathrm{d}\mathbf{x}}{\mathrm{d}t} = \frac{\partial H}{\partial \mathbf{p}} \quad \text{and} \quad \frac{\mathrm{d}\mathbf{p}}{\mathrm{d}t} = -\frac{\partial H}{\partial \mathbf{x}}\,. \tag{4.22}$$

Even though this framework is perfectly relativistic, it inherits many results from Newtonian gravity. In particular, it is possible to rigorously and unambiguously define a conserved, total linear momentum $\mathbf{P}$ and a conserved, total angular momentum $\mathbf{L}$. In the "center-of-mass" frame (in which $\mathbf{P} = \mathbf{0}$) the conservation of $\mathbf{L}$ defines a 2-dimensional orbital plane. Introducing polar coordinates (r, ϕ) in this plane, the 3PN ADM Hamiltonian is shown to only depend on $(r, p_r, p_\phi; m_{\rm i})$, where (p_r, p_ϕ) denote the conjugated momenta associated to (r, ϕ). Again, we emphasize that all these familiar, Newtonian-looking results are not straightforward and require an elaborate construction.

In this context, the first law of mechanics follows from a series of calculations based, in essence, on a (orbital-averaged) variation of the Hamiltonian $H(r, p_r, p_\phi; m_{\rm i})$ with respect to the phase-space coordinates and the individual

---

[6] We limit the present discussion to 3PN, even though this Hamiltonian is known at 4PN [462, 463]. However, at that level it brings new, non-local effects, which are precisely the content of the first law derived in the next paragraph.

masses m_i. After performing these steps (detailed in [426]), the first law of mechanics for binaries on an eccentric orbit follows from the on-shell condition $H = M$, stating that the numerical value of the Hamiltonian along any solution to Eq. (4.22) coincides with the invariant, ADM mass of the underlying spacetime [464]. The result for the first law reads [426]

$$\delta M - \Omega_\phi \delta L = \Omega_r \delta J_r + \sum_i \langle z_i \rangle \delta m_i \,, \tag{4.23}$$

where (Ω_r, Ω_ϕ) are the radial and angular Hamiltonian frequencies, J_r is the radial action, the integral invariant associated to the radial motion,[7] just like $p_\phi = L$ is that of the angular motion. Lastly, in the ADM context, the redshift z_i that enters Eq. (4.23) is defined by

$$\frac{\partial H}{\partial m_i} = \frac{d\tau_i}{dt} \equiv z_i \,, \tag{4.24}$$

where the partial derivatives is computed while keeping the phase-space variables fixed, and τ_i is the proper time along the worldline of particle i. Again, it should be emphasized that Eq. (4.24), in spite of its simplicity, is a non-trivial result of relativistic mechanics, as explained in [406]. The law for eccentric orbits (4.23) is compatible with that for circular orbits (4.21). Indeed, in the circular case, the radial action J_r vanishes (by definition), the polar angular frequency Ω_ϕ becomes that of the circular motion Ω, and finally the averaged redshift coincides with the constant redshift of the circular case.

Non-local Effects

The first law of mechanics (4.23), for eccentric orbits, was based on a calculation involving an autonomous Hamiltonian H expressed in terms of canonical coordinates $(\mathbf{x}, \mathbf{p})$. As of today, the full expression of this Hamiltonian is known at full 4PN order, and contrasts with its 3PN expansion in one key aspect: its *nonlocal*-in-time nature. Following the notations of [427] (see Sect. II for details), at 4PN order the Hamiltonian H splits into two pieces, as

$$H_{4PN} \equiv H_0(r, p_r, p_\phi; m_i) + H_{tail}[r, \phi, p_r, p_\phi; m_i] \,. \tag{4.25}$$

In this equation, H_0 denotes a "traditional" Hamiltonian, that depends algebraically on the phase space coordinates. In particular, H_0 at a given instant t is known from the value of (r, p_r, p_ϕ) at this very instant t. In contrast, H_{tail} is a non-local-in-time piece, which is a functional of the phase space coordinates (whence the bracket notation). In particular, the value of H_{tail} at a given time depends on the whole past

[7] This action is defined in the same way in Newtonian mechanics, and we shall use it extensively in the second part of this thesis, in particular in Chap. 10.

and future evolution of (r, ϕ, p_r, p_ϕ), as shown by its explicit, integral expression, provided in Eqs. (2.2)–(2.4) in [427]. This functional dependence of the Hamiltonian implies that in the Hamilton equations (4.22), the right-hand sides must be replaced by functional derivatives,

$$\frac{\mathrm{d}\mathbf{x}}{\mathrm{d}t} = \frac{\delta H}{\delta \mathbf{p}(t)} \quad \text{and} \quad \frac{\mathrm{d}\mathbf{p}}{\mathrm{d}t} = -\frac{\delta H}{\delta \mathbf{x}(t)}, \tag{4.26}$$

making it a system of integro-differential equations, and contrasting with the traditional PDE system that would be obtained with the local Hamiltonian H_0 alone.

It should be emphasized that this peculiarity arising at 4PN in binary mechanics is a feature of the conservative part of the dynamics, even though its physical interpretation does have something to do with GWs. Heuristically, the dependence of the body's motion on its own past and future can be seen as a consequence of the interaction between that body and the GWs it emitted in the past (and future): it is an effect of self-interaction, stemming from the nonlinearity of GR. Accordingly, this effect his known under the name of *gravitational-wave-tail*. In [427], the authors showed that the 3PN "eccentric" first law (4.23) was still valid at 4PN order for generic bound orbits, as long as these GW-tail effects are re-absorbed into the definition of the radial action R that enters it. This result is of importance as it shows, at once, that the first law is not limited to instantaneous, local-in-time effects, and does holds at 4PN order, which is the current state-of-the-art result (cf. Sect. 1.4 of [145]) for the motion of binary mechanics.

Spinning Particles

The effect of spin (i.e., the proper rotation of the body) on the dynamics of binary systems is known today with a satisfying precision. For example, in the PN formalism, spin-orbit couplings are known to high PN orders [145] and leading quadratic- [370] and cubic-in-spin [320] have also been calculated explicitly. We refer to Sylvain Marsat's thesis [465] for the effects of spin in compact objects binary systems.

The contributions of spin in the first law of mechanics have been first derived by Luc Blanchet, Alessandra Buonanno and Alexandre Le Tiec in [406]. They derived a linear-in-spin result that applies to circular orbits using the Hamiltonian formulation of the problem of motion for a binary system of spinning objects. More specifically, the starting point is the Lagrangian (2.13) of a massive, spinning particle, identical from the one presented in the paragraph on the Lagrangian formalism in Chap. 2, Sect. 2.2.1. From this Lagrangian they construct an ADM Hamiltonian that now depends on the canonical variables $(\mathbf{x}, \mathbf{p}, \mathbf{S})$, with the vector $\mathbf{S}$ encoding the spinning degrees of freedom of the particle. After carefully constructing all the necessary tools, they eventually obtain the following first law

$$\delta M - \Omega\, \delta J = \sum_{\mathrm{i}} z_{\mathrm{i}} \delta m_{\mathrm{i}} + (\Omega_{\mathrm{i}} - \Omega)\, \delta S_{\mathrm{i}} + O(S^2)\,, \tag{4.27}$$

where Ω_i is the spin precession frequency of the i-th particle, J is the total angular momentum of the system (containing both the orbital contribution L and the spins S_i), and Ω is the binary's angular frequency. In general, the precession frequencies Ω_i depend on the spin themselves, and the first law (4.27) only holds at linear order in the spin (whence the $O(S^2)$). It obviously reduces to the non-spinning case (4.21) when $S_i = 0$. In particular, the PDE's (4.20) derived from the non spinning first law are still valid, but two additional ones can be obtained, namely

$$\frac{\partial M}{\partial S_i} - \Omega \frac{\partial J}{\partial S_i} = \Omega - \Omega_i \,. \tag{4.28}$$

It should be noted that to derive this law, the Tulczyjew SSC $p_a S^{ab} = 0$ was chosen. However, as we saw in the Chap. 2, at linear order this is equivalent to the Pirani SSC $u_a S^{ab} = 0$, since $p^a = m u^a + O(S^2)$. We shall compare this linear-in-spin first law to our results in Chap. 5, where we derive an SSC-free first law that holds at all orders in the spin, for a dipolar particle. Finally, we note that the first law with spin presented above, valid for circular orbits, was recently generalized to eccentric orbits in [466].

4.1.4 Perturbed Systems

The derivations of the first laws of mechanics presented above all exploit, in some sense, a symmetry of the system: the helical isometry for circular-orbit binaries, or the Poincaré symmetries associated to the ADM Hamiltonian formulation of GR (see, e.g., [464]). If the mass-ratio of the binary system is small, then the lightest of the bodies can be described as moving in the background of a (Schwarzschild or Kerr) black hole, akin to the GSF framework. Again, the isometries of these metrics allow a derivation of the first laws, in a perturbative regime. We provide three examples of such derivations.

BH with a Moon

In the case of Kerr BH of mass M and spin S perturbed by an orbiting, small body, or *moon*, of mass m, Sam Gralla and Alexandre Le Tiec were able to derive in [388] a zeroth and first law of mechanics, generalizing the classical result for an isolated BH. One of the key features of their analysis is the observation that under the assumption that the moon is corotating with the BH, the helical isometry of the isolated Kerr BH is preserved (recall the Killing field for the Kerr metric in Eq. (4.3)). By "corotating", it is meant that the moon's (orbital) angular velocity coincides with that of the (unperturbed) event horizon of the BH. Under this hypothesis, they showed that the BH's perturbed horizon remains a Killing horizon, and therefore that its surface gravity is constant over it (see Sect. III in [388]). The constant value of this surface gravity is shifted with respect to the exact Kerr expression (4.5), by a small amount $c\,(m/M^2)$, where $-\sqrt{3}/2 \leq c \leq 0$ depends on the background BH parameters, thereby "cooling" the BH (recall that κ is proportional to the BH's Hawking temperature).

In addition to this zeroth law, using a calculation similar to that presented in the next chapter, the authors derive a first law of mechanics for the (Kerr BH + moon) system, valid at linear order in the mass ratio, in the form

$$\delta M_{\flat} - \tilde{\Omega}\,\delta J_{\flat} = \frac{\tilde{\kappa}}{8\pi}\delta\tilde{A} + z\delta m \tag{4.29}$$

where $M_{\flat}$, $J_{\flat}$ are the Bondi mass and angular momentum of the spacetime, $\tilde{\Omega}$, $\tilde{\kappa}$, $\tilde{A}$ are the angular frequency, surface gravity and area of the *perturbed* BH respectively, and z, m are the moon's redshift and mass. Remarkably, this first law can be derived geometrically, without explicitly knowing how the perturbations $(\tilde{\Omega}, \tilde{\kappa}, \tilde{A})$ relate to the background Kerr values (Ω, κ, A). It turns out, as the authors show (in Sect. V of [388]), that these perturbation can all be expressed simply in terms of the Kerr parameters and the Hamiltonian of the moon $\mathcal{H}$ described in the test-particle approximation (i.e., the Hamiltonian generating geodesic motion in Kerr).

Test Particle

As mentioned many times by now, in the test-particle approximation a small body of mass m orbiting a Kerr BH of parameters (M, S) follows a geodesic γ of the Kerr spacetime. This particle is characterized by its four-momentum vector $p^a \equiv m u^a$, where u^a is its velocity. The motion of this particle can be encoded in a Hamiltonian $\mathcal{H}$ given by

$$\mathcal{H}(z, p) = \frac{1}{2}\mathring{g}_{\alpha\beta}(z)p^{\alpha}p^{\beta}\,, \tag{4.30}$$

where z^α is the phase-space coordinate of the system and $\mathring{g}_{\alpha\beta}$ are the components of the Kerr metric $\mathring{g}_{ab}$. This Hamiltonian is related to the rest mass of the particle, since, *on shell*, i.e., when evaluating (4.30) anywhere on γ, we have $\mathcal{H} = -m^2/2$. The Hamilton equation then relate the position z^α and the components p^α through

$$\frac{\mathrm{d}z^{\alpha}}{\mathrm{d}\lambda} = \frac{\partial\mathcal{H}}{\partial p_{\alpha}} \quad \text{and} \quad \frac{\mathrm{d}p_{\alpha}}{\mathrm{d}\lambda} = -\frac{\partial\mathcal{H}}{\partial z^{\alpha}}\,, \tag{4.31}$$

using $\lambda \equiv \tau/m$ as an affine parameter along γ. In [467], the author used this basic setup to derive a first law of mechanics for this particular system, namely a test-particle orbiting a Kerr BH. The simplest way to derive such a first law is to exploit the isometries of the Kerr spacetime, as follows. Along with the Hamiltonian $\mathcal{H}$, two other constants of motion can be found by contracting the four-momentum p^a with the two Killing vectors of the Kerr metric (recall the discussion above Eq. (4.3)). These are the energy $\mathcal{E}$ and angular momentum $\mathcal{J}$ of the particle. Since the Hamiltonian system (4.31) is 8-dimensional, these three constants of motion $(\mathcal{H}, \mathcal{E}, \mathcal{J})$ are not sufficient to make that system *integrable*. However, as shown by Brandon Carter in his formidable paper [468], there exists a fourth constant of motion $\mathcal{K}$ (following Carter's notation), which is associated to a Killing rank-2 tensor K_{ab}, and such that $\mathcal{K} \equiv K_{ab} p^a p^b$ is conserved. These four constants of motion, or any four functions thereof, can be promoted to the level of canonical variables, as part of a system of angle-action coordinates $(J_t, J_r, J_\theta, J_\phi)$.[8] The first law of mechanics then reads, in terms of these actions

$$\delta\mathcal{E} - \Omega_\phi\, \delta\mathcal{L} = \Omega_r \delta J_r + \Omega_\theta J_\theta + z\delta m \tag{4.32}$$

where Ω_j are the Hamiltonian frequencies associated to the tri-periodic motion the Kerr BH, z is the redshift of the particle (linked to the Hamiltonian frequency for in the "time" direction), and the t-and ϕ-actions J_t, J_ϕ are nothing but $\mathcal{E}$, $\mathcal{L}$. This particular form of the first law, although only valid for a test-particle around a Kerr BH, looks the same as that of eccentric orbits (4.23) or spinning particles (4.27) when taken in the appropriate limit, as discussed in [467] (see also Sect. 3.3.4 of [469]).

Self-force

As a final example of first law of mechanics, let us now consider extending the previous test-particle result to a first-order GSF calculation, taking into account the effect of the small body on its own motion. This was done in [407], with a calculation that we now outline. As we have seen in Chap. 2, Sect. 2.2.4, even at first order the worldline of the particle can be considered a geodesic, provided that the Kerr metric $\mathring{g}_{ab}$ is replaced by the effective, regular metric $\mathring{g}_{ab} + h^R_{ab}$, with h^R_{ab} the regular piece of the metric perturbation h_{ab} (cf. the discussion in Sect. 2.2.4). Consequently, the Hamiltonian of the particle that accounts for the first-order perturbation is simply

$$\mathcal{H}(z, p; \gamma) = \frac{1}{2}\big(\mathring{g}_{\alpha\beta}(z) + h^R_{\alpha\beta}(z; \gamma)\big)\, p^\alpha p^\beta \,, \tag{4.33}$$

where the dependence on γ is here to remind the reader that the perturbation h_{ab} is worldline-dependent. In particular, Eq. (4.33) is not simply the Hamiltonian (4.30) to which we simply add $h^R_{\alpha\beta} p^\alpha p^\beta$, in particular because the four-momentum is now defined in the effective spacetime, and not the background, as in (4.30). As a consequence of the perturbation h^R_{ab}, the background actions $(\mathcal{E}, J_r, J_\theta, \mathcal{L})$ are not constants anymore in the real spacetime, but undergo slow oscillations (as expected for a perturbed Hamiltonian system), compared to the orbital period timescale. However, the authors show in [407] that it is possible to find a gauge in which these oscillations

[8] These are nothing but the Poincaré invariants, used already in the ADM context, for the Hamiltonian first laws. They will also be used in the second part of this thesis, cf. Chap. 10. However, see [378] for details on such a construction of actions in a relativistic context.

vanish, and thus that $(\mathcal{E}, J_r, J_\theta, \mathcal{L})$ can be promoted to action-variables even in the real spacetime. From there, an effective Hamiltonian can be defined, and calculations similar to those performed in the exact Kerr case lead to the first law given by Eq. (4.34), albeit with the replacement rule

$$J_\alpha \to J_\alpha \left(1 + \tfrac{1}{4}\langle h^{R}_{\alpha\beta} u^\alpha u^\beta \rangle\right) \quad \text{and} \quad z \to z\left(1 - \tfrac{1}{2}\langle h^{R}_{\alpha\beta} u^\alpha u^\beta \rangle\right) \tag{4.34}$$

with $\langle\cdot\rangle$ denoting an infinite long time average, and we recall that $J_\alpha = (\mathcal{E}, J_r, J_\theta, \mathcal{L})$. Evidently, this form of the first law is compatible with previously established results and shows, once again, the great variety of contexts to which it applies.

4.2 Details and Applications

Having reviewed the literature on the first laws of mechanics, we now turn to their applications. Most of these are based on the fact that the first law of mechanics is equivalent to a set of PDE's, as presented in Eq. (4.20). First, we start with two general comments on the interpretation of the various first laws in Sect. 4.2.1. Then, a detailed example of and application of Eq. (4.21) in Sect. 4.2.2, while a number of other applications are succinctly reviewed in Sect. 4.2.3.

4.2.1 *How to Interpret the First Law?*

All the first laws presented above have in common one general feature: through a variational identity, they relate *global, binary quantities*, e.g., binding energy, total angular momentum, orbital frequency; to the *intrinsic properties of the individual bodies*, e.g., masses, spins, redshifts, etc. However, in spite of their common general structure, there may be subtle, yet important, differences between them. We clarify the definition (and interpretation) of the δ-variation used in the formulation of the first laws, and make a comment on the global quantities that appear on their left-hand side.

What is δ?

The δ-variation that appears in the first laws of mechanics is defined as the first-order variation of the quantity it is adjoined to, with respect to a parameter indexing a family of different spacetimes. More precisely, let $(\mathscr{E}_\lambda, g_{ab}(\lambda))$ be a family of spacetimes indexed by a real, continuous parameter λ. The δ-variation is defined by (following [382, 470])

$$\delta g_{ab} \equiv \left.\frac{\mathrm{d}g_{ab}}{\mathrm{d}\lambda}\right|_{\lambda=0}, \tag{4.35}$$

where the derivative is computed while fixing any other fields present in the spacetime. The δ-variation of other quantities of interest in spacetime are defined similarly. Equivalently, δg_{ab} is simply the first term in the Taylor expansion $g_{ab}(\lambda) = g_{ab}(0) + \lambda \delta g_{ab} + o(\lambda)$, with respect to a given, arbitrary, reference spacetime $(\mathscr{E}_0, g_{ab}(0))$. This definition implies that the first laws relate the variation of quantities defined in *two different*, but arbitrarily λ-close, spacetimes.

It is also possible to interpret the first law as describing a *physical process*. However, one would generally lose information along the way. Consider, for example, the first law (4.21), $\delta M - \Omega \delta J = \sum_{\mathrm{i}} z_{\mathrm{i}} \delta m_{\mathrm{i}}$. One could argue that, for a typical binary system, the masses of the individual bodies are constant, and thus readily discard the δm_{i} contribution to the first law. This would be perfectly reasonable, and, in fact, the resulting equation $\delta M = \Omega \delta J$ was well-known long before (4.21), and has already been used to numerically evolve quasi-equilibrium sequences of corotating BH binaries in [380, 399]. However, we stress that imposing $\delta m_{\mathrm{i}} = 0$ is generally an early, simplifying assumption that prevents further insight on the problem at hand. As we shall see, even if the masses of the individual bodies do remain constant physically, the contribution δm_{i} to the first law is where most of its success comes from.

The Global Quantities

The global quantities involved in the left-hand side of the first law of mechanics always include (1) a notion of mass or energy and (2) a notion of angular momentum. Because of the many different definitions of mass, energy and angular momentum in GR, this can be confusing, especially to compare two different first laws, or even to apply them in a given context. Already in the first law presented above, several notions of mass have been involved: ADM mass, Bondi mass, binding energy, on-shell value of some Hamiltonian, etc. Our goal here is not to give a detailed review of how these quantities are defined and how they differ. Rather, we want to point the reader to valuable information on this topic. In particular: reference [471], which provides a detailed overview of the different global notions of mass, energy and angular momentum; Sect. II.E of [387] which discusses thoroughly the equality between these masses in the PN context, and reference [472] which discusses and defines conserved, global quantities in the context of diffeomorphism covariant, Lagrangian theories (such as GR).

In the present work, the combination of an approximation scheme and the helical isometry provides, just as in [387], no ambiguity as to the notion of mass and angular momentum used in the first law. In particular, it is always possible to write the left-hand side of the first law as δQ, as we shall do in Sect. 4.3 below, where Q is the unambiguously well-defined, conserved Noether charge associated to the helical isometry introduced in Chap. 3. Expressing δQ into physical quantities such as the mass and angular momentum is then another problem in itself, which is discussed at the end of Sect. 4.3.

4.2.2 A Detailed Example

Let us start with an application from the first law of mechanics at monopolar order for a binary of point particles, on a circular orbit. As we have seen previously, the first law, in this case, reads

$$\delta M - \Omega\,\delta J = z_1 \delta m_1 + z_2 \delta m_2\,, \tag{4.36}$$

where M and J are the ADM mass and the total angular momentum, both defined as integrals at spatial infinity. Physically, these two quantities, as well as the redshifts z_i of the particles, depend on the orbital frequency Ω of the binary, and on the masses m_i of the particles. In what follows, it will be more convenient to work with the "variables" (m, ν, x) instead of (Ω, m_1, m_2), where $m \equiv m_1 + m_2$ is the total mass, $\nu \equiv m_1 m_2 / m^2$ is the symmetric mass ratio and $x \equiv (m\Omega)^{2/3}$ is the usual dimensionless, invariant PN parameter adapted to circular orbits. Similarly, instead of the pair (M, J), let us introduce

$$\hat{E}(m,\nu,x) \equiv \frac{M - m}{m\nu} \quad \text{and} \quad \hat{J}(m,\nu,x) \equiv \frac{J}{m\nu^2}\,, \tag{4.37}$$

in which $\hat{E}$ is the specific (dimensionless) binding energy of the system and $\hat{J}$ its dimensionless, total angular momentum. Now consider the same system in the framework of the GSF. For example, let us consider a small NS of mass m_1 orbiting a large, Schwarzschild BH of mass m_2. The mass ratio $q \equiv m_1/m_2$ is small, and so is the symmetric mass ratio ν, since they are related by $\nu = q/(1+q)^2 = q + O(q^2)$. Consequently, in the limit $\nu \to 0$, the system approaches that of a test particle of mass m_1 orbiting the large Schwarzschild BH m_2. Therefore, if we expand the functions $\hat{E}$, $\hat{J}$ and z_i in powers of ν, we should find

- at zeroth order in ν, the contribution from a test-particle following a circular Schwarzschild geodesic (known in closed-form, cf. Sect. 7.3.3 of [82]);
- at first order in ν, the correction induced by the (conservative piece of) the first-order gravitational self-force.

Putting these results together, we find that $\hat{E}$, $\hat{J}$ and $z \equiv z_1$ (the redshift of the lightest body), have the following expansions

$$\hat{E} = -1 + \frac{1-2x}{\sqrt{1-3x}} + \nu \hat{E}_{\rm SF}(x) + O(\nu^2)\,, \tag{4.38a}$$

$$\hat{J} = \frac{1}{\sqrt{x(1-3x)}} + \nu \hat{J}_{\rm SF}(x) + O(\nu^2)\,, \tag{4.38b}$$

$$z = \sqrt{1-3x} + \nu z_{\rm SF}(x) + O(\nu^2)\,, \tag{4.38c}$$

where the leading terms are the classical results for the test-particle limit (namely the circular Schwarzschild geodesic [82]), and $\hat{E}_{\rm SF}$, $\hat{J}_{\rm SF}$, $z_{\rm SF}$ are the unknown, first order self-force corrections. A priori, if one wants to control all the dynamics at linear order in ν, this requires three different calculations in the GSF framework, one for each of correction in $\hat{E}$, $\hat{J}$ and z. But thanks to the first law, these three quantities must satisfy a system of PDE's, namely Eq. (4.20), discussed above. When we insert Eqs. (4.38) into this set of PDE's and expand the result at first order in ν, we find (1) that the leading-orders pieces (test-particle limit) verify the first law identically, and (2) that the first-order corrections for $\hat{E}$ and $\hat{J}$ satisfy

$$\hat{E}_{\rm SF}(x) = \frac{z_{\rm SF}(x)}{2} - \frac{x}{3} z'_{\rm SF}(x) - 1 + \sqrt{1-3x} + \frac{x}{6}\frac{7-24x}{(1-3x)^{3/2}}\,, \tag{4.39a}$$

$$\hat{J}_{\rm SF}(x) = -\frac{z'_{\rm SF}(x)}{3\sqrt{x}} + \frac{1}{6\sqrt{x}}\frac{4-15x}{(1-3x)^{3/2}}\,, \tag{4.39b}$$

where a prime denotes a derivative with respect to x. There are several ways in which this first law result could be used. For example, it could be used to check the consistency of a GSF calculation: if one computes all three of $(\hat{E}, \hat{J}, z)$, then the resulting expression should satisfy Eqs. (4.39). Alternatively, it can be used to infer from z the values of $(\hat{E}, \hat{J})$: as we can see from Eqs. (4.39), the sole knowledge of the first-order correction to the redshift (the function $z_{\rm SF}(x)$, and thus its derivative), immediately fixes the corrections for $\hat{E}$, $\hat{J}$.

The previous calculation was first performed in [408]. At the time of this publication, the authors could find, in the literature, 55 data points for the function $x \mapsto z_{\rm SF}(x)$, obtained from numerical, GSF calculations, with relative errors smaller than 10^{-6}. Thanks to this data, they reconstructed the function $x \mapsto \hat{E}(x)$ and $x \mapsto \hat{J}(x)$, and used them as a parametric representation for the coordinate-invariant relation $\hat{E}(\hat{J})$. This diagram, depicted in Fig. 4.1, allowed to compare their GSF result to the relation $\hat{E}(\hat{J})$ obtained using other methods, namely a 3PN expansion [473], a result from the EOB(3PN) adiabatic model [474], and an exact (up to numerical errors) result from NR [475]. The result showed remarkable agreement between GSF and NR, both in the weak and strong field regime, for all mass ratios considered.

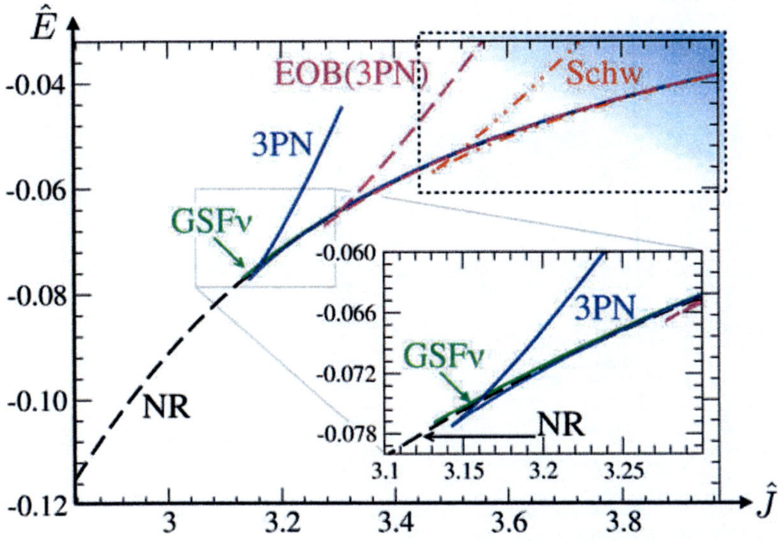

Fig. 4.1 A comparison between different approximation schemes (GSF in green, 3PN in blue, EOB in pink) to the exact result of numerical relativity (NR, in black). This comparison is made for an equal mass, nonspinning black hole binary, by plotting the invariant relation $\hat{E}(\hat{J})$. The GSF result was inferred from the first law of mechanics, as explained in the text. When the GSF is shut down, a zoom of the top-right, blue rectangular section is given in Fig. 4.2. *Source* [408]

4.2.3 Other Applications

The first law of mechanics has been used extensively for a wide variety of problems. We now present a (non-exhaustive) *tour d'horizon* of these applications. Some of them have been done separately by different groups and/or presented in a series of papers: we only refer to a selection of them, and point the reader to [427] for a more exhaustive list references. Lastly, we also mention reference [469] (Sect. 3.4 in particular), which contains an account on most applications of the laws of binary mechanics, with a greater level of details than what is presented below.

ISCO/IBCO Frequency Shift

In the test-particle limit, a small body of mass m orbiting a Schwarzshild BH of mass M on a circular orbit moves on a timelike geodesic. For a given value of the BH mass M, any circular geodesic is characterized by two parameters: the particle's specific angular momentum $\hat{J}_0$ and energy $\hat{E}_0$, associated to the invariance of the Schwarzschild spacetime under spatial rotations and time translations, respectively. Circular orbits are bound orbits, by definition. Therefore, they have negative energy, $\hat{E}_0 \leq 0$. Moreover, these circular orbits only exist if the angular momentum satisfies

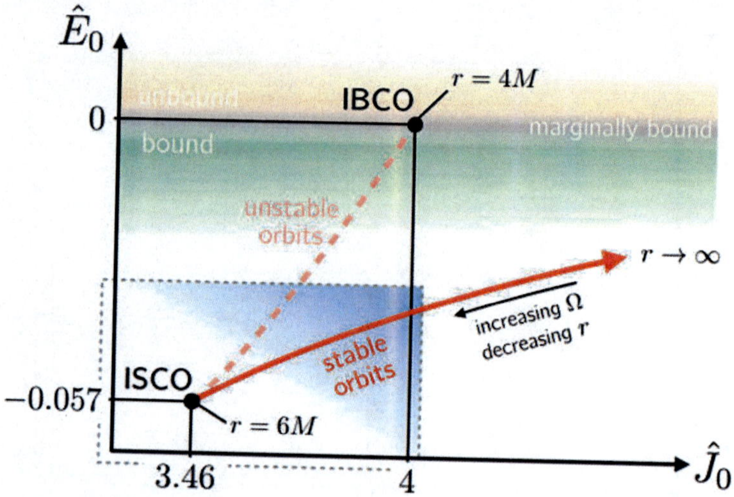

Fig. 4.2 The parameter space $(\hat{E}_0, \hat{J}_0)$ of circular orbits in the Schwarzschild spacetime. Orbits on the upper branch (dashed curve) are unstable, while those on the lower branch are stable. The IBCO is located on the $\hat{E}_0 = 0$ line of marginally bound orbits, while the ISCO is located at the cusp of the two branches, making it marginally stable. All values given in the diagram are classical results obtained by solving the Schwarzshild geodesic equation (see Chap. 7 of [82] for a nice presentation). The rectangular, blue rectangle corresponds to the one in Fig. 4.1, allowing for an easy visualization of the effect of the GSF on this diagram

$\hat{J}_0 \geq \hat{J}_0^{\rm crit}$ where the critical value can be shown to be $\hat{J}_0^{\rm crit} \equiv 2\sqrt{3} \simeq 3.46$ [82]. For a given value of $\hat{J}_0 > \hat{J}_0^{\rm crit}$, there exists exactly two distinct circular orbits: one, closer to the BH, is unstable; while the other one, farther out, is stable. As $\hat{J}_0$ varies between $\hat{J}_0^{\rm crit}$ and $+\infty$, these pairs of circular orbits draw two distinct branches in a $\hat{E}_0$ versus $\hat{J}_0$ diagram, as depicted in Fig. 4.2.

These two branches of circular orbits define two unique and important orbits. The first one, at the cusp point where the branches meet, is where $\hat{J}_0 = \hat{J}_0^{\rm crit}$ and $\hat{E}_0 = 2\sqrt{2}/3 - 1 \simeq -0.057$. This is the so-called *innermost stable circular orbit* (ISCO). It is only marginally stable as it belongs to both branches, and is located at $r = 6M$ in Schwarzschild-Droste coordinates, i.e., three times further than the BH's horizon at $r = 2M$. The other important circular orbits is found where the unstable branch (dashed line in Fig. 4.2) intersects the line $\hat{E}_0 = 0$, where orbits become unbound. This orbit is the IBCO, for *innermost bound circular orbit*. It corresponds to $\hat{J}_0 = 4$, and is located closer to the BH, at $r = 4M$. It can be shown [82] that the ISCO and the IBCO have a uniquely defined, invariant orbital frequency given by

$$\Omega^{(0)}_{\rm ISCO} \equiv \frac{1}{6\sqrt{6}M} \quad \text{and} \quad \Omega^{(0)}_{\rm IBCO} \equiv \frac{1}{8M} \,. \tag{4.40}$$

The above discussion and formulae holds in the exact Schwarzschild spacetime. Beyond the test-particle limit, when GSF effects are taken into account, this picture will change and the ISCO and IBCO orbital frequency will slightly change. For example, under the GSF a particle may venture closer to the BH while still remaining on a stable circular orbit. This orbit, of smaller radius than the (unperturbed) ISCO, is associated with an increased angular velocity. In other words, the GSF causes a *frequency shift* of the ISCO. This shift was first computed numerically in [476] for circular orbits, a result confirmed a few years later by the authors of [408], using a first law calculation similar to that of Sect. 4.2.2. Their result, valid at linear order in the mass ratio $q \equiv m/M$, reads $\Omega_{\rm ISCO} = \Omega^{(0)}_{\rm ISCO}[1 + q\Omega^{(1)}_{\rm ISCO} + O(q^2)]$, with the test-particle contribution $\Omega^{(0)}_{\rm ISCO}$ given in Eq. (4.40), and the first order GSF correction by

$$\Omega^{(1)}_{\rm ISCO} \equiv \frac{1}{2} + \frac{1}{12\sqrt{2}}\Big[z'_{\rm SF}(x) - 3z''_{\rm SF}(x)\Big]_{x=1/6}, \tag{4.41}$$

where $x = 1/6$ corresponds to the (unperturbed) ISCO frequency in the dimensionless, coordinate-invariant variable $x \equiv (M\Omega)^{2/3}$. We see that, as for the binding energy and angular momentum in Sect. 4.2.2, the sole knowledge of the redshift $z_{\rm SF}$ allows one to compute the GSF correction of the ISCO. Other works of importance studied the GSF-induced ISCO frequency shift in a Schwarzschild background, including [422, 477, 478]. Moreover these calculations have been extended to the case of a Kerr BH, first in [479], and more recently in [480].

Regarding the IBCO, the authors of [481] recently performed a similar comparison, between the GSF/first law prediction, and a direct, numerical evaluation of the GSF. Just like the ISCO calculation, at first order in q they found that $\Omega_{\rm IBCO} = \Omega^{(0)}_{\rm IBCO}[1 + q\Omega^{(1)}_{\rm IBCO} + O(q^2)$, where this time the first-order correction is

$$\Omega^{(1)}_{\rm IBCO} \equiv -\frac{1}{2} + \frac{1}{4}\Big[z'_{\rm SF}(x) - 6z_{\rm SF}(x)\Big]_{x=1/4}. \tag{4.42}$$

Notice the similarity between the ISCO (4.41) and the IBCO (4.42) results, located at $x = 1/6$ and $x = 1/4$ respectively. Again, the work presented in [481] showed agreement between the above result (which only requires the numerical calculation of the redshift of a circular orbit) and that obtained by a full numerical integration. This agreement was found for both the frequency shift, as well as the particle's orbital angular momentum shift. These agreements between numerical/analytical and full-numerical results shows two of the most important applications of the first law of mechanics. First, on the practical side, it allows one to benchmark GSF calculations and can be used as consistency-checks for GSF numerical codes. Second, on a more fundamental level, it provides evidence that different notions of energy and angular momentum (in this case, mechanical versus ADM-type) can be identified, at least in this particular GSF context.

Second-Order GSF Benchmark

The above comparisons between first-law/GSF and numerical GSF calculations were performed at linear order in the mass ratio $\nu = m_1 m_2/m^2$ (or equivalently $q = m_1/m_2$, since $\nu = q + O(q^2)$). However, even if first-order GSF theory should be sufficient to build templates accurate enough for GW *detection*, it is known that a precise *parameter estimations* from the GW signal will require (at least pieces of) the second-order GSF effects. Consequently, pushing GSF calculations to second-order is an intense, ongoing program of research. Although substantial obstacles are still in the way, the first results are slowly becoming available. In particular, the first full, self-consistent, second-order (metric) perturbation calculation of a physical effect, was recently reported in [369]. There, the authors computed the binding energy E of a quasi-circular EMRI made of a nonspinning object of mass m around a large BH of mass $M_{\rm BH}$.

As described in Chap. 2, Sect. 2.2.4, in GSF theory such an EMRI is described by extending the test-particle approximation of a small object (mass m) orbiting (on a geodesic of) a Schwarzschild BH (mass M). The natural small parameter of the problem is therefore $\epsilon \equiv m/M$. Beyond this geodesic approximation, the central BH is not Schwarzschild anymore, but only approximately so. Similarly, the total (Bondi) mass of the system M_B is not M either, as would be the case for $\epsilon = 0$. More precisely, it can be shown [369] that $M_{\rm BH}$ and M_B admit a second-order expansion in ϵ of the form

$$M_{\rm BH} = M + \epsilon\,(1 + \hat{E}_0 + \delta M) + \epsilon^2 M_{\rm BH}^{(2)} + O(\epsilon^3)\,, \tag{4.43a}$$

$$M_{\rm B} = M + \epsilon\,\delta M + \epsilon^2 M_{\rm B}^{(2)} + O(\epsilon^3) \tag{4.43b}$$

where $\hat{E}_0$ is the zeroth-order particle's (orbital) energy (second term in (4.38a)) and δM is the change in the background Schwarschild's black hole mass, which increases due to gravitational radiation falling into it. The second-order pieces in Eqs. (4.43) are given by complicated expressions involving the metric perturbation h_{ab}, evaluated at the BH's horizon (for $M_{\rm BH}^{(2)}$) and at future null infinity (for $M_{\rm B}^{(2)}$). The (specific) binding energy $\hat{E}^{\rm bind}$ of the EMRI is then defined as the difference between its Bondi mass $M_{\rm B}$ and the individual masses of the bodies $M_{\rm BH} + m$, normalized by the reduced mass $\mu \equiv M_{\rm BH} m/(M_{\rm BH} + m)$. Expanding in powers of ϵ thanks to Eqs. (4.43), it is found that

$$\hat{E}^{\rm bind} = \hat{E}_0(y) + q\,\hat{E}_{\rm SF}^{\rm bind} + (q^2) \tag{4.44}$$

where $q \equiv m/M_{\rm BH}$ and $y \equiv (M_{\rm BH}\Omega)^{2/3}$ is the x-variable introduced earlier adapted to the present context (in fact, one has $x = y + \frac{2}{3}qy + O(q^2)$), and $\hat{E}_0$ is the leading-order piece in (4.38a), independent of ν. Lastly, the second-order[9] correction $\hat{E}_{\rm SF}^{\rm bind}$

[9] Even though $\hat{E}_{\rm SF}$ may appear like a first-order term (multiplied by q), it does contains all second-order perturbations through the masses (4.43).

is given by

$$\hat{E}_{\rm SF}^{\rm bind} = \hat{M}_{\rm B}^{(2)} + \hat{M}_{\rm BH}^{(2)} + \hat{E}_0(y) + \frac{\delta M}{M}\frac{y(1-6)y}{(1-3y)^{3/2}}, \tag{4.45}$$

with $\hat{M}_{\rm B}^{(2)} \equiv M_{\rm B}^{(2)}/M$ and similarly for $\hat{M}_{\rm BH}^{(2)}$. Eq. (4.45), along with a numerical implementation of the computation of the first two-terms (which depend on the second-order metric perturbation) is the real tour-de-force of [369]. The first law of mechanics was then used to check their calculation, by comparing it to the result

$$\hat{E}_{\rm SF}(y)^{\rm 1st\ law} = \frac{z_{\rm SF}(y)}{2} - \frac{y}{3}z'_{\rm SF}(y) - 1 + \sqrt{1-3y} + \frac{y}{6}\frac{5-12y}{(1-3y)^{3/2}}, \tag{4.46}$$

which is simply obtained by combining Eqs. (4.38a) and (4.39), as well as the link $x = y + \frac{2}{3}qy + O(q^2)$ between the variables of the two problems. The comparison between Eqs. (4.45) and (4.46) is shown in Fig. 1. of [369] and shows a very good overall agreement, although non-negligible discrepancies are present, and are (most probably) due to the difference between the notions of mass and angular momentum used in the two contexts. After all, the first law was derived in a non-radiative context, whereas the GSF frameworks account for the GWs of the system. In spite of this, and quite remarkably, the first law provided a valuable comparison tool, even in the strong-field regime of an EMRI.

Cosmic Sensor

The cosmic censorship conjecture (CCC), first formulated by Roger Penrose in [92], is the statement that no naked singularity exists in the Universe. This implies, for example, that any spacetime singularity formed by gravitational collapse must be cloaked behind an event horizon. As an elementary example of such a naked singularity, consider the Kerr BH, given by the metric (4.1). In this metric, the BH parameters (M, a) are required to satisfy the inequality $a \leq M$, with the upper limit corresponding to a so-called *extremal* Kerr BH. Now, for illustrative purposes, consider what would happen for $a > M$: the location of the event horizon $\mathscr{H}$, given by $r = M + \sqrt{M^2 - a^2}$ in BL-coordinates (t, r, θ, ϕ), becomes complex-valued, and the horizon thus "vanishes", leaving the ring-singularity of the Kerr BH naked.

In light of this heuristics, consider now the more realistic case of a *near-extremal* Kerr BH, i.e., with $a - M \ll 1$, undergoing a physical process making its spin increase. This could be the capture of a small body, for example. If the CCC is true, then the capture of the small body *cannot* increase a to arbitrarily large values. If it led to an overspinning of the BH (i.e., $a > M$), then the CCC would not be true. A particularly well-suited arena to address this problem is that of GSF theory, in which the effect of a small body on the evolution of the BH parameters (M, a) can be explicitly computed. Such an analysis was performed by Marta Colleoni, Leor Barrack and their collaborators in [482, 483]. In [482], the authors showed that, without GSF corrections, overspinning was possible for a non-negligible region of the parameter space for the problem. Then, accounting for the first-order GSF, they used[10] the first law of mechanics in the perturbed Kerr spacetime (given by combining Eqs. (4.34) and (4.32)) to establish a very simple criterion for whether overspinning was possible or not, in terms of the particle's energy and the GSF correction to its redshift (see Eq. (104) there). Finally, in [483] a subsequent numerical analysis was led using this criterion. The conclusion is best put by quoting the authors: "*Within the first-order GSF approximation (and excluding deeply bound orbits), equatorial captures generically result in a subextremal post-capture geometry. One can at best achieve extremality, through weak fine-tuning, but overspinning is not possible.* This result, evidently, is a strong argument in favor of the CCC. The same result was obtain in [484] in the general case, i.e., accounting for all types of orbits.

PN Theory

The fourth post-Newtonian order equations of motion for a binary system of compact objects have been obtained recently and using different, independent methods. First, using the ADM Hamiltonian formalism, the authors of [463, 485, 486] arrived at conclusive 4PN result, albeit with a so-called "ambiguity parameter", a undetermined constant coefficient whose value could not be fixed by their methods. This parameter was finally computed explicitly in [462] (see also [420]), by comparing the on-shell value of the 4PN Hamiltonian to the binding energy of the system, as computed from a 4PN-accurate GSF calculation that involved the first law of mechanics. Soon after, the Fokker Lagrangian approach was used by a second group [487, 488], arriving at the same result with, this time, the need for two ambiguity parameters.[11] To find the value of these coefficients, the authors used again a comparison from 4PN GSF calculations, this time requiring the use of two first laws, namely that for circular (4.21) and for eccentric (4.23) orbits. It is worth noting that an ambiguity-free derivation (eliminating the need for GSF information) of the full 4PN dynamics has recently been obtained in [396, 489] (see also [196]).

Informing EOB

The first law has been of particular importance to "calibrate" the potentials that enter the EOB formalism. It was first performed in [422, 490], just after the first publication of the first law for binary mechanics (4.21). It was then extended a few years later to

[10] Strictly speaking, they stumbled upon integral that could not be computed with available GSF codes and data at the time. This integral provided a key information between the asymptotic quantities and the local properties of the particle, which is precisely the type of information the first law provides, *by good fortune*, to quote the authors of [482].

[11] These ambiguity parameters find their origin in the use of regularization schemes for the calculation. Typically, the so-called "UV" divergences arise from the use of point-particle models (introducing a singularity), while "IR" divergences come from the infinite temporal range of the tail effects, discussed in Sect. 4.1.3.

account for spin-orbit couplings [491], as well as eccentric orbits [428, 429, 492]. To illustrate how the first law has helped in this context, we provide a very simple example based on [426].

As we mentioned in Chap. 1, Sect. 1.3.1, the EOB formalism is a semi-analytical framework in which the geometry of a two-body system are mapped to an effective metric which takes the form (here for nonspinning binaries)

$$\mathrm{d}s^2 = -A(r)\,\mathrm{d}t^2 + \frac{D(r)}{A(r)}\,\mathrm{d}r^2 + r^2(\mathrm{d}\theta^2 + \sin^2\theta\mathrm{d}\phi^2)\,, \tag{4.47}$$

where the two potentials A, D encode the Schwarzschild "deformation". These are two of the so-called EOB potentials [218], and they depend on the binary's characteristics, the two masses, in the nonspinning case. Now consider the case where one of the two bodies, say m_2, is much more massive than the other, m_1. In the limit case $q \equiv m_1/m_2 \to 0$, m_1 should follow a geodesic in a Schwarzschild spacetime of mass parameter m_2. Consequently, in the small ratio limit, the potentials A and D must behave like

$$A = 1 - \frac{2m_2}{r} + \nu a(r) + O(\nu^2) \quad \text{and} \quad D = 1 + \nu d(r) + O(\nu^2)\,, \tag{4.48}$$

where we used the exact Schwarzschild limit $A = 1 - 2m_2/r$ and $D = 1$ for the leading order. The functions $a(r)$ and $d(r)$ contain all the information on the EOB dynamics at linear order in $\nu = q + O(q)$. It is to inform these functions that the first law can be used. In particular, using the first law (4.23) for eccentric orbits, it was shown in [426] that the potentials $a(r), d(r)$ (and another one entering the EOB Hamiltonian) could be expressed entirely in terms of the sole quantity z_{SF}, the leading order GSF correction to the (eccentric orbit) redshift, see Eqs. (5.25)–(5.26) there.

Finally, let us mention the recent works [493, 494], in which the authors extended the above calculation (still based on the "eccentric" first law) to compute coefficients of the 5PN and 5.5PN Hamiltonian of a two-body binary system.[12]

4.3 A Variational Identity

In this section, we derive a general identity that relates the first-order variations of conserved asymptotic quantities in a diffeomorphism invariant theory of gravity—such as GR—to those of hypersurface integrals over the SEM tensor of a generic distribution of matter with compact support. Following a short recollection of some preliminary results in Sect. 4.3.1, a gravity-matter split is performed in Sect. 4.3.2, out of which the variational identity is derived in Sect. 4.3.3. The link to conserved

[12] To reach this high PN order, their method uses at once PN, post-Minkowskian, multipolar-post-Minkowskian, GSF and EOB, and is therefore coined the *tutti-frutti* method.

asymptotic quantities is discussed in Sect. 4.3.4, and the arbitrariness of the hypersurface of integration over the SEM tensor is proven in Sect. 4.3.5.

Throughout this section we shall use boldface symbols to denote differential forms defined over a 4-dimensional spacetime manifold. Given an arbitrary differential p-form $\boldsymbol{X} = X_{a_1 \cdots a_p}$, its exterior derivative will be denoted $\mathrm{d}\boldsymbol{X} = (\mathrm{d}X)_{a_1 \cdots a_{p+1}}$.

4.3.1 Preliminaries

Iyer and Wald [470, 495] gave a general derivation of the first law of black hole mechanics for arbitrary vacuum perturbations of a stationary black hole that are asymptotically flat at spatial infinity and regular on the event horizon. This derivation was extended to arbitrary electro-nonvacuum perturbations of charged black holes by Gao and Wald [496], who further derived a "physical process" version of the first law. Here we follow their general strategy, while making appropriate modifications for our nonvacuum perturbations of a nonstationary spacetime with a generic, compactly supported SEM tensor. The following analysis will follow closely that of Iyer [497], except that our background spacetime will not be assumed to be a stationary-axisymmetric black hole solution. For simplicity and definiteness, we shall restrict our analysis to the classical theory of GR in four spacetime dimensions, but most of the calculations hold for a general diffeomorphism invariant theory of gravity in any dimension [497].

Our starting point is the Lagrangian of the theory, taken to be a diffeomorphism invariant 4-form $\boldsymbol{L}$ on spacetime, which depends on the metric g_{ab} and other dynamical "matter" fields ψ, denoted collectively as $\phi \equiv (g, \psi)$. Let us consider a one-parameter family of spacetimes with metric $g_{ab}(\lambda)$. The first-order variation of $g_{ab}(\lambda)$ is defined as $\delta g_{ab} \equiv \mathrm{d}g_{ab}/\mathrm{d}\lambda|_{\lambda=0}$, and similarly for other dynamical fields. The first-order variation of the Lagrangian $\boldsymbol{L}$ can always be written in the form [470, 495–498]

$$\delta \boldsymbol{L} = \boldsymbol{E}\,\delta\phi + \mathrm{d}\boldsymbol{\Theta}(\phi, \delta\phi)\,, \tag{4.49}$$

where summation over the dynamical fields (and contraction of the associated tensor indices) is understood in the first term on the right-hand side, and the Euler-Lagrange equations of motion can be read off as $\boldsymbol{E}(\phi) = 0$. The *symplectic potential 3-form* $\boldsymbol{\Theta}$ is a linear differential operator in the field variations $\delta\phi$. Because the Lagrangian is uniquely defined only up to an exact form, $\boldsymbol{L} \to \boldsymbol{L} + \mathrm{d}\boldsymbol{\mu}$, the symplectic potential is defined only up to $\boldsymbol{\Theta} \to \boldsymbol{\Theta} + \delta\boldsymbol{\mu} + \mathrm{d}\boldsymbol{Y}$, for some arbitrary 3-form $\boldsymbol{\mu}(\phi)$ and 2-form $\boldsymbol{Y}(\phi, \delta\phi)$.

Now, let ξ^a denote an arbitrary smooth vector field on the unperturbed spacetime and $\mathcal{L}_\xi$ the Lie derivative along ξ^a. From the Lagrangian $\boldsymbol{L}$ and its associated symplectic potential $\boldsymbol{\Theta}$, we define the *Noether current 3-form* $\boldsymbol{J}$ relative to ξ^a according to

$$\boldsymbol{J}[\xi] \equiv \boldsymbol{\Theta}(\phi, \mathcal{L}_\xi\phi) - \xi \cdot \boldsymbol{L}\,, \tag{4.50}$$

where $\boldsymbol{\Theta}(\phi, \mathcal{L}_\xi \phi)$ stands for the expression of $\boldsymbol{\Theta}(\phi, \delta\phi)$ with each occurrence of $\delta\phi$ replaced by $\mathcal{L}_\xi \phi$, and "·" denotes the contraction of a vector field with the first index of a differential form, so that $\xi \cdot \boldsymbol{L} \equiv \xi^d L_{dabc}$. The key property of the Noether current (4.50) is that it is closed ($\mathrm{d}\boldsymbol{J} = 0$) if the field equations are satisfied ($\boldsymbol{E} = 0$) or if the vector field ξ^a Lie derives all of the dynamical fields ($\mathcal{L}_\xi \phi = 0$). Indeed, taking the exterior derivative of Eq. (4.50) readily gives [498]

$$\begin{aligned} \mathrm{d}\boldsymbol{J}[\xi] &= \mathrm{d}\boldsymbol{\Theta}(\phi, \mathcal{L}_\xi \phi) - \mathrm{d}(\xi \cdot \boldsymbol{L}) \\ &= \mathcal{L}_\xi \boldsymbol{L} - \boldsymbol{E}\, \mathcal{L}_\xi \phi - (\mathcal{L}_\xi \boldsymbol{L} - \xi \cdot \mathrm{d}\boldsymbol{L}) \\ &= -\boldsymbol{E}\, \mathcal{L}_\xi \phi\,, \end{aligned} \tag{4.51}$$

where in the second equality we used the Lagrangian variation $\mathcal{L}_\xi \boldsymbol{L} = \boldsymbol{E}\, \mathcal{L}_\xi \phi + \mathrm{d}\boldsymbol{\Theta}(\phi, \mathcal{L}_\xi \phi)$ [formally analogous to Eq. (4.49)] and Cartan's magic formula,[13] and $\mathrm{d}\boldsymbol{L} = 0$ in the third, since $\mathrm{d}\boldsymbol{L}$ is a 5-form on a 4D manifold.

The form of the Noether current (4.50) can be further specified thanks to the identity (4.51). Indeed, it can be shown [497, 500] that there exists a 3-form (with an extra dual vector index) $\boldsymbol{C}_a(\phi)$ that is locally constructed out of the dynamical fields ϕ in a covariant manner, such that the rightmost term in (4.51) reads $\boldsymbol{E}\, \mathcal{L}_\xi \phi = \mathrm{d}(\boldsymbol{C}_a \xi^a)$, thus implying $\mathrm{d}(\boldsymbol{J}[\xi] + \boldsymbol{C}_a \xi^a) = 0$. Consequently, according to the Poincaré lemma, there exists a 2-form $\boldsymbol{Q}[\xi]$ such that the Noether current (4.50) can locally be written in the form

$$\boldsymbol{J}[\xi] = -\boldsymbol{C}_a \xi^a + \mathrm{d}\boldsymbol{Q}[\xi]\,. \tag{4.52}$$

Crucially, $\boldsymbol{C}_a = 0$ whenever the equation of motion, $\boldsymbol{E} = 0$, are satisfied. One may view $\boldsymbol{C}_a = 0$ as being the constraint equations of the theory which are associated with its diffeomorphism invariance [498]. The ambiguity in $\boldsymbol{\Theta}$ discussed below (4.49) implies that the Noether current is uniquely defined only up to $\boldsymbol{J}[\xi] \to \boldsymbol{J}[\xi] + \mathrm{d}\big[\boldsymbol{Y}(\phi, \mathcal{L}_\xi \phi) - \xi \cdot \boldsymbol{\mu}\big]$ and the Noether charge up to $\boldsymbol{Q}[\xi] \to \boldsymbol{Q}[\xi] + \boldsymbol{Y}(\phi, \mathcal{L}_\xi \phi) - \xi \cdot \boldsymbol{\mu}$. As shown in Ref. [470], those ambiguities will not affect the results stated in the following paragraphs, so from now on we shall omit them.

Next, we define the *symplectic current 3-form* by [495]

$$\boldsymbol{\omega}(\phi, \delta_1\phi, \delta_2\phi) \equiv \delta_2[\boldsymbol{\Theta}(\phi, \delta_1\phi)] - \delta_1[\boldsymbol{\Theta}(\phi, \delta_2\phi)]\,, \tag{4.53}$$

which depends on two linearly independent first-order variations $\delta_1\phi$ and $\delta_2\phi$ of the fields ϕ. It can be shown that this differential form is closed ($\mathrm{d}\boldsymbol{\omega} = 0$) when ϕ is a solution of the field equations and $\delta_1\phi$ and $\delta_2\phi$ are solutions of the linearized field equations [498]. As discussed in Sect. 4.3.4 below, the symplectic current (4.53) is used to define the notion of a Hamiltonian, which, in turn, gives rise to the notions of total energy and angular momentum.

Now, set $\delta_1\phi \equiv \mathcal{L}_\xi \phi$ and let $\delta_2\phi \equiv \delta\phi$ correspond to a nearby solution for which $\delta\xi^a = 0$, as allowed by the diffeomorphism gauge freedom of GR. Then

[13] By Cartan's magic formula we mean $\mathcal{L}_\xi \boldsymbol{L} = \mathrm{d}(\xi \cdot \boldsymbol{L}) + \xi \cdot \mathrm{d}\boldsymbol{L}$. It is due to Élie Cartan (in [499]) and not his son Henri, as it is sometimes claimed in textbooks.

$$\begin{aligned}\omega(\phi, \mathcal{L}_\xi\phi, \delta\phi) &= \delta\boldsymbol{\Theta}(\phi, \mathcal{L}_\xi\phi) - \mathcal{L}_\xi\boldsymbol{\Theta}(\phi, \delta\phi) \\ &= \delta \boldsymbol{J}[\xi] + \xi \cdot \delta \boldsymbol{L} - (\xi \cdot \mathrm{d}\boldsymbol{\Theta} + \mathrm{d}(\xi \cdot \boldsymbol{\Theta})) \\ &= \mathrm{d}\big(\delta \boldsymbol{Q}[\xi] - \xi \cdot \boldsymbol{\Theta}\big) + \xi \cdot \boldsymbol{E}\,\delta\phi - \delta(\boldsymbol{C}_a \xi^a)\,, \end{aligned} \tag{4.54}$$

where we used the definition (4.50) and Cartan's magic formula in the second equality, as well as the identities (4.49) and (4.52) in the last equality. When the equations of motion are satisfied, $\boldsymbol{E} = 0$ and $\boldsymbol{C}_a = 0$ imply that the symplectic current (4.54) is exact and thus closed, as mentioned above. Integrating the resulting identity over a hypersurface $\mathscr{S}$ transverse to ξ^a, with boundary $\partial\mathscr{S}$, and using Stokes' theorem, we obtain the general formula [497]

$$\int_{\mathscr{S}} \omega(\phi, \mathcal{L}_\xi\phi, \delta\phi) = \int_{\partial\mathscr{S}} \delta \boldsymbol{Q}[\xi] - \xi \cdot \boldsymbol{\Theta}(\phi, \delta\phi)\,. \tag{4.55}$$

The symplectic current in Eq. (4.54) is a linear differential operator in the field variation $\mathcal{L}_\xi\phi$. Consequently, if ξ^a Lie derives *all* of the dynamical fields in the background ($\mathcal{L}_\xi\phi = 0$), then the boundary integral on the right-hand side of Eq. (4.55) vanishes identically.

4.3.2 Gravity-Matter Split

To derive the general variational identity of interest, we shall further split the Lagrangian 4-form $\boldsymbol{L}$ of the theory into a purely gravitational (vacuum) part and a matter part:

$$\boldsymbol{L}(g, \psi) \equiv \boldsymbol{L}_\mathrm{g}(g) + \boldsymbol{L}_\mathrm{m}(g, \psi)\,. \tag{4.56}$$

The vacuum GR Lagrangian $\boldsymbol{L}_\mathrm{g}$ depends on the metric g_{ab} and its derivatives and is explicitly given by $16\pi \boldsymbol{L}_\mathrm{g} = R\,\boldsymbol{\varepsilon}$, where R is the Ricci scalar and $\boldsymbol{\varepsilon}$ the canonical volume form associated with g_{ab}. The matter part $\boldsymbol{L}_\mathrm{m}$ is left unspecified, but is required to depend only on the metric g_{ab} and the other dynamical "matter" fields ψ. Following Eq. (4.49), the first-order variation of each Lagrangian in Eq. (4.56) can then be split into a total derivative and a part related to the field equations, according to

$$\delta \boldsymbol{L}_\mathrm{g} = -\frac{1}{16\pi}\,\boldsymbol{\varepsilon}\, G^{ab}\delta g_{ab} + \mathrm{d}\boldsymbol{\Theta}_\mathrm{g}(g, \delta g)\,, \tag{4.57a}$$

$$\delta \boldsymbol{L}_\mathrm{m} = \frac{1}{2}\,\boldsymbol{\varepsilon}\, T^{ab}\delta g_{ab} + \boldsymbol{E}_\mathrm{m}(g, \psi)\,\delta\psi + \mathrm{d}\boldsymbol{\Theta}_\mathrm{m}(\phi, \delta\phi)\,. \tag{4.57b}$$

Here, $G_{ab} \equiv R_{ab} - \frac{1}{2} R g_{ab}$ is the Einstein tensor, T_{ab} is the SEM tensor and the matter field equations read $\boldsymbol{E}_\mathrm{m}(\phi) = 0$. Repeating the analysis performed above in Sect. 4.3.1, separately for the (vacuum) gravity and matter sectors, one can easily show that the gravity and matter Noether currents take the form

$$\boldsymbol{J}_{\mathsf{g}}[\xi] \equiv \boldsymbol{\Theta}_{\mathsf{g}}(g, \mathcal{L}_\xi g) - \xi \cdot \boldsymbol{L}_{\mathsf{g}} = -\boldsymbol{C}^a_{\mathsf{g}}\xi_a + \mathrm{d}\boldsymbol{Q}_{\mathsf{g}}[\xi] \,, \tag{4.58a}$$

$$\boldsymbol{J}_{\mathsf{m}}[\xi] \equiv \boldsymbol{\Theta}_{\mathsf{m}}(\phi, \mathcal{L}_\xi \phi) - \xi \cdot \boldsymbol{L}_{\mathsf{m}} = -\boldsymbol{C}^a_{\mathsf{m}}\xi_a + \mathrm{d}\boldsymbol{Q}_{\mathsf{m}}[\xi] \,. \tag{4.58b}$$

On the one hand, the vacuum GR contributions are well known and read [470, 495]

$$\Theta^{\mathsf{g}}_{abc}(g, \delta g) = -\frac{1}{16\pi}\, \varepsilon_{abcd}\, g^{de} g^{fh} (\nabla_f \delta g_{eh} - \nabla_e \delta g_{fh}) \,, \tag{4.59a}$$

$$J^{\mathsf{g}}_{abc}[\xi] = -\frac{1}{8\pi}\, \varepsilon_{abcd} \nabla_e \nabla^{[e}\xi^{d]} \,, \tag{4.59b}$$

$$Q^{\mathsf{g}}_{ab}[\xi] = -\frac{1}{16\pi}\, \varepsilon_{abcd} \nabla^c \xi^d \,, \tag{4.59c}$$

$$C^{\mathsf{g}}_{abce} = -\frac{1}{8\pi}\, \varepsilon_{abcd}\, G^d{}_e \,. \tag{4.59d}$$

On the other hand, explicit forms of the matter contributions $\boldsymbol{\Theta}_{\mathsf{m}}$, $\boldsymbol{C}^a_{\mathsf{m}}$, $\boldsymbol{Q}_{\mathsf{m}}$ and $\boldsymbol{J}_{\mathsf{m}}$ depend on the particular choice of matter Lagrangian $\boldsymbol{L}_{\mathsf{m}}$. Typical examples include perfect fluids and electromagnetic fields [497, 501]. Importantly, whenever the matter field equations $\boldsymbol{E}_{\mathsf{m}} = 0$ are satisfied, the matter constraint simply reduces to [497]

$$C^{\mathsf{m}}_{abce} = \varepsilon_{abcd}\, T^d{}_e \,, \tag{4.60}$$

in such a way that the total constraint $\boldsymbol{C}^a \equiv \boldsymbol{C}^a_{\mathsf{g}} + \boldsymbol{C}^a_{\mathsf{m}}$ in Eq. (4.52) vanishes identically when the Einstein field equation $G_{ab} = 8\pi\, T_{ab}$ are satisfied as well.

4.3.3 *Variational Identity*

So far we considered an arbitrary smooth vector field ξ^a defined on a background geometry g_{ab}. From now on, we shall further assume that ξ^a is a *Killing field* of the background, such that $\mathcal{L}_\xi g_{ab} = 0$. Moreover, as allowed by the diffeomorphism gauge freedom of GR, we shall consider first-order variations for which

$$\delta\xi^a = 0 \,, \quad \text{implying} \quad \mathcal{L}_\xi \delta g_{ab} = 0 \quad \text{and} \quad \mathcal{L}_\xi \delta\xi_a = 0 \,. \tag{4.61}$$

Consequently, the vacuum GR contribution $\boldsymbol{\omega}_{\mathsf{g}}$ to the symplectic current (4.53) can easily be shown to vanish identically. Indeed, from the explicit expression (4.59a) for the pure gravity part $\boldsymbol{\Theta}_{\mathsf{g}}$ of the symplectic potential 3-form, we have

$$\boldsymbol{\omega}_{\mathsf{g}}(g, \mathcal{L}_\xi g, \delta g) = -\mathcal{L}_\xi \boldsymbol{\Theta}_{\mathsf{g}}(g, \delta g) = -\boldsymbol{\Theta}_{\mathsf{g}}(g, \mathcal{L}_\xi \delta g) = 0 \,, \tag{4.62}$$

where we used the Lie-dragging along ξ^a of g_{ab}, δg_{ab} and ε_{abcd}, as well as the commutation of the covariant derivative ∇_a and the Lie derivative $\mathcal{L}_\xi$, as shown in Eq. (A.9). Consequently, only the matter part $\boldsymbol{\Theta}_\mathrm{m}$ contributes to (4.53), and we have

$$\begin{aligned}\boldsymbol{\omega}(\phi, \mathcal{L}_\xi\phi, \delta\phi) &= \delta\boldsymbol{\Theta}_\mathrm{m}(\phi, \mathcal{L}_\xi\phi) - \mathcal{L}_\xi\boldsymbol{\Theta}_\mathrm{m}(\phi, \delta\phi)\\ &= \delta\boldsymbol{J}_\mathrm{m}[\xi] + \xi\cdot\delta\boldsymbol{L}_\mathrm{m} - (\xi\cdot\mathrm{d}\boldsymbol{\Theta}_\mathrm{m} + \mathrm{d}(\xi\cdot\boldsymbol{\Theta}_\mathrm{m}))\\ &= \mathrm{d}\big(\delta\boldsymbol{Q}_\mathrm{m} - \xi\cdot\boldsymbol{\Theta}_\mathrm{m}\big) + \xi\cdot\big(\tfrac{1}{2}\,\boldsymbol{\varepsilon}\,T^{ab}\delta g_{ab} + \boldsymbol{E}_\mathrm{m}\,\delta\psi\big) - \delta(\boldsymbol{C}^a_\mathrm{m}\xi_a)\,,\end{aligned} \tag{4.63}$$

where we used the definition (4.50) and Cartan's magic formula in the second equality, as well as the Lagrangian variation (4.57b) and (4.52) in the last equality. Equating the expressions (4.54) and (4.63), in which $\boldsymbol{Q} = \boldsymbol{Q}_\mathrm{g} + \boldsymbol{Q}_\mathrm{m}$ and $\boldsymbol{\Theta} = \boldsymbol{\Theta}_\mathrm{g} + \boldsymbol{\Theta}_\mathrm{m}$, and imposing the equations of motion ($\boldsymbol{E} = 0$, $\boldsymbol{E}_\mathrm{m} = 0$ and $\boldsymbol{C} = 0$) implies[14]

$$\mathrm{d}\big(\delta\boldsymbol{Q}_\mathrm{g}[\xi] - \xi\cdot\boldsymbol{\Theta}_\mathrm{g}\big) = \frac{1}{2}\,\xi\cdot\boldsymbol{\varepsilon}\,T^{ab}\delta g_{ab} - \delta(\boldsymbol{\varepsilon}\cdot T\cdot\xi)\,, \tag{4.64}$$

where we introduced the shorthand $\boldsymbol{\varepsilon}\cdot T\cdot\xi = \varepsilon_{abcd}T^{de}\xi_e$. Finally, integrating this equation over a hypersurface $\mathscr{S}$ transverse to ξ^a, with boundary $\partial\mathscr{S}$, and using Stokes' theorem, we obtain the simple identity

$$\int_{\partial\mathscr{S}} \delta\boldsymbol{Q}_\mathrm{g}[\xi] - \xi\cdot\boldsymbol{\Theta}_\mathrm{g} = \frac{1}{2}\int_{\mathscr{S}} \xi\cdot\boldsymbol{\varepsilon}\,T^{ab}\delta g_{ab} - \delta\int_{\mathscr{S}} \boldsymbol{\varepsilon}\cdot T\cdot\xi\,. \tag{4.65}$$

This formula is very closely related to Eq. (32) of Ref. [497], valid for nonvacuum perturbations of stationary-axisymmetric black hole solutions in a general diffeomorphism invariant theory of gravity. Equation (4.65) was also written down (without a detailed derivation) in Ref. [388], by adapting Refs. [470, 472, 497], and applied to nonvacuum, nonstationary, nonaxisymmetric perturbations of a Kerr-black-hole-with-a-corotating-moon solution that is asymptotically flat at future null infinity.

4.3.4 *Asymptotic Conserved Quantities*

General Case

Up to a numerical prefactor, the Noether (scalar) charge Q_ξ relative to ξ^a is defined as the integral of the Noether 2-form (4.59c) over a topological 2-sphere $\mathscr{U}$ that includes all the matter fields:

---

[14] If ξ^a Lie derives all the dynamical fields in the background ($\mathcal{L}_\xi\phi = 0$) and not merely the metric g_{ab}, then the right-hand side of (4.54) and (4.63) vanish identically, yielding $\mathrm{d}\big(\delta\boldsymbol{Q}_\mathrm{g}[\xi] - \xi\cdot\boldsymbol{\Theta}_\mathrm{g}\big) = -\mathrm{d}\big(\delta\boldsymbol{Q}_\mathrm{m}[\xi] - \xi\cdot\boldsymbol{\Theta}_\mathrm{m}\big)$. This condition is stronger than merely equating the right-hand sides of Eqs. (4.54) and (4.63), but it gives rise to the same identity (4.64).

$$Q_\xi \equiv \int_S \boldsymbol{Q}_{\rm g}[\xi] \,. \tag{4.66}$$

This charge is conserved in the sense that it does not depend on the choice of integration 2-surface S. Indeed, if $\mathscr{U}$ and $\mathscr{U}'$ denote two such topological 2-spheres, and $\mathscr{S}$ any hypersurface bounded by $\mathscr{U}$ and $\mathscr{U}'$, then

$$\int_{\mathscr{U}} \boldsymbol{Q}_{\rm g}[\xi] - \int_{\mathscr{U}'} \boldsymbol{Q}_{\rm g}[\xi] = \int_{\mathscr{S}} {\rm d}\boldsymbol{Q}_{\rm g}[\xi] = -\frac{1}{8\pi} \int_{\mathscr{S}} \varepsilon_{abcd} \nabla_e \nabla^{[e} \xi^{d]} = 0 \,, \tag{4.67}$$

where we successively used Stokes' theorem, Eqs. (4.58a), (4.59b) and (4.59d), the identity $\nabla_e \nabla^{[e}\xi^{d]} = -R^{cd}\xi_c$ following from Eqs. (3.8) and (A.3), as well as the Einstein equation $R_{ab} = 0$ over the vacuum region $\mathscr{S}$ (cf. [502, 503]).

For an asymptotically flat spacetime with no isometry, the formula (4.66) can be evaluated on a topological 2-sphere at spatial infinity. For instance, if t^a and ϕ^a denote the *asymptotic* Killing vectors associated with the invariance of an asymptotically Minkowskian spacetime under time translations and spatial rotations, then the Noether charge (4.66) gives rise to the notions of Komar mass and Komar angular momentum

$$M_{\rm K} \equiv 2 \int_\infty \boldsymbol{Q}_{\rm g}[t] \quad \text{and} \quad J_{\rm K} \equiv - \int_\infty \boldsymbol{Q}_{\rm g}[\phi] \,. \tag{4.68}$$

For an asymptotically flat spacetime, it can easily be established (see e.g. Ref. [464]) that the Komar angular momentum J_K is equal to the ADM-like angular momentum J, also defined as a surface integral at spatial infinity, namely $J_{\rm K} = J$. Under the additional assumption of stationarity, the equality $M_{\rm K} = M$ of the Komar mass and the ADM mass was proven long ago [504, 505]. This equality is closely related to a general relativistic generalization of the Newtonian virial theorem [506], and was used as a criterion to compute quasi-equilibrium sequences of initial data for binary black holes [380, 399, 507–510]. Shibata et al. [386] showed that the equality $M_{\rm K} = M$ holds for a much larger class of spacetimes; in particular, they could relax the restrictive hypothesis of stationarity.

When evaluated at infinity, the boundary term on the left-hand side of the identity (4.65) has the natural interpretation of being the variation of the "conserved quantity" canonically conjugate to the asymptotic symmetry generated by ξ^a. Indeed, according to the analysis of Refs. [470, 472, 495] (also Sect. 2 in [500]), if a Hamiltonian H_ξ exists for the dynamics generated by the vector field ξ^a, then there exists a 3-form $\mathbf{B}_{\rm g}$ such that

$$\int_\infty \delta \boldsymbol{H}_{\rm g}[\xi] - \xi \cdot \boldsymbol{\Theta}_{\rm g} = \delta H_\xi \,, \quad \text{with} \quad H_\xi \equiv \int_\infty \boldsymbol{Q}_{\rm g}[\xi] - \xi \cdot \mathbf{B}_{\rm g} \,. \tag{4.69}$$

It can be shown that a necessary and sufficient condition for the existence of a Hamiltonian H_ξ conjugate to ξ^a on $\mathscr{S}$ is that for all solutions ϕ and all pairs of linearized solutions $\delta_1\phi$ and $\delta_2\phi$ we have [472]

$$\int_{\partial \mathscr{S}} \xi \cdot \omega(\phi, \delta_1 \phi, \delta_2 \phi) = 0 \,. \tag{4.70}$$

Finally, by combining (4.65) and (4.69), we conclude that if the hypersurface Σ has no inner boundary (corresponding to the intersection of Σ with the horizon of a black hole), then the variation of the conserved Noether charge associated with ξ^a is related to the energy-momentum content through

$$\delta H_\xi = \delta \int_{\mathscr{S}} \varepsilon_{abcd} \, T^{de} \xi_e - \frac{1}{2} \int_{\mathscr{S}} \varepsilon_{abcd} \, \xi^d \, T^{ef} \, \delta g_{ef} \,. \tag{4.71}$$

This variational formula is valid for a generic "matter" source with compact support. This is the most important (new) result of this chapter, as it will be the basis of all calculations for the derivation of the first law in the next chapter. In particular, we shall be interested in applying this general result to a binary system of spinning compact objects, modeled within the multipolar gravitational skeleton formalism reviewed in Chap. 2, up to dipolar order.

Helical Isometry

As mentioned earlier, in general neither t^a nor ϕ^a in (3.7) is a Killing vector. If the spacetime is asymptotically flat, however, then t^a and ϕ^a are *asymptotically* Killing, with the normalization $t^a t_a \to -1$ at infinity, and the surface integral at spatial infinity of the 2-form $\boldsymbol{Q}_\mathrm{g}[k]$ yields the conserved charge associated with the generator (3.7), namely (cf. Eq. (4.68))

$$\int_\infty \boldsymbol{Q}_\mathrm{g}[k] = \frac{1}{2} M_\mathrm{K} - \Omega J_\mathrm{K} = \frac{1}{2} M_\mathrm{K} - \Omega J \,. \tag{4.72}$$

The curious relative factor of two is related to the famous Komar "anomalous factor" entering the definitions of the Komar mass and angular momentum [470, 511].

Moreover, according to the general result (4.69), the boundary term at spatial infinity in the identity (4.65) yields a linear combination of the conserved charges associated with the asymptotic symmetry generators t^a and ϕ^a, namely[15]

$$\int_\infty (\delta \boldsymbol{Q}_\mathrm{g}[k] - k \cdot \boldsymbol{\Theta}_\mathrm{g}) = \delta \int_\infty (\boldsymbol{Q}_\mathrm{g}[k] - k \cdot \mathbf{B}_\mathrm{g}) = \delta H_k = \delta M - \Omega \, \delta J \,, \tag{4.73}$$

The formula (4.73) is consistent with the results of Ref. [382], obtained from a related analysis.

---

[15] Asymptotically, the Killing vector field (3.7) reduces to a linear combination of the generators t^a and ϕ^a of time translations and spatial rotations, such that Ω should be treated as a constant while evaluating the surface integral (4.73).

Finally, combining (4.65) and (4.73), we conclude that for a helically symmetric spacetime, the variations of the total mass and angular momentum are related to the energy-momentum content through

$$\delta M - \Omega\,\delta J = \delta \int_{\mathscr{S}} \varepsilon_{abcd}\, T^{de} k_e - \frac{1}{2}\int_{\mathscr{S}} \varepsilon_{abcd}\, k^d\, T^{ef}\, \delta g_{ef}\,. \tag{4.74}$$

The generalized first law (4.74) is valid for a generic "matter" source with compact support, whose energy-momentum tensor T^{ab} must be compatible with the helical isometry, $\mathcal{L}_k T^{ab} = 0$. The variational formula (4.74) holds, in particular, for perfect fluids. In this work, we shall be interested in applying this general formula to a binary system of spinning compact objects, modeled within the multipolar gravitational skeleton formalism, up to dipolar order, as described in Chap. 2.

4.3.5 Arbitrariness of the Hypersurface

Before closing this chapter, let us show that the integrals appearing in the right-hand side of (4.71) are independent of the choice of hypersurface $\mathscr{S}$, and hence of "time." It will prove convenient to introduce special notations for these hypersurface integrals, namely

$$I(\mathscr{S}) \equiv -\int_{\mathscr{S}} \varepsilon_{abcd}\, T^{de} \xi_e = \int_{\mathscr{S}} T^{ab} \xi_b\, \mathrm{d}\Sigma_a\,, \tag{4.75a}$$

$$K(\mathscr{S}) \equiv -\int_{\mathscr{S}} \varepsilon_{abcd}\, \xi^d\, T^{ef}\, \delta g_{ef} = \int_{\mathscr{S}} T^{ab} \delta g_{ab}\, \xi^c \mathrm{d}\Sigma_c\,, \tag{4.75b}$$

where $\mathrm{d}\Sigma_a$ is the (hyper)surface element normal to $\mathscr{S}$. The integral I is simply the flux across $\mathscr{S}$ of the conserved Noether current $T^{ab}\xi_b$ associated with the Killing field ξ^a. The integral K, which involves the perturbed metric δg_{ab}, has no such simple physical interpretation. To show that these integrals are invariant with respect to $\mathscr{S}$, consider $\Delta I \equiv I(\mathscr{S}_1) - I(\mathscr{S}_2)$ and $\Delta K = K(\mathscr{S}_1) - K(\mathscr{S}_2)$, where $\mathscr{S}_1, \mathscr{S}_2$ are two arbitrary spacelike hypersurfaces. Let also $\mathscr{V}$ denote the volume bounded by $\mathscr{S}_1$ and $\mathscr{S}_2$ and $\mathscr{T}$ a worldtube that includes the support of T^{ab}. Then, by using Stokes' theorem and the Leibniz rule, we readily find

$$\Delta I = \int_{\mathscr{V}} \nabla_a (T^{ab}\xi_b)\, \mathrm{d}V = \int_{\mathscr{V}} \left[(\nabla_a T^{ab})\xi_b + T^{ab}\nabla_{(a}\xi_{b)}\right] \mathrm{d}V = 0\,, \tag{4.76a}$$

$$\Delta K = \int_{\mathscr{V}} \nabla_c (\xi^c T^{ab} \delta g_{ab})\, \mathrm{d}V = \int_{\mathscr{V}} \left[(\mathcal{L}_\xi T^{ab})\delta g_{ab} + T^{ab}\mathcal{L}_\xi \delta g_{ab}\right] \mathrm{d}V = 0\,, \tag{4.76b}$$

where $\mathrm{d}V$ is the invariant volume element. We used the local conservation of energy and momentum, $\nabla_a T^{ab} = 0$, together with Killing's equation $\nabla_{(a}\xi_{b)} = 0$, which implies $\nabla_c \xi^c = 0$, the gauge choice $\delta k^a = 0$, and finally $\mathcal{L}_\xi T^{ab} = 0$.

As a final note, we mention that both integrals (4.75) are also invariant under Lie-dragging of the hypersurface $\mathscr{S}$ in the direction of ξ^a. Given a spacelike hypersurface $\mathscr{S}$ and a small positive number ϵ, let $\mathscr{S}_\epsilon$ denote the hypersurface obtained by Lie-dragging $\mathscr{S}_0$ along the direction $\epsilon\xi^a$. With the shorthands $I_0 \equiv I(\mathscr{S}_0)$, $I_\epsilon \equiv I(\mathscr{S}_\epsilon)$ and $\dot{I} \equiv \lim_{\epsilon\to 0} (I_\epsilon - I_0)/\epsilon$, we have

$$\dot{I} = -\int_{\mathscr{S}_0} \mathcal{L}_\xi (\varepsilon_{abcd}\, T^{de} \xi_e) = 0\,, \tag{4.77a}$$

$$\dot{K} = -\int_{\mathscr{S}_0} \mathcal{L}_\xi (\varepsilon_{abcd}\, \xi^d\, T^{ef} \delta g_{ef}) = 0\,, \tag{4.77b}$$

as a consequence of the Leibniz rule, $\mathcal{L}_\xi \varepsilon_{abcd} = 0$ and $\mathcal{L}_\xi T^{ab} = 0$ and Eq. (4.61). The fact that the right-hand side of Eq. (4.71) is invariant under Lie-dragging in the direction of ξ^a is consistent with the invariance (4.76) of I and K on the choice of hypersurface.

Chapter 5
The First Law at Dipolar Order

Contents

What distinguishes the straight line and circle more than anything else? Where, then, did Euclid fail? In not introducing the third curve, which has the same property: the screw. The representations of translation, rotation, and the two combined ought to have been the instruments of geometry.

A. De Morgan,
Letter to W.R. Hamilton (1852)

This chapter is dedicated to the derivation of the first law of mechanics, for a system of two spinning, massive points particles. It relies on a number of results that have been obtained in the previous chapters, in particular: (1) the skeleton stress-energy-momentum tensor of a dipolar particle discussed in Chap. 2; (2) a number of key,

P. Ramond, *The First Law of Mechanics in General Relativity & Isochrone Orbits in Newtonian Gravity*, Springer Theses,
https://doi.org/10.1007/978-3-031-17964-8_5

geometrical identities derived from the helical isometry in Chap. 3, and (3) the generalized first law of mechanics valid for an arbitrary distribution of matter obtained in Chap. 4. First, we provide a detailed and explicit computation of the first law in Sect. 5.1 in terms of the particle's multipoles. The derivation is fully covariant, and the result free of any spin supplementary condition. Then, we rewrite it in terms of geometrical scalars associated to the helical Killing field, and scalar multipoles in Sect. 5.2. Lastly, in Sect. 5.3 we discuss how our result compares (and reduces) to the Hamiltonian first law of mechanics found in the literature.

5.1 Derivation

In this first section we derive the first law of mechanics, for a binary system of dipolar particles moving along an exactly circular orbit. We begin in Sect. 5.1.1 by choosing the spacelike hypersurface of integration in order to simplify the calculations. Then, from the integral form of the first law derived in Sect. 4.3, we compute in Sect. 5.1.2 the two integrals that appear in the right-hand side of (4.74). The resulting expressions are simplified algebraically in Sect. 5.1.3, and those results are combined in Sect. 5.1.4 to establish the final formula. Throughout this section we do *not* impose the SSC (2.64) to keep the results as general as possible.

Note: *the derivation presented in this chapter relies heavily on a large number of identities established in Chap. 2, in particular Sect. 2.3. There, these formulae were written for a quadrupolar particle. For the remaining of this chapter, in which we limit ourselves to the dipolar case, any quadrupolar term (i.e., one that includes a contribution with* J^{abcd}*) imported from Chap. 2 will be systematically set to zero.*

5.1.1 Choice of Spacelike Hypersurface

We proved in Chap. 3, Sect. 4.3 that the right-hand side of the identity (4.74) does not depend on the choice of spacelike hypersurface $\mathscr{S}$. We may thus conveniently choose this hypersurface such that, for each particle, the Killing field k^a is orthogonal to $\mathscr{S}$ at the intersection point $\mathscr{P} \equiv \mathscr{S} \cap \gamma$, i.e.

$$k^a \stackrel{\mathscr{P}}{=} |k| n^a \,, \tag{5.1}$$

where n^a is the future-directed, unit normal to $\mathscr{S}$. Since $k^a|_\gamma = z u^a$, this implies $n^a|_\mathscr{P} = u^a$ and $|k|_\mathscr{P} = z$, where the redshift parameter z was shown in Chap. 3, Sect. 3.2.1 to be constant along γ. More generally, one may introduce a foliation of the spacetime manifold $\mathscr{E}$ by a family of spacelike hypersurfaces $(\mathscr{S}_t)_{t\in\mathbb{R}}$ such that Eq. (5.1) holds at *any* point $\mathscr{P} \in \gamma$, so that

$$k^a \overset{\gamma}{=} |k| n^a \,, \tag{5.2}$$

where n^a is the future-directed, unit normal to the family $(\mathscr{S}_t)_{t\in\mathbb{R}}$ of spacelike hypersurfaces, such that $n^a|_\gamma = u^a$. Along γ, the norm $|k|_\gamma = z$ of the helical Killing field then plays the role of a constant lapse function, and k^a that of the normal evolution vector (see, e.g., Ref. [512]). According to our findings in Chap. 3, we then have $\mathcal{L}_k n^a|_\gamma = 0$, such that the extrinsic curvature vanishes at any point along γ:

$$K_{ab} \equiv -\frac{1}{2}\mathcal{L}_n \gamma_{ab} \overset{\gamma}{=} -\frac{1}{2|k|}\mathcal{L}_k \gamma_{ab} \overset{\gamma}{=} -\frac{1}{|k|} n_{(a}\mathcal{L}_k n_{b)} \overset{\gamma}{=} 0 \,, \tag{5.3}$$

where $\gamma_{ab} \equiv g_{ab} + n_a n_b$ is the induced metric on any spacelike hypersurface of the foliation, so that $n^a \gamma_{ab} = 0$. By combining (5.3) with the equalities $n_a|_\gamma = u_a$ and $n'_a|_\gamma \equiv n^c \nabla_c n_a|_\gamma = \dot{u}_a$, the usual 3+1 formula for the gradient of the unit normal to $\mathscr{S}$ becomes [512]

$$\nabla_a n_b = -K_{ab} - n_a n'_b \overset{\gamma}{=} -u_a \dot{u}_b \,. \tag{5.4}$$

We emphasize that Eqs. (5.2)–(5.4) are valid only along the worldline γ, and thus in particular at the intersection point $\mathscr{P} = \mathscr{S} \cap \gamma$ with a given hypersurface $\mathscr{S}$. Most importantly, one cannot choose the hypersurface $\mathscr{S}$ such that (5.1) holds in an open neighborhood of $\mathscr{P}$. This is closely related to the helical nature of the Killing vector field (3.7), whose twist (3.54) does not vanish everywhere, and which, by Frobenius' theorem, cannot be hypersurface orthogonal [355] (Fig. 5.1).

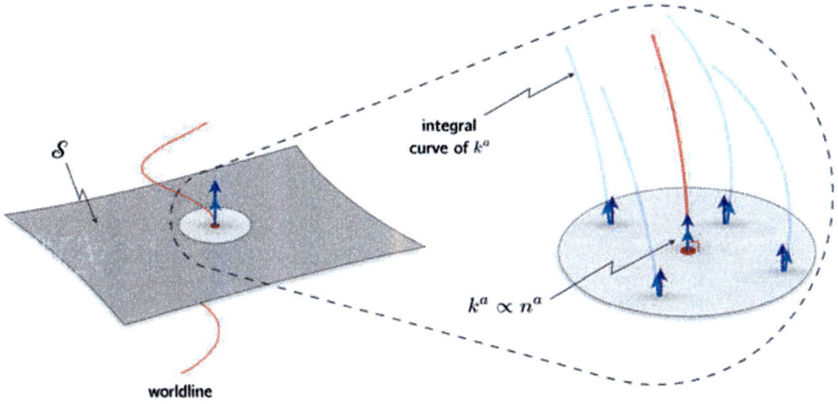

Fig. 5.1 The normal n^a of the integration hypersurface $\mathscr{S}$ is parallel to the four-velocity u^a (and thus to $k^a = zu^a$) only at the intersection point $\mathscr{P} \equiv \gamma \cap \mathscr{S}$. Elsewhere, the vector fields k^a and n^a are not tangent, due to the helical nature of the former

5.1.2 Hypersurface Integrals

In this subsection, we provide integrated expressions for the integrals I and K that appear in the right-hand side of the first law, as given in the form (4.74). To avoid being repetitive, we shall detail the calculation for I only, and merely quote the final result for K.

We begin by substituting the dipolar SEM tensor (2.58) into the definition (4.75a) of I to obtain

$$I = \int_{\mathscr{S}} \int_{\gamma} u^{(a} p^{b)} k_b \, \delta_4 \, \mathrm{d}\tau \, \mathrm{d}\Sigma_a + \int_{\mathscr{S}} \int_{\gamma} \nabla_c \big(u^{(a} S^{b)c} \delta_4 \big) k_b \, \mathrm{d}\tau \, \mathrm{d}\Sigma_a \,. \tag{5.5}$$

To evaluate those two integrals, it is convenient to consider the foliation $(\mathscr{S}_t)_{t\in\mathbb{R}}$ introduced above, and to choose for $\mathscr{S}$ one of the leaves of this foliation, say $\mathscr{S}_{t_0}$ for some fixed $t_0 \in \mathbb{R}$. We recall that n^a denotes the future-directed, unit normal to any leaf $\mathscr{S}_t$ and $\gamma_{ab} \equiv g_{ab} + n_a n_b$ the induced metric on $\mathscr{S}_t$. After performing the change of variable $\tau \to t$, while using standard 3+1 formulae [512] $\mathrm{d}\tau = N \mathrm{d}t$, $\mathrm{d}\Sigma_a = -n_a \sqrt{\overline{\gamma}} \, \mathrm{d}^3 x$ and $N\sqrt{\overline{\gamma}} = \sqrt{-g}$, where N is the lapse function, $\overline{\gamma} \equiv \det(\gamma_{ij})$ and $g \equiv \det(g_{\alpha\beta})$, we obtain

$$I = -\int_{\mathscr{E}} u^{(a} p^{b)} k_b n_a \, \delta_4 \, \mathrm{d}V - \int_{\mathscr{E}} \nabla_c \big(u^{(a} S^{b)c} \delta_4 \big) k_b n_a \, \mathrm{d}V \,, \tag{5.6}$$

with $\mathrm{d}V = \sqrt{-g} \, \mathrm{d}t \, \mathrm{d}^3 x$ the invariant 4-volume element over $\mathscr{E} = \bigcup_{t\in\mathbb{R}} \mathscr{S}_t$. The first integral in (5.6) can readily be evaluated by using the defining property (A.15) of the invariant Dirac distribution δ_4. Using the Leibniz rule and applying Stokes' theorem to the second integral yields

$$\int_{\mathscr{E}} \nabla_c \big(u^{(a} S^{b)c} \delta_4 \big) k_b n_a \, \mathrm{d}V = \int_{\partial\mathscr{E}} u^{(a} S^{b)c} k_b n_a \, \delta_4 \, \mathrm{d}\Sigma_c - \int_{\mathscr{E}} u^{(a} S^{b)c} \nabla_c (k_b n_a) \, \delta_4 \, \mathrm{d}V \,. \tag{5.7}$$

The boundary term vanishes because its support is restricted to the single point $\mathscr{P} = \gamma \cap \mathscr{S}$, which does not intersect the boundary $\partial\mathscr{E}$ of the manifold $\mathscr{E}$. The remaining integral over $\mathscr{E}$ in (5.7) can be evaluated once again by means of the property (A.15). Finally, we obtain the expression

$$I = -u^{(a} p^{b)} k_b n_a + u^{(a} S^{b)c} \nabla_c (k_b n_a) \,. \tag{5.8}$$

For the integral K defined in Eq. (4.75b) we follow the exact same steps, i.e., we perform a 3+1 decomposition, integrate by parts, apply Stokes' theorem, and lastly we use Eq. (A.15). We obtain the integrated formula

$$K = p^a k^b \delta g_{ab} + u^b S^{cd} \nabla_d (k^a n_a \delta g_{bc}) \,. \tag{5.9}$$

It should be understood that Eqs. (5.8)–(5.9) are to be evaluated at the point $\mathscr{P}$. Therefore, in the first term in the right-hand side of Eq. (5.8), one may freely use (5.1), which implies in particular $n^a|_{\mathscr{P}} = u^a$. However, those relations cannot be used in the second term in the right-hand side of Eqs. (5.8) and (5.9), because the formula (5.1) is only valid at $\mathscr{P}$, and not in an open neighborhood of it. Finally, we note that the expressions (5.8) and (5.9) hold irrespective of a particular choice of SSC.

5.1.3 Algebraic Reduction of I and K

We shall now simplify algebraically the expressions (5.8)–(5.9) for the integrals I and K. We start with the result (5.8). First, as $u^a n_a = -1$ and $k_b n_a = k_a n_b$ at $\mathscr{P}$ by virtue of (5.1), the first term is simply $p^a k_a$. For the second term, we expand the symmetry in $u^{(a}S^{b)c}$ and the gradient $\nabla_c(k_b n_a)$ by the Leibniz rule. This gives

$$I = p^a k_a + \frac{1}{2}(u^a S^{bc} + u^b S^{ac})(n_a \nabla_c k_b + k_b \nabla_c n_a)\,. \tag{5.10}$$

By substituting the formula (5.4) into Eq. (5.10), while using Killing's equation $\nabla_{(a}k_{b)} = 0$, the helical constraint $k^a|_\gamma = zu^a$, which implies $u^a k_a = -z$ and $\dot{k}^a = z\dot{u}^a$, as well as $n^a|_{\mathscr{P}} = u^a$ and the orthogonality $u^a \dot{u}_a = 0$, we readily obtain the simple expression

$$I = p^a k_a - D^a \dot{k}_a + \frac{1}{2} S^{ab} \nabla_a k_b\,, \tag{5.11}$$

where we recall that $D^a = -S^{ab}u_b$ is the mass dipole moment with respect to γ. Lastly, we can simplify the first term on the right-hand side of (5.11) by using the equality $k_a|_{\mathscr{P}} = zu_a$ and the definition (2.60) of the rest mass m, and use Leibniz rule combined with the constraint $D^a u_a = 0$, yielding

$$I = -mz + \dot{D}^a k_a + \frac{1}{2} S^{ab} \nabla_a k_b\,. \tag{5.12}$$

Next we turn to the simplification of K. We start with equation (5.9) by expanding the covariant derivative in the second term by the Leibniz rule. This gives

$$K = p^b k^c \delta g_{bc} - \big[(\nabla_d k^a) n_a \delta g_{bc} - k^a (\nabla_d n_a) \delta g_{bc} - k^a n_a \nabla_d \delta g_{bc}\big] S^{db} u^c\,, \tag{5.13}$$

of which we will consider the three terms in brackets separately. For the first term, we use Killing's equation $\nabla_{(a}k_{b)} = 0$, as well as $n^a|_{\mathscr{P}} = u^a$, so that we can write $(\nabla_d k^a) n_a|_{\mathscr{P}} = -z\dot{u}_d$. The second one vanishes, since by Eqs. (3.30) and (5.4) it is proportional to $u^a \dot{u}_a = 0$. In the third term, we use $k^a n_a|_{\mathscr{P}} = -z$. Renaming some indices and using $k^c|_{\mathscr{P}} = zu^c$ in the remaining terms yields

$$K = (p^a - S^{ab}\dot{u}_b)\delta k_a + S^{ab}k^c\nabla_a\delta g_{bc}\,, \tag{5.14}$$

where we have also used $k^c\delta g_{bc} = \delta k_b$, this last equality coming from Eq. (4.61) with $\xi^a = k^a$. Now let us focus on the first term in the right-hand side of (5.14). First we substitute the formula (2.61) and simplify the result by using the antisymmetry of S^{ab} and the Leibniz rule, so that

$$(p^a - S^{ab}\dot{u}_b)\delta k_a = m u^a \delta k_a + \dot{D}^a \delta k_a\,. \tag{5.15}$$

Second, we use in the right-hand side of (5.15) the identity $u^a\delta k_a = -2\delta z$, which is derived by applying Eqs. (4.61) and (3.30) to the equality $u^a k_a = -z$. Substituting all this into (5.14) gives the following final formula for the integral (4.75b):

$$K = -2m\delta z + \dot{D}^a\delta k_a + S^{ab}k^c\nabla_a\delta g_{bc}\,. \tag{5.16}$$

The formulae (5.12) and (5.16) for I and K are consistent with the results established in Sect. 4.3, for their integral forms (4.75). Indeed, the right-hand sides of (5.12) and (5.16) are independent of the normal vector n^a, and thus of the choice of hypersurface of integration. Moreover, these expressions only involve Lie-dragged quantities: the tensors k^a, δk_a and δg_{ab} were shown to be Lie-dragged in Sect. 4.3, while the velocity u^a and the multipoles (p^a, S^{ab}) were shown to be Lie-dragged in Chap. 3, Sect. 3.2.2, and z and m are constants of motion. The only term for which we did not prove conservation thus far is $\dot{D}^a k_a$. Fortunately, this can be done thanks to a combination of identities established in Chap. 2. Indeed, we have

$$D^a\ddot{k}_a = D^a(\dot{u}^b\nabla_b k_a) = -(D^a\nabla_a k_b)\dot{u}^b = (k^a\nabla_a D_b)\dot{u}^b = -\dot{D}_b\dot{k}^b\,, \tag{5.17}$$

where we used successively $\ddot{k}_a = \dot{u}^b\nabla_b k_a$ (which follows from the Kostant formula (3.9)), Killing's equation (3.8), the Lie-dragging of the mass dipole (3.50), and the equality $\dot{k}_a = z\dot{u}_a$ (which follows from (3.30)). Equation (5.17) readily implies that $(D^a\dot{k}_a)^{\cdot} = 0$, thanks to the Leibniz rule, and therefore each term on the right-hand side of (5.13) is separately conserved along γ.

Combining all those results with the commutation of the Lie and covariant derivatives (3.10), we find, as expected,

$$\dot{I} = z^{-1}\mathcal{L}_k I = 0 \quad \text{and} \quad \dot{K} = z^{-1}\mathcal{L}_k K = 0\,. \tag{5.18}$$

As a final note, let us mention that up to the conserved dipolar term $D^a\dot{k}_a$, the conserved integral (4.75a) is found to coincide with the Killing energy of a pole-dipole particle, a constant of motion in the dynamics of dipolar particles (see, e.g., [513, 514]) whose constancy is naturally encoded in the GKF formalism [346, 347] (recall the discussion in Chap. 2, Sect. 2.2.3).

5.1.4 Linear Combination of I and K

We are finally ready to combine the previous results for I and K in order to compute the quantity $\delta W \equiv -\delta I + K/2$ which appears in the right-hand side of the variational identity (4.74). Combining the variation of Eq. (5.12) with (5.16) readily gives

$$\delta W = \delta(mz) - \delta(\dot{D}^a k_a) - \frac{1}{2}\delta(S^a{}_b \nabla_a k^b) - m\delta z + \frac{1}{2}\dot{D}^a \delta k_a + \frac{1}{2}S^{ab}k^c \nabla_a \delta g_{bc} \, . \tag{5.19}$$

Combining the first and fourth terms yields the monopolar contribution $z\delta m$. By expanding the third term and factorizing by the spin tensor S^{ab} we obtain, after renaming some indices,

$$\delta W = z\delta m - k_a \delta \dot{D}^a - \frac{1}{2}\dot{D}^a \delta k_a - \frac{1}{2}\nabla_a k^b \delta S^a{}_b - \frac{1}{2}S^{ab}\big[g_{bc}\delta(\nabla_a k^c) - k^c \nabla_a \delta g_{bc}\big] \, . \tag{5.20}$$

The last step is to show that the last term in the right-hand side vanishes identically. The easiest way is to compute the commutator of δ and ∇_a applied to both g_{bc} and k^c, while expressing ∇_a in terms of partial derivatives and Christoffel symbols (any coordinate system will do). Then, using (4.61) and the condition $\nabla_a g_{bc} = 0$ of metric compatibility, one gets $g_{bc}\delta(\nabla_a k^c) - k^c \nabla_a \delta g_{bc} = k_c \, \delta\Gamma^c{}_{ab}$, which is symmetric under $a \leftrightarrow b$. But since it is contracted with $S^{ab} = S^{[ab]}$, the whole term vanishes identically. At last, we find

$$\delta W = z\delta m - k_a \delta \dot{D}^a - \frac{1}{2}\dot{D}^a \delta k_a - \frac{1}{2}\nabla_a k^b \delta S^a{}_b \, . \tag{5.21}$$

To obtain the first law of binary mechanics in its final form, we note that the result (5.21) is valid for a single dipolar particle in the binary system. Given the definitions (4.75) of I and K, as well as the linearity of the first law (4.74) with respect to the SEM tensor, we have $\delta M - \Omega \, \delta J = \delta W_1 + \delta W_2$, where δW_i is given by Eq. (5.21) and corresponds to the contribution of particle $\mathrm{i} \in \{1, 2\}$ to the binary system. Our final result thus reads

$$\delta M - \Omega \, \delta J = \sum_\mathrm{i} \Big(z_\mathrm{i} \, \delta m_\mathrm{i} - \frac{1}{2}\nabla_a k^b \, \delta S^a_{\mathrm{i}\,b} - k_a \, \delta \dot{D}^a_\mathrm{i} - \frac{1}{2}\dot{D}^a_\mathrm{i} \, \delta k_a \Big), \tag{5.22}$$

where z_i, m_i, $S^a_{\mathrm{i}\,b}$ and D^a_i are the redshift, the rest mass, the spin tensor and the mass dipole of the ith particle, respectively.

The variational formula (5.22) is one of the most important results of this chapter. Let us comment on this particular form of the first law of binary mechanics. First, in the simplest case of a binary system of nonspinning particles, for which $S^{ab}_\mathrm{i} = 0$, which implies $D^a_\mathrm{i} = 0$, Eq. (5.22) reduces to the standard result already established in Ref. [387], albeit by following a different route. Second, Eq. (5.22) is *exact* to dipolar order, in the sense that no truncation in the spin tensor has been performed.

Third, by imposing the SSC (2.64), the last two terms in the right-hand side of (5.22) vanish identically, and the first law takes the simple form

$$\delta M - \Omega\,\delta J \overset{\text{SSC}}{=} \sum_{\mathrm{i}} |k|\,\delta m_{\mathrm{i}} - \frac{1}{2}\sum_{\mathrm{i}} (\nabla_a k^b)\,\delta S^a_{\mathrm{i}\,b}\,, \tag{5.23}$$

where we used the fact that the redshift z_{i} coincides with the norm of the Killing field along the wordline γ_{i}, and we recall that $m_{\mathrm{i}} = -p^a_{\mathrm{i}} u^{\mathrm{i}}_a$. Equation (5.23) naturally suggests that, at higher multipolar order, the right-hand side of the first law may take the form of a multipolar expansion, with multipole index ℓ, that reads schematically (getting rid of spacetime indices)

$$\delta M - \Omega\,\delta J \sim \sum_{\mathrm{i}} \sum_{\ell \geqslant 0} \underbrace{(\nabla \cdots \nabla k)}_{\ell \text{ derivatives}}\,\delta Q^{(\ell)}_{\mathrm{i}}\,. \tag{5.24}$$

5.2 Spin Precession

In this section, we will focus on a *single* dipolar particle of the binary system and introduce an orthonormal tetrad along its worldline γ. Combined with the SSC (2.64), this will allow us to define an Euclidean 3-vector associated with the covariant spin S^a of the particle. We then discuss the evolution along γ of this 3-vector, with respect to a preferred frame that is Lie-dragged along γ. This will allow us, in the next section, to formulate the first law (5.23) in terms of scalar quantities.

5.2.1 *Orthonormal Tetrad*

We introduce an orthonormal tetrad $(e^a_\alpha) = (e^a_0, e^a_i)$, where $e^a_0 = u^a$ is taken to coincide with the 4-velocity along γ, while the italic Roman subscript $i \in \{1, 2, 3\}$ labels the spacelike vectors of the triad (e^a_1, e^a_2, e^a_3). By construction, those four vectors satisfy the orthonormality conditions

$$g_{ab} e^a_i e^b_j = \delta_{ij} \quad \text{and} \quad g_{ab} e^a_i u^b = 0\,, \tag{5.25}$$

and $u^a u_a = -1$, where δ_{ij} is the Kronecker symbol. We are interested in the evolution of this tetrad along the worldline γ. It is natural to first expand the vectors $\dot{u}^a$ and $\dot{e}^a_i$ along the tetrad. Using the fact that $\dot{u}^a u_a = 0$ and $\dot{u}_a e^a_i = -u_a \dot{e}^a_i$, these expansions take the form[1]

$$\dot{u}^a = a^i e^a_i \quad \text{and} \quad \dot{e}^a_i = a_i u^a + \omega_{ij} e^a_j\,, \tag{5.26}$$

[1] Since the labels $(i, j, k, \ldots)$ correspond to internal Euclidean indices, we may raise them and lower them indistinctly.

with the tetrad components $a_i \equiv \dot{u}_a e_i^a = -u_a \dot{e}_i^a$ and $\omega_{ij} \equiv g_{ab} \dot{e}_i^a e_j^b$. Those are closely related to the so-called Ricci rotation coefficients of the tetrad formalism in general relativity [6]. Notice that the orthogonality relations (5.25) and the metric compatibility $\nabla_c g_{ab} = 0$ imply the antisymmetry of ω_{ij}:

$$\omega_{ij} = g_{ab} \dot{e}_i^a e_j^b = -\omega_{ji} \,. \tag{5.27}$$

Consequently, ω_{ij} may be viewed as a 3×3 antisymmetric matrix with 3 degrees of freedom. It is then natural to introduce a dual 3-vector $\boldsymbol{\omega}$ whose components ω^i are given by

$$\omega^i \equiv -\frac{1}{2} \epsilon^{ijk} \omega_{jk} \quad \Longleftrightarrow \quad \omega_{ij} = -\epsilon_{ijk} \omega^k \,, \tag{5.28}$$

where ϵ_{ijk} is the totally antisymmetric Levi-Civita symbol, such that $\epsilon_{123} = 1$. Thus far, ω_i and a_i merely encode the evolution of the tetrad vectors along γ. As we shall see in the next subsections, for a geometrically motivated class of tetrads, they can be given a fairly simple physical interpretation.

5.2.2 Spin Precession

In Sect. 2.3.2 we showed how the SSC (2.64) implies the existence of a spacelike spin vector S^a, defined in Eq. (2.69). Let us now expand this vector over the triad. By Eq. (2.70a) we have $S^a u_a = 0$, so the expansion only involves spatial components S^i, such that

$$S^a = S^i e_i^a \quad \text{with} \quad S_i \equiv S_a e_i^a \,. \tag{5.29}$$

An Euclidean spin vector $\mathbf{S}$ can be defined from the three components S^i. Those are related to the tetrad components $S_{ij} \equiv S_{ab} e_i^a e_j^b$ of the spin tensor S^{ab} by an equation analogous to Eq. (5.28), namely

$$S^i = -\frac{1}{2} \epsilon^{ijk} S_{jk} \quad \Longleftrightarrow \quad S_{ij} = -\epsilon_{ijk} S^k \,, \tag{5.30}$$

where we used the definition (2.69) and the formula $\varepsilon_{abcd} u^b e_i^c e_j^d = \epsilon_{ijk} e_a^k$, which follows from the expansion $\varepsilon^{abcd} = -4!\, u^{[a} e_1^b e_2^c e_3^{d]}$ of the volume form ε_{abcd} on the orthonormal basis (u^a, e_i^a). Equation (5.30) allows us to compute the Euclidean norm of the spin vector $\mathbf{S}$, which is found to be conserved, in the sense that [recall Eqs. (2.60c) and (2.66)]

$$\delta_{ij} S^i S^j = \frac{1}{2} S_{ab} S^{ab} = S^2 \,. \tag{5.31}$$

Next, we look for an equation of evolution along γ for the Euclidean spin vector $\mathbf{S} = (S^i)$. To do so, we simply compute the proper time derivative of the equality

$S_i = S_a e^a_i$ while using Eqs. (2.72) and (5.26). With the help of Eq. (5.25), the resulting formula can be turned into an evolution equation for **S** that reads

$$\dot{S}_i = \omega_{ij} S^j \iff \dot{\mathbf{S}} = \boldsymbol{\omega} \times \mathbf{S}\,, \tag{5.32}$$

where we used (5.28) to introduce a cross product. With this Newtonian-looking (but exact) equation of precession for the spin vector **S**, the vector $\boldsymbol{\omega}$ can be interpreted as the precession frequency vector for **S**. This spin vector processed in the (e^a_i) frame with an angular frequency ω given by

$$\omega^2 \equiv \delta_{ij}\omega^i\omega^j = \frac{1}{2}\omega_{ij}\omega^{ij} = \frac{1}{2}\left(\dot{e}^a_i \dot{e}^i_a + \dot{u}^a \dot{u}_a\right), \tag{5.33}$$

where the last two equalities follow from the relations (5.28) and (5.27), respectively, together with the identity $\delta^{ij} e^a_i e^b_j = g^{ab} + u^a u^b$. Despite the natural interpretation of (5.32) as a spin precession equation for the 3-vector **S**, the precession frequency 3-vector $\boldsymbol{\omega}$ depends on the choice of triad [as Eq. (5.33) illustrates most clearly], and as such has no invariant meaning.

5.2.3 A Geometrically-Motivated Class of Tetrads

In order to give an invariant meaning to the spin precession frequency, thereafter we shall restrict ourselves to the geometrically-motivated class of tetrads (u^a, e^a_i) that are Lie-dragged along the helical Killing field k^a, or equivalently along u^a. Because we already have $\mathcal{L}_k u^a = 0$ (cf. Chap. 3), we additionally require that

$$\mathcal{L}_k e^a_i = 0\,. \tag{5.34}$$

Since $k^a|_\gamma = z u^a$, the formula (5.34) implies that e^a_i evolves along γ according to $z\dot{e}^a_i = e^c_i \nabla_c k^a$. Using the expansions (5.26) and projecting on the tetrad readily implies

$$z\omega_{ij} = e^a_i e^b_j \nabla_a k_b\,, \tag{5.35a}$$

$$z a_i = u^a e^b_i \nabla_a k_b\,, \tag{5.35b}$$

which shows that ω_{ij} is manifestly antisymmetric via Killing's equation. This provides a new interpretation of ω_{ij} and a_i which are, up to a factor of z, the space-space and space-time components of the Killing 2-form $\nabla_a k_b$ in the Lie-dragged tetrad, respectively.[2] In other words,

$$\nabla^a k^b \overset{\gamma}{=} 2 z a^i e^{[a}_i u^{b]} + z\omega^{ij} e^a_i e^b_j\,. \tag{5.36}$$

---

[2] The time-time component $u^a u^b \nabla_a k_b$ vanishes by virtue of Killing's equation $\nabla_{(a} k_{b)} = 0$.

This formula will allow us to compute, in Sect. 5.2.4, the norm of $\nabla_a k_b$ along γ. Notice also that, since $\mathcal{L}_k \nabla_a k_b = \nabla_a \mathcal{L}_k k_b = 0$ (recall Eq. (3.10)) and $\mathcal{L}_k z = z\dot{z} = 0$, both right-hand sides of Eqs. (5.35) are Lie-dragged, such that

$$\dot{\omega}_{ij} = z^{-1} \mathcal{L}_k \omega_{ij} = 0\,, \tag{5.37a}$$

$$\dot{a}_i = z^{-1} \mathcal{L}_k a_i = 0\,. \tag{5.37b}$$

Combining this result with the definition (5.28) gives the important result that the precession frequency vector is constant along the worldline, in this Lie-dragged frame:

$$\dot{\boldsymbol{\omega}} = \mathbf{0}\,. \tag{5.38}$$

On the other hand, we showed in Chap. 3, Sect. 3.2.2 that $\mathcal{L}_k S^a = 0$. From the definition (5.29) and the property (5.34), it readily follows that

$$\dot{S}_i = z^{-1} \mathcal{L}_k S_i = 0\,. \tag{5.39}$$

Therefore each component S^i is conserved along the worldline, and so is the vector $\mathbf{S}$. By the equation of spin precession (5.32), the spin $\mathbf{S}$ must thus satisfy $\boldsymbol{\omega} \times \mathbf{S} = \mathbf{0}$; as a consequence it must be aligned or anti-aligned with the precession frequency vector $\boldsymbol{\omega}$. Introducing the notation $\mathbf{n} = (n^i)$ for their common constant direction, such that $\mathbf{n} \cdot \mathbf{n} = 1$, we thus have

$$\boldsymbol{\omega} = \omega\, \mathbf{n}\,, \tag{5.40a}$$

$$\mathbf{S} = S\, \mathbf{n}\,, \tag{5.40b}$$

where ω, S and n_i are constant. We emphasize that although the results (5.37)–(5.40) express conservation laws, they have little dynamical contents by themselves, in the sense that they rely crucially on the particular choice of a Lie-dragged triad obeying Eq. (5.34).

5.2.4 *Precession Frequency and Vorticity*

In this subsection, we shall express the Euclidean norm ω of the spin precession frequency vector $\boldsymbol{\omega}$ in terms of geometrically-defined quantities related to the helical Killing field (3.7). This scalar will, in particular, be shown to be closely related to the vorticity V^a associated to k^a. Let us consider the norm $|\nabla k|$ of the 2-form $\nabla_a k_b$ along the worldline γ. By making use of the orthogonality properties (5.25), the formula (5.36) readily implies

$$|\nabla k|^2 \equiv \frac{1}{2} \nabla^a k^b \nabla_a k_b \overset{\gamma}{=} z^2 (\omega^2 - a^2)\,, \tag{5.41}$$

where we introduced $a^2 \equiv \boldsymbol{a} \cdot \boldsymbol{a} = \dot{u}^a \dot{u}_a$ by virtue of (5.26), and we recall that $\omega^2 \equiv \boldsymbol{\omega} \cdot \boldsymbol{\omega} = \frac{1}{2} \omega_{ij} \omega^{ij}$. All the scalar fields appearing in this formula are constant along γ, as was shown in Chap. 3 and Eq. (5.38). The second term in the right-hand side of Eq. (5.41) can alternatively be written as

$$z^2 a^2 \overset{\gamma}{=} \nabla_a |k| \nabla^a |k| \,, \tag{5.42}$$

where we made use of Eq. (3.74) derived in Chap. 3. Combining Eqs. (5.41) and (5.42) yields an exact, coordinate-invariant and frame-independent expression for the (redshifted) norm $z\omega$ of the precession frequency $\boldsymbol{\omega}$, namely

$$z^2 \omega^2 \overset{\gamma}{=} \frac{1}{2} \nabla_a k_b \nabla^a k^b + \nabla_a |k| \nabla^a |k| \,. \tag{5.43}$$

This simple formula generalizes—for a spinning particle that follows a nongeodesic motion driven by the Mathisson-Papapetrou spin force (2.57a)—the result of Refs. [436, 437, 515], which was established in the particular case of a small but massive test spin that follows a geodesic motion in a (properly regularized) helically-symmetric perturbed black hole spacetime.

To linear order in the spin, the motion is not geodesic because $\dot{u}^a = O(S)$, as established in Eq. (2.77). Therefore, the second term in the right-hand side of Eq. (5.43) is of quadratic order in the spin, such that

$$z\omega \overset{\gamma}{=} |\nabla k| + O(S^2) \,. \tag{5.44}$$

Comparing this formula with Eq. (3) of Ref. [436] or Eqs. (2.11)–(2.12) of Ref. [437], we notice an extra factor of the redshift z. This can easily be understood because the helical Killing field considered in Ref. [436] is normalized such that $k^a|_\gamma = u^a$ (equivalent to $z = 1$), while the spin precession frequency considered in Ref. [437] is defined with respect to the coordinate time t, and not with respect to the proper time τ (while $z = \mathrm{d}\tau/\mathrm{d}t$ in adapted coordinates).

We now establish a simple relation between the Lorentzian vorticity V^a and the Euclidean spin precession frequency $\boldsymbol{\omega}$, through the basis vectors of the Lie-dragged tetrad introduced in Sect. 5.2.3. By substituting Eq. (5.36) into the definition (3.57) of the vorticity and by using the identity $\varepsilon_{abcd} u^b e_i^c e_j^d = \epsilon_{ijk} e_a^k$ and the definition (5.28) of the precession 3-vector, we obtain

$$V^a = z\omega^i e_i^a \,. \tag{5.45}$$

Equation (5.45) establishes that the components of the vorticity V^a with respect to the Lie-dragged frame (e_i^a) coincide—up to a redshift factor—with the Euclidean components ω^i of the spin precession frequency vector $\boldsymbol{\omega}$. Moreover, by comparing Eq. (5.29) and (5.45), we conclude that the Euclidean colinearity (5.40) of $\boldsymbol{\omega}$ and $\mathbf{S}$ implies the Lorentzian colinearity of V^a and S^a, in agreement with the conclusion (3.62) reached in Sect. 3.3.

Finally, using the orthonormality of the triad (e_i^a), we find that (5.45) implies the following simple relationship between the Lorentzian norm V of the vorticity V^a and the Euclidean norm ω of spin precession frequency $\boldsymbol{\omega}$:

$$V = z\omega\,. \tag{5.46}$$

Alternatively, this conclusion could be reached by computing the norm of the vorticity (3.57) and comparing it with Eq. (5.43). The conservation of z and ω along γ [recall Eq. (5.38)] is of course compatible with that of V, as established in Sect. 3.3.

5.3 Hamiltonian First Law of Mechanics

In this section, we shall compare our variational formula (5.23) to the canonical Hamiltonian first law of mechanics established in Ref. [406], for a binary system of spinning particles moving along circular orbits, for spins aligned or anti-aligned with the orbital angular momentum. In Sect. 5.3.1 we first rewrite (5.23) in term of the scalars ω and S, to linear order in the spins, by using a Lie-dragged tetrad as introduced in the previous section. Then, in Sect. 5.3.2 the scalars S and ω are shown to be related to the Euclidean norms of the canonical spin variable and the spin precession frequency used in Ref. [406], allowing us to prove the equivalence of the differential geometric first law (5.23) to the Hamiltonian first law of Ref. [406].

5.3.1 Alternative Form of the First Law

In this first subsection, we write the dipolar contribution $\nabla_a k^b \, \delta S^a{}_b$ in Eq. (5.22) in terms of the conserved scalars ω (or $|\nabla k|$) and δS that were defined in Eqs. (5.33) [or Eq. (5.41)] and (5.31), respectively. We will do so at linear order in the spins, since this is all we need in order to compare to the Hamiltonian first law of Ref. [406].

By using the formulas (2.69) and (3.58) for $S^a{}_b$ and $\nabla_a k^b$, as well as the Leibniz rule and the antisymmetry of ε^{abcd}, we readily obtain

$$\begin{aligned}\nabla_a k^b \delta S^a{}_b &= \varepsilon^{abcd} \varepsilon_{abef} u^e V^f \delta(u_c S_d) + \varepsilon_a{}^b{}_{ef} u^e V^f u_c S_d \, \delta\varepsilon^a{}_b{}^{cd} \\ &\quad + 2\varepsilon^{abcd} k_a \dot{u}_b \, \delta(u_c S_d) + (k_a \dot{u}^b - \dot{u}_a k^b) u_c S_d \, \delta\varepsilon^a{}_b{}^{cd}\,. \end{aligned} \tag{5.47}$$

Let us consider those four terms successively. Using the Leibniz rule and the orthogonality of u^a to both V^a and S^a, the first term reduces to $2V^a \delta S_a - 2V^c S_c u^a \delta u_a$. With the help of the identity $\delta\varepsilon^{abcd} = -\frac{1}{2}\varepsilon^{abcd} g^{ef} \delta g_{ef}$, the second term simplifies to $-V^a S^b \delta g_{ab} + V^c S_c u^a u^b \delta g_{ab}$. Similarly, the third and fourth terms yield $2S^{ab} \dot{k}_a \delta u_b$ and $-S^{ab} \dot{u}_a k^c \delta g_{bc}$, respectively. Using $2u^a \delta u_a = u^a u^b \delta g_{ab}$, as a consequence of $u^a u_a = -1$, as well as $\delta k^c = 0$, we then obtain

$$\nabla_a k^b \delta S^a{}_b = 2V^a \delta S_a - V^a S^b \delta g_{ab} + 2S^{ab} \dot{k}_a \delta u_b - S^{ab} \dot{u}_a \delta k_b\,. \tag{5.48}$$

Next, we use the colinearity (3.62) of V^a and S^a, as a consequence of the SSC (2.64), so that the first two terms combine to give $2V\delta S$. The last two terms can be simplified thanks to the SSC (2.64) and the helical constraint (3.30). Using the relation (5.46) we finally obtain the simple result

$$\nabla_a k^b \delta S^a{}_b = 2z\omega\delta S + \dot{u}_a S^{ab} \delta k_b\,. \tag{5.49}$$

Therefore, the dipolar contribution in Eq. (5.23) involves a term $z\omega\delta S$ that is linear in spin and a term proportional to $\dot{u}_a S^{ab} \delta k_b$, which is quadratic in spin. Indeed, by virtue of (2.77) we have

$$\dot{u}_a S^{ab} \delta k_b = \frac{1}{m} S^a B_{ab} S^{bc} \delta k_c + O(S^3)\,. \tag{5.50}$$

We are at last ready to write down the first law of compact binary mechanics in terms of scalar quantities, to *linear* order in the spin amplitudes. To do so, we substitute Eqs. (5.49)–(5.50) into (5.23) for each particle, and obtain the simple variational formula

$$\delta M - \Omega\,\delta J \overset{\text{SSC}}{=} \sum_{\text{i}} z_{\text{i}}\,(\delta m_{\text{i}} - \omega_{\text{i}}\,\delta S_{\text{i}}) + O(S_{\text{i}}^2)\,. \tag{5.51}$$

This is another important result of our work. Interestingly, for a helically symmetric spacetime that contains one/two black holes, the necessary conditions of vanishing expansion and shear (i.e. Killing horizon) imply that each black hole must be in a state of co-rotation [206, 382, 388, 406]. By contrast, for a binary system of dipolar particles, the helical isometry merely constrains each spin vector $\mathbf{S}_{\text{i}}$ to be aligned with the precession frequency vector $\boldsymbol{\omega}_{\text{i}}$ [recall Eqs. (5.40)], while the spin amplitude S_{i} of each particle in Eq. (5.51) is left entirely free.

Now, recalling that $z_{\text{i}} = |k|_{\text{i}}$ and $z_{\text{i}}\omega_{\text{i}} = |\nabla k|_{\text{i}} + O(S_{\text{i}}^2)$, from Chap. 3 and Eq. (5.44) above, the variational formula (5.51) looks explicitly like an expansion in powers of (the norms of) the covariant derivatives of the helical Killing vector field k^a, namely

$$\delta M - \Omega\,\delta J \overset{\text{SSC}}{=} \sum_{\text{i}} (|k|_{\text{i}}\,\delta m_{\text{i}} - |\nabla k|_{\text{i}}\,\delta S_{\text{i}}) + O(S_{\text{i}}^2)\,. \tag{5.52}$$

This naturally suggests that, at the next quadrupolar order, one might obtain an additional contribution of the form $\sum_{\text{i}} |\nabla\nabla k|_{\text{i}}\,\delta Q_{\text{i}}$, where the double covariant derivative of the helical Killing field can be related to the curvature tensor through the Kostant formula (3.9), and Q_{i} would be the spacetime norm of the quadrupole moment tensor of each particle.

5.3.2 Comparison to the Hamiltonian First Law

By using the canonical Arnowitt-Deser-Misner (ADM) Hamiltonian framework of general relativity applied to spinning point particles, the authors of Ref. [406] derived a first law of mechanics for binary systems of compacts objects with spins aligned (or anti-aligned) with the orbital angular momentum, to *linear* order in the spins. Our goal is to relate this earlier result to the scalar version (5.51) of the first law, which also holds to linear order in spins.

In Sect. 5.2 we have established the geometrical precession, with respect to an orthonormal frame (e_i^a) orthogonal to γ and Lie-dragged along it, of a Euclidean spin vector $\mathbf{S}$ orthogonal to γ. Using $z = \mathrm{d}\tau/\mathrm{d}t$, which holds in any coordinate system adapted to the helical isometry (cf. Chap. 3), Eqs. (5.32) and (5.44) can be rewritten, for each particle, in the equivalent form

$$\frac{\mathrm{d}\mathbf{S}_\mathrm{i}}{\mathrm{d}t} = z_\mathrm{i}\boldsymbol{\omega}_\mathrm{i} \times \mathbf{S}_\mathrm{i}\,, \quad \text{where} \quad z_\mathrm{i}\omega_\mathrm{i} \overset{\gamma_\mathrm{i}}{=} |\nabla k|_\mathrm{i} + O(S_\mathrm{i}^2)\,. \tag{5.53}$$

As shown in Refs. [437, 516], one can relate this well established kinematical result to *dynamical* properties of spin-orbit coupling in a binary system of spinning compact objects.

This can be done in the context of the canonical ADM Hamiltonian framework of general relativity, applied to a binary system of spinning point masses with canonical positions $\mathbf{r}_\mathrm{i}(t)$, momenta $\mathbf{p}_\mathrm{i}(t)$ and spins $\bar{\mathbf{S}}_\mathrm{i}(t)$. In the center-of-mass frame, the dynamics depends on the relative position $\mathbf{r} \equiv \mathbf{r}_1 - \mathbf{r}_2$, the relative momentum $\mathbf{p} \equiv \mathbf{p}_1 = -\mathbf{p}_2$ and the individual spins. The evolution of the canonical variables is then governed, to linear order in the spins, by a canonical Hamiltonian

$$H(\mathbf{r}, \mathbf{p}, \bar{\mathbf{S}}_\mathrm{i}) = H_\mathrm{orb}(\mathbf{r}, \mathbf{p}) + \sum_\mathrm{i} \boldsymbol{\Omega}_\mathrm{i}(\mathbf{r}, \mathbf{p}) \cdot \bar{\mathbf{S}}_\mathrm{i}\,, \tag{5.54}$$

where the pseudo-vectors $\boldsymbol{\Omega}_\mathrm{i} = \sigma_\mathrm{i}\,\mathbf{L}$ are both proportional to the orbital angular momentum $\mathbf{L} = \mathbf{r} \times \mathbf{p}$, while σ_i is closely related to the "gyro-gravitomagnetic" ratio g_i of the i-th particle [517]. The usual Poisson bracket structure of the Cartesian components $\bar{S}_\mathrm{i}^j$ of the canonical spin $\bar{\mathbf{S}}_\mathrm{i}$ ensures that the spin-orbit (linear-in-spin) part of the canonical Hamiltonian (5.54) implies Newtonian-looking, but exact precession equations of the form [516]

$$\frac{\mathrm{d}\bar{\mathbf{S}}_\mathrm{i}}{\mathrm{d}t} = \boldsymbol{\Omega}_\mathrm{i} \times \bar{\mathbf{S}}_\mathrm{i}\,. \tag{5.55}$$

Consequently, the Euclidean norm $\|\bar{\mathbf{S}}_\mathrm{i}\|$ of each canonical spin variable is conserved. For that reason, the canonical variables $\bar{\mathbf{S}}_\mathrm{i}$ are known as the "constant-in-magnitude" spins. Those variables are, however, by no means unique. In particular, it can be shown that the gauge freedom (local rotation group) to perform an infinitesimal rotation of $\bar{\mathbf{S}}_\mathrm{i}$ can be seen as being induced by an infinitesimal canonical transformation in the full phase space [516].

Despite the striking similarity between Eqs. (5.53) and (5.55), the canonical spin variable $\bar{\mathbf{S}}_{\rm i}$ need not coincide with the Euclidean spin vector $\mathbf{S}_{\rm i}$ constructed from the components of the 4-vector $S_{\rm i}^a$ along an orthonormal triad (e_i^a), and the precession frequency $z_{\rm i}\boldsymbol{\omega}_{\rm i}$ need not coincide with the coefficient $\boldsymbol{\Omega}_{\rm i}$ appearing in the spin-orbit piece of the canonical Hamiltonian (5.54). In fact, several definitions of a globally SO(3)-compatible, canonical spin variable $\bar{\mathbf{S}}$ constructed from a spacelike, covariant 4-vector S^a are possible [437, 516]. For instance, given the spatial components S_i of the covector $S_a = g_{ab} S^b$ with respect to an ADM coordinate system compatible with the Hamiltonian formulation, one may define a particular canonical spin variable according to

$$\bar{S}^i = H^{ij} S_j \,, \tag{5.56}$$

where H^{ij} is the unique symmetric and positive definite square root of the effective metric $G^{ij} \equiv g^{ij} - 2g^{0(i}v^{j)} + g^{00} v^i v^j$, which is constructed from the components $g^{\alpha\beta}$ of the inverse metric and the components $v^\alpha = u^\alpha / u^0$ of the coordinate 3-velocity of the particle [516, 518]. The key point is that the Euclidean norm $\bar{S} \equiv \|\bar{\mathbf{S}}\|$ of the canonical spin variable (5.56) is numerically equal to the norm (5.31) of the spin vector $\mathbf{S}$ (and is thus conserved):

$$\bar{S}^2 \equiv \delta_{ij} \bar{S}^i \bar{S}^j = S^2 \,. \tag{5.57}$$

Moreover, the global SO(3) symmetry of the Hamiltonian dynamics generated by (5.54) implies, in particular, that the conserved total angular momentum vector $\mathbf{J}$ has the simple, additive form $\mathbf{J} = \mathbf{L} + \sum_{\rm i} \bar{\mathbf{S}}_{\rm i}$ [516]. For spins aligned or anti-aligned with the orbital angular momentum $\mathbf{L}$, this reduces to the algebraic equality

$$J = L + \sum_{\rm i} \bar{S}_{\rm i} \,. \tag{5.58}$$

Finally, in the particular case of circular orbits of angular velocity Ω, the analysis of Ref. [437] (see also [436]) shows that, for each particle, the Euclidean norm $\Omega_{\rm i} \equiv \|\boldsymbol{\Omega}_{\rm i}\|$ of the precession frequency vector of the canonical spin variable $\bar{\mathbf{S}}_{\rm i}$ is related to the Lorentzian norm (5.41) of the helical Killing 2-form $\nabla_a k_b$ by the simple numerical link

$$\Omega_{\rm i} = \Omega - |\nabla k|_{\rm i} \,. \tag{5.59}$$

The occurrence of the circular-orbit angular velocity Ω can be understood from the fact that the precession frequency $\Omega_{\rm i}$ is defined with respect to the coordinate time of the global chart associated with the 3+1 split required to construct the canonical Hamiltonian (5.54), while the (redshifted) precession frequency $z_{\rm i}\omega_{\rm i} = |\nabla k|_{\rm i}$ is defined with respect to a local spatial frame that is Lie-dragged along the spinning particle's worldline. Importantly, the relation (5.59) was established to linear order in the spin amplitudes $S_{\rm i}$.

Substituting the relationships (5.57)–(5.59) into the geometric, scalar first law (5.51), we readily find

$$\delta M - \Omega\,\delta L = \sum_{\rm i} (z_{\rm i}\,\delta m_{\rm i} + \Omega_{\rm i}\,\delta \bar{S}_{\rm i}) + O(\bar{S}_{\rm i}^2)\,. \tag{5.60}$$

Provided that the total mass M coincides with the on-shell value of the Hamiltonian (5.54), the variational formula (5.60) precisely agrees with the Hamiltonian first law of mechanics derived in Ref. [406] [see Eq (4.6) there], for binary systems of spinning compact objects moving along circular orbits, with spins aligned or anti-aligned with the orbital angular momentum, to linear order in the spins. The first integral associated with the variational formula (5.60) reads [406]

$$M - 2\Omega L = \sum_{\rm i} (z_{\rm i} m_{\rm i} + 2\Omega_{\rm i}\,\bar{S}_{\rm i}) + O(\bar{S}_{\rm i}^2)\,. \tag{5.61}$$

Combined with Eqs. (5.58) and (5.59), together with the formula (5.44), this agrees with the "geometrical" first integral relation (6.43), as we shall see in the next chapter.

Chapter 6
Extensions at Quadrupolar Order

Contents

Dear Mr Poincaré, I am sorry to say it is too true that there are, as you tell me in your letter which I have received this morning, several mistakes in respect to magnitude and to sign in my statements regarding the nutation which would exist if the earth consisted of a rigid ellipsoidal shell filled with frictionless liquid.

W. Thomson
Letter to H. Poincaré (1901).

In this last chapter, the aim is to extend the results of Chap. 5, obtained at dipolar order, to account for several finite-size effects. In particular, we detail two quadrupolar models, that account for the deformation of the physical body described by the particle: a spin-induced quadrupole, which encodes the information on the response of a body to its proper rotation, and a tidally-induced quadrupole, which accounts

P. Ramond, *The First Law of Mechanics in General Relativity & Isochrone Orbits in Newtonian Gravity*, Springer Theses,
https://doi.org/10.1007/978-3-031-17964-8_6

for the tidal effects. These models are reviewed in Sect. 6.1 along with some useful identities and remarks that will help afterwards. In Sect. 6.2, we derive an integral first law of mechanics at quadrupolar order, valid for any quadrupolar model. This result extends the integral first law that can be found in the literature for spinning particles, with which we make contact for a comparison. This section includes many calculations that are covariant and free of any spin supplementary condition. Lastly, in (the very short) Sect. 6.3, we provide a number of preliminary results for the variational first law at quadrupolar order. Although its derivation is still underway, we motivate and discuss the anticipated final result.

6.1 Quadrupolar Particles

We reviewed in Chap. 2, Sect. 2.3 the main results regarding quadrupolar particles. In particular, we saw there that at the quadrupolar level, a particle is endowed with three multipoles: the four-momentum p^a, the spin tensor S^{ab} and the quadrupole tensor J^{abcd}. Associated to this particle is a SEM tensor, which takes the form of the dipolar part T^{ab}_{dipo} and a quadrupolar part T^{ab}_{quad} (given in Chap. 2, Eqs. (2.58)–(2.59)) which we rewrite here for convenience

$$T^{ab}_{\mathsf{dipo}} = \int_\gamma u^{(a} p^{b)} \, \delta_4 \, \mathrm{d}\tau + \nabla_c \int_\gamma u^{(a} S^{b)c} \, \delta_4 \, \mathrm{d}\tau \,, \tag{6.1a}$$

$$T^{ab}_{\mathsf{quad}} = \frac{1}{3} \int_\gamma R_{cde}{}^{(a} J^{b)cde} \, \delta_4 \, \mathrm{d}\tau - \frac{2}{3} \nabla_{cd} \int_\gamma J^{c(ab)d} \, \delta_4 \, \mathrm{d}\tau \,. \tag{6.1b}$$

We also pointed out in Chap. 2 that multipolar schemes do not provide an evolution equation for J^{abcd}, and that closing the system of evolution equations requires (an SSC and) an expression for J^{abcd} in terms of the other unknowns of the problem, namely the four-velocity u^a, the multipoles (p^a, S^{ab}), and the geometry (g_{ab}, R_{abcd}) evaluated at the location of the particle. In this first section, we give the two types of quadrupolar models that will be considered in this work.

Preliminaries

As we mentioned already in Chap. 2, the quadrupole tensor J^{abcd} possesses the same algebraic symmetries as the Riemann tensor R_{abcd}. This is reminiscent of its definition in the Lagrangian formalism, as the conjugated momentum of the Riemann tensor. Using these symmetries, we can decompose any quadrupole into a set of three tensors of smaller rank, as [318, 371, 519]

$$J^{abcd} = \hat{J}^{abcd} + 2\big(u^{[a} J^{b]cd} + u^{[d} J^{c]ba}\big) - 4 u^{[a} J^{b][c} u^{d]} \,, \tag{6.2}$$

where $\hat{J}^{abcd}$ is the stress quadrupole, J^{abc} the momentum quadrupole, and J^{ab} the mass quadrupole, respectively. These names come from the fact that these tensors reduce to their Newtonian equivalent (when they have one) in the right limit, as discussed in [520]. They are built from various space/time contractions of J^{abcd}, as

$$\hat{J}^{abcd} \equiv J^{\hat{a}\hat{b}\hat{c}\hat{d}}\,, \quad J^{acd} \equiv u_b J^{\hat{a}b\hat{c}\hat{d}} \quad \text{and} \quad J^{ab} \equiv u_c u_d J^{\hat{a}c\hat{b}d}\,, \tag{6.3}$$

reminding the reader that a hatted index denotes a projection orthogonal to the four-velocity. Consequently, these tensors are orthogonal to u_a with respect to all their indices by construction, and verify the following algebraic symmetries, as can be checked easily

$$J^{[ab]} = 0\,, \quad J^{a(bc)} = 0 \quad \text{and} \quad J^{[abc]} = 0\,. \tag{6.4}$$

In terms of components, J^{abcd} contains 20 independent ones, with 6 in J^{ab} + 6 in J^{abc} + 8 in $\hat{J}^{abcd}$. Each of the three terms in the right-hand side of Eq. (6.2) verifies the algebraic symmetries of the Riemann tensor, and, therefore, any quadrupolar model should be of one of these three forms. This will be the case for the two models considered below.

The previous "3 + 1" decomposition of J^{abcd} depends only on its algebraic symmetries and could very well be applied to the Riemann tensor R_{abcd} itself. Instead of this 3 + 1 expansion, we shall introduce the classical Hodge, electric-magnetic decomposition of the Riemann tensor, much more convenient for our purpose. These two rank-2 tensors are defined as follows

$$\begin{cases} E_{ab} \equiv R_{acbd} u^c u^d\,, \\ B_{ab} \equiv \star R_{acbd} u^c u^d\,, \end{cases} \quad \text{where} \quad \star R_{abcd} \equiv \tfrac{1}{2} \varepsilon_{ab}{}^{ef} R_{efcd}\,, \tag{6.5}$$

is the self-dual of R_{abcd}. The two models of quadrupoles that will be considered will be easily expressed in terms of these electric-part E_{ab} and magnetic-part B_{ab} of the Riemann tensor. For a physical interpretation of these fundamental quantities, see [515] and references therein.

6.1.1 Quadrupole Models

Generally, the explicit form of a quadrupole tensor J^{abcd} cannot be obtained[1] from the point-particle models presented in Chap. 2. However, if the quadrupole is expected to depend on the pole-dipole properties of the body (and geometric quantities evaluated at its location), then it is possible to infer the generic form that J^{abcd} must have, if it is to be compatible with a number of natural assumptions. The most natural way

[1] Except for its 4-indices nature, its algebraic symmetries, and the actual number of degrees of freedom that do enter the evolution equations, which is ten (i.e., 10 less than the total number of independent degrees of freedom in J^{abcd}), see for example [368] for details.

of doing this is via some kind of effective-field-theory method, extending ideas from the pole-dipole Lagrangian formalism presented in Chap. 2, Sect. 2.2.1. Giving details about these constructions would be much beyond the scope of this chapter, and, for the overall scheme of EFT we refer the reader to the recent review [195], as well as references therein. Although a study of the compatibility between EFT assumptions and those of our work is necessary (and underway), we shall simply use the EFT results "as is" in what follows, as explicit expressions for the quadrupole. However, we shall work with formulae as general as possible as long as we can. In particular, most of the following results will be valid for any J^{abcd}, thanks to the generic decomposition in Eq. 6.2.

Spin-Induced Quadrupole

As in Newtonian gravity, we shall consider the deformation of a body resulting from its proper rotation, or *spin-induced* quadrupole, denoted J^{abcd}_{spin}. Its expression is well-known [195, 318, 320, 521–525], and, in a nutshell, results from the addition of a scalar in the dipolar Lagrangian (2.13) that is (1) built from the Riemann tensor (evaluated at the particle's location) and the spin tensor S^{ab} of the body, and (2) preserves the aforementioned symmetry/conservation conditions for the Lagrangian. The unique possibility[2] is a term $\propto E_{ab} S^a{}_c S^{cb}$. Introducing a dimensionless, coupling constant, the spin-induced quadrupole readily follows [320]

$$J^{abcd}_{\text{spin}} \equiv \frac{\kappa_2}{2m} u^{[a} S^{b]e} S_e{}^{[c} u^{d]} , \tag{6.6}$$

where, in the prefactor $1/2m$ is conventional, and the dimensionless constant κ_2 measures the quadrupolar "polarisability" of the body induced by its proper rotation. In practice, one has $\kappa_2 = 1$ for an isolated (i.e. Kerr) black hole [155, 526] and $\kappa_2 \sim 4 - 8$ for a neutron star [310, 527, 528], depending on its equation of state. In the EFT formalism, this constant is simply the (non-minimal) coupling constant between gravity and the extended object, and is written C_{ES^2} [195]. Indeed, when S^{ab} is expressed into a spin vector S^a (and thus under a given SSC), the right-hand side of Eq. (6.6) is simply $\propto E_{ab} S^a S^b$. If no SSC is imposed, then from the general (SSC-free) decomposition (cf. Chap. 2, Eq. (2.63)) of S^{ab} in terms of a spin vector S^a and a mass dipole D^a, an easy calculation shows that L_{S} also contains a term $\propto E_{ab} D^a D^b$. Since we try to impose the SSC as late as possible, we keep the general expression (6.6) for now. Lastly, we note that the spin-induced quadrupole gives a mass-type contribution to the total quadrupole J^{abcd}, according to the decomposition Eq. (6.2).

---

[2] Under suitable redefinition (or re-absorption of terms) in the other multipoles [320, 523, 524].

Tidally-Induced Quadrupole

The spin-induced quadrupole arises naturally in the case of spinning particles. However, there are also quadrupolar deformations to an extended body in the case of massive but non-spinning bodies, in particular for binary systems, through tidal interactions. In this case, the Lagrangian and/or EFT formalism can be applied and also give explicit formulae for the tidally-induced quadrupole. Once again, the idea is to add a scalar term to the (non-spinning) point-particle Lagrangian that (1) depends on the Riemann tensor (evaluated at the particle's location), and (2) preserves the Lagrangian covariance and parametrization invariance of the worldline. From these constraints, two possible contributions emerge, and read [529–532] (see also Sect. 3.1 in [195]) one $\propto E_{ab}E^{ab}$ and another $\propto B_{ab}B^{ab}$. From these expressions follows the tidally-induced quadrupole, made of two contributions, an "electric" one and a "magnetic" one, given by

$$J^{abcd}_{\text{elec}} = \frac{3\mu_2}{2}\, u^{[a} E^{b][c} u^{d]}\,, \tag{6.7a}$$

$$J^{abcd}_{\text{mag}} = -2\sigma_2\big(u^{[a} B^{b]f} \varepsilon^{cd}{}_{ef} u^e + u^{[c} B^{d]f} \varepsilon^{ab}{}_{ef} u^e\big)\,, \tag{6.7b}$$

where, again, the numerical prefactors are conventional in the literature, and this time the constants μ_2 and σ_2 (so-called quadrupolar "tidal Love numbers") have dimension [length]^5. Remarkably, μ_2 and σ_2 vanish for a nonspinning black hole [533–538]. For a nonspinning neutron star of areal radius R,

$$\mu_2 = \frac{2}{3} R^5 k_2 \quad \text{and} \quad \sigma_2 = \frac{1}{48} R^5 j_2\,, \tag{6.8}$$

with $k_2 \sim 0.05 - 0.15$ and $|j_2| \lesssim 0.02$, depending on the equation of state [533, 534]. The effect of the spin of the compact objects on the tidal Love numbers was explored in Refs. [539–543]. The quadrupole model (6.7) assumes that each compact object responds *adiabatically* and *linearly* to the tidal field induced by the orbiting companion. A more realistic model would include a dynamical response of the body to the applied tidal field [544, 545] and would account for nonlinear tidal effects (see e.g. footnote 2 in [530]). Interestingly, we note that the electric part of the tidally-induced quadrupole (6.7a), just like the spin-induced part, induces a mass-type contribution to the total quadrupole J^{abcd}, according to the decomposition Eq. (6.2). It contrasts with the magnetic part (6.7b) corresponding to a momentum-type quadrupole.

6.2 Integral First Law at Quadrupolar Order

As shown e.g. in Refs. [387, 388, 406, 426, 427, 467], each variational first law of binary mechanics implies the existence of an associated "first integral" relationship. This is the case for the laws derived in this work too, and is the topic of discussion for this section. This "integral first law" does not necessarily involve integrals; rather, it is an algebraic equation that does not contain any δ's anymore, hence the wording *integral*. It contains less information than the variational law, in the sense that the former follows from the latter, but not the opposite. However, it is an interesting relation for two reasons: (1) it already provides a checking tool for comparisons between results from different approximations schemes [387], and (2) its derivation will provide intermediary results for the quadrupolar order, variational first law calculations. In what follows, we provide some preliminary, geometric results in Sect. 6.2.1, before explicitly deriving the dipolar contribution in Sect. 6.2.2 and quadrupolar contributions in Sect. 6.2.3, and stating our final result in Sect. 6.2.4.

6.2.1 Preliminaries

Before performing the calculation, we state a general integral first law that applies for arbitrary matter sources in helically isometric spacetimes, and motivate the choice of integration hypersurface that we make, expanding on the discussion in Chap. 5, Sect. 5.1.1.

Komar Quantities

Our starting point is Eq. (4.72) derived in Chap. 4 and employs the notations introduced in Sect. 4.3 there. Using the Noether theorem, Eqs. (4.59b) and (4.59c), the Kostant formula Eq. (A.3) and the Einstein equation, we readily obtain

$$M_{\mathrm{K}} - 2\Omega J_{\mathrm{K}} = 2\int_{\infty} \mathcal{Q}_{\mathrm{g}}[k] = -2\int_{\mathscr{S}} \left(T^{ab}k_b - \tfrac{1}{2}Tk^a\right)\mathrm{d}\Sigma_a\,, \tag{6.9}$$

where $T \equiv g_{ab}T^{ab}$ is the trace of the energy-momentum tensor and $\mathscr{S}$ any spacelike hypersurface bounded by the 2-sphere at spatial infinity. This formula was previously written down in Eq. (3.18) of Ref. [387]. It is closely related to the Tolman formula for the mass and angular momentum of a stationary-axisymmetric star [546].

Let $F(\mathscr{S})$ denote the integral in the right-hand side of the identity (6.9). This integral can be shown to be independent of the choice of hypersurface $\mathscr{S}$, as was done in Chap. 4 for the integrals (4.75). Let $\mathscr{V}$ denote a volume bounded by two spacelike hypersurfaces $\mathscr{S}_1$ and $\mathscr{S}_2$ and by a worldtube $\mathscr{T}$ that includes the support of T^{ab}. Then by using Stokes' theorem we readily find

$$F(\mathscr{S}_1) - F(\mathscr{S}_2) = \int_{\mathscr{V}} \nabla_a\left(T^{ab}k_b - \tfrac{1}{2}Tk^a\right)\mathrm{d}V = 0\,, \tag{6.10}$$

as a consequence of the local conservation law $\nabla_a T^{ab} = 0$, of Killing's equation $\nabla_{(a}k_{b)} = 0$, which implies $\nabla_a k^a = 0$, and of the Lie-dragging $\mathcal{L}_k T = k^a\nabla_a T = 0$ along k^a of the trace T, itself a consequence of the Lie-dragging of T^{ab}, as established in Chap. 3.

Choice of Integration Hypersurface

The hypersurface $\mathcal{S}$ involved in Eq. (6.9) will be the same as that used for the derivation of the variational first law at dipolar order, in Sect. 5.1. Following the notations used there, $\mathcal{S}$ is chosen so that its future-directed, unit normal n^a coincides with the four-velocity u^a of the particle at the intersection $\mathscr{P} \equiv \gamma \cap \mathcal{S}$. Consequently, extending this requirement at all points along γ just as in Sect. 5.1, we will be able to use the result

$$\nabla_a n_b \overset{\gamma}{=} -u_a \dot{u}_b \tag{6.11}$$

to simplify any geometric, $\mathcal{S}$-dependent term involving $\nabla_a n_b$. For the same reason, we will have access to $k^a n_a|_\gamma = -z$, according to the general result $k^a|_\gamma = z u^a$. These two results were enough for the dipolar case, with the SEM integrals only involving one covariant derivative. At quadrupolar order, we will have to consider double covariant derivatives, such as $\nabla_a \nabla_b n_c$. This term will arise frequently and it is convenient to have a reduced formula for it, which we derive as follows. Taking the covariant derivative of the general 3 + 1 formula $\nabla_b n_c = -K_{bc} - n_b n'_c$ (recall Sect. 5.1), we readily find

$$\begin{aligned}\nabla_a \nabla_b n_c &= -\nabla_a K_{bc} - n'_c \nabla_a n_b - n_b (\nabla_a n^e) \nabla_e n_c - n_b n^e \nabla_{ae} n_c \\ &= -\nabla_a K_{bc} + \left(K_{ab} n'_c + n_a n'_b n'_c\right) - \left(K_a{}^d n_b K_{cd} + n_a n_b K_{cd} n'^d\right) \\ &\quad + \left(K'_{ac} n_b + n'_a n_b n'_c + n_a n_b n''_c - R_{aecd} n_b n^d n^e\right),\end{aligned} \tag{6.12a}$$

where each of the four terms in the first line yields the corresponding parenthesis in the second line. To obtain the latter, we have commuted the second covariant derivative in the last term of the first equality, and used the properties $n^a K_{ab} = 0$ and $n^a n'_a = 0$. By evaluating this general formula at the point $\mathscr{P} = \gamma \cap \mathcal{S}$, where $n_a = u_a$, $n'_a = \dot{u}_a$ and $K_{ab} = 0$, so that $\nabla_a n_b = -u_a \dot{u}_b$, as well as $K'_{ab} = \dot{K}_{ab} = 0$, we obtain

$$\nabla_a \nabla_b n_c \overset{\gamma}{=} -\nabla_a K_{bc} - E_{ac} u_b + 2 u_{(a} \dot{u}_{b)} \dot{u}_c + u_a u_b \ddot{u}_c \,. \tag{6.13}$$

This formula will be used in the following section, for the quadrupolar contributions to Eq. (6.9).

6.2.2 *Dipolar Contribution*

First, we rewrite Eq. (6.9) as follows

$$M_{\mathsf{K}} - 2\Omega J_{\mathsf{K}} = W_{\mathsf{dipo}} + W_{\mathsf{quad}} \tag{6.14}$$

where W_{dipo} is the integral on the right-hand side of Eq. (6.9) obtained by inserting in place of T^{ab} the dipolar contribution T^{ab}_{dipo} given in Eq. (6.1a), and similarly for

W_{quad} with Eq. (6.1b). For now we focus on W_{dipo} and relegate the calculation of W_{quad} to the next section. To evaluate W_{dipo}, we note that it can be written as

$$W_{\mathsf{dipo}} \equiv -2 \int_{\mathcal{S}} \left(T^{ab}_{\mathsf{dipo}} k_b - \tfrac{1}{2} T_{\mathsf{dipo}} k^a\right) \mathrm{d}\Sigma_a \equiv -2 I_{\mathsf{dipo}} + L_{\mathsf{dipo}} \,, \tag{6.15}$$

where I_{dipo} is precisely the integral I computed in the previous chapter, Sect. 4.3.3 (defined in Chap. 1, Eq. (4.75a)), and only the trace-term $L_{\mathsf{dipo}} \equiv \int T_{\mathsf{dipo}} k^a \mathrm{d}\Sigma_a$ needs to be computed. To this end, we proceed as in Sect. 4.3.3 and write explicitly the trace T_{dipo} of the dipolar SEM tensor (6.1a), to obtain

$$\begin{aligned} L_{\mathsf{dipo}} &= - \int_{\mathcal{V}} \left[p^a u_a \, \delta_4 + \nabla_a (D^a \delta_4)\right] k^b n_b \, \mathrm{d}V & \text{(6.16a)} \\ &= - \, p^a u_a k^b n_b + D^a \nabla_a (k^b n_b) \,, & \text{(6.16b)} \end{aligned}$$

where $D^a = -S^{ab} u_b$ is the mass dipole defined in Eq. (2.63) and n^a is the unit vector normal to $\mathcal{S}$. In the second equality we used the defining property of the invariant Dirac distribution and Stokes' theorem to integrate by parts the dipolar term (see Sect. 5.1 for details). Equation (6.16) can be further simplified by combining formula (5.4) with the formula (5.11) for the integral I_{dipo}. We find

$$W_{\mathsf{dipo}} = -p^a k_a - S^{ab} \nabla_a k_b + D^a \dot{k}_a \,. \tag{6.17}$$

This formula can be given an alternative form by using the decomposition (2.63) of the spin tensor S^{ab} in terms of the spin vector S^a and the mass dipole D^a, as well as the definition (3.57) of the vorticity V^a. In particular, with the identity $\varepsilon^{abgh} \varepsilon_{abcd} = -4 \delta^{[g}_c \delta^{h]}_d$ (see Appendix 2.2 in [6]), we readily find

$$S^{ab} \nabla_a k_b = 2 S_a V^a + 2 D^a \dot{k}_a \,, \tag{6.18}$$

where the orthogonality relations $D^a u_a = 0$ and $V^a u_a = 0$ were used. Inserting the latter into Eq. (6.17) gives the final, integral formulation of the first law of mechanics for dipolar particles

$$W_{\mathsf{dipo}} = -p^a k_a - 2 \, S^a \, V_a - D^a \dot{k}_a \,. \tag{6.19}$$

Notice the couplings of the particle's multipoles to the Killing field and its derivatives in the right-hand side, reminiscent of the general structure discussed around Eq. (5.24) in Chap. 5. Lastly, we stress that this result is independent of any SSC.

6.2.3 Quadrupolar Contributions

Now let us turn to the quadrupolar contribution W_{quad} in Eq. (6.14). It will closely follow the previous one for a dipolar particle. In fact, in view of the fact that the right-hand side of Eq. (6.9) is linear in the SEM tensor, we need only compute the quadrupolar contribution, and add them to the pole-dipole part already done and given in Eq. (6.17). Consequently, we want to compute

$$W_{\mathsf{quad}} \equiv -2\int_{\mathcal{S}}(T^{ab}_{\mathsf{quad}}k_b - \frac{1}{2}T_{\mathsf{quad}}k^a)\mathrm{d}\Sigma_a \equiv -2I_{\mathsf{quad}} + L_{\mathsf{quad}}\,, \tag{6.20}$$

where $T_{\mathsf{quad}} \equiv g_{ab}T^{ab}_{\mathsf{quad}}$ is the trace of the quadrupolar piece of the SEM tensor, given in Eq. (6.1b), and on the right we defined $I_{\mathsf{quad}} \equiv \int T^{ab}_{\mathsf{quad}}k_b\mathrm{d}\Sigma_a$, as well as $L_{\mathsf{quad}} \equiv \int T_{\mathsf{quad}}k^a\mathrm{d}\Sigma_a$. Let us start with the trace-contribution L_{quad}. We take the trace of Eq. (6.1b), insert the result into the definition of L_{quad}, integrate by means of Stokes theorem (applied twice) and use the δ_4 distribution, just as in the dipolar case. We proceed in the same way for I_{quad}. We find

$$L_{\mathsf{quad}} = -\frac{1}{3}R_{cdeb}J^{bcde}k^a n_a + \frac{2}{3}g_{eb}J^{cebd}\nabla_{dc}(k^a n_a)\,, \tag{6.21a}$$

$$I_{\mathsf{quad}} = -\frac{1}{3}R_{cde}{}^{(a}J^{b)cde}k_b n_a + \frac{2}{3}J^{c(ab)d}\nabla_{dc}(k_b n_a)\,, \tag{6.21b}$$

where, again, n_a denotes the normal vector field to the integration hypersurface $\mathcal{S}$, and the right-hand side of Eqs. (6.21) is to be evaluated at the position of the particle $\mathcal{P} \equiv \gamma \cap \mathcal{S}$.

Next, we simplify Eqs. (6.21). If we write them (schematically) as $RJ \cdot kn + gJ \cdot \nabla kn$, the calculation is best done by splitting it in two steps: first the "quadrupole" terms (RJ, gJ), and then the "Killing" terms (kn, ∇kn). The former depends on the specific quadrupole, while the latter does not. Let us first compute it, and then we will combine it with the three quadrupoles types (mass, momentum and stress), in three separate parts.

Killing Terms

We start with the "Killing" terms $k^a n_a$ and $\nabla_{dc}(k^a n_a)$ in Eq. (6.21a). The first one is easy: we chose the hypersurface $\mathcal{S}$ such that its normal n^a coincides with the four-velocity u^a. Hence we have $k^a n_a = -z$, from the (now) classical result $k^a|_\gamma = zu^a$. The second term, with the second covariant derivative, is more complicated. We start by applying the Leibniz rule twice, to get

$$\nabla_{dc}(k^a n_a) = n_a\nabla_{dc}k^a + (\nabla_d k^a)\nabla_c n_a + (\nabla_d n_a)\nabla_c k^a + k^a\nabla_{dc}n_a\,. \tag{6.22}$$

The first two terms are easily dealt with: the first one can be simplified by the Kostant formula (A.3), and in the second one we can use Eq. (6.11) as well as the conservation $(\nabla_a k_b)^{\cdot} = 0$, established in Chap. 3, Eq. (3.78). In details:

$$n_a \nabla_{dc} k^a = n^a R_{acde} k^e = -z E_{cd} \,, \tag{6.23a}$$

$$(\nabla_d k^a) \nabla_c n_a = -u_c \dot{u}_a \nabla_d k^a = u_c \ddot{k}_d \,, \tag{6.23b}$$

where Killing's equation $\nabla_{(a} k_{b)} = 0$, and the identities $k^a|_\gamma = z u^a$ and $\dot{z} = 0$ were also used several times. Next, the third term on the right-hand side of Eq. (6.22) is simply the second one with $c \leftrightarrow d$, and the fourth term is directly given by Eq. (6.13). All these results put together and some simplifications finally leads to

$$\nabla_{dc}(k^a n_a) = -z E_{cd} + 2 u_{(c} \ddot{k}_{d)} - k^a \nabla_d K_{ca} - \dot{u}^a \dot{u}_a k_d u_c \,. \tag{6.24}$$

In a similar fashion, we can compute the Killing terms involved in Eq. (6.21b). Again, the first one is simply $k_b n_a = z u_b u_a$ and is thus symmetric. For the other term, $\nabla_{dc}(k_b n_a)$, we provide the result only, the calculation following closely that of (6.24). We find

$$\begin{aligned} \nabla_{dc}(k_b n_a) = n_a R_{bcde} k^e - u_c \dot{u}_a \nabla_d k_b - u_d \dot{u}_a \nabla_c k_b \\ - k_b E_{da} u_c + k_b u_d u_c \ddot{u}_a - k_b \nabla_d K_{ca} + 2 k_b u_{(d} \dot{u}_{c)} \dot{u}_a \,, \end{aligned} \tag{6.25}$$

and we can check that the trace with respect to ab of Eq. (6.25) gives, as expected, exactly Eq. (6.24). Equations (6.24) and (6.25) are key identities that will be used below.

Mass-Type Quadrupole

We now compute the quantities L_{quad}, I_{quad} given in Eqs. (6.21) for the case of a mass-type quadrupole, of the form

$$J^{abcd} = u^{[a} J^{b][c} u^{d]} \,. \tag{6.26}$$

In practice, J^{ab} will be a placeholder tensor, so that the spin-induced quadrupole (6.6) is found by setting $J^{ab} = \frac{\kappa_2}{2m} S^a{}_c S^{cb}$, and the electric-type tidal quadrupole (6.7a) by setting $J^{ab} = \frac{3\mu_2}{2} E^{ab}$ (we shall treat the magnetic part further down). We also assume that the rank-2 tensor J^{ab} in Eq. (6.26) is symmetric and orthogonal to u_a, since it will be the case of both $S^a{}_c S^{cb}$ (under the assumption of the SSC) and E^{ab} (in general). With the general form (6.26) and these properties of J^{ab}, a simple calculation readily shows that

$$R_{cdeb} J^{bcde} = -\frac{1}{2} E_{ab} J^{ab} \quad \text{and} \quad g_{eb} J^{cebd} = \frac{1}{4}(J u^c u^d - J^{cd}) \,, \tag{6.27}$$

where $J \equiv g_{ab} J^{ab}$ is the trace of J^{ab}. We can now combine all these results into Eq. (6.21a), and we find a number of terms that cancel one another, in particular the electric terms $E_{ab} J^{ab}$ and four-acceleration terms (using $u^a \ddot{u}_a = -\dot{u}^a \dot{u}_a$, cf. Eq. (3.77)). Moreover, all contractions of the form $J^{ab} u_a$ vanish by assumption on J^{ab}, and, eventually, we obtain

$$L_{\mathrm{quad}}^{\mathrm{mass}} = \frac{z}{6}\big(J \dot{u}^a \dot{u}_a + u^a J^{cd} \nabla_d K_{ca}\big)\,, \tag{6.28}$$

We note that this equation depends on the extrinsic curvature K_{ca} and thus on the hypersurface of integration. Since we showed that our final result should be independent of it, we expect this term exactly cancel with the I_{quad} integral (cf. Eq. (6.20)). In fact, the extrinsic curvature term in the above equation vanishes identically, since, along γ, we have

$$u^a J^{cd} \nabla_d K_{ca} = J^{cd}\big(\nabla_d (K_{ca} n^a) - u_d K_{ca} \dot{u}^a\big) = 0\,, \tag{6.29}$$

where in the first equality we used the Leibniz rule, Eq. (6.11) and $n^a|_\gamma = u^a$, and in the second the fact that $K_{ca} n^a = 0$ by definition and $J^{cd} u_d = 0$ by assumption. A similar calculation holds for $I_{\mathrm{quad}}^{\mathrm{mass}}$, and it is also shown to be independent of K_{ab}, and thus of the foliation. In the end, the result for both quantities is (for a mass-type quadrupole J^{ab})

$$L_{\mathrm{quad}}^{\mathrm{mass}} = \frac{z}{6} g_{ab} J^{ab} \dot{u}^c \dot{u}_c\,, \tag{6.30a}$$

$$I_{\mathrm{quad}}^{\mathrm{mass}} = \frac{z}{6} J^{ab} \dot{u}_a \dot{u}_b + \frac{z}{4} J^{ab} E_{ab}\,. \tag{6.30b}$$

Momentum-Type Quadrupole

Now we turn to deriving the contribution $W_{\mathrm{quad}}^{\mathrm{mom}}$ from a momentum-type quadrupole, the middle term in (6.2). Once again, we write

$$J^{abcd} = u^{[a} J^{b]cd} + u^{[c} J^{d]ab}\,, \tag{6.31}$$

with a momentum-type quadrupole J^{abc}. Note that $J^{bcd} = -2\sigma_2 B^{bf} \varepsilon^{cd}{}_{ef} u^e$ in the case of the magnetic-type tidally-induced quadrupole Eq. (6.7b). We note that, by construction, J^{bcd} verifies $J^{bcd} = J^{b[cd]}$ and is orthogonal to the four-velocity with respect to all its indices. To get the (momentum-type) quadrupolar contribution $W_{\mathrm{quad}}^{\mathrm{mom}} = -2 I_{\mathrm{quad}}^{\mathrm{mom}} + L_{\mathrm{quad}}^{\mathrm{mom}}$, we insert Eq. (6.31) into the general expressions (6.21a) (for $L_{\mathrm{quad}}^{\mathrm{mom}}$) and (6.21b) (for $I_{\mathrm{quad}}^{\mathrm{mom}}$). After some calculations similar to those performed above, we find

$$L^{\text{mom}}_{\text{quad}} = -\frac{z}{3}\big(R_{abcu}J^{cba} + 2\,J^a\ddot{u}_a + J^a u^b u^c \nabla_a K_{bc}\big)\,, \tag{6.32a}$$

$$I^{\text{mom}}_{\text{quad}} = -\frac{z}{3}\big(R_{abcu}J^{abc} - \tfrac{1}{2}\,J^{abc}\nabla_b K_{ca}\big) - \frac{2}{3}J^{abc}\,\dot{u}_{(a}\nabla_{b)}k_c\,, \tag{6.32b}$$

where we defined $J^a \equiv g_{bc}J^{bca}$ and $R_{abcu} \equiv R_{abcd}u^d$. Now let us simplify these two formulae. Among the terms involving the extrinsic curvature K_{ab}, the one in Eq. (6.32a) is easily shown to vanish because of $u^a u^b|_\gamma = n^a n^b$, the Leibniz rule and $K_{ab}n^a = 0$. The one in Eq. (6.32b) can be simplified as follows

$$J^{abc}\nabla_b K_{ca} = J^{abc}\nabla_{[b}K_{c]a} = \frac{1}{2}J^{abc}R_{\hat{b}\hat{c}\hat{a}d}n^d = \frac{1}{2}R_{bcau}J^{abc}\,, \tag{6.33}$$

where used the antisymmetry of J^{abc} in the first equality, the Codazzi relation (cf. [464]) in the second, and the fact that $J^{abc} = J^{\hat{a}\hat{b}\hat{c}}$ as well as $u^a = n^a|_\gamma$ throughout. Using all this and the symmetries of the Riemann tensor and J^{abc} one last time, we find

$$L^{\text{mom}}_{\text{quad}} = -\frac{2z}{3}\big(R_{abcu}J^{abc} + g_{ab}J^{abc}\ddot{u}_c\big)\,, \tag{6.34a}$$

$$I^{\text{mom}}_{\text{quad}} = -\frac{z}{2}R_{abcu}J^{abc} - \frac{2}{3}J^{abc}\,\dot{u}_{(a}\nabla_{b)}k_c\,. \tag{6.34b}$$

Stress-Type Quadrupole

For the last, stress-piece of the quadrupole (first term in Eq. (6.2)) the calculation is rather easy since $\hat{J}^{abcd} \equiv J^{\hat{a}\hat{b}\hat{c}\hat{d}}$ is orthogonal to u_a with respect to any index. The calculation thus amounts to finding the (rare) terms that will not involve such contractions. Regarding $L^{\text{stress}}_{\text{quad}}$ (Eq. (6.21a)) the first term will evidently not produce any such term since $\hat{J}^{abcd}$ is contracted with Riemann tensor. The second term however contains a number of canceling contractions in view of Eq. (6.24) for $\nabla_{dc}(k^a n_a)$, and only one remains, the E_{cd} term. In regards to $I^{\text{stress}}_{\text{quad}}$ (Eq. (6.21b)), both terms vanish, since both will contain at least one contraction of the four-velocity with some index of $\hat{J}^{abcd}$. In the end, the result for the stress-type quadrupole contribution reads

$$L^{\text{stress}}_{\text{quad}} = \frac{z}{3}\big(R_{abcd} - 2g_{ab}E_{cd}\big)\hat{J}^{dabc}\,, \tag{6.35a}$$

$$I^{\text{stress}}_{\text{quad}} = 0\,. \tag{6.35b}$$

6.2.4 Results and Discussion

The previously established results for the mass-, momentum- and stress-type quadrupolar tensors given in Eqs. (6.30), (6.34) and (6.35), respectively, can be used to write the total quadrupolar contribution W_{quad} in the right-hand side of the integral first law (6.20) for a *generic* quadrupole. Indeed, since the latter can always be written in the general form (6.2), we can simply compute W_{quad} as

$$W_{\mathsf{quad}} \equiv W_{\mathsf{quad}}^{\mathsf{stress}} + 2W_{\mathsf{quad}}^{\mathsf{mom}} - 4W_{\mathsf{quad}}^{\mathsf{mass}}\,, \tag{6.36}$$

where each of these contributions are given by the combination $-2I + L$ in the aforementioned equations. Putting all this together, we find the total quadrupolar contribution as

$$\begin{aligned} W_{\mathsf{quad}} &= \frac{z}{3}\hat{J}^{acbd}\big(R_{abcd} - 2g_{ab}E_{cd}\big) && \text{(6.37a)}\\ &+ \frac{2z}{3}J^{abc}\big(R_{abcu} - 2g_{ab}\ddot{u}_c\big) + \frac{8}{3}J^{abc}\dot{u}_{(a}\nabla_{b)}k_c && \text{(6.37b)}\\ &+ \frac{2z}{3}J^{ab}\big(3E_{ab} + 2\dot{u}_a\dot{u}_b - g_{ab}\dot{u}^c\dot{u}_c\big)\,. && \text{(6.37c)} \end{aligned}$$

This relation holds for *any* quadrupolar model, and can be rewritten in terms of J^{abcd} simply by using Eq. (6.3). From it one can derive the trace-free version, which could be of interest as they are the only one affecting the motion of particles in vacuum spacetimes [368]. Lastly, let us mention that, to be self-consistent, the quadrupolar contribution (6.37) has to be added to the dipolar one, derived in Sect. 6.2.2, namely

$$W_{\mathsf{dipo}} = -p^a k_a - S^{ab}\nabla_a k_b + D^a \dot{k}_a\,. \tag{6.38}$$

Although we call Eq. (6.38) a "dipolar contribution", it does contain J^{abcd}-dependent terms, which come indirectly from the term $\dot{k}_a = z\dot{u}^a$ in (6.38), and thus the evolution equations. A similar remark can be made for Eq. (6.37) as well, which also contains p^a- and S^{ab}-dependent terms through $\dot{u}^a$. In fact, all this is simplified by a choice of SSC which allows one to consistently expand the results in powers of a given quantity, typically the spin, as we shall do subsequently.

Extended Spinning Particle

In the particular case where the quadrupolar model only assumes a spin-dependence for the (non-minimal) Lagrangian, the tidally-induced quadrupole do not enter the description of the body. Neglecting the tidal forces with respect to spin-induced ones is a reasonable assumption, at least in the inspiral stage of binary systems, where tidal effects enter at higher PN order than the spin-induced ones. In any case, the previous result allows us to write the integral first law of mechanics for a spinning particle, accounting for all quadratic-in-spin effects. To find it, we simply add the dipolar result (6.38) with the quadrupolar one (6.37) in which we take for the generic quadrupole (6.2) the spin-induced one (6.6). To be consistent, we assume the SSC $S_{ab}u^a = 0$ to ensure that[3] $J^{ab}u_a = 0$. We find

$$M - 2\Omega J \overset{\text{SSC}}{=} -p^a k_a - S^{ab}\nabla_a k_b + \frac{2z\kappa_2}{m} S^{ac} S_c{}^b E_{ab} + O(S^4)\,. \tag{6.39}$$

with the S^3 contribution containing the last two terms in the line (6.37), using $\dot{u} = O(S)$ as established in Chap. 2, Sect. 2.3.2. Equation (6.39) can be rewritten in two interesting ways. First, by recalling the definition (6.7a) of the electric-part of the Riemann tensor and the Kostant formula (A.3),

$$M - 2\Omega J \overset{\text{SSC}}{=} -p^a k_a - S^{ab}\nabla_a k_b - \tfrac{2\kappa_2}{m} S^{ac} S_c{}^b u^d \nabla_{ab} k_d + O(S^3)\,, \tag{6.40}$$

which agrees with the conjecture made in Chap. 5 where we expected to see higher derivatives of k^a contracted with higher-order multipoles. Second, we can turn Eq. (6.39) into a form that only depends on the spin vector S^a (Eq. (2.69)). Introducing the helical vorticity V^a (Eq. (3.57)), we easily show that (given the SSC) $S^{ac}S_c{}^b = S^a S^b - S^2 h^{ab}$, with $h^{ab} = g^{ab} + u^a u^b$ the projector orthogonal to u^a. Since E_{ab} is traceless, Eq. (6.39) becomes

$$M - 2\Omega J \overset{\text{SSC}}{=} -p^a k_a - 2S^a V_a + \tfrac{2\kappa_2}{m} S^a S^b E_{ab} + O(S^3)\,. \tag{6.41}$$

This equation is a self-consistent, quadratic-in-spin result for a binary systems of spinning point particles. Indeed, it is known that the next-order in the multipolar approximation brings an octupole with cubic-in-spin terms (see [320] or [195]), and any indirect quadratic-in-spin piece that would come from the quadrupolar equations of motion actually contributes at quartic-order, as already above. We also note that we did not include any tidal effect here for the sake of brevity, but our general result (6.37) along with the explicit formulation (6.7) for tidally-induced quadrupoles can be used to add the tidal contributions to Eq. 6.41 as easily as for the spin-induced contribution.

Comparison at Dipolar Order

Let us now show that the (Komar)-integral law (6.19) is perfectly compatible with the that derived from the variational law established in Chap. 5, Sect. 5.3.1, at linear order in spin

$$\delta M - \Omega\,\delta J \overset{\text{SSC}}{=} \sum_{\text{i}} z_{\text{i}}\,(\delta m_{\text{i}} - \omega_{\text{i}}\,\delta S_{\text{i}}) + O(S_{\text{i}}^2)\,. \tag{6.42}$$

[3] If we do not choose an SSC, there is simply an extra quadrupolar contribution $J^{ab} \propto D^a D^b$, in terms of the mass dipole D^a.

Indeed, by applying Euler's theorem to the function $M(\sqrt{J}, \sqrt{S_1}, \sqrt{S_2}, m_1, m_2)$, which is homogeneous of degree one, one obtains from the variational first law (5.51), the following relationship

$$M - 2\Omega J \overset{\text{SSC}}{=} \sum_{\text{i}} z_{\text{i}} \left(m_{\text{i}} - 2\omega_{\text{i}} S_{\text{i}} \right) + O(S_{\text{i}}^2) \,. \tag{6.43}$$

After imposing the SSC (2.64) and using $p_{\text{i}}^a k_a = -m_{\text{i}} z_{\text{i}}$ and $S_{\text{i}}^a V_a = S_{\text{i}} V = z_{\text{i}} \omega_{\text{i}} S_{\text{i}}$, as a consequence of Eqs. (3.30), (2.60a), (3.62) and (5.45), we readily obtain the simple algebraic formula

$$M_{\text{K}} - 2\Omega J_{\text{K}} \overset{\text{SSC}}{=} \sum_{\text{i}} z_{\text{i}} \left(m_{\text{i}} - 2\omega_{\text{i}} S_{\text{i}} \right) . \tag{6.44}$$

Assuming that our helically symmetric spacetimes would obey appropriate falloff conditions [386, 512], it can be shown that $M_{\text{K}} = M$ and $J_{\text{K}} = J$, so that the Komar-type derivation of the first integral relation is consistent with the formula (6.43). In fact, the algebraic formula (6.44) suggests that the first integral (6.43) is *exact* at dipolar order, and not merely valid to linear order in the spin amplitudes S_{i}.

6.3 Variational First Law at Quadrupolar Order

Note: as pointed out in the text, some results presented in this section are preliminary and part of ongoing calculations. The definitive, correct expressions may ultimately differ from these "anticipated" ones.

The derivation of the first law of mechanics for quadrupolar particles is one of the main problem tackled during this work. The calculation follow the same step as that presented for the dipolar case. In particular, one must insert the full quadrupolar SEM tensor (Eqs. (6.1a) and (6.1b)) into the generalized first law derived in Chap. 4 and compute the two integrals I and K (defined in Sect. (4.75). Thankfully, it suffices, at the quadrupolar order, to simply add the quadrupolar contribution, coming from the J^{abcd}-dependent part (6.1a) of the SEM tensor. We have thus two integrals to calculate, namely I_{quad}, K_{quad}.

6.3.1 Integrals

Thankfully, our previous calculations for the integral first law provided I_{quad} for an arbitrary quadrupole. We can therefore read their expression off Eqs. (6.30b), (6.34b) in which we simply substitute the explicit quadrupole models given in Eqs. (6.6) and (6.7). This calculation is done with the same method as above: we skip over the details for the spin and electric-type contributions $I_{\text{quad}}^{\text{spin}}$, $I_{\text{quad}}^{\text{elec}}$, which are relatively straightforward. For $I^{\text{mag}}_{\text{quad}}$, we mention the following key steps of the calculation

$$J^{(ab)c}\dot{u}_a\nabla_b k_c = J^{(ab)c}\dot{u}_a\varepsilon_{bcde}u^d V^e = -\sigma_2 B_{ab}\dot{u}^a V^b\,, \tag{6.45a}$$

$$R_{abcu}J^{abc} = \frac{1}{2}R_{cbau}J^{abc} = \sigma_2 B_{ab}B^{ab}\,. \tag{6.45b}$$

In Eq. (6.45a), the first equality follows from the expansion (3.58) for $\nabla_a k_b$ in terms of the vorticity V^a, and the second by replacing J^{abc} by the magnetic-type quadrupole (6.7). Regarding Eq. (6.45b), the first equality is obtained by combining the algebraic symmetries of both R_{abcd} and J^{abc}, and the second by again replacing J^{abc} and using the definition (6.5) of the self-dual of the Riemann tensor. These two formulae can readily be inserted in Eq. (6.34b) to find $I^{\text{mag}}_{\text{quad}}$. In the end, the three quadrupolar contributions are given

$$I^{\text{spin}}_{\text{quad}} \overset{\text{SSC}}{=} -\frac{z\kappa_2}{2m}S^a{}_c S^{cb}E_{ab} - \frac{z\kappa_2}{3m}S^a{}_c S^{cb}\dot{u}_a\dot{u}_b\,, \tag{6.46a}$$

$$I^{\text{elec}}_{\text{quad}} \overset{\text{SSC}}{=} -\frac{3z\mu_2}{2}E_{ab}E^{ab} - z\mu_2 E_{ab}\dot{u}^a\dot{u}^b\,, \tag{6.46b}$$

$$I^{\text{mag}}_{\text{quad}} \overset{\text{SSC}}{=} -\frac{z\sigma_2}{2}B_{ab}B^{ab} + \frac{2\sigma_2}{3}B_{ab}\dot{u}^a V^b\,, \tag{6.46c}$$

Even for dipolar particles, we showed in Chap. 2 that $\dot{u}^a = O(S)$. Therefore, the last two terms on the right-hand side of Eq. (6.46a) are $O(S^4)$. As already mentioned above, such quartic-in-spin contributions must be neglected to remain self-consistent at quadrupolar order. Indeed, we do not control the terms of order $O(S^3)$ or higher that would come from spin-induced multipoles at octupolar order and beyond. Similarly, for a nonspinning, tidally-perturbed body, one has $\dot{u} = O(R\nabla R)$, as can be checked using the equations provided in Eq. (A.61e) App. A.4.4, and inserting the tidally-induced quadrupole (6.7) into the force and torque (recall Eq. (2.54)). As a consequence, the last term in Eq. (6.46b) and in Eq. (6.46c) are $O(R^3(\nabla R)^2)$ and $O(R^2\nabla R)$, respectively. These quintic-in-curvature and cubic-in-curvature contributions can also be neglected, as we do not control the terms $O(R^3)$ or higher that would come from nonlinear tidal effects. Consequently, we find the simplified expression for the total quadrupolar integral I_{quad}

$$I_{\text{quad}} \overset{\text{SSC}}{=} -\frac{z}{2m}\kappa_2\,E_{ab}S^a S^b - \frac{3z}{2}\mu_2\mathcal{E}^2 - \frac{z}{2}\sigma_2\mathcal{B}^2 + O(R^3, S^4) \tag{6.47}$$

where, as for the integral law, we used the equality $S^{ac}S_c{}^b E_{ab} = S^a S^b E_{ab}$, and we also introduced the *tidal invariants* $\mathcal{E}^2 \equiv E_{ab}E^{ab}$ and $\mathcal{B}^2 \equiv B_{ab}B^{ab}$ (see e.g., [515, 547, 548]). Here, the spin vector S^a is uniquely determined from the spin tensor thanks to the SSC. The quadratic invariants play a key role in interfacing GSF/PN/PM theory to EOB theory, and, therefore, it is not very surprising that they would appear in the first law of mechanics.

As we mentioned earlier, having computed the integral I_{quad} is not sufficient for the derivation of the first law as we also need an expression for the following integral:

$$K_{\text{quad}} \equiv \int_{\mathcal{S}} T^{ab}_{\text{quad}} \delta g_{ab} \, \xi^c \mathrm{d}\Sigma_c \,, \tag{6.48}$$

with T^{ab}_{quad} given by Eq. (6.1b). The integration process for this integral is as easy as for the other one, and is based on the same choice of integration hypersurface. The difficulty stems from the interplay between the double covariant derivative arising from the quadrupolar term in the SEM tensor, and the δ-variations. As a consequence, these calculations are still underway, and there would be no point in giving the expression that we have obtained at the current stage, for which a simplification remains to be made.

6.3.2 *Preliminary Result*

Given the fact that the integral K_{quad} has not been obtained yet, the explicit form of the first law at quadrupolar order is still unknown. However, given the explicit results obtained in Eqs. (6.46) and the general structure of the first law as a sum of terms of the form (derivative of k^a) times the corresponding quadrupole (recall the discussion around Eq. (5.24)), we *anticipate* that the scalar version *may* take the form

$$\delta M - \Omega \delta J \overset{!!!}{=} \sum_{\mathrm{i}} z_{\mathrm{i}} \Big(\delta m_{\mathrm{i}} - \omega_{\mathrm{i}} \delta S_{\mathrm{i}} + E^S_{\mathrm{i}} \delta(\kappa_{2,\mathrm{i}} \mathcal{Q}_{\mathrm{i}}) + \mathcal{E}_{\mathrm{i}} \delta(\lambda_{2,\mathrm{i}} \mathcal{E}_{\mathrm{i}}) + \mathcal{B}_{\mathrm{i}} \delta(\sigma_{2,\mathrm{i}} \mathcal{B}_{\mathrm{i}}) \Big) \tag{6.49}$$

where $E^S \equiv E_{ab} s^a s^b$ is the eigenvalue associated with the eigendirection $s^a \equiv S^a/S$ of the electric-type tidal quadrupole, $\mathcal{Q} \equiv -S^2/m$ is the quadrupole (mass) moment of a Kerr black hole of mass m and spin S. Once again, we stress as it does not have a complete derivation yet, there should be numerical factors in front of the last three terms in Eq. (6.49), and some of the quadrupolar factors may ultimately have to be moved inside or outside of the δ-variations. This is the general form of the "expected" result, and it does not have a complete derivation yet.

Naturally the spin-induced quadrupolar contribution is proportional to the scalar quadrupole $\kappa_2 \mathcal{Q}$, but also to the eigenvalue of the electric tidal field in the axial direction s^a of the spin. For each body, only two out of the six parameters $(m, S, z, \kappa^E_2, \mathcal{E}, \mathcal{B})$ are truly linearly independent, e.g. m and S, the other four parameters, say $\boldsymbol{\theta} \equiv (z, \kappa^E_2, \mathcal{E}, \mathcal{B})$, being functions of m and S. Therefore,

$$\delta \boldsymbol{\theta} = \partial_m \boldsymbol{\theta} \, \delta m + \partial_S \boldsymbol{\theta} \, \delta S \,, \tag{6.50}$$

where the partial derivatives can be computed explicitly in certain physically relevant limits, e.g. the large-mass-ratio limit (black hole perturbation theory and the GSF framework) and the large-separation limit (PN approximation). Those finite-size contributions will effectively renormalize the redshift z and the spin precession frequency ω of each particle in the first law of binary mechanics. In particular, recent GSF results for the redshift parameter z of a small, extended, compact body on a circular orbit around a Schwarzschild black hole [417], combined with the quadrupolar first law, will give new results for the binding energy and angular momentum, which could then be used to calibrate EOB models with tidal effects.

Lastly, let us make a comment on the phenomenological parameters κ_2, μ_2 and σ_2 that appear in Eq. (6.49). An isolated black hole has $\kappa_2 = 1$, which implies $\delta\kappa_2 = 0$ under a small variation of the black hole mass m and spin S. For an isolated neutron star, $\kappa_2 \sim 4 - 8$, depending on the equation of state. Hence comparing two neutron stars with two slightly different internal properties gives $\delta\kappa_2 \neq 0$ in general. In principle, the tidal field of the companion can correct the $O(S^2)$ quadrupole of the compact object, thus effectively promoting the constant κ_2 to a function of the mass m and spin S. In particular, this would imply $\kappa_2 \neq 1$ for a black hole and thus $\delta\kappa_2 \neq 0$ in general. However, we did not account for such a small (to be quantified) effect in this work, which would require the addition of appropriate operators in the EFT description of the binary system and a proper matching to a tidally perturbed spinning black hole at $O(S^2)$. Regarding the tidal coupling constants μ_2 and σ_2, it has been shown that they vanish for *non*spinning black holes, thus implying $\delta\mu_2 = \delta\sigma_2 = 0$ while comparing two *non*spinning black holes of nearly identical masses. For a nonspinning neutron star, we have generically nonzero $\delta\mu_2$ and $\delta\sigma_2$. Those scale as the fifth power of the areal radius $R = m/\Xi$, where the value of the compactness parameter $\Xi \sim 0.2$ depends on the equation of state.

Interlude

For most physics students learning about general relativity (GR), it is fair to say that having followed an introductory course on special relativity (SR) is *recommended* at the least. Most of these students (myself included) might have felt that SR is less "complicated" than GR. Without doubts, this is primarily because of the flatness of the Minkowski spacetime, which allows one to stay within the realm of affine geometry, a world already familiar to Euclideanly-trained students. This misconception may also be due to the way SR is usually taught, with more emphasis on a matrix formulation and Lorentz transformations than on the tensorial, operational nature of the theory. But as was beautifully showed by Éric Gourgoulhon in his treatise on SR in non-inertial frames [22], the geometry of Minkowski spacetime is nothing short of complicated and beautiful. It may as well be curved if one chooses arbitrary coordinates and/or studies accelerated, rotating motions. Of course, GR forces us to leave affine spaces for manifolds, but the effort to make the SR → GR step is, in my opinion, drastically reduced when learning about SR chronogeometrically, as Gourgoulhon teaches.

In any case, these relativistic theories showed how to think of space, time and gravitation geometrically. Einstein himself talked about this matter several times.

His thoughts on the geometrization of physics are beautifully summarized in the article *Geometrie und Erfahrung* (Geometry and Experience) [549], which he had prepared for his "particularly nice lecture" given in 1921, at the Prussian Academy's in honor of Frederick the Great. During this lecture, Einstein makes the distinction between the "practical" geometry, which he refers to as the oldest branch of physics; and "purely axiomatic geometry", which he does not give any credit for his discovery of relativity. But by distinguishing between the two, Einstein only emphasizes of the geometry of space (and time) in which physical processes take place, and not on the inherent geometry of the laws themselves. Yet, this is where the geometrization of physics started, long before Einstein and his revolutionary ideas, at a time when the frontier between mathematics and physics was very much nuanced.

Indeed, it was classical mechanics that became the first geometrized branch of physics during the eighteenth century, through an ironical situation often found in books on physics history. While it is often said that Lagrange is the father of analytical mechanics (which would become the foundation of symplectic geometry), we find in his 1788 masterwork on analytical mechanics the following warning:

> *On ne trouvera point de Figures dans cet Ouvrage. Les méthodes que j'y expose ne demandent ni construction, ni raisonnemens géométriques ou méchaniques, mais seulement des opérations algébriques, assujetties à une marche réguliere et uniforme.*[4]

In English: *The reader will find no figures in this work. The methods which I set forth require no constructions nor geometrical or mechanical reasoning, but only algebraic operations, subject to a regular and uniform rule of procedure.* Here, Lagrange speaks, in this particular paragraph, of the "synthetic" geometry, which he will not be using, even though it was the method of choice among geometers and physicists of the time. Instead, Lagrange warns that he will be using new tools, that are now part of classical "analysis". Nevertheless, it is ironic that by inventing his new analysis tools, in particular, the *Lagrange parentheses*, Lagrange did invent a new geometry: *symplectic geometry*. Indeed, we may rightfully set the birthdate of symplectic geometry in 1808, when Lagrange's work on celestial mechanics showed that the equations of planetary motion could be cast in a greatly simplified form [550], using the tools he invented. The form of these equations were then extended to any mechanical problem by Hamilton, and these ideas were all simplified, extended and understood in details by Poisson, Jacobi and Liouville, to name but a few. An illustration of Lagrange's early insight can be found in the "Lagrange parentheses", equivalent to the more commonly known Poisson brackets, which are nothing but the components of the symplectic form [551].

The word "symplectic" was introduced by Hermann Weyl in his treatise on group theory [552] (see the first footnote in chapter VI there) as a replacement for (and Greek-equivalent of) *complex*, also introduced by himself in the context of group

[4] One shall not find figures in this book. Methods here exposed require no construction, no geometrical or mechanical reasoning, but only algebraic operations subject to regular and uniform steps.

theory. Symplectic geometry, at the basis of Mechanics, is more demanding, in some sense, than Lorentzian geometry, on which GR is founded. While pretty much every manifold can be endowed with a Riemannian metric, regardless of its dimension, not all of them can carry a symplectic form. Even excluding odd-dimensional manifolds is not enough. For example, the 4D and 6D sphere cannot be endowed with a symplectic structure (not even a nondegenerate antisymmetric bilinear form for the former!) The main reason behind this fundamental difference between the two is encoded in Jacobi's identity, or, to paraphrase Arnold [553], *because the three altitudes of a triangle are concurrent* [554]. We leave for the reader the pleasure of discovering why Arnold would have said such a thing.

The first part of this thesis was based on a relativistic problem, where a "symmetry" was provided by the existence of a helical Killing vector field. In the second part, we will be dealing with a non-relativistic problem, with a symplectic symmetry, hidden at the level of the Hamiltonian of the system. Since the new problem at hand s be rather independent of that studied in the first, let us, at least, keep with the tradition of letting Galileo open the introductory chapter, like in the first part of this thesis.

Part II
Isochrone Orbits in Newtonian Gravity

Chapter 7
Isochronies

Contents

This life as you now live it and have lived it, you will have to live once more and innumerable times more; and there will be nothing new in it, but every pain and every joy and every thought and sigh and everything unutterably small or great in your life will have to return to you, all in the same succession and sequence.

Nietsche
The Gay Science, Sect. 341 (1882).

The embodiment of isochrony is undoubtedly the pendulum. Galileo was surely not the first to play with pendula, but was a pioneer in relating his observations of their oscillations and their dependence on the pendulum's parameters. However, the first

P. Ramond, *The First Law of Mechanics in General Relativity & Isochrone Orbits in Newtonian Gravity*, Springer Theses,
https://doi.org/10.1007/978-3-031-17964-8_7

satisfying analysis of the isochrone pendulum is attributed to Christian Huygens, who managed to construct such a device by combining his geometrical skills and his great knowledge of planar curves. This chapter opens the second part of this thesis with a historical account on the (near-)isochrone pendulum of Galileo, the (exact) isochrone pendulum of Huygens (Sect. 7.1. Then, Sect. 7.2 is dedicated to the introduction of notations and classical results regarding periodic orbits in generic, spherically symmetric potentials. Lastly, in Sect. 7.3, we provide some astrophysical background on Hénon's isochrony, and give a precise definition of the problem that will be explored in the next chapters.

7.1 Isochrone Pendula

In this introductory section, we present a very brief, modern-day summary of Galileo and Huygens' work and ideas in the quest for the isochrone pendulum. Although fascinating in itself, this story presents strong similarities with that related in the second part of this thesis, regarding celestial isochrony.

7.1.1 Galileo's Pendulum

Attach a bob of mass m to one end of a massless string of length ℓ, and suspend that string from the other end, so that the bob is only acted upon by gravity and the string's restoring force. Remarkably, this simple mechanical system has been one of the most important physical apparatus in the history of science. There hardly is any other physical system that can, alone, demonstrate the Earth' proper rotation[1] [557], act as a gravimeter and a time keeper [558], set standards of length [559], compute air drag coefficients [560], etc. Among all its properties, the one that kept Galileo's attention was *isochrony*, or the ability of a pendulum to oscillate with a constant period, independently of its releasing height.

If a pendulum is released from rest at an angle θ_o with respect to the vertical axis meeting its fixed end, an exact formula for its oscillation period T can be obtained. Galileo made his observations on the pendulum in the first decade of the seventeenth century, but published them about thirty years later in his famous *Discorsi*. These observations led him to two central results. First, T is independent of the mass. As we know today, this is a consequence of the universal law of free fall, which he discovered himself [561]. Second, the oscillation period is proportional to the square root of the string's length. Putting these together, we obtain $T = C\sqrt{\ell}$ where C is a constant, independent of the mass m. Galileo is convinced of the isochrony of the

[1] Foucault's pendulum has a close link with general relativity, see for example this proposal [555] for an experimental check of general relativity involving a Foucault pendulum, or this article [556] explaining the precession of the oscillation plane in terms of the relativistic Thomas precession.

pendulum, even though his arguments, from a modern perspective, were not at all convincing [562]. The constant of proportionality C to which Galileo refers is known today to depend only on the strength of the local gravitational field g, resulting in the famous result

$$T = 2\pi\sqrt{\frac{\ell}{g}}\,. \tag{7.1}$$

Galileo's result was, therefore, perfectly sound, and indeed, based on Eq. (7.1), the pendulum does seem isochrone. However, soon after the publication of his results, Galileo's conclusions were rapidly examined by others, notably Marin Mersenne and René Descartes in Paris, who pointed out that Galileo's result were correct but limited to pendula with small oscillations. As a matter of fact, in Mersenne's translation [563] of Galileo's work, we find in Book I, Article XVII the comment[2]

Si l'autheur eust esté plus exact en ses essais, il eust remarqué que la chorde est sensiblement plus long-temps á descendre depuis le haut de son quart de cercle jusques á la perpendiculaire, que lorsqu'on la tire seulement dix ou quinze degrez...

In fact, Galileo did observe these small differences and, retrospectively, nothing should have prevented him to conclude as to the *near-isochrony* of the pendulum. Nothing, except his confirmation bias. Galileo, the father of experimentation and the scientific method, fell into the trap of trying to justify isochrony at any cost, instead of looking for what contradicted his isochrone hypothesis. There are a number of reasons why he did so, but this discussion would lead us too far (we refer to [562, 564] for good references on this matter). Let us simply say that his justifications of any non-isochrone behavior were not convincing from a scientific point of view. Although his conclusion was (partly) incorrect, Galileo had already done the most difficult part: convincing his successors to examine closely *the pendulum*, for it was not a mere oscillating ball anymore, but an apparatus worthy of interest for accurate time-keeping. This was the dawn of precise time measurement [558], that culminated in 1761 with John Harrisson's invention of the *marine chronometer* [565], that would revolutionize sea-travel navigation by allowing for precise longitude measurements.

As of today, we known the pendulum much better, with exact and explicit formulae for T accounting for arbitrarily large oscillations. Usually stated in terms of elliptic integrals [566] (see also [567]), let us provide another closed-form expression in terms of the mechanical energy E of the pendulum. As a conserved quantity, E may be evaluated at any point on the trajectory. Let us choose the release point, where the amplitude is θ_o and the velocity vanishes. The mechanical energy E is then pure potential energy E_p, simply given by $mg\ell(1 - \cos\theta_o)$ when we choose $E_p = 0$ when the pendulum is vertical, at rest. The exact period of oscillation T takes the form

---

2 *"If the author had made his measurements more carefully, he would have noticed that a pendulum released from* 45° *takes a longer time to reach the vertical than from* 10° or 15°.".

$$T(E) = 2\pi\sqrt{\frac{\ell}{g}}\left(1 + \sum_{n\geq 1} a_n \mathcal{E}^n\right), \quad \text{with } \mathcal{E} \equiv \frac{E}{2mg\ell}, \tag{7.2}$$

where a_n are explicit positive coefficients[3] and $\mathcal{E}$ the dimensionless energy. Since $\mathcal{E} = \frac{1}{2}(1 - \cos\theta_o)$, we find $\mathcal{E} = O(\theta_o^2)$ for small oscillations $\theta_o \ll 1$, we do recover the *approximate isochrony* of Galileo's pendulum, which holds if we neglect effects quadratic in the amplitude. To put Galileo's result into perspective, let us take an initial amplitude of $\theta_o = \pi/6$, so that $\mathcal{E} = 1/4$, thus adding a 3% correction to the small-amplitude result (7.1). With a 3% change, Galileo should have seen this effect, as he reveals using pendula swinging for up to a hundred times [562].

A hint for a way to build a truly isochrone pendulum can readily be found by inspecting the exact formula (7.2) for $T(E)$. Notice that, by definition, the dimensionless energy reads $\mathcal{E} = \frac{1}{2}(1 - \cos\theta_o)$, so that in Eq. (7.2) T depends on ℓ through the prefactor and on θ_o through the power expansion. These two mathematical dependencies correspond to two competing physical effects. If the oscillation θ_o is too wide, then the energy $\mathcal{E}$ contributes significantly and T increases. Conversely, if the length ℓ of the pendulum shortens, then T decreases. Consequently, if somehow one could constrain the string's length to decrease when the wide oscillations take place, there may be a way to make these two effects compensate one another, so that T remains, in fact, constant. In practice, this can be achieved by placing fixed, curved walls on the side of the pendulum. With this apparatus, as the pendulum swings the string is constrained to wrap around the wall, making it *effectively* shorter, as explained in Fig. 7.1. The shape of these walls can be tuned so that they only affect the motion in the wide-oscillation regime. In this case, the bob does not follow a perfect circle of radius ℓ anymore. Rather, its path draws a more complicated curve, of equation $y(x)$ on the right of Fig. 7.1. The solution to the isochrone pendulum will involve such walls and a non-circular curve $y(x)$, as we will see below.

7.1.2 *Horlogium Oscillatorium*

It is Christiaan Huygens who provided both the first correct explanation of the near-isochrony of Galileo's pendulum and the solution to the problem regarding the isochrone pendulum. His findings were published in 1673 in a famous and groundbreaking treatise, nowadays called by the title's first two words: *Horlogium Oscillatorium*. His demonstrations only use elementary geometry, as calculus would only be invented a few decades later.[4] In this section we would like to explain the key steps Huygens took for his discovery of the isochrone pendulum, rewritten here with

---

[3] Explicitly, $a_n \equiv (\frac{(2n)!}{4^n (n!)^2})^2$. Therefore $a_n \sim 1/\pi n$ for large n and the series converges for $|\mathcal{E}| < 1$, which is equivalent to $\theta_o \neq \pi$. Indeed, $\theta_o = \pi$ is the (unstable) equilibrium, associated to an infinite period. In that case, $\mathcal{E} = 1$ and the series diverges harmonically.

[4] Although, quite remarkably, reading through Huygens' geometric arguments clearly reveals hints of what would later become limits, derivatives and integration [568].

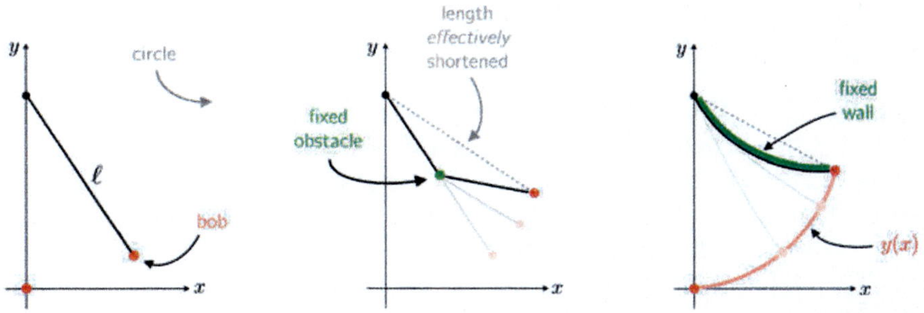

Fig. 7.1 *Left*: in the classical pendulum the bob moves on a circle of radius ℓ. *Middle*: a fixed point obstacle (green dot) acts as a secondary, temporary oscillation center, that decreases artificially the length, and thus the oscillation period. *Right*: an infinity of such point obstacles are placed on a curve (green curve) to make a "wall" around which the string wraps. This allows one to change the path $y(x)$ followed by the bob's path (in red). Such an apparatus was actually used by Huygens, found in his magnum opus *Horlogium Oscillatorium*

modern mathematics. Explaining Huygens' ideas in this fashion provides an interesting derivation seldom seen in the literature, and, crucially, it closely resembles celestial isochrony as it will be defined and explored in the next chapters.

Revisiting Galileo

The first goal of Huygens is to find the curve $y(x)$ that produces an isochrone pendulum, i.e., with an oscillating period T independent of the initial release height. Only then will he discuss the right shape for the side walls that enforces this motion (again, cf. Fig. 7.1). Huygens starts by writing, using geometry, what we would now call the master ODE for this kind of problem, namely

$$\frac{1}{2}\left(\frac{\mathrm{d}s}{\mathrm{d}t}\right)^2 = g\,(h_o - y)\,, \tag{7.3}$$

where s is the (curvilinear) arclength along the path $y(x)$ and h_o is the release height. Equation (7.3) is nothing but the conservation of energy, with the kinetic energy difference on the left equal, at all times, to the gravitational potential energy difference on the right, the points of reference being an arbitrary point on the curve and the turning point, where $(\dot{s}, y) = (0, h_o)$. Huygens considers first the case of Galileo's free pendulum. As said before, $y(x)$ is then a circle of radius ℓ, the string's length. Therefore, this path is given by $x^2 + (y - \ell)^2 = \ell^2$, with the Cartesian coordinates (x, y) erected with respect to the lowest point on the circle (as in Fig. 7.1). A classical calculation then gives $\mathrm{d}s/\mathrm{d}y = \ell/x$, so that Eq. (7.3) reads, for the *classical* pendulum

$$\frac{1}{2}\left(\frac{\mathrm{d}y}{\mathrm{d}t}\right)^2 = g\,(h_o - y)\,(1 - (y/\ell - 1)^2)\,. \tag{7.4}$$

To get the period T, it suffices to isolate $\mathrm{d}t$, express it in terms of $\mathrm{d}y$ and integrate the result. In fact, it is the cubic-in-y polynomial in the right-hand side of Equation (7.4) that will give rise to the aforementioned elliptic integral, which can then be turned into Eq. (7.2). Huygens already knows about the anisochrony of the regular pendulum, and rather wants to understand why, geometrically, in small-oscillation limit it becomes isochrone. With modern day notations, this limit is simply $y \ll \ell$, and in this case Eq. (7.4) becomes

$$\frac{1}{2}\left(\frac{\mathrm{d}y}{\mathrm{d}t}\right)^2 = 2g\,(h_o - y)\,y/\ell\,. \tag{7.5}$$

Now we may compute the oscillating period as follows. As the pendulum is released from the left-hand side of the vertical, it goes down (from height h_o to height 0), and then up (from 0 to h_o) before reaching the turning point on the right-hand side of the vertical. It then swings all the way back, symmetrically. Consequently, by symmetry, T is four times that the time taken from height h_o to height 0. Since $\dot{y} < 0$ on this portion, we obtain by inverting Equation (7.5) for $\mathrm{d}t$ and integrating the result:

$$T = 2\sqrt{\frac{\ell}{g}}\int_0^{h_o} \frac{\mathrm{d}y}{\sqrt{(h_o - y)y}}\,. \tag{7.6}$$

One can then check that the integral on the right-hand side of (7.6) is independent of h_o, as it evaluates exactly to π, giving the classical, small-angle result (7.1).

The Parabola Property

Of course, Huygens did not have access to integrals as we have shown here. His work was based on pure geometry. In geometrical terms, the Taylor approximation $y/\ell \ll 1$ we made to go from Eq. (7.3) to Eq. (7.4) is nothing but the act of replacing the original circle by its *osculating parabola* at the origin, as showed on Huygens' own drawings (cf. Fig. 7.2). With our method or Huygens', the result is the same: it allows one to get the "time of the curve", i.e., T in terms of $y(x)$ (Eq. (7.6)). The next goal is, naturally, to find a way to invert this relation. Huygens thus asks:

For what trajectory, if any, would the parabolic relationship used to construct the inverse curve of time be exact?

Here, he speaks of the osculating parabola argument that allowed an explicit calculation. Heuristically, if the effect of this osculation could be applied (somehow) to the whole curve (and not just at one point), then the resulting pendulum would be isochrone in all oscillations, not just small ones. From this point on, Huygens' geometric demonstration consists in showing that the *cycloid* is such a curve (see

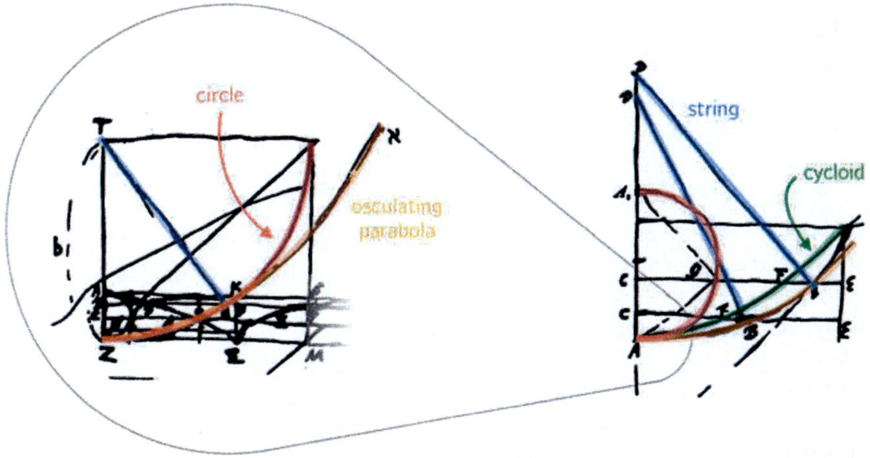

Fig. 7.2 The first (left) and last (right) figures drawn by Huygens in his 1659 *Worknote on the Pendulum*. Highlighted in color are the different curves considered in his proof: the circle (classical pendulum), its osculating parabola (small-angle approximation) and the cycloid (the isochrone pendulum). The left figure represents small-angle approximation (osculating parabola) of the classical pendulum (circle). The right figure shows that it is an approximation of the actual isochrone pendulum (cycloid). For references and details on these notes, see [570]

Fig. 7.2). His proof, in this sense, may be a bit disappointing at first: he uses the known, end result to motivate the means to prove it.[5] Nevertheless, the beauty of the geometrical arguments involved makes up for it.

To understand, with modern-day notations, how his proof works, let us summarize how we arrive at the isochrone result in the small amplitude limit:

- First, we assumed that $y(x)$ was a circle, which allowed to replace the arclength parameter s for the height y, going from (7.3) to (7.4)
- Then we replaced the circle by its osculating parabola, so as to turn the cubic into a quadratic, going from (7.4) to (7.5)
- Finally, the peculiar form of the integral (7.6), with a square root of a quadratic in the denominator, and its roots as integral limits, gives a constant value.

Clearly, isochrony comes from the second step, i.e., the right-hand side of (7.5) being *quadratic* in y. With this in mind, let us go back to the master equation (7.3) but without making the assumption of step 1, keeping a general curve $y(x)$ (or $y(s)$, equivalently). We write it as an ODE for y using the Leibniz rule as

$$\frac{1}{2}\left(\frac{\mathrm{d}y}{\mathrm{d}t}\right)^2 = g\,(h_o - y)\left(\frac{\mathrm{d}y}{\mathrm{d}s}\right)^2 . \tag{7.7}$$

[5] One might even wonder if Huygens did not even discover that the cycloid was a solution in the first place, and then checked if it yielded an isochrone pendulum. Indeed, before starting to think about this problem he already knew about the cycloid, as he was just getting out of a mathematical controversy after a challenge on this very curve posed by Blaise Pascal in 1658 [569].

Now let us suppose that the curve $y(s)$ is such that Eq. (7.7) (an exact equation) is *exactly the same ODE* as Eq. (7.5) (an approximate one). By comparing Eqs. (7.7) and (7.5), this implies that $y(s)$ satisfies

$$\left(\frac{\mathrm{d}y}{\mathrm{d}s}\right)^2 = \frac{2y}{\ell} \quad \Rightarrow \quad y(s) = \frac{s^2}{2\ell}, \tag{7.8}$$

where the integration constant has been set to 0 to ensure $y(0) = 0$. The[6] solution to the isochrone pendulum problem is right there, in Eq. (7.8). Indeed, combining Eqs. (7.7) and (7.8) to compute the period just as before readily gives $T = 2\pi\sqrt{\ell/g}$, as expected (we leave this easy integral for the reader). The associated pendulum is exactly isochrone, as no small-angle approximation was made to obtain it.

Nevertheless, it is not straightforward to infer by eye the actual shape of the curve $y(x)$ from its arclength parametrization (7.8). To obtain a more traditional, Cartesian equation, we look for a parametric equation, using s as a parameter. To this end, we derive from $y(s)$ the other coordinate function $x(s)$, using again the Euclidean line element $\mathrm{d}s^2 = \mathrm{d}x^2 + \mathrm{d}y^2$, which readily gives $\mathrm{d}x/\mathrm{d}s = \sqrt{1 - s^2/\ell^2}$. After an integration by parts, we readily obtain

$$x(s) = \frac{\ell}{2}\arcsin\frac{s}{\ell} + \frac{s}{2}\sqrt{1 - \frac{s^2}{\ell^2}}, \tag{7.9}$$

with the integration constant set to 0 to ensure $x(0) = 0$. Equations (7.8) and (7.9) provide the arclength parametrization for the Cartesian coordinates of the curve. However, there is a much more natural one. Indeed, it is clear that Eq. (7.9) calls (screams, even) for the introduction of an angle ϕ such that $s = \ell \sin\phi/2$. Replacing s by this new parameter ϕ in Eqs. (7.8) and (7.9), we find (with well-chosen trigonometric identities)

$$x(\phi) = \frac{\ell}{4}(\phi + \sin\phi) \quad \text{and} \quad y(\phi) = \frac{\ell}{4}(1 - \cos\phi), \tag{7.10}$$

the range of the parameter ϕ being $[-\phi_o; \phi_o]$, where $\phi_o \equiv \arccos(1 - 4h_o/\ell)$ is find by setting $y(\phi_o) = h_o$. Equation (7.10) is the familiar parametrization of the *cycloid*.[7] This particular expression is reminiscent of its geometrical definition as the path followed by a fixed point on the circumference of freely rolling circle (of radius $\ell/4$ here), as depicted at the top in Fig. 7.3. In this parametrization (7.10), ϕ is simply the angle by which the circle has rotated. That the cycloid is the solution to the isochrone pendulum is not really a surprise for who knows about two other equally

---

[6] Or at least, *one* solution to the problem. Just like Huygens, we did not prove its unicity. Although it turns out that the cycloid is the unique symmetric isochrone pendulum (i.e., the motion is symmetric with respect to the vertical axis), there exists an infinity of asymmetric isochrone pendula, as shown in [571].

[7] In light of the above, it is rather amusing to note that the name *cycloid*, was coined by Galileo, who claimed to have studied it for more than fifty years [572].

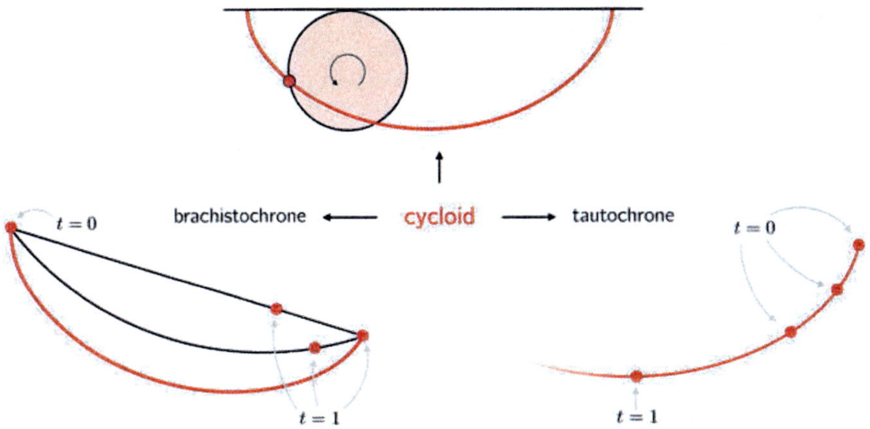

Fig. 7.3 *Top:* the traditional definition of the cycloid as the curve followed by a point on a rolling circle. *Bottom left:* the brachistochrone (cycloid, in red) and two other curves in black. Of the three rolling balls released from the top at $t = 0$, the first to arrive at the tip is the one on the cycloid ($t = 1$). *Bottom right:* the cycloid is a tautochrone: balls released at $t = 0$ from different heights all reach the bottom at the same time ($t = 1$). We encourage the reader to look at [572] which contains a nice historical perspective on these curves

famous curves: the *brachistochrone* and the *tautochrone*. The former is the curve that minimizes travel time of a freely rolling ball between any two points (left in Fig. 7.3), and the latter is the curve along which any two balls released from different heights will reach the bottom point at the same time (right in Fig. 7.3). These two curves are also cycloids, and are very closely related to the isochrone pendulum. Huygens solved the tautochrone problem,[8] while both Bernoulli brothers (and Leibniz) solved the brachistochrone problem first, in 1796 [572].

The Right Variable

One of the most important result of the previous calculation is that it allows one to rewrite the cycloid pendulum as a true harmonic system, where Galileo's circular pendulum was only nearly harmonic. This harmonic behavior is only revealed when the problem is written in term of the arclength s along the path. In terms of s, the equation of motion, obtained by combining the master ODE (7.3) and the parametrization (7.8), is simply

$$\frac{1}{2}\left(\frac{\mathrm{d}s}{\mathrm{d}t}\right)^2 = g\left(h_o - \frac{s^2}{2\ell}\right), \tag{7.11}$$

[8] Solving the tautochrone problem is equivalent to finding an isochrone pendulum, the only difference being that the string tension is replaced by the contact force from the solid curve.

This ODE, with a parabola on the right-hand side, is a hallmark of isochrony, as we shall see in this the next chapter. By differentiating Equation (7.11) with respect to time, we obtain $\ddot{s}(t) + (g/\ell)s(t) = 0$, which is the ODE of an harmonic oscillator.

The exact equation (7.7) can be used to understand the reason why Galileo's pendulum is nearly isochrone for small oscillations, and also why any pendulum is. Indeed, consider a pendulum described by the Cartesian curve $y(x)$, and expand it around $x = 0$, i.e., the downward equilibrium. There, we have $y = 0$ by construction of the coordinates, and $\mathrm{d}y/\mathrm{d}x = 0$ since the velocity must me perpendicular to the (vertical) string. Assuming a smooth curve, we have,

$$y(x) = \frac{x^2}{2\ell} + O(x^3)\,, \tag{7.12}$$

for some constant $\ell \equiv 1/y''(0)$, which is nothing but the radius of curvature at $x = 0$. Equation (7.12) is simply the second-order Taylor approximation of *any* pendulum's motion. To link this to the arclength s, we can use the Euclidean line element in the form $\mathrm{d}s^2 = \mathrm{d}x^2 + \mathrm{d}y^2$, and after a short calculation, we obtain

$$\left(\frac{\mathrm{d}y}{\mathrm{d}s}\right)^2 = \frac{2y}{\ell} + O(y^2)\,. \tag{7.13}$$

This result, although elementary, is crucial because it holds for any pendulum, and can be used to show two things. On the one hand, inserting it into the master equation (7.8) readily gives Eq. (7.5), the small amplitude equation of Galileo's circular pendulum. Consequently, as claimed, any pendulum will exhibit isochrony in the small-angle limit. On the other hand, we see that the exact cycloid ODE Eq. (7.8) is the first-order approximation to the general result (7.13). In other words, where all pendulum are nearly isochrone because they can all be approximated by a parabola (in the y or s coordinate), only one curve extends this parabolic property to the whole pendulum, the cycloid. The difference between these two cases is encapsulated in the $O(y^2)$ term in Eq. (7.13).

Now that Huygens has found that the cycloid solves the isochrone problem, his last step is to find the shape of the pendulum walls (cf. Fig. 7.1) to enforce this curve. This reverse problem leads him to introduce the notions of *involute* and *evolute* of a curve. The involute of a curve is the locus of points drawn by a pendulum's bob when it wraps around that curve, and the evolute is the converse, i.e., the evolute of the involute of a curve is the curve itself. Huygens goes on to show that the cycloid is its own involute (and evolute), as can be seen on Fig. 7.4. Consequently, by the very geometry of the problem, the walls that will enforce cycloidal motion are portions of that same, exact cycloid. This is a unique property that, once again, characterizes the cycloid among all other curves.

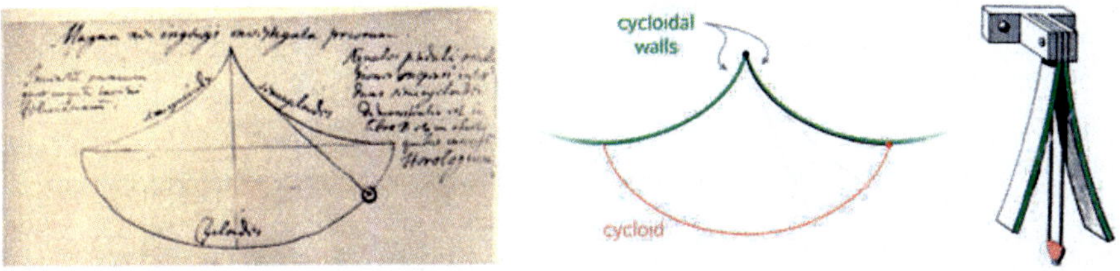

Fig. 7.4 *Left and middle*: an excerpt from Huygens' notebook on the isochrone pendulum, where we see both the cycloid path followed by the pendulum's bob, and the cycloid walls that generate the motion. This is only possible because the involute of a cycloid (the red curve in the middle) is a cycloid (the green walls). *Right*: the actual size and shape of the walls used by Huygens for his cycloid pendulum, as depicted in his Horlogium Oscillatorium, with the same color-coding as the other figures

7.1.3 Staying in Harmony

After Galileo and Huygens' investigations of the pendulum, a profusion of designs for pendulum clocks were proposed. Huygens' cycloid was perfectly isochrone on the theoretical point of view, and on the experimental side too. In 1657, him and the horologist Salomon Coster built a pendulum providing a hundred-fold increase in accuracy compared to older designs, losing only 15 s per day. However, for most purposes, it did not prove that useful, as the cycloid pendulum only provided increased accuracy for wide oscillations, whereas a rigid, classical pendulum with small-oscillations was enough for most applications.

Yet, Huygens found another way of applying his isochrone pendulum to time-keeping devices. After the mastering of regular pendulums, the advent of pocket clocks and wrist watches came with another challenge: for these devices, gravity does not always point downward with respect to the oscillator. Since for the pendulum gravity acts as a restoring force, the natural first try was to mimic gravity by replacing it with a spring. It is in this context that Robert Hooke stated the eponymous law in 1675, first publishing with an anagram (a common practice at the time)

$$\textit{ceiiinosssttuv} \quad \Leftrightarrow \quad \textit{ut tensio, sic vis}$$

which translates in modern day mathematics into $F \propto x$, where F is the restoring force and x the displacement. As we know today, this force derives from a potential proportional to the displacement squared, nowadays called the harmonic (or Hooke) potential. In other words, the time-dependent displacement of such a body under the influence of a Hooke force satisfies the ODE (7.11), just like the isochrone pendulum. Huygens only needed to find how to construct, mechanically, such a force. Thankfully, Huygens knows all about curves, and invents right around that time the *spiral balance*, an apparatus composed of a metallic spring in the shape of an Archimedean spiral.[9] This curve has a polar equation of the form $r = a + b\theta$, where a, b are constants, and as a spring, produces a restoring force that satisfies (an angular) Hooke's law even for wide amplitudes. Coupled to an oscillator, it produce isochrone motion. Just like Hooke, Huygens shares his discovery in a coded message that he sends to Royal Society's secretary Henry Oldenburg in 1675 [559]:

$$
\begin{gathered}
413537312343242 \\
abcefilmnorstux
\end{gathered}
$$

which, once deciphered, gives in Latin "*Axis Circuli mobilis affixus in Centro Volute ferrea*", or "the axis of the mobile circle is fixed at the center of an iron spiral". This is indeed the basic mechanism that can be found drawn on several letters sent by Huygens around 1675, as depicted in Fig. 7.5. The period of oscillation T for the spiral balance depends on the moment of inertia I of this "mobile circle" (the balance wheel in red in Fig. 7.5) and on the stiffness κ of the spiral spring, through the relation

$$T = 2\pi\sqrt{\frac{I}{\kappa}}, \tag{7.14}$$

We see from Eq. 7.14 the isochrone nature of this system (of course, up to nonlinearities in the spring and energy losses). This very mechanism is still present in most mechanical clocks and watches.

The quest for isochrone oscillations and precise time-keeping devices did not end with the seventeenth century, of which we provide a summarizing timeline in Fig. 7.6 for the reader's convenience. This rich and major part of scientific history is fascinating, and the interested reader will find in [558] (in French) much more about it.

This first introductory section could be seen as rather unrelated to the problem that will occupy us in the next chapters. After all, we will deal with Newtonian, celestial mechanics with the true Kepler potential ($\psi(r) \propto -1/r$), whereas Huygens considered a constant gravitational field, associated to a Hooke potential ($\psi(z) = mgz$). However, it is my hope that, at the end of the second part of this thesis, the reader will understand why Huygens' derivation is so interesting in the light of celestial isochrone mechanics. In particular, regarding the previous discussion, three key points stand out, and will be worth remembering:

- there exists an adapted variable in which the isochrone pendulum looks like a harmonic oscillator,
- isochrony is equivalent to the potential looking like a parabola in this variable,
- sometimes, insight can be found by looking at Archimedes' work.

Remarkably, these three features will be at the core of our resolution of the isochrone problem as initially formulated by Michel Hénon in 1959.

[9] Huygens, whose father nicknamed *my little Archimedes*, knew about Archimedes' work very well. Interestingly enough, we will also meet Archimedes in celestial isochrony.

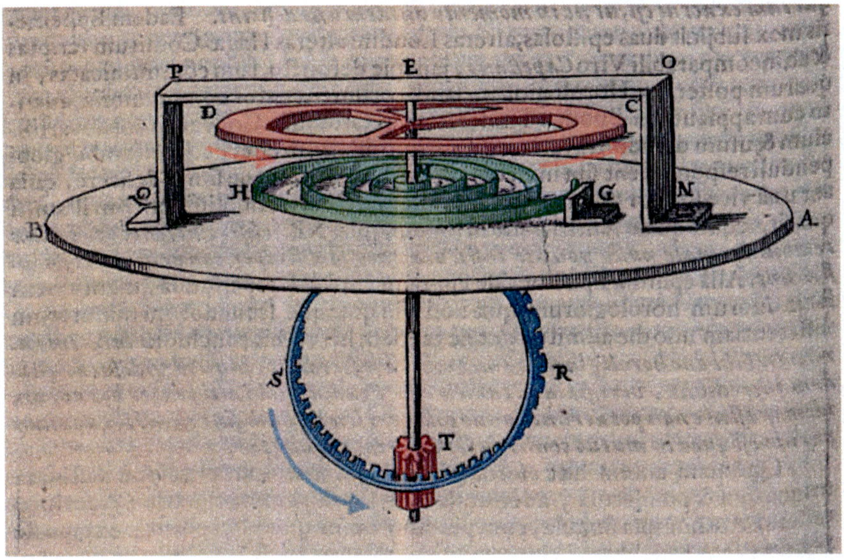

Fig. 7.5 Huygens' drawing of a spiral balance, found in a letter sent to Jean Gallois in 1675 [559]. The balance wheel (in red) connects to the spiral spring (in green) through a vertical shaft. Turning the wheel loads potential energy into the spring, and releasing it starts the back and forth motion. The shaft terminates at the bottom with a cog transmitting motion the geared wheel (blue)

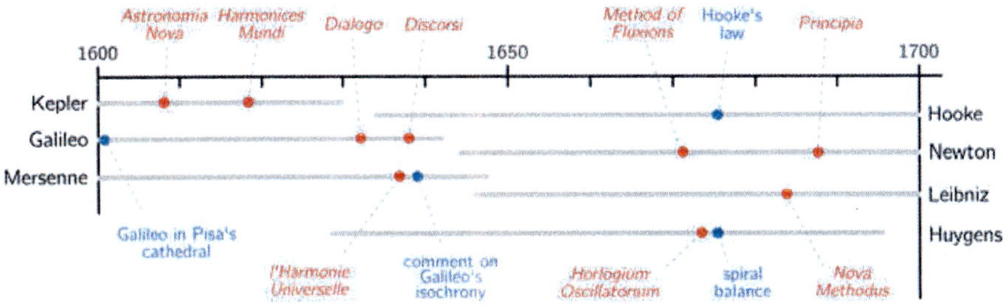

Fig. 7.6 Seventeenth century timeline of the main works and events cited in this introduction. Horizontal grey lines shows the life span of the associated scientist, on left and right (note that birth and death fit into this century for Huygens only). In red their *magnum opus* (*opera* for some of them), with their common name (in general a shorthand of the whole title). In blue historical notes that have been cited in the text. Many additional and interesting historical references can be found in the very nice account [559]

7.2 Periodic Orbits in Central Potentials

The main difference between the isochrony of the pendulum, as described above, and the isochrony defined by Michel Hénon is that (1) it deals with a 2-dimensional problem, and (2) it is defined in the context of celestial mechanics. Consequently, in this section we shall lay the general definitions that will be used throughout the next chapters. In Sect. 7.2.1, we derive some standard results regarding periodic orbits of test particles in a given central, gravitational potential. Then, in Sects. 7.2.2 and 7.2.3 we define two constants of motion that will be at the core of Hénon's isochrony, defined in the next chapter.

7.2.1 *Generalities*

Let us consider the three-dimensional Euclidean space and an inertial frame of reference equipped with a set of spherical coordinates (r, θ, φ) and the associated natural basis $(\vec{e}_r, \vec{e}_\theta, \vec{e}_\varphi)$. We assume that around the origin $O = (0, 0, 0)$ lies a spherically symmetric distribution of matter with mass density $\rho(r)$. This system generates a gravitational potential, denoted $\psi(r)$, that obeys Poisson's equation

$$\Delta\psi(r) = \frac{1}{r^2}\frac{\mathrm{d}(r^2\psi')}{\mathrm{d}r} = 4\pi G\rho(r)\,, \tag{7.15}$$

where a prime $'$ denotes a differentiation with respect to r, and G is Newton's constant. We consider a test particle of mass m orbiting this system, with position vector $\vec{r}$ and velocity vector $\vec{v} \equiv \mathrm{d}\vec{r}/\mathrm{d}t$. Owing the spherical symmetry, the angular momentum $\vec{L} \equiv m\vec{r} \times \vec{v}$ of the particle is conserved. Its norm is given by $|\vec{L}| = mr^2\dot{\theta}$, with the usual notation $\dot{\theta} = \mathrm{d}\theta/\mathrm{d}t$ for the time derivative. The total energy E of the particle, sum of a kinetic term $\frac{1}{2}m|\vec{v}|^2$ and a potential term $m\psi$, is conserved as well. Let us introduce $\xi \equiv E/m$, the (total) energy of the particle per unit mass ; and $\Lambda \equiv |\vec{L}|/m$, the (norm of the) angular momentum per unit mass. The explicit computation of the energy in terms of r yields the following energy conservation equation

$$\xi = \frac{1}{2}\left(\frac{\mathrm{d}r}{\mathrm{d}t}\right)^2 + \frac{\Lambda^2}{2r^2} + \psi(r)\,. \tag{7.16}$$

Since E, $\vec{L}$ and m are conserved quantities, ξ and Λ are two constants of motion. In a given potential ψ, the quantities (ξ, Λ) are sufficient to know everything about the dynamics of a particle, up to initial conditions in the position. Accordingly, we will abuse notation and speak of (ξ, Λ) as *a particle*. Along with these initial conditions, Eq. (7.16) forms a nonlinear ODE for the function $t \mapsto r(t)$. We are interested in orbits and therefore will consider bounded solutions to Eq. (7.16). Once $r(t)$ has been obtained by solving Equation (7.16), we can solve for the angular part $\theta(t)$ by means of conservation of angular momentum, which readily gives

$$\frac{\mathrm{d}\theta}{\mathrm{d}t} = \frac{\Lambda}{r^2} . \tag{7.17}$$

This one-way decoupling between radial and angular motion will be important for our purpose, as the definition of isochrony depends on the characteristics of the radial motion of a test particle.

Since the orbit is bounded and the function $r(t)$ is continuous, we may define r_P and r_A as the minimum and maximum values of $r(t)$. Physically, r_P is (the radius of) the *periapsis*, i.e., the point on the orbit closest to O; and r_A that of the *apoapsis*, the one farthest from O. At these turning points, the radial velocity $\dot{r}\vec{e}_r$ vanishes and $\dot{r}$ changes sign. Consequently, by Eq. (7.16), r_P and r_A are two solutions to the following algebraic equation[10]

$$\xi = \psi_e(r) \equiv \frac{\Lambda^2}{2r^2} + \psi(r) , \tag{7.18}$$

where we have introduced the *effective potential* $\psi_e(r)$, sum of the gravitational potential $\psi(r)$ and the centrifugal term $\Lambda^2/2r^2$. Note that, for some values of (ξ, Λ), there is unique solution r_C to Eq. (7.18). The associated orbit is circular, of radius $r = r_C$. This can always be seen as the degenerate case $r_P \to r_A$.

It is customary to use the effective potential to study geometrically the orbit of a particle, depending on its energy. As depicted in Fig. 7.7, one plots the function ψ_e for a given value of Λ, and then draws a line of height ξ. By construction, any choice of initial conditions will yield $\xi \geq \min \psi_e$. When there are two intersections between the line $y = \xi$ and the curve $y = \psi_e(r)$, the orbit is non-circular and one can read the periapsis r_P and apoapsis r_A as the abscissae of the intersection points. When there is only one intersection, its abscissa is the orbital radius r_C and the orbit is circular. Furthermore, at $r = r_C$ the tangent to the curve is the horizontal line $y = \xi$, and therefore, $\psi_e'(r_C) = 0$. Two examples of this classical construction are depicted in Fig. 7.7 for two particles sharing the same energy but having different angular momenta.

Lastly, it is worth noting that, by virtue of Eq. (7.16), the quantity $\xi - \psi_e(r) \propto \dot{r}^2$ should always be strictly positive when $r(t) \in]r_P, r_A[$, and vanish at r_P and r_A, by definition.

---

[10] When there are more than two solutions to Eq. (7.18)., say $(r_1, \ldots, r_n)$ for some $n \geq 3$, the orbit is selected on the graph by the initial radius r_0: its periapsis and apoapsis being $r_P = r_i$ and $r_A = r_{i+1}$, where i is such that $r_0 \in [r_i, r_{i+1}]$.

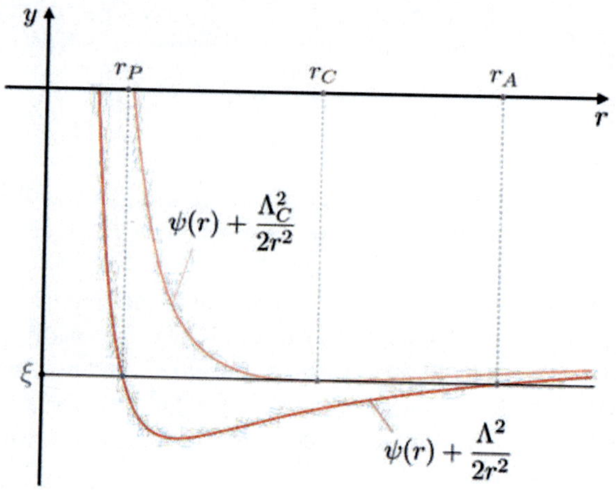

Fig. 7.7 The graph $y = \psi_e(r)$ corresponds to the effective potential $\psi(r) + \Lambda^2/2r^2$. Two ψ_e are depicted, associated with two particles with different angular momenta : Λ (bottom curve, red) and $\Lambda_C > \Lambda$ (top curve, light red). The vertical line $y = \xi$ defines two orbits associated with the same energy ξ. Particle (ξ, Λ) is on a generic, non-circular orbit $[r_P, r_A]$ and particle (ξ, Λ_C) is on a circular orbit of radius r_C. Note that they both orbit in the same potential ψ

7.2.2 Radial Period

The Hamiltonian nature of the system implies that any bounded and continuous solution to Eq. (7.16) must be periodic [345]. In other words, if an orbit is bounded in a central potential, it is necessarily radially periodic. We shall define the *radial period* (*the period* in short hereafter) as the smallest $T \in \mathbb{R}_+^\star$ such that $r(t+T) = r(t)$ for all $t \geq 0$. Note that the *radial period T* always exists for bound orbits. It should not be confused with the *period of motion* of the particle (i.e., the period of the position vector $t \mapsto \vec{r}(t) \in \mathbb{R}^3$). The later only exists if the orbit is closed in real space, as for Keplerian ellipses.

For a generic, non-circular orbit, one can get a formula for T by first isolating the variables t and r in Eq. (7.16). This yields

$$\mathrm{d}t = \pm \frac{\mathrm{d}r}{\sqrt{2\xi - 2\psi(r) - \Lambda^2/r^2}}\,. \tag{7.19}$$

In this formula, the + sign corresponds to an increasing radius $r(t)$, i.e., when the particle goes from r_P to r_A, whereas the − sign corresponds to a decreasing radius, i.e., when the particle comes from r_A back to r_P. Integrating Equation (7.19) over a full period and taking into account the two different signs provides the following integral formula for the period, which we take as a definition

$$T \equiv 2 \int_{r_P}^{r_A} \frac{\mathrm{d}r}{\sqrt{2\xi - 2\psi(r) - \Lambda^2/r^2}}. \tag{7.20}$$

Notice that the bounds of the integral r_P and r_A are precisely the zeros of the denominator vanish, by virtue of Eq. (7.18). The fact that $x \mapsto 1/\sqrt{x}$ is integrable near 0 ensures the convergence of the integral.[11] We note that strictly speaking, the integral in (7.20) does not make immediate sense for circular orbits. As we shall see with explicit formulae in the next chapters, by continuity there is a well-defined notion of radial period associated to circular orbits, in terms of ξ and/or Λ.

7.2.3 *Apsidal Angle*

Let a particle (ξ, Λ) be at some position $(r(t), \theta(t))$ on its orbit at some fixed time t (red point on the right of Fig. 7.8). The radial period T corresponds to the time taken for the particle to go back to that radius $r(t)$ (with the same sign of $\dot{r}$). This does not mean, however, that the orbit itself is a closed curve in real space. It will be the case only if, after a period T, the new angle $\theta(t + T)$ is equal to $\theta(t) + q\pi$, for some rational number $q \in \mathbb{Q}$. The orbit then closes after a finite number of radial periods that equal the denominator of q.

To quantify this, we define the quantity $\Theta \equiv \theta(t + T) - \theta(t)$. This angle is constant[12] and corresponds physically to the angle difference between any two positions, a period T apart. In orbital mechanics it is customary to take the angle difference between two successive periapsis, as depicted in Fig. 7.8. Therefore, we shall call $\Theta > 0$ the *apsidal angle*. When Θ is a rational multiple of π, the orbit depicts a closed curve in real space. Otherwise, the orbit densely fills the shell region $r \in [r_P, r_A]$.

An integral formula can be obtained for Θ, by using the conservation of angular momentum Eq. (7.17). This equation gives $\mathrm{d}\theta = \Lambda \mathrm{d}t/r^2$, which, when combined with Eq. (7.19) and integrated over one period, gives easily

$$\Theta \equiv 2\Lambda \int_{r_P}^{r_A} \frac{\mathrm{d}r}{r^2\sqrt{2\xi - 2\psi(r) - \Lambda^2/r^2}}. \tag{7.21}$$

Equation (7.21) is well-defined for any non-circular orbit, but also admits a finite value in the circular-orbits limit, just as the integral for T in Eq. (7.20). We shall come back to this in the next chapters.

[11] Indeed, using Equation (7.18), we have the Taylor expansion $\xi - \psi_e(r) = \psi_e'(r_P)(r_P - r) + o(r_P - r)$, and $\psi_e'(r_P) \neq 0$ since the orbit is non-circular. The integrand in Eq. (7.20) is thus equivalent to $(r_P - r)^{-1/2}$, which is integrable at r_P. The same holds at r_A.

[12] Since $\Lambda = r^2\dot{\theta}$ we have $\dot{\Theta} = \Lambda/r(t+T)^2 - \Lambda/r(t)^2$, which vanishes by T-periodicity of $r(t)$.

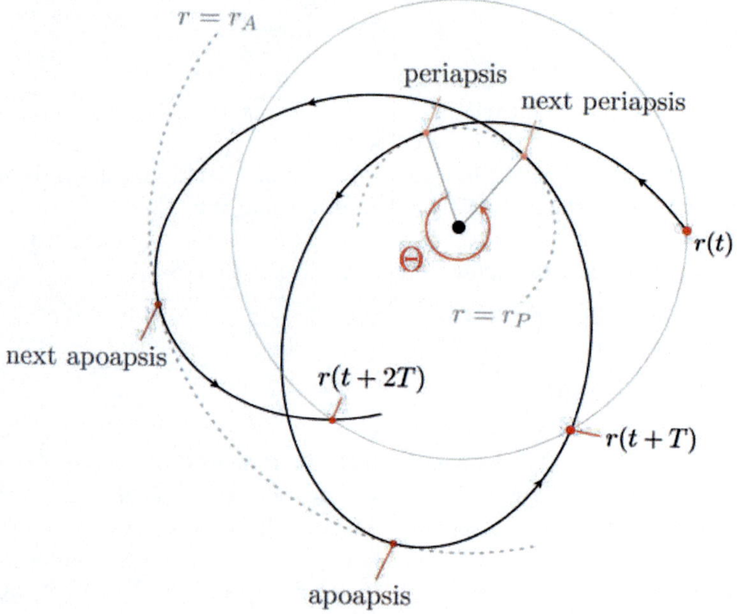

Fig. 7.8 A typical orbit in a central potential (solid black), centered on the origin O, during ~ 2 periods T. At some initial time t, the particle is at a radius $r = r(t)$ (red dot on grey circle, right). At times $t + T$ and $t + 2T$ it comes back to that same radius, crossing the grey circle with the same sign of $\dot{r}$. During the first period $[t, t + T]$, the particle reaches the periapsis (inner dashed circle) and then the apoapsis (outer dashed circle). During the second period $[t + T, t + 2T]$, the process repeats. Θ is the angle between two successive periapsis, but also between any two successive positions a period T appart

7.2.4 *Radial Action*

A Hamiltonian treatment of the problem will be thoroughly discussed in Chap. 9. For now, we will only need one result from it, the existence of a *radial action* J. Fundamentally, this is defined as the integral over one radial period of the momentum conjugated to the radial variable r. In practice, this momentum is simply $m\dot{r}$, and we will simply define this action by

$$J = \frac{\sqrt{2}}{\pi} \int_{r_P}^{r_A} \left(\xi - \frac{\Lambda^2}{2r^2} - \psi(r) \right)^{1/2} \mathrm{d}r \,. \tag{7.22}$$

The fact that it is an action (from a set of action-angle variables) is not important here. Rather, its central feature is that it acts as a *generating function* for the radial period T and the apsidal angle Θ as seen as functions of (ξ, Λ). Indeed, by comparing Eqs. (7.20) and (7.21) with Eq. (7.22), the following relations are immediate

$$\frac{T}{2\pi} = \frac{\partial J}{\partial \xi} \quad \text{and} \quad \frac{\Theta}{2\pi} = -\frac{\partial J}{\partial \Lambda}. \tag{7.23}$$

These two equations provide a fundamental result that will be used later on. To derive it, we take the partial derivative of T with respect to Λ (left of (7.23)) and the partial derivative of Θ with respect to ξ (right of (7.23)). Using Schwartz's theorem to swap the order of derivative, we readily find that

$$\frac{\partial T}{\partial \Lambda} = -\frac{\partial \Theta}{\partial \xi}. \tag{7.24}$$

Of course this differential relation could have been obtained readily by differentiating the integral definitions (7.20) and (7.21). It just more convenient through the radial action J. In any case, Eq. (7.24) will be very useful to reformulate the problem of isochrony, in the next section.

7.3 Hénon's Isochrony

With the background on isochrone pendula given in Sect. 7.1 and the prerequisites provided in Sect. 7.2, it is time to move to the statement of the isochrone problem with which we will deal in the following chapters. First we provide some context on Hénon's isochrony (Sect. 7.3.1), state the problem to be solved (Sect. 7.3.2), and make a brief summary of what can be found in the literature about this particular notion of isochrony (Sect. 7.3.3).

7.3.1 The Context

Michel Hénon (1931–2013) was a French theoretical astrophysicist. Although he touched upon many different topics in his career, it is fair to say that all of them had to do with the field of dynamical systems. In theoretical physics, he was a pioneer in numerical exploration of dynamical systems (the celebrated Hénon map is named after him [573]). Regarding astrophysics, we refer to the book [574], dedicated to Hénon's life and work, which summarizes several of his many decisive scientific contributions.

The problem of interest for us, isochrony, was introduced by Hénon at the end of the 1950s,s, at a time when he was not *Docteur és Sciences Physiques* yet. His three papers on isochrony were published in 1959 and 1960, and he defended his PhD in December 11, 1961 in front of André Danjon, director of the Paris Observatory, and Evry Shatzman, his advisor.[13] At that time, Hénon was interested in a particular type of astrophysical systems called *globular clusters*. These objects are rather peculiar, like tiny galaxies within galaxies, made of thousands to millions of stars bound together by their collective gravitational field. They were first resolved observationally as a luminous system of many individual stars at the end of the eighteenth century by William Herschel, who also introduced the term *globular* for their near-perfect spherical symmetry [575]. Today, globular clusters are of central importance for astronomy and astrophysics and are found in essentially all galaxies. They appear to contain stellar (and possibly massive) black holes in their core [576–578], and the total number of globular clusters seems to be positively correlated with the mass of the super-massive black hole of the host galaxy [579]. New literature on the dynamics and evolution of globular clusters is currently burgeoning thanks to the recent and unparalleled precision of the data collected during the ongoing GAIA mission [580–582].

In the Milky Way, around 150 globular clusters have been observed [581], the most massive and luminous of them being Omega Centauri, with around 10 million stars. In 1959–1960, Hénon publishes a three-part series of articles in French entitled *L'amas isochrone* (the isochrone cluster) [583–585]. The primary goal of this work, which he would never revisit later in his career, was to propose an analytic model for the potential generated by these globular clusters. In his seminal paper [583] (in French, for an English version see [586]) he succeeded in solving this ambitious problem and found what he called the "isochrone potential"

$$\psi(r) = -\frac{\mu}{s}, \quad \text{where} \quad s \equiv b + \sqrt{b^2 + r^2}, \tag{7.25}$$

and b is a size parameter closely related to the half-mass radius of the system. The corresponding mass density distribution, obtained by solving the Poisson equation (7.15), was in good agreement with some of the observed globular clusters available in 1959, as can be seen in Fig. 7.9.

7.3.2 *The Problem*

How did Hénon find his potential, and why did he call it "isochrone"? First, he made the following heuristic argument. In a globular cluster, the stars at the center have a nearly constant density, whereas those on the periphery see the spherical distribution beneath them. Through the Poisson equation, a constant density is associated to a quadratic (or harmonic) potential, whereas spherical distributions generate the Kepler potential, say

[13] Interestingly, the work on isochrony presented in this thesis was done in the Evry Shatzman building of the Paris observatory (Meudon site), and presented for the first time in the André Danjon room (Paris site).

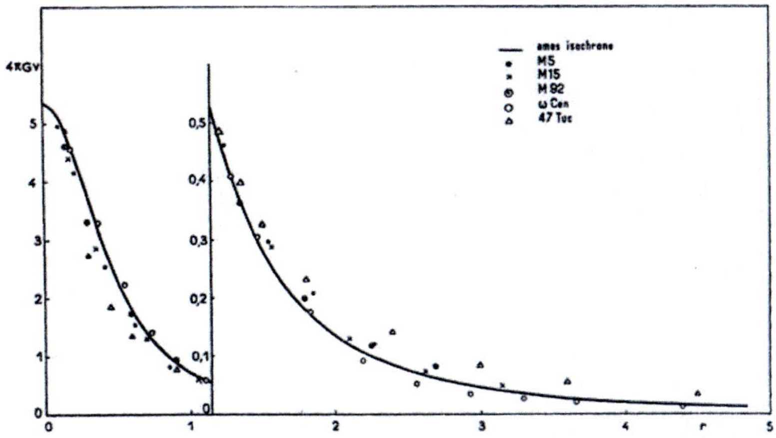

Fig. 7.9 A figure from Hénon's first article [583] on isochrony, comparing the theoretical (projected) mass density of Hénon's isochrone model and the observations for 5 globular cluster: M5, M15, M92, Omega Centauri (NGC 5139) and 47 Tucanae (NGC 104)

$$\psi_{\mathrm{Ke}}(r) = -\frac{\mu}{r} \quad \text{and} \quad \psi_{\mathrm{Ha}}(r) = \frac{1}{8}\omega^2 r^2 \,, \tag{7.26}$$

with μ and ω two constants characterizing the mass sourcing these potentials. A potential that would describe the entirety of the globular cluster should share a common feature with these two. What property do they share? A first answer is that they generate elliptic orbits. But in general, orbits in central potentials are not even closed, except for, precisely, the Kepler and harmonic potential. This is the celebrated Bertrand theorem (which we will show to be a simple consequence of isochrone theory later on). Another, less restrictive, commonality of the harmonic and Kepler potentials (7.26) can be found by examining the radial period T of orbits therein. We find

$$T_{\mathrm{Ke}} = \frac{2\pi\mu}{(-2\xi)^{3/2}} \quad \text{and} \quad T_{\mathrm{Ha}} = \frac{2\pi}{\omega} \,, \tag{7.27}$$

the first identity being the celebrated third law of Kepler, and the second the simple relationship between period and angular frequency of the harmonic oscillator. Notice that they do not depend on the other constant of motion Λ, the angular momentum. However, for a generic central potential ψ, the radial period T is a function of both ξ and Λ, as should be clear in view of Eq. (7.20). Two particles with same energy but otherwise different angular momenta will orbit the Kepler or harmonic potentials with the same (*iso*) radial period (*chrone*). Hénon then asks if there are other potentials with this property, and if, most importantly for him, these other "isochrone" potentials accurately describe globular clusters.

Our definition of isochrony in the case of celestial mechanics will thus be that of Hénon, for spherically symmetric (i.e., radial) potentials: *a radial potential $\psi(r)$ is isochrone if and only if all periodic orbits it generates have a radial period T only*

D'autre part l'angle entre deux apocentres successifs est :

$$\text{(11)} \qquad \alpha = \pi\left(1 + \frac{A}{\sqrt{A^2+2}}\right)$$

On notera le fait curieux que α dépend uniquement du moment angulaire A, et non de E.

Fig. 7.10 Hénon points out the "*curious fact*" that, in his isochrone model, the apsidal angle α (Θ in our notations) depends only the angular momentum A (Λ for us)

function of the energy ξ. Schematically, we thus have

$$\not\psi \text{ is isochrone} \quad \Leftrightarrow \quad T = T(\xi, \not\Lambda)\,, \tag{7.28}$$

with the radial period being a functional of $\not\psi$, defined by a integral in (7.20). Hénon found one particular isochrone potential, called *the* isochrone in the astrophysical literature. As we will see, there exists many other. Consequently, in the following we will use *isochrone* as a qualifier for the whole class of potentials verifying (7.28), and call the potential discovered by Hénon (7.25), the *Hénon potential.*

There is an alternative, equivalent definition of isochrony, on which, without knowing, Hénon commented. Indeed, we can find on page 3 of his second paper [584], a remark depicted in Fig. 7.10.

After having computed the apsidal angle Θ in the Hénon potential, he notices that the resulting expression depends only on Λ. Thanks to the previously established result in Eq. (7.24), we see that this is not a coincidence, but an equivalent definition of isochrony. Accordingly, we can complete the definition (7.28) as

$$\not\psi \text{ is isochrone} \quad \Leftrightarrow \quad T = T(\xi, \not\Lambda) \quad \Leftrightarrow \quad \Theta = \Theta(\not\xi, \Lambda)\,. \tag{7.29}$$

The problem that will occupy us for the remaining chapters is about this notion of isochrony. In particular, starting from the definition (7.29), we will be interested in answering as thoroughly as possible (1) the question of the existence of other isochrone potentials and (2) the properties of orbits therein. Our exploration will be made under the light of mathematical physics, and we will not discuss the applicability of the theory of isochrone potentials to the modeling of globular clusters, or other astrophysical systems. However, work remains to be done in that direction, and several prospects are provided in the thesis' conclusion section.

7.3.3 The Literature

The literature on this particular isochrone problem is rather scarce if we look at it from a mathematical-physics perspective (which is that of the present work), but rather rich from an astrophysics/dynamics point of view. Before ending this chapter, let us provide an overview of the literature that deals directly with Hénon's isochrone.

Astrophysics

Evidently, the first to provide answers to the problem is Michel Hénon himself. In his three papers, the first one [583] is dedicated to finding the explicit formula for his isochrone potential, given in Eq. (7.25) and compute the corresponding density distribution. He then proposes a heuristic argument as to why his isochrone model fits well with the observations. His argument is as follows. First, resonances are expected to be stronger between stars of equal period. Second, if the cluster is not yet isochrone, then the set F_T of stars with period T do not have the same energy ξ (typically, radial orbits (large Λ) have an energy smaller than quasi-circular orbits (small Λ). Third, mutual interactions in F_T will naturally drive it a state of energy equipartition. Consequently, all stars in F_T will have the same energy, which defines the isochrone cluster. Hénon points out that this is but a heuristic, and that a formal demonstration of this process, if it exists, remains to be given.

Hénon's second paper [584] aims at exploring the orbits in his isochrone potential. He computes the period T and apsidal angle Θ, checks that they indeed satisfy (7.29), and provide drawings of some isochrone orbits.[14] In the third and last paper [585], he goes back to the statistical physics approach and computes the distribution function of an isochrone cluster using his model. Although the recent refinement of observations has actually revealed a wider diversity, Hénon's isochrone model remains at the center of cluster modeling for at least two reasons. First, just like the harmonic and Kepler potentials, this potential is fully integrable and its action-angle formalism provides a fundamental basis for both the modeling and simulations of stellar systems (see e.g., [587]). More recently, a detailed numerical analysis [588] showed that the isochrone model could be associated with the initial state of the evolution of singular stellar systems (e.g., globular clusters and/or Low Surface Brightness Galaxies); a result that followed an involved extension of many aspects of Hénon's work on isochrone potentials [589].

Hénon's isochrone model, and more precisely his elegant (but heuristic) explanation of why it may be physically relevant to cluster modeling, does not seem to be widely known nor used. Yet, to my knowledge, no further exploration of this particular isochrone resonance mechanism can be found in the literature, except for brief mentions in [590] (see Sects. III and IV there). However, the isochrone model has been used extensively in cluster and galactic dynamics, not as an exact potential for these systems, but to develop physical intuition on otherwise very complex mechanisms. The main reason is because it provides completely analytic results while presenting much richer dynamics that perfectly closed ellipses. The authoritative book [591] contains a number of references on these applications. As an example, let us mention the extension of Hénon's isochrone to flattened potentials in [592, 593], so that it can be properly applied to galactic potentials which are (at least for the non dark matter sector) clearly not spherically symmetric like Hénon's isochrone.

---

[14] I do not know if Hénon used graphical methods to plot the complicated parametric equations of the orbit, or if he used his own mechanical calculator (he used to build is own before the advent of modern computers [574]).

Mathematical Physics

On the mathematical physics side, expect for Hénon's first paper, in which he partly solves the (restricted) problem of isochrony, not much is found in the literature. The first edition of James Binney and Scott Tremaine's *Galactic Dynamics* (1987) makes a brief mention of it, essentially re-deriving all of Hénon's results from the explicit form for the potential (7.25) (they do not discuss how and why Hénon found it in the first place). The second edition (2008) [591] contains a new section on action-angle variables, and uses for Hénon's isochrone for illustration. However, no reference is made to earlier works by Dino Boccaletti and Guiseppe Pucacco's *Theory of Orbits* published in 2003–2004, where one can already find all these action-angle results. In all these references, however, the isochrone is a mean to illustrate key ideas about the orbits of stars in radial potentials, and no particular emphasis is made on Hénon's model regarding is peculiar mathematical properties, as will be explored here. Hénon's derivation of the form (7.25) was elegantly re-visited by Donald Lyndel-Bell in [590]. There, he showed that Hénon's isochrone could be recovered by generalizing the classical mapping between the Hooke and Kepler potentials (which both produce ellipses), as was already described by Newton with his "transmutation" of the force [594].

Except for these rare references, no real and thorough discussion or revisit of the mathematical problem posed by Michel Hénon (find all isochrone potentials and discuss their properties) exists before 2018 (to my knowledge). There is a good reason for it. For physical applications, we shall see that although the class of isochrone potentials is very large, only the Kepler, harmonic and Hénon potentials ((7.26) and (7.25)) really matter, as other families are associated to more exotic mass distributions. However, when considered as a whole, the class of isochrone potentials presents an inherent, profound structure with many geometric properties and symmetries. This was revealed in 2018, with a paper by Alicia Simon-Petit, Jérôme Perez and Guillaume Duval [589]. In this work, other kinds of potentials with the isochrone property were found and classified using elements of group theory and Euclidean geometry.

The present work builds on the findings of [589], and aims to go further in three directions, one for each chapter. These extensions are organized as follows.

- First, we provide a fully geometrical solution of Hénon's main problem, finding all isochrone potentials. We shall see that with a geometrical treatment, one family of potentials was left aside in [589]. Therefore, we give a complete solution of Hénon's problem and exhaustively classify all isochrone potentials, based on their physical properties. This is the content of Chap. 8.
- Then, we study in details the shape, properties and conditions of existence of isochrone orbits. In particular we generalize Kepler's third law to all isochrone orbits, providing a synthetic analytic formula for both the radial period and the apsidal angle. We also detail and fulfill a geometrical program that leads to an analytic parameterization of any isochrone orbit, completing the program started in [589]. This is the content of Chap. 9.

- Lastly, we complete this exploration by examining isochrone potentials and isochrone orbits within the realm of Hamiltonian mechanics. We bring to light the central role and universal property of isochrone potentials, and show that fundamental properties and symmetries of the two academic isochrones (Kepler and harmonic), for example Kepler's laws of motion, the Bertrand Theorem, the Kepler Equation, eccentric orbital elements are all but special cases of the more general theory of isochrony.

Chapter 8
Isochrone Parabolae

Contents

It is remarkable how, sometimes, in the glimpse of an eye, one notices something for the first time even though it was there all along.

C. Huygens
Letter to G.W. Leibniz (1691).

Huygens' solution of the isochrone pendulum problem, as we saw in the previous chapter, reveals that the isochrone pendulum traces a cycloid curve in its plane of oscillation. Fundamentally, we pointed out that this was reminiscent of a quadratic potential energy, as expressed in terms of the arclength defined by the pendulum motion. In this chapter, we will see how this feature is also shared by celestial isochrony. In particular, by introducing what we call the Hénon variables in Sect. 8.1, it will be shown in Sect. 8.2 that any isochrone potential has the shape of a parabola in these variables. This remarkable result, called the *fundamental theorem of isochrony*, is the most important feature of the theory of isochrone potentials. Its proof involves

P. Ramond, *The First Law of Mechanics in General Relativity & Isochrone Orbits in Newtonian Gravity*, Springer Theses,
https://doi.org/10.1007/978-3-031-17964-8_8

two main ingredients: (1) a calculation based off Hénon's own ideas found in his seminal paper, and (2) a geometrical characterization of parabolae that can be traced back to Archimedes. With this theorem, we complete in Sect. 8.3 the classification of isochrone potentials, thus answering Hénon's problem once and for all.

8.1 The Hénon Variables

The effective potential method, presented in Fig. 7.7, is particularly useful to find geometrically the characteristics of an orbit of energy and angular momentum (ξ, Λ). It is based on the equation of conservation of energy (7.16), in the form

$$\frac{1}{2}\left(\frac{\mathrm{d}r}{\mathrm{d}t}\right)^2 = \xi - \left(\frac{\Lambda^2}{2r^2} + \psi(r)\right). \tag{8.1}$$

with the parenthesis on the right being the *effective potential* $\psi_e(r)$. However, this method also presents a disadvantage: since ψ_e includes the centrifugal term $\Lambda^2/2r^2$, it depends on the particle through its angular momentum. Therefore, it is impossible with this method to draw and compare the orbits of two distinct particles (with different (ξ, Λ)) in a given potential ψ. As an illustration, consider the Solar system. With this effective potential method, one needs eight pairs of different curves and lines to represent each planet. Would not it be easier to have a diagram where a single curve corresponds to the Sun's potential, and each line to a planet, instead?

One way to perform this decoupling between the particle (ξ, Λ) and the potential (ψ) is to introduce the Hénon variables. Their definition comes from the following consideration. Multiplying Eq. (8.1) by $2r^2$ on each side and regrouping the terms gives

$$\frac{1}{16}\left(\frac{\mathrm{d}(2r^2)}{\mathrm{d}t}\right)^2 = \left(\xi \cdot 2r^2 - \Lambda^2\right) - 2r^2 \cdot \psi(r). \tag{8.2}$$

As such, this equation is equivalent to the original (8.1), and not much more useful. However, if we look at it in terms of the variable $2r^2$ instead of r, then the term between parenthesis is a straight line, of which the slope is the energy ξ and the y-intercept is (minus the square root of) the angular momentum Λ. To make things even more clear, define the following "variables"

$$x \equiv 2r^2 \quad \text{and} \quad Y \equiv 2r^2\psi(r). \tag{8.3}$$

Since $Y(x)$ and $\psi(r)$ are in a one-to-one correspondence through Eq. (8.3), the x variable can still be thought of as a radius and Y as a potential, and we shall sometimes abuse and speak of the radius x and the potential Y, always referring to this duality. In terms of these, the energy conservation Eq. (7.16) reads

$$\frac{1}{16}\left(\frac{\mathrm{d}x}{\mathrm{d}t}\right)^2 = (\xi x - \Lambda^2) - Y(x)\,. \tag{8.4}$$

In his seminal paper on isochrony [583], Michel Hénon introduced this change of variables just as a mean to compute some complicated integrals. As we can see, these have a much broader use, as they decouple the particle from the potential it lives in, contrary to the effective potential method. As we shall see below, these variables are the equivalent of the arclength for the isochrone pendulum.

Plotting orbits in the (x, Y) variables is the same as with (r, ψ_e). In view of the paragraph above Eq. (7.18), the apoapsis and periapsis $x_A \equiv 2r_A^2$ and $x_P \equiv 2r_P^2$ in Hénon's variables are given by the intersection between the curve $\mathscr{C} : y = Y(x)$ and the straight line $\mathscr{L} : y = \xi x - \Lambda^2$, as can be read off the right-hand side of Eq. (8.4). Furthermore, it is clear that the line $\mathscr{L}$ should always lie *above* curve $\mathscr{C}$ since the left-hand side of Eq. (8.4) is always positive. In what follows, $\mathscr{L}$ will always denote a line of equation $y = \xi x - \Lambda^2$, associated with a particle (ξ, Λ); and $\mathscr{C}$ will always be the curve of equation $y = Y(x)$, associated with a potential $Y(x) = 2r^2\psi(r)$.

We note that, in terms of Hénon's variables, the conservation of energy provides the ODE for $x(t) = 2r(t)^2$, Eq. (8.4). We still call it the equation of radial motion. Similarly, the conservation of angular momentum reads as an ODE for $\theta(t)$:

$$\Lambda = \frac{x}{2}\frac{\mathrm{d}\theta}{\mathrm{d}t}\,. \tag{8.5}$$

To summarize, Hénon's variables $(x, Y(x))$ take advantage of the fact that a test particle in a radial potential is entirely characterized by two numbers (ξ, Λ), and is therefore in a one-to-one correspondence with a line, that has two degrees of freedom (e.g., the slope and the y-intercept). The potential $Y(x)$ corresponds to a unique curve $\mathscr{C}$, and a particle (ξ, Λ) is associated with a unique straight line $\mathscr{L}$. If $\mathscr{L}$ intersects $\mathscr{C}$ and lies above it, this particle orbits periodically, as detailed in Fig. 8.1. For the sake of completeness, we provide in Fig. 8.2 a comparison between the effective potential method and the Hénon variables, going back to the point made relative to the Solar system earlier.

8.2 Hénon Meets Archimedes

This section aims at giving a geometrical characterization of isochrony. In Sect. 8.2.1 we derive the Hénon formulae which give T and Θ explicitly for isochrone potentials. In Sect. 8.2.2, we give a geometrical proof that the Hénon formula for T implies that the potential Y must be an arc of parabola. We follow the notation introduced in the last section using Hénon's variables: A particle (ξ, Λ) is associated with a line $\mathscr{L} : y = \xi x - \Lambda^2$, and $\mathscr{C} : y = Y(x)$ is the curve of an arbitrary isochrone potential $Y(x) = 2r^2\psi(r)$.

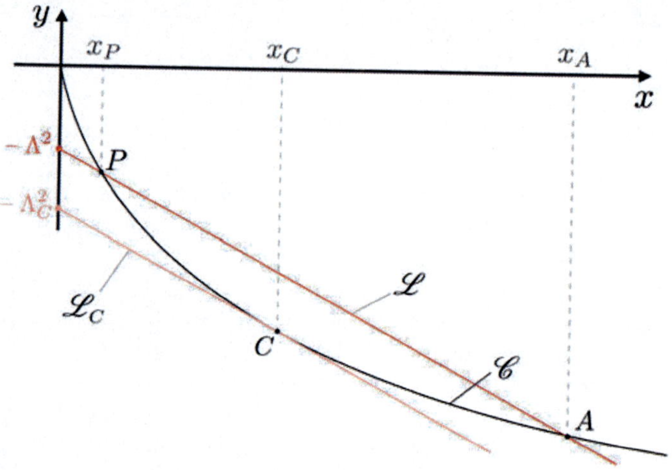

Fig. 8.1 Same situation as in Fig. 7.7 depicted here in the Hénon plane, with Hénon's variables. The curve $\mathscr{C}$ of the graph $y = Y(x)$ corresponds to the potential ψ, in Hénon's variables. Two particles are depicted as straight lines $\mathscr{L}$ (red) and $\mathscr{L}_C$ (light red). They have the same energy ξ (same slope for both lines) but different angular momenta Λ and $\Lambda_C > \Lambda$ (different y-intercept). Particle (ξ, Λ) is on a generic orbit with periapsis x_P and apoapsis x_A given by the two intersections at P and A. Particle (ξ, Λ_C) is on a circular orbit of radius x_C given the unique intersection at C. Notice that one can draw several different particles orbiting the potential without changing the curve $y = Y(x)$

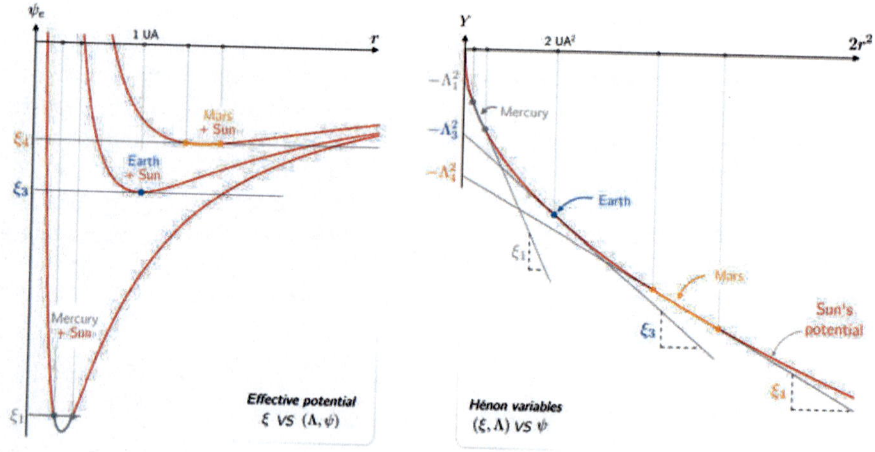

Fig. 8.2 Comparison between the two methods discussed in the text, to represent orbits of three planets (1 = Mercury, 3 = Earth and 4 = Mars) in the gravitational potential of the Sun. On the left, 3 curves, one for each planet. They contain information about both the Sun and the planet (angular momentum). On the right, thanks to the decoupling permitted by the Hénon variables, a single curve is enough (the Sun), on which one can depicts the orbits of the three planets. This gives access to their energy (slope) and angular momentum (y-intercept). Both diagrams are to scale. The units are adapted to Earth (*left:* unit of Earth's energy VS unit of Earth's mean distance to the sun (1 AU); *right:* unit of Earth's angular momentum squared VS AU squared)

8.2.1 *Hénon's Formulae*

This subsection is split into three parts. In the first, we integrate Eq. (7.20) explicitly, for any central potential, following a method of Hénon [583]. Assuming isochrony, we simplify in the second part this result to get the Hénon formula for T. The third part presents more briefly this computation for Θ.

Computing the Integral for T

As we motivated below Eq. (8.4), we start by performing in Eq. (7.20) the change of variables $r \to x = 2r^2$ and we introduce the potential $Y(x) = 2r^2\psi(r)$. We readily obtain the following expression

$$T = \frac{1}{2}\int_{x_P}^{x_A} \frac{\mathrm{d}x}{\sqrt{D(x)}}, \quad \text{with} \quad D(x) \equiv (\xi x - \Lambda^2) - Y(x). \tag{8.6}$$

The bounds of the integral are $x_P \equiv 2r_P^2$ and $x_A \equiv 2r_A^2 \geq x_P$. In the (x, y) plane, the quantity D appearing in Eq. (8.6) is the vertical distance between the curve $\mathscr{C}$ and the line $\mathscr{L}$. The fact that $D(x) \geq 0$ is ensured by the very existence of the orbit, or equivalently by Eq. (8.4), as discussed in the last section.

Since the curve $\mathscr{C}$ is smooth and lies below $\mathscr{L}$ on $[x_P, x_A]$, there exists a line $\mathscr{L}_C$ that is both parallel to $\mathscr{L}$ and tangent to $\mathscr{C}$ at some point C of abscissa $x_C \in [x_P, x_A]$. This line intersects $\mathscr{C}$ exactly once, and corresponds to a particle with a circular radius r_C, such that $x_C = 2r_C^2$. Moreover, $\mathscr{L}$ and $\mathscr{L}_C$ are parallel and therefore associated with particles that share the same energy ξ. Consequently we may write $\mathscr{L}_C : y = \xi x - \Lambda_C^2$, where Λ_C is the angular momentum of the other particle, on the circular orbit.

With the help of this secondary line $\mathscr{L}_C$, we may rewrite the distance D of Eq. (8.6) as the difference $\ell^2 - z(x)^2$, where $\ell^2 \equiv \Lambda_C^2 - \Lambda^2 > 0$ is the vertical distance between $\mathscr{L}$ and $\mathscr{L}_C$, and $z(x)^2 \equiv Y(x) - (\xi x - \Lambda_C^2) > 0$ is the vertical distance between $\mathscr{C}$ and $\mathscr{L}_C$ (see Fig. 8.3). We denote the latter by a squared quantity $z(x)^2$, so that we may conveniently choose $z(x) \leq 0$ on $[x_P, x_C]$ and $z(x) \geq 0$ on $[x_C, x_A]$. We stress that this is nothing but a convention: the positive distance is still $z(x)^2 \geq 0$, but the sign of $z(x)$ depends on where we are on $[x_P, x_A]$. These new quantities are depicted in Fig. 8.3, and upon insertion in Eq. (8.6), we obtain

$$T = \frac{1}{2}\int_{x_P}^{x_A} \frac{\mathrm{d}x}{\sqrt{\ell^2 - z(x)^2}}, \tag{8.7}$$

Now, by construction, $z(x)$ varies monotonically on $[x_P, x_A]$: it is negative and increasing on $[x_P, x_C]$, it hits zero at x_C and it is positive and increasing again on $[x_C, x_A]$. We can therefore perform the change of variables $x \to z(x)$ in Eq. (8.7). We readily obtain

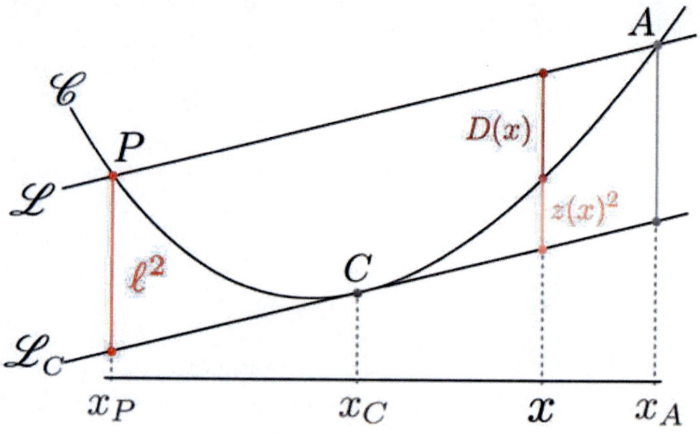

Fig. 8.3 Summary of the geometrical quantities used to compute the integral for the period T. Both lines $\mathscr{L}$ and $\mathscr{L}_C$ define an orbit with the same period, and ℓ^2 is the vertical distance between $\mathscr{L}$ and $\mathscr{L}_C$. The distance between $\mathscr{L}_C$ and $\mathscr{C}$ is $z(x)^2$, such that $D(x) + z(x)^2 = \ell^2$

$$T = \frac{1}{2}\int_{-\ell}^{\ell} \frac{f'(z)\mathrm{d}z}{\sqrt{\ell^2 - z^2}}, \quad \text{with} \quad x \equiv f(z). \tag{8.8}$$

It is now natural to perform in Eq. (8.8) one last change of variables, namely $z \to \ell \sin\phi$, with ϕ varying between $-\pi/2$ and $\pi/2$, corresponding to $z = -\ell$ and $z = \ell$, respectively. We then get

$$T = \frac{1}{2}\int_{-\pi/2}^{\pi/2} f'(\ell \sin\phi)\mathrm{d}\phi. \tag{8.9}$$

We cannot, in general, compute explicitly the integral in Eq. (8.9), for f' is but a generic, unspecified function that depends on the potential and the particle. However, assuming that the potential is regular enough, we can expand the function f' as a Taylor expansion at zero, i.e., write $f'(z) = a_0 + \sum_{n\geq 1} a_n z^n$. Inserting this in Eq. (8.9) and integrating term by term gives

$$T = \frac{\pi}{2}a_0 + \sum_{n\geq 1} a_{2n} W_{2n} \ell^{2n}, \quad \text{with} \quad W_n \equiv \int_0^{\pi/2} \sin^n\phi \,\mathrm{d}\phi. \tag{8.10}$$

Note that only the even terms a_{2n} remain since the integral of the odd function $\sin^{2n+1}$ vanishes over the symmetric interval $[-\pi/2, \pi/2]$. The integral W_n is the celebrated Wallis integral, and can be given explicitly. Notice that Eq. (8.10) is valid for any potential and any particle orbiting within it. Both the coefficients a_n and ℓ depend on (ξ, Λ) and the properties of the potential. As such, it is not that useful. However, for isochrone potentials, it is of considerable interest.

Explicit Formula for T

So far, what we have done does not take advantage of the isochrony property, and Eq. (8.10) is valid for any particle (ξ, Λ) in any central potential $Y(x)$. In particular, we insist that the coefficients a_n appearing in Eq. (8.10) are all function of ξ and Λ, a priori. Now let us fix the energy ξ of the particle. If the potential is isochrone, then by definition T is independent of Λ, and so is the right-hand side of Eq. (8.10). We may therefore choose $\Lambda = \Lambda_C$, i.e., $\ell = 0$, so that the last term on the right-hand side of Eq. (8.10) vanishes, and we readily find that

$$T(\xi) = \frac{\pi}{2} a_0 \,. \tag{8.11}$$

Now Eq. (8.11) is true for any Λ. (Actually, it is independent of Λ.) The combination of Eqs. (8.10) and (8.11) implies that the sum on the right-hand side of Eq. (8.10) is a power series in ℓ that vanishes for any ℓ. By a classical result on power series, this is true if and only if all the coefficients of the power series vanish, i.e., $a_{2n} W_{2n} = 0$ for all $n \geq 1$. Since the Wallis integrals W_n are all nonzero, we conclude that, if the potential is isochrone, $a_{2n} = 0$ for all $n \geq 1$. In particular, the Taylor expansion of f' now reads

$$f'(z) = a_0 + \sum_{n \geq 1} a_{2n+1} z^{2n+1} \,. \tag{8.12}$$

The last step consists in finding explicitly the coefficient a_0 appearing in Eq. (8.11). To this end, we integrate Eq. (8.12) over $[z(x_P), z(x_A)] = [-\ell, \ell]$. On the left-hand side, we use $f(z_A) = x_A$ and $f(z_P) = x_P$ (which follows from the definition $x = f(z)$). On the right-hand side, the first term is a mere constant, and the second term is an odd function of z: Its integral over $[-\ell, \ell]$ will vanish. Consequently, the integrated result is simply $x_A - x_P = 2\ell a_0$. With the help of $\ell^2 = \Lambda_C^2 - \Lambda^2$ and Eq. (8.11), we obtain the following explicit formula for $T(\xi)$

$$T(\xi) = \frac{\pi}{4} \frac{x_A - x_P}{\sqrt{\Lambda_C^2 - \Lambda^2}} \,. \tag{8.13}$$

This is what we call Hénon's formula for T. Let us make a few remarks on Eq. (8.13). First, and quite remarkably, we stress that although both $x_A - x_P$ and $(\Lambda_C^2 - \Lambda^2)^{1/2}$ depend explicitly on Λ, their ratio does not, since T is independent of Λ by assumption. Second, if we square both sides of Eq. (8.13), we observe that the horizontal distance $x_P - x_A$ squared is proportional to the vertical one $\Lambda_C^2 - \Lambda^2$, and that the constant of proportionality, namely $16T^2/\pi^2$, is independent of Λ. We shall use this geometrical result to prove that the curve $\mathscr{C}$ must be a parabola in the (x, y) plane. Third, we insist that this relation is valid for all isochrone potentials, even though their explicit form is unknown at this stage. In particular, given an isochrone potential, the radial period of any orbit can be read simply by drawing the line $\mathscr{L}$ intersecting the curve $\mathscr{C}$, and then finding the secondary line $\mathscr{L}_C$ that is both parallel to $\mathscr{L}$ and tangent to $\mathscr{C}$.

Lastly, let us explain why we call formula (8.13) Hénon's formula. In fact, it can be found as an intermediate equation in the seminal paper of Hénon [583], (with a missing factor of 1/2 there). Since our method here is similar to his (although it is more detailed here) this is not unexpected. However, Hénon did not seem to be interested in this particular equation, probably because his main goal was not to obtain a formula for the period. Yet, we shall see that this equation is rather central in the context of isochrony.

Explicit Formula of Θ

In the last paragraphs, we were able to obtain the explicit formula (8.13) for $T(\xi)$. The recipe for the computation went in five steps that can be summarized as follows:

- fix ξ and rewrite the integrand in (7.20) as $1/\sqrt{D(x)}$ using Hénon's variables,
- rewrite $D(x)$ as $\ell^2 - z(x)$ using the line $\mathscr{L}_C$ associated with the circular orbit of same energy (and thus same period),
- introduce $x = f(z)$, perform the change of variables $x \to z$ and then $z \to \phi$,
- perform a Taylor expansion of $f\prime$ around 0 and integrate explicitly,
- assume that ψ is isochrone and use $T = T(\xi, \not{\Lambda})$ to constrain f and conclude.

Ultimately, the effectiveness of this recipe can be traced back to the Hénon variable $x = 2r^2$ which *isolates* Λ from the denominator of the integrand in Eq. (7.20), as can be seen on Eq. (8.6). Knowing this, it is possible to try and adapt the recipe to find an explicit formula for Θ, starting from its integral definition in Eq. (7.21). As we argued earlier, examining the radial action (7.22) shows that $T = T(\xi, \not{\Lambda})$ is equivalent to $\Theta = \Theta(\not{\xi}, \Lambda)$. Therefore, one can apply the same recipe provided that one uses a variable that *isolates* ξ in the denominator in Eq. (7.21). The Binet variable $u \equiv 1/r$ turns out to be the appropriate variable this time. Indeed, in terms of u, Eq. (7.21) becomes

$$\Theta = \sqrt{2}\Lambda \int_{u_A}^{u_P} \frac{du}{\sqrt{D_\Theta(u)}}, \quad \text{with} \quad D_\Theta(u) \equiv \xi - \Psi_e(u)\,. \tag{8.14}$$

From there, the same computation can be made to obtain an explicit formula for Θ. The detailed computation is given in Appendix C.2.2. More precisely, with the Binet variable $u = 1/r$ and the Binet effective potential $\Psi_e(u) \equiv \psi_e(1/u)$, it is possible to make a one-to-one dictionary between what was used for T and what can be used for Θ. The latter is presented in Table 8.1. At the end of the computation, for any given Λ we obtain the following formula in the case of isochrone potentials

$$\Theta(\Lambda) = \frac{\pi\Lambda}{\sqrt{2}} \frac{u_P - u_A}{\sqrt{\xi - \xi_C}}, \tag{8.15}$$

with $u_{P,A} \equiv 1/r_{P,A}$. This is *Hénon's formula for* Θ. The value ξ_C depends only on Λ and is the energy to be given to a particle of angular momentum Λ to obtain a

Table 8.1 Dictionary between the geometrical quantities involved in the derivation of the Hénon formulae: (8.13) for $T(\xi)$ and (8.15) for $\Theta(\Lambda)$

Isochrony	Variable	Curve $\mathscr{C}$	Line $\mathscr{L}$
$T(\xi, \not{\Lambda})$	$x = 2r^2$	$y = x\psi(\sqrt{x/2})$	$y = \xi x - \Lambda^2$
$\Theta(\not{\xi}, \Lambda)$	$u = 1/r$	$y = \psi_e(1/u)$	$y = \xi$

circular orbit. Moreover, as we argued earlier for T in Eq. (8.13), despite appearances the right-hand side of Eq. (8.15) is independent of ξ.

Just as Eq. (8.13) will be used in Chap. 9 to write a generalized Kepler's third law for all isochrone orbits, Eq. (8.15) will be used to find a similar law for the apsidal angle of any isochrone orbit. We shall not use it directly and present a more astute computation, but it is possible to derive, without any trick, this periapsis law directly from Eq. (8.15).

This apsidal angle law can, in turn, be used to give a proof of Bertrand's theorem, a well-known result of classical mechanics that states that the only two potentials in which all periodic orbits are closed are the Kepler and the harmonic potentials. In fact, since these two are also isochrone potentials, it should come as no surprise that Bertrand's theorem is closely related to isochrony. As demonstrated in [589], the theorem actually follows from the examination of Eq. (8.15), once the latter is expressed in terms of Λ. Let us mention that the equivalent of Eq. (8.10) for Θ (Eq. (C.11)) can be used to give a proof a Bertrand's theorem with brute force as in [595] [compare Eq. (20) of [595] to Eq. (C.11)]. We shall revisit Bertrand's theorem ourselves with the Hamiltonian treatment of isochrony, in Chap. 10.

8.2.2 Geometry of Parabolae

In this section, we provide a geometrical proof that the curve $\mathscr{C} : y = Y(x)$ must be a parabola[1] in order for the associated potential ψ to be isochrone. This result, which we will refer to as the fundamental theorem of isochrony, was first established by Hénon in [583], using a very technical argument (see also [589] for proof using complex analysis). The present proof is based only on Eq. (8.13) and a characterization of parabolae that can be traced back to Archimedes.

[1] Strictly speaking, we will show that $\mathscr{C}$ is *an arc of parabola*, as it is the graph of a function. However, any given arc of parabola defines a unique parabola, so there will be no possible confusion.

Archimedean Characterization

Archimedes, in his treatise *Quadrature of the Parabola*,[2] proved in a series of 24 propositions the following remarkable property shared by all parabolae. On a given parabola $\mathscr{P}$, take two points A and B defining a chord AB and a third point C where the tangent to $\mathscr{P}$ is parallel to the chord AB. Then the area enclosed by $\mathscr{P}$ and AB is four thirds that of the triangle ABC. Although it was not known to Archimedes, it turns out that this property uniquely characterizes parabolae [598]. In other words, we have the following theorem:

Theorem. *Let $\mathscr{C}$ be an arbitrary smooth curve in the plane, and $\mathscr{L}$ any line that intersects $\mathscr{C}$ exactly twice, say at points P and A. Let C be the point where the tangent to $\mathscr{C}$ is parallel to $\mathscr{L}$. Then, if $\mathscr{T}$ denotes the triangle PAC and $\mathscr{S}$ the region enclosed by $\mathscr{L}$ and $\mathscr{C}$, the following equivalence holds:*

$$\text{Area}\,(\mathscr{T}) = \frac{3}{4}\text{Area}\,(\mathscr{S}) \quad \Leftrightarrow \quad \mathscr{C} \text{ is an arc of parabola}\,. \tag{8.16}$$

Notice that in order for $\mathscr{C}$ to be a parabola, the area ratio should be $4/3$ for *all of its chords*. As stated above, the $\Leftarrow$ result was the aim of Archimedes' work.

Rewriting of Hénon's Formula

Consider the curve $\mathscr{C} : y = Y(x)$ associated with an isochrone potential $Y(x) = 2r^2\psi(r)$. Following the notation used so far, let us take a line $\mathscr{L} : y = \xi x - \Lambda^2$, such that P and A correspond to the periapsis and apoapsis of the orbit of particle (ξ, Λ). Accordingly, the parallel line that passes through C is $\mathscr{L}_C : y = \xi x - \Lambda_C^2$, and defines a circular orbit with the same energy ξ, and thus the same period $T(\xi)$, given by Eq. (8.13). If for any line $\mathscr{L}$ the areas involved in the theorem are in proportion $4/3$, we will have shown that $\mathscr{C}$ is a parabola. Therefore, the goal is to find an expression for these areas, using Eq. (8.13).

First let us do a bit of geometry. We define B to be the orthogonal projection of C on $\mathscr{L}$, and take M to be an arbitrary point on CB. We parameterize the length CM by $h \geq 0$, with the convention $h = 0$ when $M = C$, and $h = CB$ when $M = B$. Next we define a chord $P'A'$ that is parallel to $\mathscr{L}$ and passes through M. We denote by $L(h)$ the length of that chord A' and P'. Note that $L(h)$ varies between 0 (when $h = 0$) and PA (when $h = CB$). All these quantities are depicted in Fig. 8.4.

Now let us rewrite Eq. (8.13) in terms of these geometrical quantities. For the numerator, $x_A - x_P$ is but the horizontal projection of PA, and thus $x_A - x_P = PA \cos\varphi$, where φ is the angle that $\mathscr{L}$ makes with the horizontal axis, i.e., $\varphi = \arctan\xi$. Similarly, for the denominator, $\Lambda_C^2 - \Lambda^2$ is simply the vertical

---

[2] We refer the interested reader to pp. 233–252 of Heath [596] for a modern English translation of this work, and to pp. 51–62 of Stein [597], for a pedagogical version.

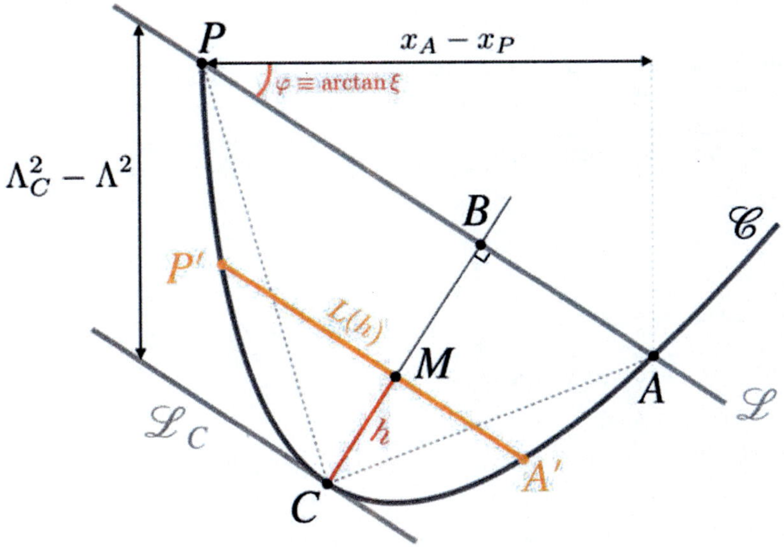

Fig. 8.4 Initially, an arbitrary curve $\mathscr{C}$ and an intersecting line $\mathscr{L}$ of slope ξ are drawn. They define the intersection points P and A. The horizontal distance between P and A is $x_A - x_P$ (top). The line $\mathscr{L}_C$, parallel to $\mathscr{L}$ and tangent to $\mathscr{C}$ at C, defines a circular orbit with energy ξ (bottom). The vertical distance between $\mathscr{L}$ and $\mathscr{L}_C$ is $\Lambda_C^2 - \Lambda^2$ (left). The intermediary chord $P'A'$ defined in the text is parallel to $\mathscr{L}$ and defines yet another orbit with energy ξ

projection of CB; consequently, we also have $\Lambda_C^2 - \Lambda^2 = CB/\cos\varphi$. Inserting these two identities in Eq. (8.13) gives its geometrical variant

$$T(\xi) = \frac{\pi}{4(1+\xi^2)^{3/4}} \frac{PA}{\sqrt{CB}}, \tag{8.17}$$

where we used the trigonometric identity $\cos(\arctan\xi) = (1+\xi^2)^{-1/2}$. Now, formula (8.17) has been obtained for any chord PA of the curve $\mathscr{C}$, corresponding to a particle of energy ξ. However, by construction, for any h the chord $P'A'$ is parallel to PA and thus corresponds to an orbit with the same energy ξ. Therefore, the potential being isochrone, all parallel chords $P'A'$ generated by varying h correspond to orbits with the same energy ξ, and therefore the same period $T(\xi)$. The conclusion is that Eq. (8.17), which corresponds to the case $h = CB$, is also verified for any value of h when the potential is isochrone. In other words, for any $h \in]0, CB]$, we have

$$T(\xi) = \frac{\pi}{4(1+\xi^2)^{3/4}} \frac{L(h)}{\sqrt{h}}. \tag{8.18}$$

Of course, Eq. (8.17) is just a particular case of Eq. (8.18), when $h = CB$ and $L(h) = PA$.

Computing the Areas

With Eq. (8.18) at hand, we can now turn to the computation of the areas involved in the theorem. For the triangle PAC, we have the basis PA and the height CB. For the area between $\mathscr{C}$ and $\mathscr{L}$, we can simply integrate *à la Lebesgue* the infinitesimal area $L(h)\mathrm{d}h$ while h varies between 0 and CB. We thus have, respectively

$$\text{Area}\,(\mathscr{T}) = \frac{1}{2} PA \times CB \quad \text{and} \quad \text{Area}\,(\mathscr{S}) = \int_0^{CB} L(h)\mathrm{d}h\,. \tag{8.19}$$

Now we compute these areas and we show that they are in proportion 4/3. For the area of the triangle $\mathscr{T}$, we use Eq. (8.17) to express PA in terms of CB and plug the result in Eq. (8.19). We obtain the following expression

$$\text{Area}\,(\mathscr{T}) = \frac{2T}{\pi}(1+\xi^2)^{3/4} CB^{3/2}\,. \tag{8.20}$$

In a similar manner, the area of the region $\mathscr{S}$ can be found by isolating $L(h)$ from Eq. (8.18) and expressing it in terms of h. Plugging the result in the area formula for $\mathscr{S}$ in Eq. (8.19) and computing the integral explicitly give easily

$$\text{Area}\,(\mathscr{S}) = \frac{8T}{3\pi}(1+\xi^2)^{3/4} CB^{3/2}\,. \tag{8.21}$$

Comparing Eqs. (8.20) and (8.21) shows that, indeed, Area $(\mathscr{S})/$Area $(\mathscr{T}) = 4/3$. The conclusion from this calculation is so central that we give it its own paragraph, below.

Conclusion

We have just shown, thanks to the Hénon formula (8.13) for $T(\xi)$ that if ψ is isochrone, then the curve $\mathscr{C} : y = Y(x)$ of that potential in Hénon's variables satisfies the 4/3-area condition of Theorem (8.16). Since the latter uniquely characterizes parabolae, we obtain what we call the *fundamental theorem of isochrony*:

$$\psi \text{ is isochrone} \quad \Rightarrow \quad \mathscr{C} \text{ is a parabola}\,. \tag{8.22}$$

The converse result holds as well, although will require a discussion on the conditions under which a parabola actually corresponds to a physically well-posed isochrone potential. Nevertheless, we may already check the result (8.22) for potentials that we know to be isochrone. Recall in Chap. 7 that three of them were already mentioned: the Kepler and harmonic potentials (Eq. (7.26)) and the Hénon potential (Eq. (7.25)). Let us look at the first two, which we rewrite here for convenience

$$\psi_{\mathrm{Ke}}(r) = -\frac{\mu}{r} \quad \text{and} \quad \psi_{\mathrm{Ha}}(r) = \frac{1}{8}\omega^2 r^2\,, \tag{8.23}$$

where μ, ω and b are real constants. Let us re-express them in terms of the Hénon variables using the definitions (8.3). We find

$$Y_{\mathrm{Ke}}(x) = -\mu\sqrt{2x} \quad \text{and} \quad Y_{\mathrm{Ha}}(x) = \frac{1}{16}\omega^2 x^2\,, \tag{8.24}$$

and as we can see, both of these are arcs of parabolae in the (x, y) plane, and thus satisfy the Theorem (8.22). The same computation can be done for Hénon's isochrone potential (7.25), which also looks like a parabola in the (x, Y) variables. This theorem drastically reduces the size of the set of potentials that needs to be searched to find isochrone ones. This will be the goal of the next section.

Let us make a last comment on the derivation of theorem (8.22), which has been derived thanks to the Hénon formula (8.13) for $T(\xi)$. Following the same logic, one could ask whether the Hénon formula for $\Theta(\Lambda)$ (Eq. (8.15)) could not be used to reach the same result. The answer is *no*, unfortunately. The reason is that the Θ-formula is to be read in the (u, y) plane, where $u = 1/r$ is the Binet variable, and orbits correspond to horizontal lines $\mathscr{L} : y = \xi$; whereas Eq. (8.13) is to be read in the (x, y) plane, where orbits correspond to straight lines $\mathscr{L} : y = \xi x - \Lambda^2$ (recall Table 8.1). Yet, the Archimedean characterization of parabolas (8.16) requires the areas ratio to be $4/3$ for *any* chord, not just horizontal ones. Therefore, Eq. (8.15) cannot be used to conclude that $\mathscr{C}$ should be a parabola, at least not with the Archimedean characterization.

We end this section by making contact, as promised in Chap. 6, with Huygens' isochrony of the pendulum. In particular, let us ask ourselves the question: Why is it that $\mathscr{C}$ should be a parabola, and not any other type of curve, when the potential is isochrone ? What is so special about parabolae ? To understand this, let us focus our attention on a point M of a generic curve $\mathscr{C}$ (i.e., non-necessarily a parabola). Close enough to M, $\mathscr{C}$ always looks like a parabola, as can be seen by writing its Taylor expansion around M. To see this, consider a Cartesian coordinate system (x, y) centered on M where the tangent to $\mathscr{C}$ at M is horizontal. These two conditions indicate that $y(0) = 0$ and $y'(0) = 0$ respectively. Therefore the curve has an implicit equation of the type

$$y = \alpha x^2 + o(x^2) \quad \text{for some constant } \alpha\,. \tag{8.25}$$

For a generic curve, the $o(x^2)$ in Eq. (8.25) corrects the local *parabolicness* of the curve as one moves away from M. However, Eq. (8.18) shows that the $o(x^2)$ terms vanishes identically in the case of isochrony. Indeed, in Fig. 8.4, this particular frame (x, y) we are considering is precisely the one centered on C equipped with coordinates $(x, y) = (L, h)$. Now Eq. (8.18) may be rewritten as

$$h = \alpha L^2\,, \quad \text{with } \alpha = \pi^2/16T^2(1+\xi^2)^{3/2}\,. \tag{8.26}$$

Table 8.2 The derivation of the isochrone pendulum (by Huygens) and isochrone potentials (by Hénon) follow common steps and the results are similar: the potential driving the motion is a parabola in terms of a well-chosen proper variable.

Isochrony	System	Variable	Potential
Huygens	Pendulum	Radius squared	Quadratic
Hénon	Orbit	Arc length	Parabolic

Since this should be true for all h, or equivalently any x, comparing Eqs. (8.25) and (8.26) shows that in the case of isochrony the $o(x^2)$ vanishes identically as claimed. In other words, isochrony constrains the curve $\mathscr{C}$ to *be* a parabola, in addition to locally *look like* one. Remarkably, this is exactly like the osculating parabola argument that Huygens used in his derivation of the isochrone pendulum, as we covered in Chap. 7. We have therefore seen all three of the take away results provided in the list at the very end of Sect. 7.1: the change of variable, the parabolic nature of the potential, and an insight from Archimedes insight, as summarized in Table 8.2.

8.3 Isochrone Potentials

The fundamental theorem of isochrony (8.22) implies that the curve $\mathscr{C}$ of an isochrone potential Y corresponds to (at least an arc of) a parabola. However, not all parabolae will *contain* the potential of a physically realistic system. The first aim of this section is to classify, based on their geometrical properties, the *isochrone parabolae*, i.e., those that contain the curve $\mathscr{C}$ of a well-defined, isochrone potential Y (Sect. 8.3.1). Once this is done, we provide explicit formulae and show that there are five families of isochrone potentials (Sect. 8.3.2). We then conclude in Sect. 8.3.3 by examining the conditions on (ξ, Λ) for which a particle has a bounded orbit in these five families of isochrone potentials. Inspired by Arnold [345], these result are written as bifurcation diagrams in the (ξ, Λ^2) plane.

8.3.1 Isochrone Parabolae

In this section, we start by a potpourri of algebraic and geometrical properties of parabolae, and derive general results that shall be used throughout the next sections. Then, we examine under which conditions a parabola is associated to an isochrone potential well-defined physically.

Geometric Preliminaries

Let us give an overview of parabolae in the plane, equipped with Cartesian coordinates (x, y). A generic parabola is given by an implicit equation, of the form

$$\mathscr{P} : (ax + by)^2 + cx + dy + e = 0 \quad \text{with} \quad \delta \equiv ad - bc \neq 0\,, \tag{8.27}$$

and where (a, b, c, d, e) are five real numbers. The quantity δ is the discriminant of $\mathscr{P}$ and is taken to be nonzero, otherwise $\mathscr{P}$ degenerates into a pair of parallel lines. Without loss of generality, we will assume from now on that $\delta > 0$.[3]

As an algebraic curve, a parabola is not, in general, the graph of a function. It is, however, always the union of such graphs. To see this, let us take a point (x, y) on the parabola $\mathscr{P}$ given by Eq. (8.27). Its ordinate y can be given as a function of its abscissa x by solving Eq. (8.27) for y. There are two cases depending on the parameter b:

- when $b = 0$, the whole parabola $\mathscr{P}$ is the graph of a function. Its equation is obtained by isolating y in Eq. (8.27):

$$\mathscr{P} : y = -\frac{e}{d} - \frac{c}{d}x - \frac{a^2}{d}x^2\,, \tag{8.28}$$

where $d \neq 0$ since the discriminant $\delta = ad \neq 0$. For $d < 0$ this parabola opens upwards and we shall say that it is *top-oriented.* When $d > 0$, it opens downwards and we will say *bottom-oriented.* Any top- or bottom-oriented parabola crosses the y-axis once, and the ordinate of this point is

$$\lambda \equiv -\frac{e}{d}\,. \tag{8.29}$$

- when $b \neq 0$, the curve $\mathscr{P}$ is the union of two branches, which are actual graphs of two functions. Indeed, for a fixed x, Eq. (8.27) is a quadratic in y. Its solutions are easily found to be

$$\mathscr{P}_{\pm} : y = -\frac{ax}{b} - \frac{d}{2b^2} \pm \frac{1}{2b^2}\sqrt{4b(\delta x - be) + d^2}\,, \tag{8.30}$$

where the condition $4b(\delta x - be) + d^2 \geq 0$ ensures the positivity inside the square root. The *support* of the parabola is the set of x such that $4b(\delta x - be) + d^2 \geq 0$. In particular, there is a unique value

$$x_v \equiv \frac{4b^2e - d^2}{4b\delta}\,, \tag{8.31}$$

[3] If $\delta < 0$ we can always replace (a, b) by $(-a, -b)$. This leaves Eq. (8.27) unchanged and thus corresponds to the same parabola, but changes the sign of δ.

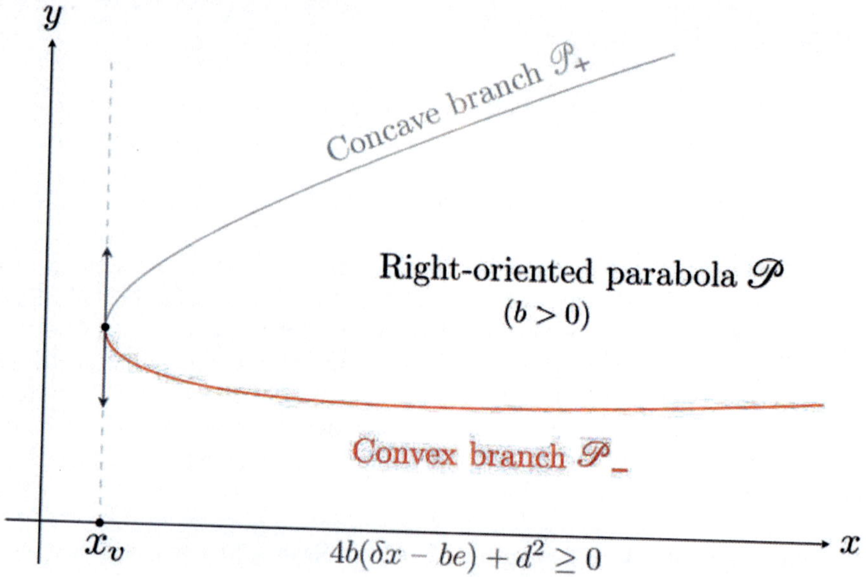

Fig. 8.5 Geometrical properties of a right-oriented parabola $\mathscr{P}$, with its two branches $\mathscr{P}_\pm$ that are actual graphs of functions. The parabola itself is $\mathscr{P} = \mathscr{P}_- \cup \mathscr{P}_+$. On the left, the point of abscissa x_v given by Eq. (8.31) belongs to both $\mathscr{P}_+$ and $\mathscr{P}_-$, and the domain of the parabola is $[x_v, +\infty[$ (highlighted in grey on the x-axis)

that makes the square root in Eq. (8.30) vanish. The quantity x_v is the abscissa of the common point between $\mathscr{P}_+$ and $\mathscr{P}_-$, where the branches meet and the tangent to $\mathscr{P}$ is vertical, as shown in Fig. 8.5. Note that $\mathscr{P}_-$ is always convex and always below $\mathscr{P}_+$ which is concave.

The sign of b controls the orientation of the parabola. If $b > 0$, we will say that the parabola is *right-oriented.* Its support is $[x_v, +\infty]$, and the parabola crosses the y-axis if and only if $x_v \leq 0$. If $b < 0$, we say that it is *left-oriented.* Its support is $[-\infty, x_v]$ and it crosses the y-axis if and only if $x_v \geq 0$.

A left or right-oriented parabola may not always cross the y-axis. When it does, the ordinates of the intersection points are obtained by setting $x = 0$ in Eq. (8.30). In particular, the convex branch $\mathscr{P}_-$ crosses the y-axis at ordinate

$$\lambda \equiv -\frac{d + \sqrt{d^2 - 4b^2 e}}{2b^2}. \tag{8.32}$$

This implies that $d^2 - 4eb^2$ should always be positive for parabolae crossing the y-axis. In particular, we have $d^2 - 4b^2 e > 0$ when there are two intersections. The case $d^2 - 4eb^2 = 0 \Leftrightarrow d^2 = 4eb^2$ happens when the two intersections degenerate into one, and its ordinate is simply $-d/2b^2$.

To summarize, the graph $\mathscr{C}$ of an isochrone potential $Y(x)$ must be contained within a parabola $\mathscr{P}$ in the plane. From the preceding generalities on parabolae, it follows that $\mathscr{C}$ is to be looked for in any of the following families:

- top- and bottom-oriented parabolae, whose whole curve $\mathscr{P}$ is that of a function defined on $\mathbb{R}$, cf. Eq. (8.28).
- left-oriented parabolae, whose curve $\mathscr{P}$ is the union of a convex branch $\mathscr{P}_-$ and a concave branch, $\mathscr{P}_+$, each of the two being the graph of a function defined on $]-\infty, x_v]$, cf. Eq. (8.30) with $b < 0$.
- right-oriented parabolae, whose curve $\mathscr{P}$ is the union of a convex branch $\mathscr{P}_-$ and a concave branch, $\mathscr{P}_+$, each of the two being the graph of a function defined on $[x_v, +\infty[$, cf. Eq. (8.30) with $b > 0$.

Geometrical Hypotheses

Now, we examine under which conditions a curve $\mathscr{C}$ of the potential $Y(x) = 2r^2\psi(r)$ is a physically well-posed, isochrone potential, in the sense that it does contain bounded orbits. This will be done in three steps, each consisting in imposing a geometrical hypothesis (H_i, with $i = 1, 2, 3$), on a candidate curve. These three hypotheses come from, respectively: the fundamental theorem of isochrony (8.22), the very definition of the Hénon variable (8.3), the existence of orbits (8.4). They are as follows:

- H_0: *$\mathscr{C}$ must be an arc of parabola.* This geometrical requirement ensures that the potential ψ is isochrone, regardless of its mathematical and physical properties. The next hypotheses are therefore concerned with the parabola $\mathscr{P}$ that contains the curve $\mathscr{C}$.
- H_1: *$\mathscr{C}$ must lie on the right half plane.* From a purely mathematical perspective, an isochrone potential is a function $\psi(r)$ defined on some subset of $\mathbb{R}_+$ (since r is a positive radius). Since $x = 2r^2 > 0$, *we only keep parabolae that exhibit a portion on the right half plane $x > 0$.* The only parabolae that do not have this property, are the left-oriented ones crossing the y-axis at most once (Eq. (8.30) with $b < 0$ and $x_v \leq 0$).
- H_2: *$\mathscr{C}$ must be convex.* Indeed, a particle orbits periodically when *$\mathscr{C}$ is below $\mathscr{L}$*, since the (directed) distance is positive (recall Eq. (8.4)). The geometrical equivalent of this is that $\mathscr{C}$ should lie under its chords, and therefore be convex. Since this is not possible on the concave branch of a parabola, *we only keep parabolae that exhibit a convex branch.* These that do not are the bottom-oriented ones[4] (Eq. (8.28) with $d < 0$.). Therefore, we discard the bottom-oriented parabolae, and stress that on the right- and left-oriented ones, the curve $\mathscr{C}$ will be located on the convex branch $\mathscr{P}_-$.

Equipped with these three hypotheses H_0, H_1 and H_2, it is now a simple matter to examine each parabola and discard any that does not satisfy one of them. Instead of

[4] Strictly speaking, these bottom-oriented parabolae contain a unique, circular, unstable orbit.

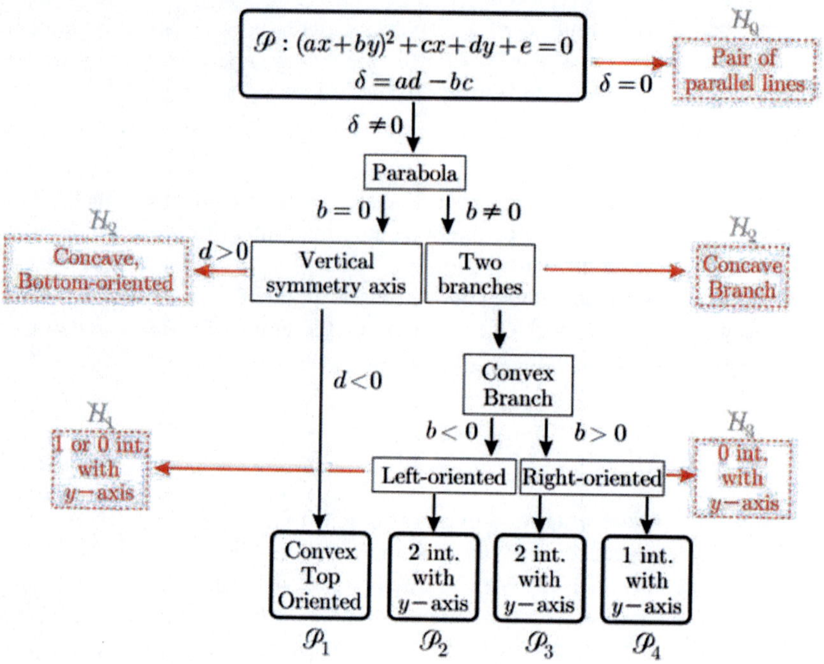

Fig. 8.6 Tree showing the reduction process of isochrone parabolae. Starting at the top with the implicit Eq. (8.27), the reduction consists in exploring the properties of parabolae associated with the sign of δ, b, d. Red horizontal lines correspond to a discarded parabola (that does not satisfy one of the hypotheses H_i). Black, downwards arrows lead naturally to the five families $\mathscr{P}_i$ that remain at the bottom. Any of these is associated with an isochrone potential

presenting these calculations, we summarize the result in Fig. 8.6. This figure shows the reduction process: starting at the top with the general Eq. (8.27), parabolae are discriminated with respect to their parameters, and some are discarded, as indicated by horizontally red arrows, and the corresponding hypothesis they do not fulfill. At the bottom, by construction, all remaining parabolae satisfy H_0, H_1 and H_2, and are the last candidate for isochrone potentials.

After this reduction process, the resulting set of parabolae is split into five families, according to their orientation in the (x, y) plane and their number of intersections with the y-axis:

- $\mathscr{P}_1$: *top*-oriented (Eq.(8.28), with $d < 0$),
- $\mathscr{P}_2$: *left*-oriented, crossing the y-axis *twice* (Eq. (8.30) with $b < 0$, $x_v > 0$),
- $\mathscr{P}_3$: *right*-oriented, crossing the y-axis *twice* (Eq. (8.30) with $b > 0$, $x_v < 0$).
- $\mathscr{P}_4$: *right*-oriented, crossing the y-axis *once* (Eq. (8.30) with $b > 0$, $x_v = 0$).
- $\mathscr{P}_5$: *right*-oriented, *not* crossing the y-axis (Eq. (8.30) with $b > 0$, $x_v > 0$).

We see that the last four of these families are simply given by the relative signs of the pair (b, x_v), where we recall that x_v is given in Eq. (8.31) and corresponds to the

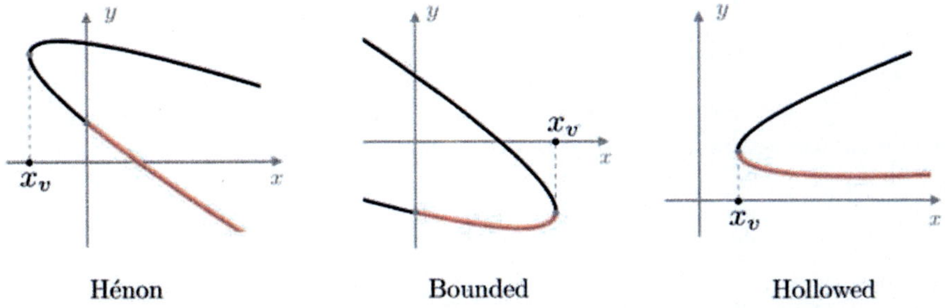

Fig. 8.7 Three isochrone parabolae, of the Hénon (left), Bounded (middle) and Hollowed (right) family. Left-orientation ($b < 0$) corresponds the Bounded family $\mathscr{P}_2$. Right-orientation $b > 0$ encompasses the Hénon class ($x_v < 0$) and the Hollowed class ($x_v > 0$). For each parabola, the physical part (i.e., where the associated $\psi(r)$ is well-defined) is highlighted in red

abscissa of the point where two branches meet (or equivalently where the parabola has a vertical tangent).

Combining the reduction process depicted in Fig. 8.6 and the generic equation of a parabola (8.27) allows us to obtain the equation of an isochrone parabola, in the form $\mathscr{P}_i : y = Y_i(x)$, for $i \in [\![1; 5]\!]$. Three of these are depicted in Fig. 8.7. We can then invert this equation to obtain the expression of the associated potential $\psi_i(r)$, using the definition of the Hénon variables (8.3). This will be much easier in terms of new parameters instead (of the Latin ones (a, b, c, d, e) that we have used up to now), as we show below.

8.3.2 *Complete Set of Isochrone Potentials*

In this section we derive the explicit form of each and every isochrone potential, completing Michel Hénon's program. We express our result in Eqs. (8.34)–(8.39) in terms of new, more adapted parameters, allowing us to determine the fine structure of the set of isochrone potentials. Finally we make a comment on the notion of "gauge" for isochrone potentials and their associated mass distribution.

New Parameters

When working in the Hénon plane, doing geometry with parabolae, the Latin parameters (a, b, c, d, e) are useful. In order to work with simpler expressions when dealing with the potentials, and prepare for the next steps, we rewrite the equations of the isochrone parabola ($\mathscr{P}_i$) with more adapted, Greek parameters $(\epsilon, \lambda, \omega, \mu, \beta)$.

We first consider the top-oriented parabolae $\mathscr{P}_1$ given by Eq. (8.28) with $d < 0$. We may combine the constants in the first two terms and define $\epsilon \equiv -c/d \in \mathbb{R}$, $\lambda \equiv -e/d \in \mathbb{R}$. Moreover, since $d < 0$ in this case, we may always write

$$\omega^2 \equiv -\frac{16a^2}{d} > 0\,. \tag{8.33}$$

Inserting the new parameters $(\epsilon, \lambda, \omega)$ in Eq. (8.28) and using $\psi(r) = Y(2r^2)/2r^2$, we obtain for the first family of potentials ψ_1 and/or parabolae $\mathscr{P}_1$, associated with top-oriented parabolae,

- *the Harmonic family*:

$$\psi_1(r) = \epsilon + \frac{\lambda}{2r^2} + \frac{1}{8}\omega^2 r^2\,. \tag{8.34}$$

The name *Harmonic* comes from the fact that ψ_1 is a harmonic potential, up to a constant ϵ and a centrifugal-like term $\lambda/2r^2$. The normalizing factor $1/16$ is chosen such that the radial period T will coincide exactly with the angular frequency ω, i.e., $T = 2\pi/\omega$, as we shall find later on.

Now consider the $b \neq 0$ case, i.e., the convex branch $\mathscr{P}_-$ of Eq. (8.30). Once again, we may define $\epsilon \equiv -a/b \in \mathbb{R}$. With a bit of rewriting, we can also introduce the λ parameter of Eq. (8.32) and move the square root of the rightmost term down, using the usual conjugate trick. Furthermore, independently of the sign of $b \neq 0$, we may always define

$$\mu \equiv \sqrt{\frac{\delta|b|}{2b^4}} > 0 \quad \text{and} \quad \beta \equiv \sqrt{\frac{d^2 - 4b^2 e}{8\delta|b|}} \geq 0\,. \tag{8.35}$$

When $\beta \neq 0$, the parabola crosses the y-axis twice. Then either $b > 0$ (right-oriented parabola) or $b < 0$ (left-oriented parabola), as depicted in Fig. 8.6. Inserting the new parameters $(\epsilon, \lambda, \mu, \beta)$ in Eq. (8.30) and using $\psi(r) = Y(2r^2)/2r^2$, we obtain the second family of potentials ψ_2 and/or parabolae $\mathscr{P}_2$, associated with left-oriented parabolae,

- *the Bounded family*:

$$\psi_2(r) = \epsilon + \frac{\lambda}{2r^2} + \frac{\mu}{\beta + \sqrt{\beta^2 - r^2}}\,. \tag{8.36}$$

The name *Bounded* comes from the fact that ψ_2 is defined only on the bounded interval $[0, \beta]$. Finally, the same parameters can be used to define the third family of potentials ψ_3 and/or parabolae $\mathscr{P}_3$, associated with right-oriented parabolae crossing the y-axis twice, namely

- *the Hénon family*:

$$\psi_3(r) = \epsilon + \frac{\lambda}{2r^2} - \frac{\mu}{\beta + \sqrt{\beta^2 + r^2}}\,. \tag{8.37}$$

The name *Hénon* comes from the fact that ψ_3 is, up to a constant and a centrifugal-like term, the potential found by Michel Hénon [583]. Finally, the case $\beta = 0$ defines a fourth family of potentials ψ_4 and/or parabolae $\mathscr{P}_4$, associated with right-oriented parabolae crossing the y-axis once, namely

- *the Kepler family*:

$$\psi_4(r) = \epsilon + \frac{\lambda}{2r^2} - \frac{\mu}{r}. \tag{8.38}$$

The name *Kepler* comes from the fact that ψ_4 is, up to a constant and a centrifugal-like term $\lambda/2r^2$, the usual Kepler potential. For the final family $\mathscr{P}_5$, the quantity λ, defined as the y-intercept of the convex branch of a parabola does not make sense anymore since, by definition, $\mathscr{P}_5$ has no intersection with the y-axis (recall Fig. 8.6). We thus have to go back to the general Eq. (8.30) and define the constants μ, β in terms of x_v (cf. Eq. (8.31)), which is well defined for any (non-harmonic) parabola. A quick calculation shows that the parameter μ can be taken as in Eq. (8.35), whereas β is simply $\sqrt{x_v/2}$. Lastly, by setting $\lambda = -d/2b^2$, we define the fifth and last family of isochrone potentials, namely:

- *the Hollowed family*:

$$\psi_5(r) = \epsilon + \frac{\lambda}{2r^2} - \frac{\mu}{r}\sqrt{1 - \frac{\beta^2}{r^2}}. \tag{8.39}$$

The qualifier "Hollowed" comes from the fact that ψ_5 is defined only on the open interval $[\beta, +\infty[$, i.e., outside a sphere of radius β in the three-dimensional physical space. For $r < \beta$, the potential is undefined and it turns out that no particle can enter this region, leaving it "hollow". In this sense, the Hollowed potential ψ_5 is complementary to the Bounded one ψ_2, since the latter is defined on $[0; \beta]$ and the former on $[\beta; +\infty[$.

With the new, Greek parameters, the result of this section can be summarized easily: *If a potential ψ is isochrone, then it must be equal to one of the $\psi_{i \in [\![1;5]\!]}$ above*. In other words, if ψ is isochrone, then there exists some constants (ϵ, λ) and (μ, β, ω) such that $\psi(r) = \psi_i(r)$ for some $i \in [\![1; 5]\!]$. We stress that, by definition, $\omega \neq 0$, $\mu > 0$ and $\beta \geq 0$. The introduction of these parameters is important since it allows one to structure the set of isochrone potentials into 5 families, and give the dimension of each. In particular, we see that both the Harmonic (ψ_1) and Kepler (ψ_4) potentials form a (2+1)-parameters family (2 because of (ϵ, λ), and 1 because of ω or μ); whereas the Bounded (ψ_2), Hénon (ψ_3) and Hollowed (ψ_5) are (2+2)-parameters families. We have now found the fine structure of the set of isochrone potentials, and thus completed Hénon's program, from a mathematical physics perspective. The main result are summarized in Fig. 8.8.

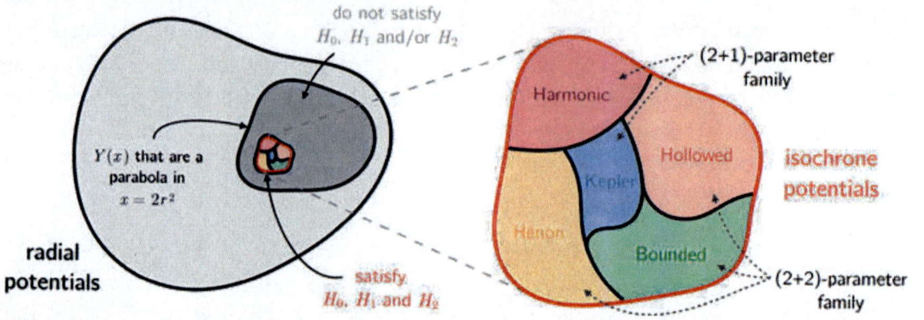

Fig. 8.8 *Left:* isochrone potentials form a subset of the larger class of radial potentials that look like a parabola in Hénon's variables, which is itself a subset of the class of radial potentials. Some radial potentials correspond to parabolae but are not physically well-defined (as explained in Table 8.6). *Right:* the set of isochrone potentials is structured into 5 subfamilies (Eqs. (8.34)–(8.39))

Gauged and Reduced Potentials

Now, for the sake of completeness, let us make a few comments about the notion of "gauged" isochrone potentials, as introduced in [589]. Notice from the explicit forms of the five potential families (8.34)–(8.39) that all isochrone potentials are the sum of three terms, the first two being identical for each of them:

$$\psi(r) = \epsilon + \frac{\lambda}{2r^2} + \cdots , \tag{8.40}$$

where ... denotes the third term, that differs from one family to another. Physically, the first, constant term is expected, since it represents a shift in the potential energy of the particle. The second term looks like it derives from a centrifugal force, although it is not dependent on the particle. Now let us see how these terms contribute to the potential in the Hénon variables. Multiplying (8.40) by $2r^2$ and remembering the definition $Y = 2r^2\psi(r)$, we find

$$Y(x) = \epsilon x + \lambda + \cdots \tag{8.41}$$

In terms of the associated parabola, these two terms make a simple affine contribution. In particular, the parameters (ϵ, λ) control the orientation of the symmetry axis and the height of the parabola respectively. Similarly, we can show that the other parameters (ω for $\mathscr{P}_1$, and (μ, β) for the others) control the aperture and the horizontal position of the parabola. The other consequence of the general form (8.41) is the partition of isochrone potentials between gauged ones and reduced ones, as introduced in [589]. Without going into the details (this is well-explained in that article), the gauged potentials are obtained by applying an affine transformation to the reduced ones, as follows.

- For any real ϵ, the affine transformation $I_\epsilon : (x, y) \mapsto (x, y + \epsilon x)$ is a *transvection*. Geometrically, this map adds the value ϵ to the slope of any straight line in the plane. In particular, when applied to $\mathscr{L}$ it gives the line $I_\epsilon(\mathscr{L}) : y = (\xi + \epsilon)x - \Lambda^2$

and when applied to $\mathscr{P}$ it adds ϵ to the slope of its symmetry axis. Moreover, vertical lines are invariant under this map, and therefore the number of intersections between $\mathscr{P}$ with $\mathscr{L}$ is preserved. In other words, the action of I_ϵ does not change the fact that a parabola is isochrone.

- For any $\lambda \leq 0$ the map $J_\lambda : (x, y) \mapsto (x, y + \lambda)$ is a downward, vertical translation. If a parabola $\mathscr{P}_i$ is intersected twice by a line $\mathscr{L}$, this intersecting configuration remains after the action of J_λ. Moreover, since negativity of the y-intercept of $\mathscr{L}$ is preserved by J_λ, the action of J_λ does not change the fact that a parabola is isochrone.

As a consequence, if $\mathscr{P}$ is an isochrone parabola, then for any $(\epsilon, \lambda) \in \mathbb{R} \times \mathbb{R}_-$, the curve $I_\epsilon \circ J_\lambda(\mathscr{P})$ is also an isochrone parabola. In topological terms, this induces an equivalence relation on the set of isochrone potentials. If one takes the quotient of that set by this equivalence relation, one can define a set of reduced isochrone potentials, that are defined up to an additional term $\epsilon + \lambda/2r^2$. This topological structure has been explored thoroughly in [589], we refer to Sect. 2.5 of that article for more details.

Finally, we mention that from the explicit formulae of the isochrone potentials $\psi_i(r)$, it is possible to compute and analyze the associated mass density $\rho_i(r)$, through the Poisson equation (7.15). This was already presented in Sect. 4.1 of [589], albeit for only a subset of isochrone potentials. Speaking of mass density, we relegated to Appendix C.1 a simple construction in the Hénon plane that allows one to compute, geometrically, the mass contained in a sphere of radius r for a given mass density. This construction works for any radial potential that is defined around the origin (not necessarily isochrone). It shows, once again, the usefulness of the Hénon variables.

8.3.3 *Bifurcation Diagrams*

In the last section, we have isolated five families of potentials (ψ_i) and their associated parabolae $(\mathscr{P}_i)$, and discussed some of their properties. We have shown that if a potential is isochrone, then its curve $\mathscr{C} : y = Y(x)$ is the portion of a parabola $\mathscr{P}_i$ that is convex and lies on the right half-plane $x > 0$. What remains to be shown is the reciprocal of this statement, i.e., that any such $\mathscr{C}$ is the curve of an isochrone potential, in the sense that it contains periodic orbits. To this end, we just need to show that $\mathscr{C}$ can always be intersected twice by some line $\mathscr{L} : y = \xi x - \Lambda^2$. We shall do this by finding explicitly which lines $\mathscr{L}$ can intersect $\mathscr{C}$. In so doing, we will find two important results:

- the set (ψ_i) is complete (it contains all and only the isochrone potentials),
- necessary and sufficient conditions on (ξ, Λ) for the existence of periodic orbits.

In what follows, we consider a curve $\mathscr{C}$ and a line $\mathscr{L} : y = \xi x - \Lambda^2$. By assumption, $\mathscr{C}$ is on the right half plane and on the convex portion of a parabola $\mathscr{P}_i$ given by Eqs. (8.34)–(8.39). When they exist, we denote by x_P and $x_A > x_P$ the abscissa of P and A, the two intersections of $\mathscr{L}$ with $\mathscr{C}$.

Harmonic Family $\mathscr{P}_1$

Let us start with the family $\mathscr{P}_1$ given in Eq. (8.34). We fix the parameters $(\epsilon, \lambda, \omega)$ and look for the conditions on (ξ, Λ) under which the line $\mathscr{L} : y = \xi x - \Lambda^2$ intersects $\mathscr{P}_1$ twice. By definition, P, A belong to both $\mathscr{L}$ and $\mathscr{P}_1$; therefore, x_P, x_A are solutions to $\xi x - \Lambda^2 = \epsilon x + \lambda + \omega^2 x^2/16$. We may equivalently write this equation in the following evocative form

$$x^2 - sx + p = 0\,, \qquad \text{where} \quad \begin{cases} s \equiv 16(\xi - \epsilon)/\omega^2\,, \\ p \equiv 16(\Lambda^2 + \lambda)/\omega^2\,, \end{cases} \tag{8.42}$$

with the sum $s = x_P + x_A$ and product $p = x_P x_A$ of the two roots. Note that x_P, x_A and therefore s and p are all functions of ξ and Λ^2. For a generic quadratic equation such as Eq. (8.42), two solutions exist if and only if $\Delta \equiv s^2 - 4p > 0$, and are given by $(s \pm \sqrt{\Delta})/2$, with a minus sign for x_P and a plus sign for x_A. We want these solutions to lie on the convex branch of the parabola, in order for an orbit to actually exist. This is always satisfied since the parabola $\mathscr{P}_1$ is everywhere convex. Furthermore, we want them to be strictly positive, in order for P and A to be in the right half plane $x \geq 0$. It is sufficient to require $x_P > 0$, because then $x_A > x_P > 0$.

The parameters of the potential are fixed; therefore, the condition $x_P(\xi, \Lambda^2) > 0$, along with $\Lambda^2 \geq 0$, defines a region in the (ξ, Λ^2) plane that contains every pair (ξ, Λ^2) such that the orbit is radially periodic. Using the formula for x_P in terms of s and p, this domain is explicitly delimited by the two inequalities

$$\Lambda^2 \geq 0 \quad \text{and} \quad x_P(\xi, \Lambda^2) > 0\,, \tag{8.43}$$

with $x_P = (s - \sqrt{s^2 - 4p})/2$ and s, p given by Eq. (8.42). Outside this region, there may be collision orbits (the particle avoids the origin and goes to infinity), or no orbit at all (for instance in the $\Lambda^2 < 0$ region). This is depicted in Fig. 8.9.

Bounded Family $\mathscr{P}_2$

We proceed similarly for the left-oriented parabolae $\mathscr{P}_2$, associated with a Bounded potential. In particular, we fix the parameters $(\epsilon, \lambda, \mu, \beta)$ and look for a domain in the (ξ, Λ^2) plane that contains all and only the periodic orbits. The condition $P, A \in \mathscr{L} \cap \mathscr{P}_2$ translates algebraically into

$$\xi x - \Lambda^2 = \epsilon x + \lambda + 2\mu\beta - 2\mu\sqrt{\beta^2 - x/2}\,. \tag{8.44}$$

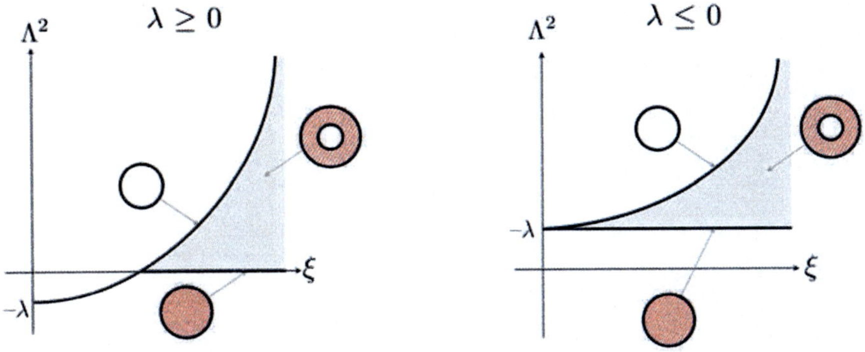

Fig. 8.9 Bifurcation diagram for a harmonic potential ψ_1, with $\lambda \geq 0$ (left) and $\lambda \leq 0$ (right). The axes are $\xi = \epsilon$ and $\Lambda^2 = 0$. The light grey region defined by the inequalities (8.43) contains the (ξ, Λ^2) associated with bounded motion. Following the convention of Arnold (Fig. (2.3) of [345]), the region of possible motion in the physical space is depicted as a light-red region in the orbital plane. For a generic orbit, $0 < r_P < r_A$ and the motion takes place in an annulus. The black boundaries correspond to degeneracies: The top one to circular motion ($r_P \to r_A$) and the bottom one to trajectories spiraling toward the center ($r_P \to 0$)

As for $\mathscr{P}_1$, with a bit of algebra we may write Eq. (8.44) as $x^2 - sx + p = 0$ where $s \equiv x_P + x_A$ and $p \equiv x_P x_A$. In terms of (ξ, Λ^2), we have explicitly

$$s \equiv 2\,\frac{(\Lambda^2 + \lambda + 2\mu\beta)(\xi - \epsilon) - \mu^2}{(\xi - \epsilon)^2} \quad \text{and} \quad p \equiv \frac{(\Lambda^2 + \lambda + 4\mu\beta)(\Lambda^2 + \lambda)}{(\xi - \epsilon)^2}\,. \tag{8.45}$$

As before, the solutions x_P, x_A must be strictly positive and this is ensured by the condition $x_P > 0$. At this point, choosing (ξ, Λ^2) in the region $x_P(\xi, \Lambda^2) > 0$ ensures that there are two intersections between $\mathscr{P}_2$ and $\mathscr{L}$. Adding the condition $\Lambda^2 \geq 0$ ensures that x_P is on the convex branch. However, a third condition is needed, namely that A belongs to the convex branch. To this end, notice that in Eq. (8.44), the minus sign in front of 2μ came from selecting the convex branch of the parabola. Therefore, to ensure that A is on this branch, it is sufficient to impose that $\xi x_A - \Lambda^2 \leq \epsilon x_A + \lambda + 2\mu\beta$. All in all, three conditions are sufficient to draw the bifurcation diagram. They read explicitly

$$\Lambda^2 \geq 0\,, \quad x_P(\xi, \Lambda^2) > 0 \quad \text{and} \quad (\xi - \epsilon)\, x_A(\xi, \Lambda^2) \leq \Lambda^2 + \lambda + 2\mu\beta\,. \tag{8.46}$$

Once we express x_P and x_A in terms of s, p, and thus in terms of (ξ, Λ^2) via Eq. (8.45), the three inequalities (8.46) delimit a region with all and only the periodic orbits. This region is depicted for a typical bounded potential ψ_2 in Fig. 8.10.

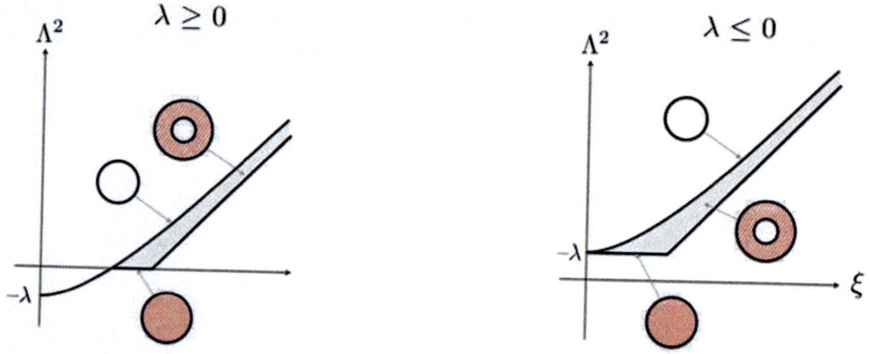

Fig. 8.10 Bifurcation diagram for a bounded potential ψ_2, with $\lambda \geq 0$ (left) and $\lambda \leq 0$ (right). The axes are $\xi = \epsilon + \mu/2\beta$ and $\Lambda^2 = 0$. All information is encoded the same way as in Fig. 8.9

Hénon and Kepler Families $\mathscr{P}_3$ and $\mathscr{P}_4$

For the right-oriented parabolae, we proceed the same way. Noticing that a parabola $\mathscr{P}_4$ can be obtained as the $\beta \to 0$ limit of a parabola $\mathscr{P}_3$, we may focus on the latter. As before, the condition $P, A \in \mathscr{P}_3$ translates into $x^2 - sx + p = 0$, where s and p are have the same expression as in Eq. (8.45), albeit with a plus sign in front of μ^2 for the former.

The conditions to be imposed to have a well-defined orbit are as before, $x_P > 0$, $\Lambda^2 \geq 0$. These two ensure that P is in the right half plane and on the convex branch. However, this does not imply that A is on the convex branch, so we must, again, add a third condition. Therefore, all three requirements are the same as in the $\mathscr{P}_2$ case, and the bifurcation diagram can be depicted using the inequalities Eq. (8.46) (again, with a plus sign in front of μ^2). For the Hénon family, $\beta \neq 0$ and the bifurcation diagram is depicted in Fig. 8.11. The bifurcation diagram for the Kepler family $\mathscr{P}_4$ is the $\beta \to 0$ limit of Fig. 8.11 and coincides precisely with Fig. (2.3) of [345].

Hollowed Family $\mathscr{P}_5$

Finally, we can do the same calculation for the fifth family $\mathscr{P}_5$. Equating $\xi x - \Lambda^2$ to $Y_5(x) = x\psi_5(\sqrt{x/2})$ and squaring the result provides a quadratic-in-x equation, from which we find the sum s and product p to be

$$s \equiv 2\,\frac{(\Lambda^2+\lambda)(\xi-\epsilon)+\mu^2}{(\xi-\epsilon)^2} \quad \text{and} \quad p \equiv \frac{(\Lambda^2+\lambda)^2+4\beta^2\mu^2}{(\xi-\epsilon)^2}. \tag{8.47}$$

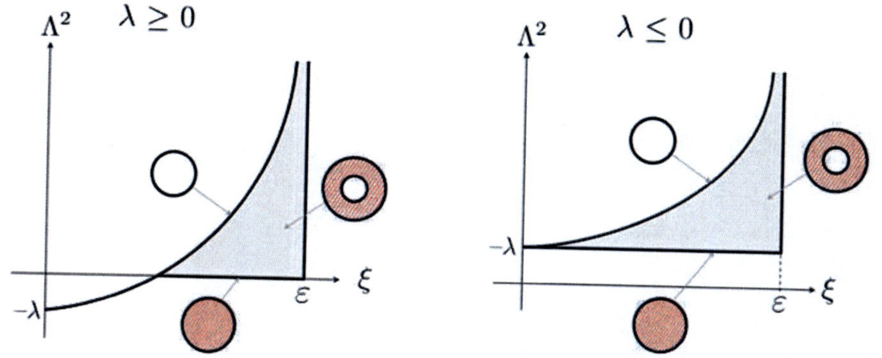

Fig. 8.11 Bifurcation diagram for a Hénon potential ψ_3, with $\lambda \geq 0$ (left) and $\lambda \leq 0$ (right). The axes are $\xi = \epsilon - \mu/2\beta$ and $\Lambda^2 = 0$. All information is encoded the same way as in Fig. 8.9

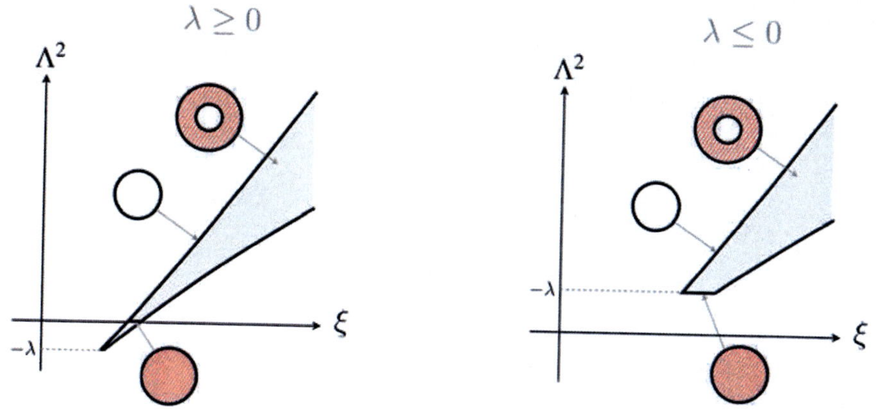

Fig. 8.12 Bifurcation diagram for a Hollowed potential ψ_5, with $\lambda \geq 0$ (left) and $\lambda \leq 0$ (right). The axes are $\xi = \epsilon$ and $\Lambda^2 = 0$. All information is encoded the same way as in Fig. 8.9

The bifurcation diagram is then given in Fig. 8.12, and is built by requiring the same conditions as before, except that this time, A lies on the convex branch if $\xi x_A \leq \epsilon x_A + \lambda$. The conditions are therefore just as Eq. (8.46), without the term $2\mu\beta$ on the right-hand side.

Chapter 9
Isochrone Mechanics

Contents

But it is absolutely certain and exact that the proportion between the periodic times of any two planets is precisely the sesquialterate proportion of their mean distances.

J. Kepler
Harmonices Mundi (1619).

After years of meticulous analysis of Tycho Brahe's observations of Mars, Johannes Kepler discovered, at the beginning of the seventeenth century, three laws that now bear his name. The first two, published in the *Astronomia Nova* (1609), state that (1) planets follow elliptical orbits with the Sun at one of their focii, and (2) the time

P. Ramond, *The First Law of Mechanics in General Relativity & Isochrone Orbits in Newtonian Gravity*, Springer Theses,
https://doi.org/10.1007/978-3-031-17964-8_9

elapsed between two points on an orbit is proportional to the swept area in that time. Kepler's third law, was published later in his *Harmonices Mundi* (1619). It states that the cube of the semi-major axis of these ellipses is proportional to the square of the planet's orbital period. This chapter is dedicated to a generalization of Kepler's laws to all isochrone orbits.[1] In particular, we will show that any periodic orbit in any isochrone potential satisfies Kepler's third law (Sect. 9.1) for the radial period T. We will interpret this law in various geometrical contexts and will also point out a similar and unified law for the apsidal angle Θ. In Sect. 9.2, we devise a geometrical method for solving the equations of motion in an isochrone potential, and show that all isochrone orbits can be parametrized by a Keplerian ellipse. Finally, we shall use all these results to exhibit and classify *isochrone orbits* in Sect. 9.3.

9.1 Isochrone Kepler's Third Laws

Kepler's third law is arguably one of the most celebrated and useful equations in astronomy and astrophysics. It arises in the context of a test particle orbiting a spherical body of mass M. Under the Newtonian law of gravitation, the particle follows a perfectly elliptical orbit. Kepler's third law expresses the relation that exists between the semi-major axis α of this ellipse[2] and the orbital period T_{orb}, as $T_{\text{orb}} \propto \alpha^{3/2}$, where the proportionality constant, $4\pi^2/GM$, does not depends on the particle. Let us re-write Kepler's third law using our notations adapted to the study of isochrony. First, notice that for a Keplerian ellipse, the orbital period (time taken for the particle to come back to the same position in the orbital plane) coincides with the radial period T that we have used up to know. Second, a classical result of celestial mechanics (see Appendix B.2 for details) shows that the semi-major axis α depends only on the particle's energy ξ, through $\alpha = -\mu/2\xi$. Consequently, Kepler's third law reads

$$T^2 = \frac{\pi^2}{2} \frac{\mu^2}{(-\xi)^3} \tag{9.1}$$

This first section is dedicated to a derivation of an equation identical to (9.1) for any *isochrone orbit*.[3] It is divided into three parts. First, we derive the generalized Kepler's third law for the period T in any isochrone potential, by solving quadratic equations (Sect. 9.1.1). Then we use the *circular orbit trick* to get a similar law for Θ (Sect. 9.1.2). Lastly we provide an alternative formulation of these two laws, in terms of purely geometrical quantities that can all be inferred solely from the parabola $\mathscr{P}$ and the line $\mathscr{L}$ (Sect. 9.1.3).

---

[1] Fortunately, we only have to generalize two of them. Indeed, Kepler's second law is a mere consequence of spherical symmetry: it holds for any radial potential, and thus for isochrone ones.

[2] We use α, since the more common notation is a and this corresponds to one of the Latin parameters for our parabolae (8.27).

[3] By *isochrone orbit*, we mean an orbit followed by a test particle in an isochrone potential.

9.1.1 *Third Law for the Radial Period T*

Let us consider a generic isochrone potential ψ and a particle (ξ, Λ) that orbits within it. As we did many times above, in the Hénon plane ψ is associated with a parabola $\mathscr{P}$, and the particle is associated with a line $\mathscr{L}$ that intersects $\mathscr{P}$ twice (or once for circular orbits). For convenience, we rewrite the equations for $\mathscr{P}$ and $\mathscr{L}$ here:

$$\mathscr{P} : (ax + by)^2 + cx + dy + e = 0 \quad \text{and} \quad \mathscr{L} : y = \xi x - \Lambda^2 . \tag{9.2}$$

Keeping the parabola $\mathscr{P}$ fixed, we take a line (ξ, Λ) with two intersections P and A, both functions of (ξ, Λ). Now since A and P belong to both $\mathscr{P}$ and $\mathscr{L}$, we can eliminate y from the two equations in (9.2) and get an equation on x whose solutions are x_P and x_A, the abscissa of P and A. Re-arranging the result gives the following quadratic equation:

$$(a + b\xi)^2 x^2 + [(c + d\xi) - 2b\Lambda^2(a + b\xi)]x + [b^2\Lambda^4 + e - d\Lambda^2] = 0 , \tag{9.3}$$

whose discriminant Δ is given by

$$\Delta(\xi, \Lambda) \equiv (c + d\xi)^2 - 4b\Lambda^2(c + d\xi)(a + b\xi) - 4(a + b\xi)^2(e - d\Lambda^2) . \tag{9.4}$$

We will now compute the period T with the help of Hénon's formula, proved in Sect. 9.1 and given in Eq. (8.13). First, we need the difference between x_A and x_P. By writing the solutions to the quadratic equation (9.3), we obtain easily

$$(x_A - x_P)^2 = \frac{\Delta(\xi, \Lambda)}{(a + b\xi)^4} . \tag{9.5}$$

Second we need a formula for Λ_C, the angular momentum of the circular orbit with energy ξ. To find it, we note that for a fixed ξ, the discriminant $\Delta(\xi, \Lambda)$ is strictly positive when $\mathscr{L}$ intersects $\mathscr{P}$ twice, but vanishes when it intersects once, for some value Λ_C. This corresponds to a circular orbit, obtained by translating $\mathscr{L}$ downward until P and A degenerate into a single point C (recall Fig. 8.1). Therefore, we have $\Delta(\xi, \Lambda_C) = 0$, which may be solved for Λ_C. With the help of Eq. (9.3) and some elementary algebra, we obtain

$$\Lambda_C^2 = \frac{4e(a + b\xi)^2 - (c + d\xi)^2}{4\delta(a + b\xi)} , \tag{9.6}$$

where δ is the discriminant of the parabola, cf. Eq. (8.27). Now we can combine Hénon's formula (8.13) with Eqs. (9.5) and (9.6). Once again, after some easy algebra we obtain the following generalization of Kepler's third law

$$T^2 = -\frac{\pi^2}{4} \frac{\delta}{(a + \xi b)^3} . \tag{9.7}$$

A few remarks are in order. First of all, we stress that Eq. (9.7) is valid for *any* particle (ξ, Λ) orbiting periodically in *any* isochrone potential, including Hollowed ones and those discarded during the reduction process[4] showed in Fig. 8.6. As long as there is a periodic orbit in a isochrone potential, physical or not, there is an associated parabola $\mathscr{P}$ given by Eq. (9.2) and its radial period T verifies Eq. (9.7). Second, we see that it involves in the numerator δ which is strictly positive. Therefore, Eq. (9.7) implies that the denominator is strictly negative and thus that $a + b\xi < 0$. This is a general property that can be traced back to the very existence of solutions to the quadratic equation (9.3). We shall use this result later in Sect. 9.2 to find a parametrization of isochrone orbits. Moreover, speaking of the parameters, we recover the two well-known cases: When $a = 0$ the parabola has horizontal symmetry and we have $T^2 \propto |\xi|^{-3}$, as in the Kepler potential. Similarly, when $b = 0$, the parabola has vertical symmetry and we have $T = \text{cst}$, i.e., T is independent of the properties of the particle, as for the harmonic potential.

9.1.2 Third Law for the Apsidal Angle Θ

All the results presented in the last paragraph regarding T are also true for the apsidal angle Θ. In particular, we can use the Hénon formula (8.15) in order to write the apsidal angle for any orbit solely in terms of Λ and the parameters (a, b, c, d, e). To this end, we start, as usual, with some geometry and Fig. 8.1 in mind. Consider a line $\mathscr{L}$ intersecting a generic isochrone parabola $\mathscr{P}$, both given by Eq. (9.2). Since Θ is independent of ξ, we may choose a value of ξ such that the orbit is circular. This can be done as follows. Keeping Λ fixed, decreasing ξ defines other lines with the same Λ and thus the same apsidal angle $\Theta(\Lambda)$ for the associated orbits. In particular, ξ can reach a critical value ξ_C such that the line $\mathscr{L}$ becomes tangent to $\mathscr{P}$, at some point of abscissa x_C. It is important to notice that ξ_C and x_C are function of Λ only.

Let us focus on this very line $\mathscr{L}_C : y = \xi_C x - \Lambda^2$ and the associated circular orbit. Its orbital radius is r_C, such that $2r_C^2 = x_C$. The period $T(\xi_C)$ of this orbit is given by Eq. (9.7). Now by definition of the angular momentum, we have, for this circular orbit $\Lambda = r_C^2\dot{\theta}$. Since $x_C = 2r_C^2$, this can be turned into a differential equality $2\Lambda \mathrm{d}t = x_C \mathrm{d}\theta$. Now, by definition of Θ, integrating the latter over a period $T(\xi_C)$ readily gives

$$\Theta = 2\Lambda \frac{T(\xi_C)}{x_C}. \tag{9.8}$$

Let us stress again that both ξ_C and x_C are functions of Λ, thus we just need a formula for these in terms of (a, b, c, d, e) and Λ. Now we can apply the results of the last section, regarding the intersections of $\mathscr{P}$ and $\mathscr{L}$, below Eq. (9.2). In particular, in the

[4] For these potentials, one must venture into a world where the orbital radius and other physical quantities become imaginary numbers. Yet, the law (9.7) still holds in these (strange) case.

present context ξ_C is such that $\Delta(\xi_C, \Lambda) = 0$, with Δ given by Eq. (9.4). After some algebra, the solution for ξ_C is easily found to be

$$\xi_C(\Lambda) = -\frac{a\Xi_C - c}{b\Xi_C - d} \quad \text{where} \quad \Xi_C \equiv 2b\Lambda \pm 2\sqrt{b^2\Lambda^4 - d\Lambda^2 + e}\,, \tag{9.9}$$

and the $+/-$ sign should be used for the left-/right-oriented parabolae, respectively. Regarding the quantity x_C, it can be found by writing the solution to Eq. (9.3) when $\Delta = 0$. We obtain easily

$$x_C(\Lambda) = \frac{b\Lambda^2}{a + b\xi_C} - \frac{c + d\xi_C}{2(a + b\xi_C)^2}\,, \tag{9.10}$$

where ξ_C is given in terms of Λ by Eq. (9.9). Now it is just a matter of inserting Eqs. (9.10) and (9.9) into Eq. (9.8) and do some algebra to obtain a formula for Θ. After a rather lengthy but simple computation, we obtain the following law, valid for any particle orbiting in any isochrone potential

$$\frac{\Theta^2}{\pi^2\Lambda^2} = \frac{2b^2\Lambda^2 - d}{b^2\Lambda^4 - d\Lambda^2 + e} + \frac{2b}{\sqrt{b^2\Lambda^4 - d\Lambda^2 + e}}\,. \tag{9.11}$$

As for Eq.(9.7), we stress that Eq. (9.11) is valid for any orbit in any isochrone potential, even the gauged and hollow ones. As a corollary of this general formula, one may insert the Greek parameters $(\epsilon, \lambda, \omega, \mu, \beta)$ introduced earlier, and find agreement with the results of [589].

9.1.3 Geometrical Reading of the Third Laws

The computation of the period T and apsidal angle Θ via Eqs. (9.7) and (9.11) involves the parameters (a, b, c, d, e), and can thus be made only if we know the algebraic equation of the parabola. Here, we show that it is also possible to express T and Θ entirely in terms of geometrical quantities, i.e., compute them solely with Euclidean geometry, once a parabola $\mathscr{P}$ and a line $\mathscr{L}$ is drawn in the Hénon plane. For the period T, we need to define three geometrical quantities

- $\vec{\xi} \equiv (1, \xi)$, generator of the line $\mathscr{L}$,
- $\vec{N} \equiv (a, b)$, generator of the symmetry axis of $\mathscr{P}$ (controls its orientation),
- $R \equiv \delta/2|\vec{N}|^3$, the curvature radius of $\mathscr{P}$ at its apex (controls its aperture).

Expressions for these quantities can be easily derived with the help of Sect. 8.3.1. We can now rewrite Eq. (9.7) without the (a, b, c, d, e) parameters, using the unit vector $\vec{n} \equiv \vec{N}/|\vec{N}|$, simply as

$$T^2 = \frac{\pi^2}{2} \frac{R}{|\vec{n} \cdot \vec{\xi}\,|^3} . \tag{9.12}$$

This formula should be compared to Kepler's third law for the radial period as given by Eq. (9.1). In a similar fashion, we can make a geometrical construction for the law of the apsidal angle Θ. In particular, let $\ell_\pm$ be the ordinate of the intersection points between the branch $\mathscr{P}_\pm$ and the y-axis (we thus discard the Hollowed potentials for the moment, for which the parabola does not intersect the y-axis, and the Harmonic ones which have only one). As we have seen already in Sect. 8.3.1, we have

$$\ell_\pm = -\frac{d \pm \sqrt{d^2 - 4b^2 e}}{2b^2} \tag{9.13}$$

In terms of the Greek parameters, we can show easily that the quantity ℓ_+ is nothing but λ, and that ℓ_- is $\lambda + 4\mu\beta$. Now we can easily turn Eq. (9.13) into $b^2\Lambda^4 - d\Lambda^2 + e = b^2(\ell_+ + \Lambda^2)(\ell_- + \Lambda^2)$ and notice that the left-hand side appears precisely in the denominator in Eq. (9.11). Therefore, we insert this result in Eq. (9.11), make a partial fraction decomposition for the first term and obtain

$$\Theta = \frac{\pi}{\sqrt{1 + \ell_+/\Lambda^2}} + \frac{\iota\pi}{\sqrt{1 + \ell_-/\Lambda^2}} , \tag{9.14}$$

where $\iota \in \{-1, 1\}$ is simply the sign of b and determines the orientation (resp. left or right) of the parabola. As for the period T, we see that the apsidal angle Θ can be found with only geometrical quantities that can be read off the parabola. In particular, $\ell_\pm/\Lambda^2$ is simply the ratio of the vertical distances between the y-intercept of $\mathscr{L}$ and the branches[5] $\mathscr{P}_\pm$, as depicted in Fig. 9.1.

Finally, at the end of Appendix C.2.1 we provide yet another form for the third law $T(\xi)$ and $\Theta(\Lambda)$. For T, this alternative form involves the slope and curvature of the parabola at the circular point C uniquely associated with the orbit. For Θ, this alternative involves the curvature of the effective potential, when looked at in the Binet variable $u = 1/r$. The interested reader may find these results useful in order to go further into the geometrical properties of isochrony, e.g., for academic purposes. Along these lines, we stress that what led to the mathematical equalities (9.7), (9.11) and other third laws in Appendix C.2.1 can be seen as a geometrical method to compute the rather complex-looking integrals (8.6) and (8.15), with $Y(x)$ given by Eq. (8.28) or Eq. (8.30). In particular, the fact that these rather complex-looking integrals do not depend on Λ for T, and on ξ for Θ, is quite remarkable.

[5] For the Kepler family, the two intersections degenerate into one and $\ell_+ = \ell_-$. For the Harmonic family, ℓ_+ goes to $+\infty$ (think of a $\pi/2$-rotation turning $y = -\sqrt{x}$ into $y = x^2$).

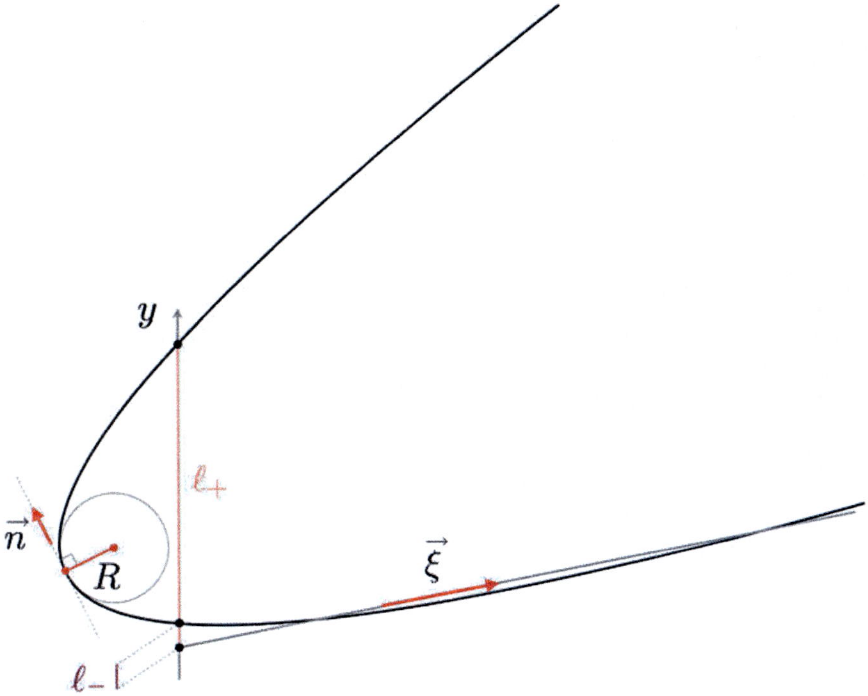

Fig. 9.1 Geometrical quantities involved in the geometrical laws for T (Eq. (9.12)) and for Θ (Eq. (9.14)). In this figure, $\vec{n}$ is a unit vector and gives the asymptotic direction of the parabola, R is the curvature radius at the apex, and $\ell_\pm$ are the vertical distances between the y-intercept of $\mathscr{L}$ and the y-intercept of $\mathscr{P}_\pm$

9.2 Isochrone Orbits Transformations

In this section, we provide a geometrical analysis that leads naturally to an explicit and analytic parametrization of any isochrone orbit in polar coordinates. The essential idea is the following: An isochrone orbit is associated with an arc of parabola in the Hénon plane. There is one isochrone orbit for which we know an analytic expression: the Keplerian ellipse. Using linear transformations in the Hénon plane, we show how to map any arc of parabola to a Keplerian one, and therefore establish a one-to-one correspondence between any isochrone orbit and a Keplerian ellipse, the latter being used to parametrize the former.

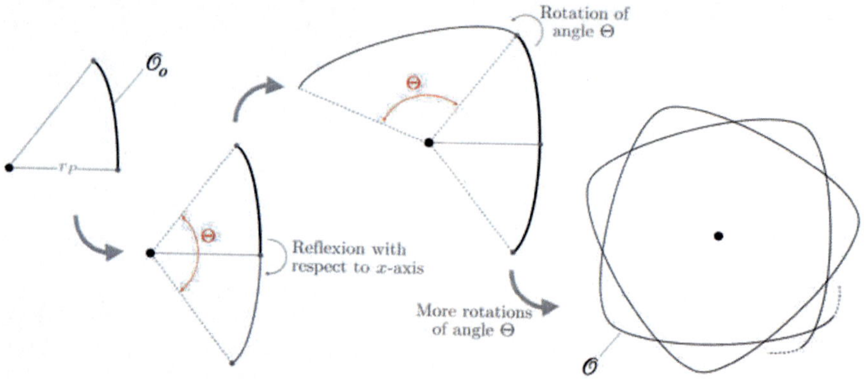

Fig. 9.2 The construction of a full orbit $\mathcal{O}$ (right) from a reduced orbit $\mathcal{O}_o$ (left). All points of $\mathcal{O}_o$ have a different radius r varying increasingly between periapsis $r = r_P$ and apoapsis $r = r_A$. By reflexion with respect to $\theta = 0$, one obtains the portion travelled in one radial period T, and the opening angle is therefore Θ. By successive rotations of angle Θ, the full orbit $\mathcal{O}$ can be constructed

9.2.1 *Reduced Orbit*

As we have seen many times before, an arc of parabola $\mathscr{A}$ in the Hénon plane (x, y) is associated with an isochrone orbit in the physical space that will be denoted by $\mathcal{O}$. By conservation of angular momentum, the particle orbits within a plane, equipped with the usual polar coordinates (r, θ). We shall always choose the angle θ such that $\theta = 0$ at periapsis $r = r_P$.

When the particle moves on an isochrone orbit $\mathcal{O}$, its radius r changes periodically and can be mapped to a point M that travels back and forth on the arc $\mathscr{A}$. However, the converse is not true: A point $M \in \mathscr{A}$ of abscissa $x = 2r^2$ corresponds to an infinite number of points on $\mathcal{O}$, namely the points $(r, \theta + k\Theta)_{k\in\mathbb{Z}}$, precisely because of the radial periodicity. To get a one-to-one correspondence, we can quotient the full orbit $\mathcal{O}$ by reflexions and rotations, to get the *reduced orbit* $\mathcal{O}_o$, as depicted in Fig. 9.2. The full orbit $\mathcal{O}$ can be constructed from $\mathcal{O}_o$, which acts as a generator of the orbit and which, contrary to the *full* orbit $\mathcal{O}$, is in a one-to-one correspondence with the arc $\mathscr{A}$: a point $(r, \theta) \in \mathcal{O}_o$ is uniquely linked to a point $(x, y) \in \mathscr{A}$ via $x = 2r^2$.

9.2.2 *Kepler Parabola*

We will need in this section a few results about the Kepler parabola $y^2 = 2\mu^2 x$. It is associated with the usual Kepler potential $\psi(r) = -\mu/r$. The bifurcation diagram in terms of (ξ, Λ) for the Kepler potential is given in Fig. 8.11 with $\beta = \epsilon = \lambda = 0$, or equivalently in Fig. (2.3) of [345]. Requiring that the values of (ξ, Λ) generate a periodic orbit (i.e., that they are in the grey region of Fig. 8.11) is equivalent to the following algebraic inequalities:

$$\Lambda^2 \geq 0 \quad \text{and} \quad 0 > \xi \geq -\frac{\mu^2}{2\Lambda^2}\,. \tag{9.15}$$

In the Kepler potential, any bounded orbit is an ellipse whose focus is at the center of polar coordinates. An elliptic orbit $\mathcal{O}$ is made of one periapsis and one apoapsis, joined by two symmetrical portions, as can be seen in the bottom left of Fig. 9.3. We will take the reduced orbit $\mathcal{O}_o$ to be the upper portion. From the classical solution to the Kepler problem (see [345] and/or Appendix B.2), the reduced orbit $\mathcal{O}_o$ can be given the following parametric representation[6]

$$\begin{cases} r(s) = p\,(1+\varepsilon\cos s)^{-1} \\ \theta(s) = s \end{cases}, \quad s \in [0,\pi]\,. \tag{9.16}$$

In these equations, ε is the eccentricity of the ellipse, and p is its semi-latus rectum (again, see [345] and/or Appendix B.2 for details). They depend explicitly on the energy ξ and angular momentum Λ of the particle, as well as the central mass $\mu \equiv GM$ of the Keplerian potential. They are given by

$$\varepsilon \equiv \sqrt{1+\frac{2\Lambda^2\xi}{\mu^2}} \quad \text{and} \quad p \equiv \frac{\Lambda^2}{\mu}\,, \tag{9.17}$$

and one can see that Eq. (9.15) is actually equivalent to $\varepsilon \in [0, 1[$ and $p \geq 0$.

9.2.3 *Linear Transformation*

We consider a reduced isochrone orbit $\mathcal{O}_o$ in a central isochrone potential with finite mass at the origin. Recall that $\mathcal{O}_o$ is in a one-to-one correspondence with an arc of parabola $\mathscr{A}$ in the Hénon plane that passes through the origin. This arc $\mathscr{A}$ is defined as the portion of a parabola $\mathscr{P}$ that lies below a given line $\mathscr{L}$, both given by Eq. (9.2) with $e = 0$

$$\mathscr{P} : (ax+by)^2 + cx + dy = 0 \quad \text{and} \quad \mathscr{L} : y = \xi x - \Lambda^2\,. \tag{9.18}$$

Now let us apply the following linear transformation F to the Hénon plane

$$F : (x, y) \mapsto (-cx - dy, ax + by) \equiv (\bar{x}, \bar{y})\,. \tag{9.19}$$

Following Eq. (9.19), any quantity X that has been mapped by F will be denoted with a bar as $\bar{X} \equiv F(X)$. For instance, a point M of coordinates (x, y) on $\mathscr{A}$ will be mapped to the point $\bar{M}$ with coordinates $(\bar{x}, \bar{y}) = F(x, y)$ given by Eq. (9.19).

[6] Note that if the periapsis is at $(r, \theta) = (r_P, 0)$, the reduced orbit $\mathcal{O}_o$ is simply the upper half of $\mathcal{O}$.

Since the set of parabolae and the set of lines are stable under affine transformations (and thus linear ones), $\bar{\mathscr{P}} = F(\mathscr{P})$ is still a parabola and $\bar{\mathscr{L}} = F(\mathscr{L})$ still a line. The parameters (a, b, c, d) in Eq. (9.19) are precisely these of the parabola $\mathscr{P}$ in Eq. (9.18) and are not chosen randomly. It is straightforward to find its implicit equation which reads

$$\bar{\mathscr{P}} : \bar{y}^2 = \bar{x}\,. \tag{9.20}$$

In view of the previous section, it is clear from Eq. (9.20) that F maps $\mathscr{P}$ to a Keplerian parabola with mass parameter $\mu = 1/\sqrt{2}$. Regarding the image $\bar{\mathscr{L}}$ of the line $\mathscr{L}$, a quick computation gives the following equation

$$\bar{\mathscr{L}} : \bar{y} = \bar{\xi}\bar{x} - \bar{\Lambda}^2\,, \quad \text{with} \quad \bar{\xi} = -\frac{a+b\xi}{c+d\xi} \quad \text{and} \quad \bar{\Lambda}^2 = -\frac{\delta\Lambda^2}{c+d\xi}\,. \tag{9.21}$$

The image arc $\bar{\mathscr{A}}$ is a portion of $\bar{\mathscr{P}}$, although we do not know yet if it lies below $\bar{\mathscr{L}}$. If it does, then $\bar{\mathscr{A}}$ is a Keplerian arc and the associated orbit is an ellipse. Let us first ensure that $\bar{\mathscr{A}}$ indeed corresponds to a well-defined elliptic orbit. According to the inequalities (9.15) with $\mu = 1/\sqrt{2}$, a Keplerian orbit is periodic provided that $\bar{\Lambda}^2 \geq 0$, $\bar{\xi} < 0$ and $\bar{\xi} > -1/4\bar{\Lambda}^2$. We now argue that these three conditions are always satisfied, in the three following steps:

- From Kepler's generalized third law (9.7) and $\delta > 0$, we have $a + b\xi < 0$. Consequently, by Eq. (9.21), the two conditions $\bar{\xi} < 0$ and $\bar{\Lambda}^2 \geq 0$ hold if and only if $c + d\xi < 0$.
- The condition $c + d\xi < 0$ is a geometrical consequence of all hypotheses (H_i) that are, by assumption, verified since the initial orbit $\mathscr{O}$ is isochrone. (The proof is easy but not central here; it can be found in Appendix C.2.4). At this stage, we thus have $\bar{\xi} < 0$ and $\bar{\Lambda}^2 \geq 0$.
- Linear transformations preserve the existence of intersection points; therefore, $\bar{\mathscr{L}}$ intersects the Kepler parabola $\bar{\mathscr{P}}$ twice. Along with $\bar{\xi} < 0$ and $\bar{\Lambda}^2 \geq 0$, we can check easily that these intersections are necessarily on the convex branch. Consequently, the orbit is an ellipse with eccentricity $\bar{\varepsilon} = \sqrt{1 + 4\bar{\Lambda}^2\bar{\xi}} \in [0, 1[$, and therefore, $\bar{\xi} > -1/4\bar{\Lambda}^2$ holds.

To summarize, we can map any isochrone arc $\mathscr{A}$ to a Keplerian one $\bar{\mathscr{A}}$ with mass $\mu = 1/\sqrt{2}$ using an appropriate linear transformation F given by Eq. (9.19). According to Keplerian dynamics, the orbit $\bar{\mathscr{O}}$ associated with $\bar{\mathscr{A}}$ is an ellipse whose polar equation is

$$\bar{r}(\bar{\theta}) = \frac{\bar{p}}{1 + \bar{\varepsilon}\cos\bar{\theta}} \quad \text{with} \quad \bar{\varepsilon} = \sqrt{1 + 4\bar{\Lambda}^2\bar{\xi}} \quad \text{and} \quad \bar{p} = \sqrt{2}\bar{\Lambda}^2\,. \tag{9.22}$$

We thus have a mapping between the generic isochrone arc $\mathscr{A}$ and the Keplerian one $\bar{\mathscr{A}}$, i.e., we have established the upper part of Fig. 9.3. The next step is to extend this to the lower part of Fig. 9.3, i.e., link the polar coordinates of each orbit.

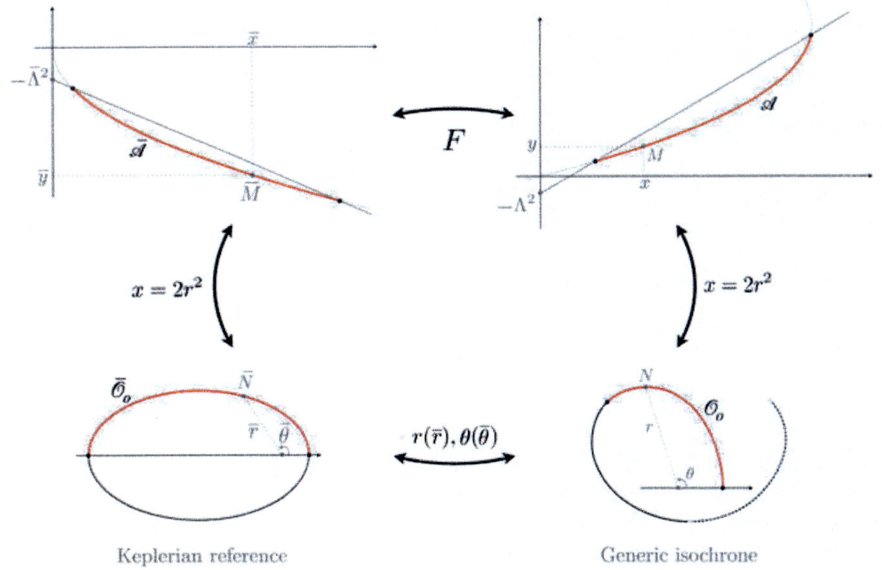

Fig. 9.3 Any reduced isochrone orbit $\mathscr{O}_o = (r, \theta)$ (red, bottom right) can be put in correspondence with a reduced Keplerian ellipse $\bar{\mathscr{O}}_o = (\bar{r}, \bar{\theta})$ (red, bottom left). The mapping $r(\bar{r}), \theta(\bar{\theta})$ can be obtained by going into the Hénon plane and making a linear transformation F to relate the arcs of parabolae associated with $\mathscr{O}_o$ (red, top right) and $\bar{\mathscr{O}}_o$ (red, top left) (see details in the text)

9.2.4 General Isochrone Orbit

Consider a point $N = (r, \theta)$ on a reduced isochrone orbit $\mathscr{O}_o$. It is in a one-to-one correspondence with the point $M \in \mathscr{A}$ of abscissa $x = 2r^2$. The point $M = (x, y)$ is in turn associated with a unique point $\bar{M} = (\bar{x}, \bar{y})$ on the Keplerian parabola (9.20) and thus with a unique point $N = (\bar{r}, \bar{\theta})$ on the reduced elliptic orbit $\bar{O}_o$. This is all depicted in Fig. 9.3. The goal now is to express (r, θ) of the generic isochrone orbit in terms of $(\bar{r}, \bar{\theta})$.

We start with the radius. By inverting Eq. (9.19) we write $x = \delta^{-1}(b\bar{x} + d\bar{y})$. Now $\bar{M}$ is on the convex branch of the Keplerian parabola (9.20); therefore, $\bar{y} = -\sqrt{\bar{x}}$. Combining these two equations readily gives a relation between the abscissa of the two points M and $\bar{M}$, namely

$$x = \delta^{-1}(b\bar{x} - d\sqrt{\bar{x}}) \quad \Rightarrow \quad r = \sqrt{\frac{\bar{r}}{\delta}\left(b\bar{r} - \frac{d}{\sqrt{2}}\right)}, \tag{9.23}$$

where we simply used the definition of Hénon variables for each orbit, i.e., $\bar{x} = 2\bar{r}^2$ and $x = 2r^2$ to get the second equation, and $\delta = ad - bc \neq 0$ (see Sect. 8.3.1). Now if we take a point $(\bar{r}, \bar{\theta})$ on the Keplerian reduced ellipse, then while $\bar{\theta}$ varies in $[0, \pi]$,

$\bar{r}$ changes according to Eq. (9.22), and r changes as well according to Eq. (9.23). Therefore, we can use $\bar{\theta}$ as a parameter, denoted s, to track the radius r on $\mathscr{O}_o$. To this end, we insert Eq. (9.22) into Eq. (9.23) and perform some algebraic manipulations to find that *the radius r of the particle on a generic reduced isochrone orbit $\mathscr{O}_o$ can be parametrized by $r = \varrho(s)$, $s \in [0, \pi]$, where*

$$\varrho(s) \equiv \frac{\sqrt{c_1 + c_2 \cos s}}{1 + c_3 \cos s}, \tag{9.24}$$

for some constants c_1, c_2 and c_3 that depend on (a, b, c, d) and (ξ, Λ), given by

$$c_1 = \frac{2b\delta\Lambda^4}{(c + d\xi)^2} + \frac{c_2}{c_3}, \quad c_2 = \frac{\Lambda^2 d}{c + d\xi} c_3 \quad \text{and} \quad c_3 = \sqrt{1 + \frac{4\delta\Lambda^2(a + b\xi)}{(c + d\xi)^2}}. \tag{9.25}$$

We note incidentally that c_3 is actually the eccentricity $\bar{\varepsilon}$ of the parametrizing ellipse (9.22). Now we seek to find a similar parametrization for the angle θ. Since $\bar{\theta} = s$ is our parameter, all we need is an expression of θ in terms of $\bar{\theta}$. To this end, we start by writing the following chain rule

$$\frac{\mathrm{d}\theta}{\mathrm{d}\bar{\theta}} = \frac{\mathrm{d}\theta}{\mathrm{d}t}\frac{\mathrm{d}t}{\mathrm{d}x}\frac{\mathrm{d}x}{\mathrm{d}\bar{x}}\frac{\mathrm{d}\bar{x}}{\mathrm{d}\bar{t}}\frac{\mathrm{d}\bar{t}}{\mathrm{d}\bar{\theta}}. \tag{9.26}$$

In this equation, the time $\bar{t}$ is the one associated with the dynamics of the Keplerian orbit $\bar{\mathscr{O}}$, i.e., such that energy and angular momentum conservation read

$$\frac{1}{16}\left(\frac{\mathrm{d}\bar{x}}{\mathrm{d}\bar{t}}\right)^2 = \bar{\xi}\bar{x} - \bar{\Lambda}^2 - \bar{y} \quad \text{and} \quad \bar{\Lambda} = \frac{\bar{x}}{2}\frac{\mathrm{d}\bar{\theta}}{\mathrm{d}\bar{t}}. \tag{9.27}$$

We now express each factor on the right-hand side of Eq. (9.26) in terms of $\bar{\theta}$, in order to integrate a first order ODE. For the first and last terms, we use angular momentum conservation in each orbit: Eq. (8.5) for $\mathscr{O}$ and Eq. (9.27) for $\bar{\mathscr{O}}$. Similarly, in the second and second-to-last terms we use energy conservation: Eq. (8.4) for $\mathscr{O}$ and Eq. (9.27) for $\bar{\mathscr{O}}$. Inserting these results in Eq. (9.26) readily gives

$$\frac{\mathrm{d}\theta}{\mathrm{d}\bar{\theta}} = \frac{\bar{x}}{x}\frac{\mathrm{d}x}{\mathrm{d}\bar{x}}\frac{\Lambda}{\bar{\Lambda}}\left(\frac{\bar{\xi}\bar{x} - \bar{\Lambda}^2 - \bar{y}}{\xi x - \Lambda^2 - y}\right)^{1/2}. \tag{9.28}$$

To simplify this equation, we express $(\bar{x}, \bar{y})$ in terms of (x, y) using Eq. (9.19) and $(\bar{\xi}, \bar{\Lambda})$ in terms of (ξ, Λ) using Eq. (9.21). When doing so, the last two terms on the right-hand side compensate each other exactly. The only terms contributing on the right-hand side of Eq. (9.28) are the first two, and they can be simplified with the help of Eq. (9.23). In the end, we find

$$\frac{d\theta}{d\bar{\theta}} = \frac{1}{2} + \frac{1}{2}\frac{b}{b - d/\sqrt{2\bar{r}}}. \tag{9.29}$$

The final step is to insert Eq. (9.22) into Eq. (9.29). We then obtain a first-order ODE that can then be integrated using the usual change of variables $u = \tan\bar{\theta}/2$. Once this integration is done and the initial condition is chosen,[7] we find that *the angle θ of the particle on a generic reduced isochrone orbit $\mathscr{O}_o$ can be parametrized by $\theta = \vartheta(s)$, with $s \in [0, \pi]$, where*

$$\vartheta(s) \equiv \frac{s}{2} + c_4 \arctan\left(c_5 \tan\frac{s}{2}\right), \tag{9.30}$$

for some constants c_4 and c_5 that depend on (a, b, c, d) and (ξ, Λ); given by

$$c_4 = \frac{b\Lambda}{\sqrt{b^2\Lambda^2 - d}} \quad \text{and} \quad c_5 = \sqrt{1 - \frac{2d(c + d\xi)c_3}{2b\delta\Lambda^2 + d(c + d\xi)(1 + c_3)}}. \tag{9.31}$$

9.2.5 Summary and Remarks

To summarize, we have found a parametrization for any reduced orbit $\mathscr{O}_o$ in an isochrone potential ψ with finite central mass. To get the full orbit $\mathscr{O}$ from $\mathscr{O}_o$, we follow Fig. 9.2. In particular, we add to $\mathscr{O}_o$ its symmetric with respect to the x-axis, by extending the range of the parameter s from $[0, \pi]$ to $[-\pi, \pi]$.[8] We then obtain a piece $\mathscr{O}_T$ of the orbit that spans a full radial period T, or equivalently a full apsidal angle Θ. The full orbit $\mathscr{O}$ is then obtained by copying and pasting the piece $\mathscr{O}_T$, albeit rotated anti-clockwise by an angle $n\Theta$, for all $n \in \mathbb{Z}$. In particular, any orbit $\mathscr{O}$ in a potential with finite central mass can be parametrized by

$$\mathscr{O} = \bigcup_{n\in\mathbb{Z}} \mathscr{O}_n \quad \text{where} \quad \mathscr{O}_n \equiv \{(r, \theta) = (\varrho(s), \vartheta(s) + n\Theta)\,,\, s \in [-\pi, \pi]\}\,, \tag{9.32}$$

with $\varrho(s)$ and $\vartheta(s)$ given by Eqs. (9.24) and (9.30), respectively. By construction, when $s = 0$, the particle is at periapsis $(r_P, 0)$ and when $s = \pi$ it is at apoapsis $(r_A, \Theta/2)$. The latter implies that $\Theta = \pi(1 + c_4)$, a result that can be checked by comparing Eqs. (9.11) and (9.31). The special case of the Keplerian ellipse of eccentricity ε and semi-latus rectum p corresponds to $(c_1, c_2, c_3, c_4, c_5) = (p^2, 0, \varepsilon, 1, 1)$.

This parametrization covers *any* isochrone orbit in a potential associated with a finite mass at the center. However, it can be extended easily to orbits in gauged

[7] We choose θ such that $\theta = 0$ at initial $r = r_P$. Since $\bar{r}_P$ is sent to r_P, we require $\theta = 0$ when $\bar{\theta} = 0$.

[8] Indeed, from Eqs. (9.24) and (9.30), two points $(\varrho(s), \vartheta(s))$ and $(\varrho(s), \vartheta(-s))$ are symmetric with respect to $\theta = 0$ for $s \in [0, \pi]$ simply because cos is even and $s \mapsto s$, tan and arctan are odd.

potentials (with $\lambda \neq 0$, i.e., infinite central mass) and to hollow potentials (with $x_v > 0$, i.e., undefined around the origin), by considering affine transformations instead of only linear ones. Indeed, starting from the appropriate parabola crossing the origin, with a vertical (resp. horizontal) translation, one can reach any orbit in a gauged (resp. hollow) potential. In particular, one can follow the previous method and send any parabola $\mathscr{P} : (ax + by)^2 + cx + dy + e = 0$ to the Keplerian parabola $\bar{\mathscr{P}} : \bar{y}^2 = \bar{x}$ by applying to $\mathscr{P}$ the affine transformation $F \circ G$, composed of the linear map F given by Eq. (9.19) and the translation $G : x \mapsto x - e$. The computation can be done to find an analytic parametrization, with a little more work in the integration of the ODE expressing θ in terms of $\bar{\theta}$, cf. Eq. (9.28). We leave this as an exercise to the interested reader.

Speaking of Eq. (9.28), we have seen that the last two terms on the right-hand side cancel each other. Notice that it would also have been the case if the following assumptions had been made: $\xi x - y = \bar{\xi}\bar{x} - \bar{y}$ and $\Lambda = \bar{\Lambda}$. In [589], the authors precisely make these assumptions and the consequence was twofold. On the one hand, not all orbits can be reached from the Keplerian one (only the ones associated with arcs that verify theses two geometrical constraints). Therefore, the so-called bolst transformations cannot bridge between any two isochrone orbits. On the other hand, these bolsts form a subgroup of the linear transformations, whose additive representation exhibits similarities with the Lorentz group (to some extent), allowing for some analogies with special relativity, for a particular subclass of bolsts (the so-called '$\boldsymbol{i}$−bolst). However, as we have seen, the integration of Eq. (9.28) is tractable without any additional assumption, and considering linear transformations is the only way to describe all isochrone orbits. To summarize, although we believe that the special relativistic analogies presented in [589] may be of pedagogical interest, the fundamental group associated with isochrony is that of parabola arcs equipped with affine transformation. Any other subgroup will necessarily miss the description of some isochrone orbits.

One last comment, to motivate the need for a Hamiltonian treatment of the problem, as presented in Chap. 10. The above calculation is, in a sense, a geometrical way of solving the equations of motion (7.16), under the assumption that the potential $\psi(r)$ is isochrone. Given the general form of these potentials (the five equations (8.34)-(8.39)), there was little hope that such an analytic solution existed. Fortunately, we can bypass this complexity by going to the Hénon plane and, do some geometry, and come back, to get the solution. This solution is given as a parametric system (Eqs. (9.24) and (9.30), with the parameters s following the polar angle of a Keplerian ellipse. However, no insight can be given as to the nature of this parameter, except perhaps by examining the polar angle, Eq. (9.30), which does look familiar to the eccentric anomaly of the Kepler problem with the term $\arctan(\ldots \tan s/2)$. In Chap. 10, we will show that this parameter s is actually the generalization to any isochrone potential of the Keplerian eccentric anomaly.

9.3 Classification of Isochrone Orbits

Now that we have an analytic expression for any isochrone orbit, we will classify each of them according to the isochrone potential in which it exists. In classical textbooks, the two academic orbits turn out to be ellipses: In the Kepler problem, these ellipses have the origin at one of their focii, and in the harmonic problem, the ellipse is centered on the origin. We will of course recover these results here, and try to exhibit the plethora of orbits arising from all five families of isochrone potentials, one by one.

9.3.1 *General Properties of Isochrone Orbits*

It is well known that in gravitational mechanics, a periodic orbit in a generic radial potential consists in a *rosette* [345]. Although no clear definition of a rosette exists, all have in common a generally not-closed, flower-shaped plane curve that may wrap numerous times around the origin while oscillating between an periapsis and apoapsis. Of course, we will recover all these results here. In Figs. 9.4, 9.5 and 9.7, we depict the orbits of particles in each of the five families of isochrone potentials, such as defined in Sect. 8.3.2. However, all isochrone orbits exhibit similar properties, due to the following fact: Any isochrone potential $\psi(r)$ can be written as

$$\psi(r) = \epsilon + \frac{\lambda}{2r^2} + f(r)\,, \tag{9.33}$$

where f is an increasing function or r, as can be checked on the definitions of the potentials ψ_i in Sect. 8.3.2. This is true for all families but the Hollowed one, the study of which we relegate to Sect. 9.3.5 below.

In the case $\lambda \leq 0$, it is immediate from Eq. (9.33) that ψ will be increasing and therefore be associated with a *gravitational* potential, i.e., with positive mass density ρ. Consequently, the orbiting particle will feel an ever-attracting force and its orbit will be some kind of rosette [345], as is well known in classical mechanics. In particular, the apsidal angle in such a case will always verify $\Theta \geq \pi$ since the particle, when approaching the origin, *misses* it as its angular velocity $v_\theta = \Lambda/r$ increases, while $r \to 0$. Given that the harmonic and Kepler potentials are isochrone, it is no surprise that all isochrone orbits with $\lambda \leq 0$ will be some kind of precessing ellipses, as we shall see in the following sections.

We mention that in the $\lambda \leq 0$ case, any particle (ξ, Λ) with $\Lambda^2 = -\lambda$ will fall toward the center without stopping, and will take an infinite amount of time to reach it, as on the innermost orbit depicted in Fig. 9.4. This is because the $\lambda/2r^2$ term will balance the centrifugal term $\Lambda^2/2r^2$ and the particle will not feel that centrifugal wall anymore. In terms of the parabola, this happens when the line $\mathscr{L}$ intersects the parabola $\mathscr{P}$ once, on the y-axis. This is a generalization of the well-known radial orbits [345], i.e., these with $\Lambda = 0$. When $\lambda = 0$, only particles with $\Lambda = 0$ travel on radial orbits. As their angular velocity $v_\theta = \Lambda/r$ vanishes they go in a straight line to the center. When $\lambda < 0$, although the centrifugal wall is not here anymore, they still have $v_\theta \neq 0$ and will thus inspiral toward the center. These orbits are all depicted as the innermost ones in Figs. 9.4, 9.5 and 9.6.

In the case $\lambda > 0$, Eq. (9.33) shows that ψ will be decreasing around the origin $r = 0$, and therefore be associated with a repulsive force in this region, associated with a negative mass density. The particle will therefore be repelled at periapsis. Far from the origin, however, the potential is always decreasing, and the particle will be attracted at apoapsis. This situation is closer to electrostatics than to gravitational dynamics, and shows that isochrony is not unique to gravitational systems, and can be found in the motion of charged, test particles in central electrostatic potentials. This special property of $\lambda > 0$ potentials will imply that Θ can take value in $[0, \pi]$, and isochrone orbits will be drastically different.

We now turn to the analysis of orbits in each of the five families of isochrone potentials. We stress that the general shape of the orbits can be classified only by the value of the λ parameter. In particular, we will set $\epsilon = 0$ for each potential as it just amounts to re-scaling the potential energy, and with a good choice of units for time and space, we may always set $\omega = 1$ for the harmonic family and $\mu = \beta = 1$ in the Bounded, Hénon, Kepler and Hollowed families. These choices do not change the general characteristics of the orbit. A dynamical system formulation of the problem (detailed in Appendix C.1.1) has been integrated numerically and used to check (and found perfect agreement with) all the isochrone formulae: (9.7) for the radial period T, (9.11) for the apsidal angle Θ and the parametrization (9.32) for the shape of the orbit.

9.3.2 *Harmonic Family*

According to Sect. 8.3.2, a potential ψ in the Harmonic family is given by

$$\psi(r) = \epsilon + \frac{\lambda}{2r^2} + \frac{1}{8}\omega^2 r^2 , \tag{9.34}$$

for some $\omega > 0$ and $(\epsilon, \lambda) \in \mathbb{R}^2$. Given the potential, the values of (ξ, Λ) that yield periodic orbits are given by the inequalities (8.43). With the help of Sect. 8.3.2, we may insert the Greek parameters in place of the Latin ones into Eq. (9.7) and (9.11) to find the period $T(\xi)$ and apsidal angle $\Theta(\Lambda)$ in terms of $\epsilon, \lambda, \omega$

$$T = \frac{2\pi}{\omega}, \quad \text{and} \quad \Theta = \frac{\pi\Lambda}{\sqrt{\Lambda^2 + \lambda}} . \tag{9.35}$$

- In the case $\lambda = 0$, we have $\Theta = \pi$ for all orbits. Up to the additive constant ϵ, ψ_1 is the well-known harmonic (or Hooke) potential. The dynamics can be solved

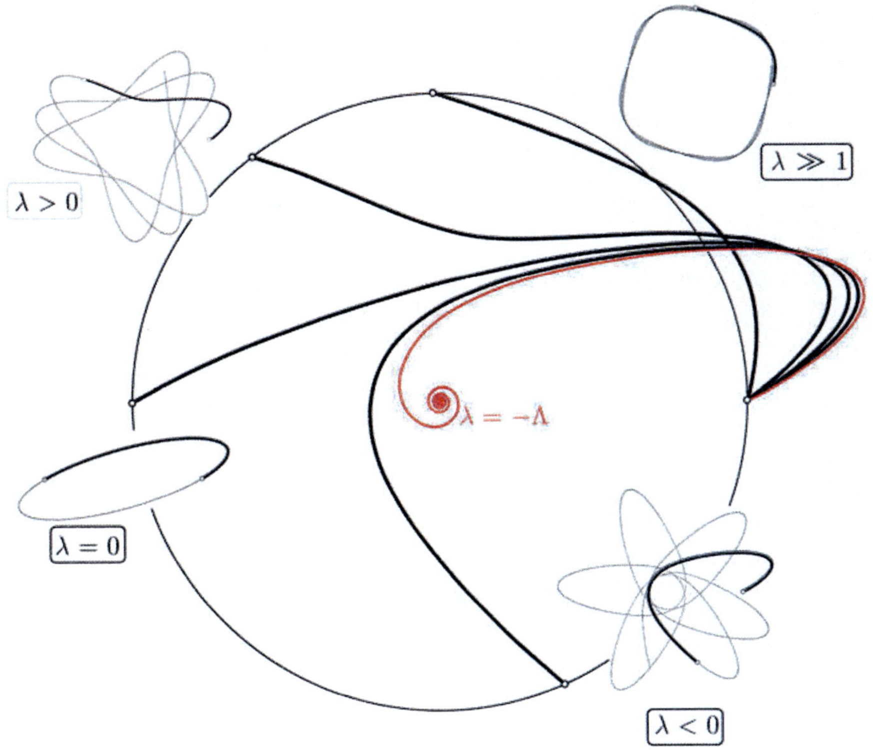

Fig. 9.4 Five isochrone orbits, generated by the same particle (ξ, Λ) orbiting in five different potentials of the harmonic family ψ_1, with varying λ. In the corners, the four orbits are depicted over several periods T. The curves in the middle, aside each orbit in the corner, is the highlighted, first period $[0, T]$ of each orbit, allowing for an easier comparison. Each of these orbits has the same initial position $(r, \theta) = (r_0, 0)$ (black dot on the right) and same initial velocity. The innermost orbit (in red) spirals toward the origin, and the four outer ones are found at $r = r_0$ again (black dots on the dashed circle $r = r_0$) after one radial period T, by definition. The spiraling one corresponds to the $\lambda = -\Lambda^2$ case, discussed in the text

analytically, and the shape of the orbit is an ellipse centered on the origin (bottom left of Fig. 9.4).

• In the case $\lambda < 0$, from Eq. (9.35) we have $\Theta > \pi$. When $\pi < \Theta < 2\pi$, the particle makes less than one turn in one radial period T. One such orbit is depicted on the bottom-right of Fig. 9.4. When $\Theta > 2\pi$ the orbit winds up at least once around the origin, and the winding number can become arbitrarily large with λ.

• In the case $\lambda > 0$, we have $\Theta \in]0, \pi[$. In this peculiar case, the orbits need many periods in order to make a complete turn around the origin. This is because particles are repelled when reaching their periapsis. Such orbits are depicted at the top of Fig. 9.4. When the orbit is close to a circular one this gives rather odd shapes, such as the top-right of Fig. 9.4.

9.3.3 Bounded Family

According to Sect. 8.3.2, a potential ψ in the Bounded family is given by

$$\psi(r) = \epsilon + \frac{\lambda}{2r^2} + \frac{\mu}{\beta + \sqrt{\beta^2 - r^2}}, \tag{9.36}$$

for some $\mu > 0$, $\beta > 0$ and $(\epsilon, \lambda) \in \mathbb{R}^2$. Given the potential, the values of (ξ, Λ) that yield periodic orbits are given by the inequalities (8.46). As established in Sect. 9.1, the period $T(\xi)$ and apsidal angle $\Theta(\Lambda)$ of the orbit are given by

$$T = \frac{\pi}{\sqrt{2}} \frac{\mu}{(\xi - \epsilon)^{3/2}} \quad \text{and} \quad \Theta = \frac{\pi\Lambda}{\sqrt{\Lambda^2 + \lambda}} - \frac{\pi\Lambda}{\sqrt{\Lambda^2 + \lambda + 4\mu\beta}}. \tag{9.37}$$

The most striking feature of orbits in Bounded potentials is the angular, almost non-differentiable, turning point at the apoapsis, as depicted in Fig. 9.5. In fact, these orbits are smooth and we provide some insight as to why they *seem* pointy in Appendix C.2.5. Regarding the classification of orbits, it will be very similar to that of the harmonic family, by examining the function $\Theta(\Lambda)$ given by Eq. (9.37), the properties of which can be found in Appendix C.2.3. We set $\epsilon = 0$ and $\mu = \beta = 1$ in Eq. (9.36) by a good choice of units, and the shape of the orbits depends on the sign of λ.

- In the case $\lambda = 0$, Θ decreases with Λ, but varies in $]0; \pi]$. The particle needs many periods to make a full rotation around the center. If the apoapsis are peaked, then this can lead to peculiar, star-shaped orbits, such as the bottom left one in Fig. 9.5. It is even possible to tune Λ so that Θ is commensurable with π in order to obtain any regular polygon whose vertices are the apsides of the orbit.
- In the case $\lambda < 0$, Θ decreases with Λ and can take arbitrary values in $]0, +\infty[$. As we said for the harmonic family, the orbit may wrap around the origin numerous times in one period, as depicted at the bottom of Fig. 9.5.
- In the case $\lambda > 0$, Θ is not monotonous with respect to Λ. It is increasing from 0 to some maximum value $\Theta_{\max} < \pi$ when Λ equals some critical value Λ_o, and then, it decreases to zero for Λ_o. In particular, all orbits have a maximum apsidal angle that is less than π. However, we are again in the case where the particle is repelled at periapsis, giving the orbits a different look than the $\lambda = 0$ case. Two examples are depicted at the top of Fig. 9.5.

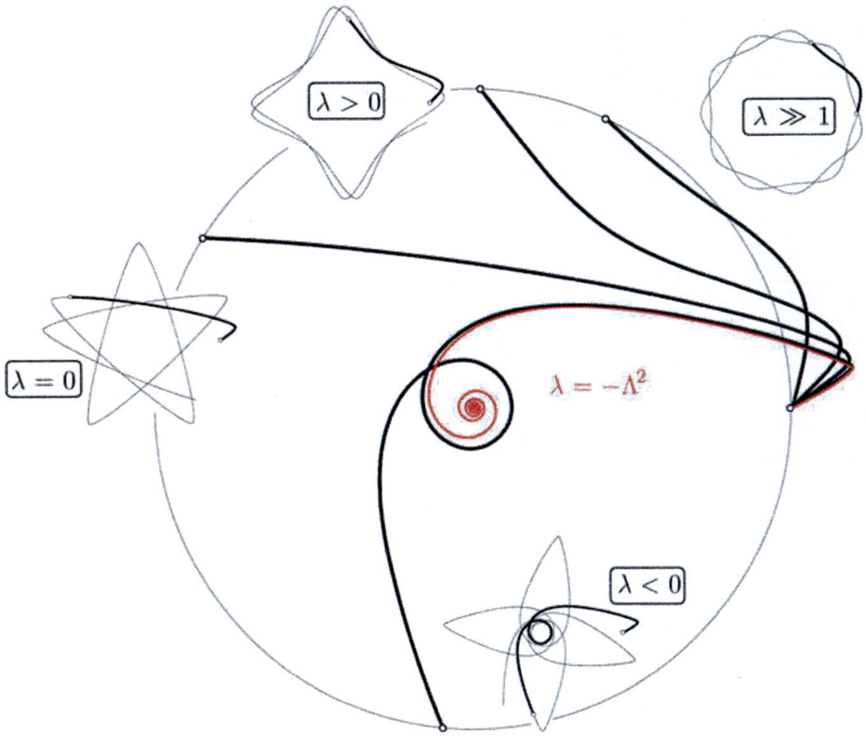

Fig. 9.5 Some orbits in various potentials of the Bounded family ψ_2, with varying λ. See Fig. 9.4 for the explanation of the figure

9.3.4 Hénon Family

According to Sect. 8.3.2, a potential ψ in the Hénon family is given by

$$\psi(r) = \epsilon + \frac{\lambda}{2r^2} - \frac{\mu}{\beta + \sqrt{\beta^2 + r^2}}, \tag{9.38}$$

for some $\mu > 0$, $\beta \geq 0$ and $(\epsilon, \lambda) \in \mathbb{R}^2$. Given the potential, the values of (ξ, Λ) that yield periodic orbits are given by the inequalities (8.46) (with a minus sign in front of μ^2). As established in Sect. 9.1, the period $T(\xi)$ and apsidal angle $\Theta(\Lambda)$ of the orbit are given by

$$T = \frac{\pi}{\sqrt{2}} \frac{\mu}{(\epsilon - \xi)^{3/2}} \quad \text{and} \quad \Theta = \frac{\pi\Lambda}{\sqrt{\Lambda^2 + \lambda}} + \frac{\pi\Lambda}{\sqrt{\Lambda^2 + \lambda + 4\mu\beta}}. \tag{9.39}$$

Regarding the classification of orbits, we apply the same method that we used for the two other families. In particular, the variations of the function $\Theta(\Lambda)$ by Eq. (9.39)

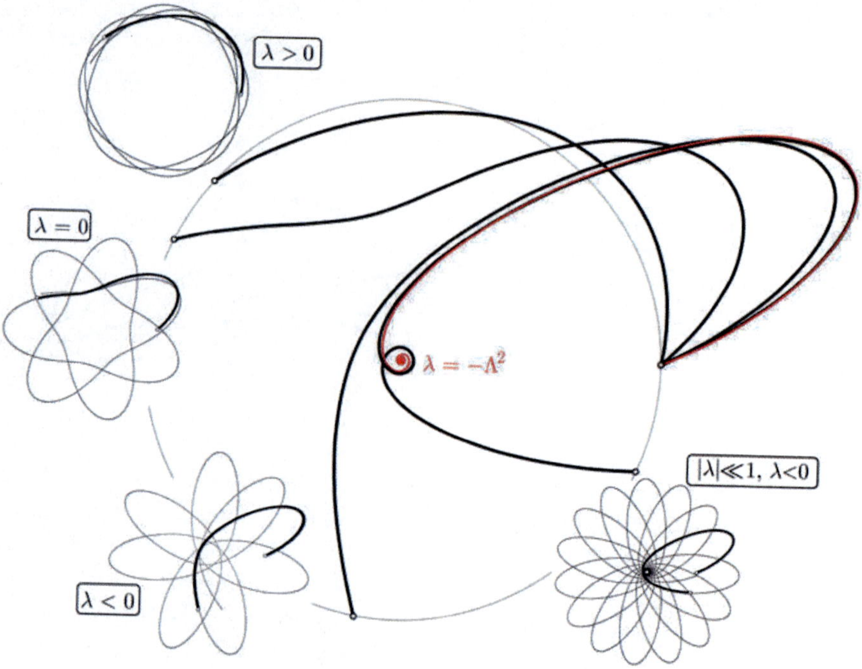

Fig. 9.6 Some orbits in various potentials of the Hénon family ψ_3, with varying λ. See Fig. 9.4 for the explanation of the figure. The Hénon family contains the usual Kepler potential when $(\beta, \lambda) = (0, 0)$, with closed ellipses. Consequently, for $(\beta, \lambda) \neq (0, 0)$ the orbits are similar to precessing ellipses

are given in Appendix C.2.3 and we set $\epsilon = 0$ and $\mu = \beta = 1$ in Eq. (9.36) by a good choice of units. The shape of the orbits depends on the sign of λ.

• In the case $\lambda = 0$, Θ increases with Λ and varies in $[\pi; 2\pi[$. The particle needs at least two periods to make a full rotation around the center. The Kepler potential belongs to the Hénon family with $\lambda = \beta = 0$, and has $\Theta = 2\pi$, recovering the elliptic orbit. It is thus not a surprise that most orbits in the Hénon family resemble precessing ellipses. One such orbit is depicted at the bottom of Fig. 9.6.

• In the case $\lambda < 0$, Θ is, in general, not monotonous with respect to Λ. The precise shape of the function $\Theta(\Lambda)$ can be found in Appendix C.2.2, but generally speaking, Θ is decreasing from $+\infty$ to some minimum value $\Theta_{\text{min}} < \pi$ when Λ equals some critical value Λ_o, and it increases to reach 2π for $\Lambda > \Lambda_o$. For some values of λ, the critical angular momentum Λ_o goes to $+\infty$, and $\Theta(\Lambda)$ is then strictly decreasing, varying between 2π and $+\infty$. In either case, $\Theta > \pi$ and the periapsis can be at an arbitrarily large radius, leading to an orbit with numerous windings around the center, as depicted on the bottom-right of Fig. 9.6.

• In the case $\lambda > 0$, the apsidal angle Θ is strictly increasing between 0 and 2π. This case is peculiar because the shape of the orbit will depend on the location of the periapsis. Indeed, note that since $\lambda > 0$, the potential is always decreasing in some region surrounding the origin. If the periapsis is in this region, then the particle will be repelled, and we will have necessarily $0 < \Theta < \pi$, as usual for repelled orbits. On the contrary, if the periapsis is outside the region where the potential decreases, the particle is always attracted and $\pi < \Theta < 2\pi$.

9.3.5 Hollowed Family

According to Sect. 8.3.2, a potential ψ in the Hollowed family is given by

$$\psi(r) = \epsilon + \frac{\lambda}{2r^2} - \frac{\mu}{r}\sqrt{1 - \frac{\beta^2}{r^2}}, \tag{9.40}$$

for some $\mu > 0$, $\beta \geq 0$ and $(\epsilon, \lambda) \in \mathbb{R}^2$. Given the potential, the values of (ξ, Λ) that yield periodic orbits are given by the inequalities (8.46) (with a minus sign in front of μ^2). The formulae for the period $T(\xi)$ and the apsidal angle $\Theta(\Lambda)$ can be found using the definition of the Greek parameters in terms of the Latin ones, and then inserting these relations in Kepler's third laws (9.7) and (9.11). We find

$$T = \frac{\pi}{\sqrt{2}} \frac{\mu}{(\epsilon - \xi)^{3/2}} \quad \text{and} \quad \Theta = \frac{\pi\Lambda}{\sqrt{\Lambda^2 + \lambda + 2\mathrm{i}\mu\beta}} + \frac{\pi\Lambda}{\sqrt{\Lambda^2 + \lambda - 2\mathrm{i}\mu\beta}}, \tag{9.41}$$

where the imaginary unit i is just here to provide a concise formula (any branch cut of the complex square root can be taken, cf. what is shown in Appendix B.3.1), the result being a real number as can be checked easily. The two conjugate complex numbers under the square root in Eq. (9.41) are reminiscent of the fact that the Hollowed potential has a parabolae does not cross the y-axis. Consequently the equation "Hollowed parabola $= y$-axis" has two solutions that are complex conjugate, whence these two terms. We note that, just as in the Hénon case, we recover the classical relation $\Theta = 2\pi$ for the Kepler potential, corresponding to $\lambda = 0 = \beta$. When written as a purely real number (for example by squaring it), the function $\Lambda \mapsto \Theta(\Lambda)$ is easily shown to be non-monotonous. Again, a proof of this like for the other four cases follows from the same arguments used for the other potentials (as in Appendix C.2.2). Physically, the repulsion of the particle by the $r \leq \beta$ inner region follows the same principle as that of the outer region $r \geq \beta$ of the Bounded potential, owing to the symmetry of this two cases, as previously mentioned.

Consider a particle orbiting the Hollowed potential (9.40), far away from the hollow region, i.e., such that $\beta \ll r$. Then, the potential is approximately Keplerian, so that near the apoapsis the orbit looks like an ellipse. This is what we see in Fig. 9.7, on the right. Now the particle comes back. If it does not go near the $r \simeq \beta$ region, nothing really happens to it: it passes the origin from a distance. But if it approaches $r \simeq \beta$, then we see from Eq. (9.40) that the term on the right behaves as $-2\mu(r - \beta)$, which corresponds to a centrifugal force linear in the distance to the sphere. Since

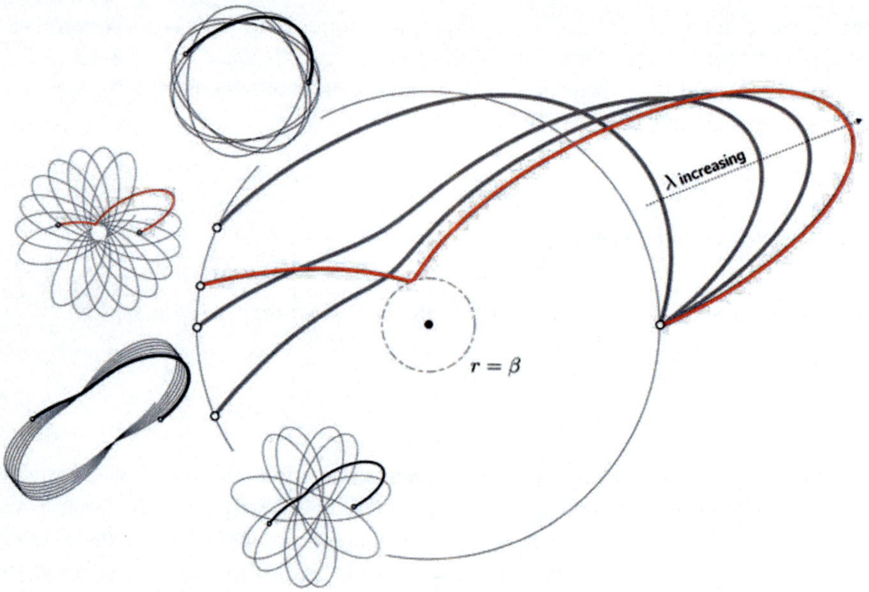

Fig. 9.7 Some orbits in various potentials of the Hollowed family ψ_5, with varying λ, as indicated in the top right corner. See Fig. 9.4 for the explanation of the figure. We clearly see the "wall" at $r = \beta$ where no particle can enter, leaving a hollow center. Particles that orbit close to it are rapidly deflected, so much so that the apsidal angle becomes smaller

all other terms become mere constants, the only thing the particle feels is a repulsive force from the sphere $r = \beta$: it is repelled. This looks like a bounce when the particle comes very close to $r = \beta$, and the turn is very sharp (although still smooth). This is the same thing that happens at the boundary $r = \beta$ from within, in the Bounded potential, as discussed in Sect. 9.3.3. All the qualitative behavior just discussed can be seen on Fig. 9.7.

Chapter 10
Hamiltonian Analysis

Contents

One must tread lightly when asserting that something is not in Poincaré's work.

E. Ghys.

In the previous chapters, we used geometry as a tool to solve a differential system, namely the motion of a test particle in an isochrone potential. From the Archimedean characterization of parabolae to the expression of Kepler's third law in terms of geometric invariants, there does not seem to be a single feature in isochrony that is not expressible in terms of *Euclidean* geometry. However, the motion of particles in isochrone potentials is also (and primarily) a Hamiltonian problem. In fact, one can check that the starting point of all derivations in the last chapter was the conservation

P. Ramond, *The First Law of Mechanics in General Relativity & Isochrone Orbits in Newtonian Gravity*, Springer Theses,
https://doi.org/10.1007/978-3-031-17964-8_10

of the particle's energy. In this tenth and last chapter, we will revisit Hénon's isochrone problem in the light of Hamiltonian mechanics, and thus switch from Euclidean to *symplectic* geometry. First, in Sect. 10.1, we extend the generalization of Keplerian mechanics to all isochrone potentials, namely with the derivation of an isochrone Kepler equation. Then, in Sect. 10.2, we introduce the notion of Birkhoff normal form of a Hamiltonian, and compute the Birkhoff invariants of the isochrone problem. These results are then exploited in Sect. 10.3, where we use these invariants to derive, at once, and in a fully self-consistent way, the fundamental theorem of isochrony, Bertrand's theorem, and the isochrone Kepler's third law.

10.1 New Methods of Isochrone Mechanics

Even though all credit must be given to Lagrange for pioneering the field of analytical mechanics, there is a strong argument for associating Henri Poincaré to its modern formulation. In his celebrated *New Methods of Celestial Mechanics*[1] [599], and the following "Mémoires" [600], Poincaré developed and explored the revolutionary idea of using geometrical and topological techniques to determine the qualitative behavior and global properties of solutions to differential systems, rupturing with the ancestral methods devised to find exact solutions. His ideas were then developed during all the twentieth century, with applications that have proved useful to solve problems well beyond the sole scope of celestial mechanics [345]. Perhaps the most ambitious problem that Poincare's methods was able to tackle was the qualitative resolution of the classical N-body problem, which ultimately gave birth to KAM-theory [601] and the stability analysis of quasi-integrable Hamiltonian systems.

At the core of Hamiltonian mechanics lies the notion of periodic motion. The two canonical examples are the two-body (or Keplerian) problem, and the harmonic oscillator, which we have discussed a lot in the previous chapters. In the gravitational two-body problem of classical mechanics (see Chap. 2 in [602] for a nice presentation), the orbit is a perfect ellipse. Therefore, an explicit, analytic polar equation $r(\theta)$ can be found. However, no analytic solution can be found in the form $(r(t), \theta(t))$, where t is the time. However, it is possible to find a parametric solution for all three, namely $(r(E), \theta(E), t(E))$, in terms of the so-called *eccentric anomaly* E (these classical results are collected in Appendix B.2). In this first section, we derive a series of formulae that are closely related to this parametric solution of the Keplerian problem, but that is actually true of any isochrone orbit. Quite remarkably, all these formulae can be derived analytically in terms of (1) the properties of the particle (ξ, Λ) and (2) the properties of the isochrone potential (a, b, c, d, e). We derive these formulae, and compare them to the Keplerian case to motivate generalized definitions. It should be noted that some of the following results were proposed in slightly different forms as "useful formula for numerical methods" in Appendix A of

[1] To which the title of this section humbly pays tribute.

[587], and in Sect. 5.3 of [602]. In both cases, this concerns only Hénon's potential, and not the whole class of isochrone.

10.1.1 Hamiltonian and Action-Angle Variables

From now on, we consider the isochrone problem from the point of view of Hamiltonian mechanics. But first, let us take a step back and let H be the Hamiltonian a test particle in a generic radial potential $\psi(r)$, not necessarily isochrone. In terms of the polar coordinates adapted to the orbital plane (r, θ), the canonical momenta (p_r, p_θ) simply read $(\dot{r}, \Lambda)$, as is well-known (we set the mass of the particle to 1). The constancy of the angular momentum Λ then follows from the fact that θ is a cyclic variable. In these variables the Hamiltonian reads

$$H(r, \theta, p_r, p_\theta) = \frac{1}{2}\left(p_r^2 + \frac{p_\theta^2}{r^2}\right) + \psi(r)\,, \tag{10.1}$$

Now we consider the problem in terms of action-angle variables. One way of doing this is to consider the two conserved quantities that we already have at hand: the energy ξ and the angular momentum Λ. However, the former is simply the numerical value of the Hamiltonian H. Instead, let us follow Poincaré[2] and define some actions as the integral invariants:

$$J_i \equiv \frac{1}{2\pi}\oint p_i \mathrm{d}q_i\,, \tag{10.2}$$

where $i \in \{r, \theta\}$ and the integral is performed over any closed curve in phase space followed during one orbital transfer (e.g., from one periastron to the following) [345, 591]. For the angular part, this is almost tautological: $J_\theta \equiv \frac{1}{2\pi}\oint p_\theta \mathrm{d}\theta = \Lambda$, which is a constant of motion. For the radial part, a quick computation provides

$$J_r \equiv \frac{1}{2\pi}\oint p_r \mathrm{d}r \quad \Rightarrow \quad J_r = \frac{\sqrt{2}}{\pi}\int_{r_P}^{r_A}\left(\xi - \frac{\Lambda^2}{2r^2} - \psi(r)\right)^{1/2} \mathrm{d}r\,, \tag{10.3}$$

where r_P and r_A are the periastron and apoastron radii, respectively, and to get the second identity we simply integrated p_r as given by inverting Eq. (10.1), and setting $H = \xi$ and $p_\theta = \Lambda$ there. We now have a set of actions which we will denote $(J_r, J_\theta) = (J, \Lambda)$ from now on, for simplicity and without risk of confusion. By definition of these action variables, the Hamiltonian H of the system is independent of their associated coordinates, the "angles", which we will call (z_J, z_Λ). In general, if one want to compute explicitly the Hamiltonian in terms of (J, Λ), the integral (10.3) needs to be inverted. However, for generic radial potentials $\psi(r)$, this integral is seldom expressible in terms of elementary functions. We will now show that, under

[2] Poincaré first introduced these eponymous invariants in Sect. 255 of his New Methods [603].

the assumption that $\psi(r)$ is isochrone, an explicit expression for $H = H(J, \Lambda)$ can be obtained.

One way of showing this would be to (1) take one of the five families of isochrone potentials (given in the Chap. 8, Eqs. (8.34)–(8.39)), (2) compute the periastron and apoastron in terms of (ξ, Λ) by solving the algebraic equation (7.18), (3) insert all these in (10.3) and compute the integral. And then, start over with the other isochrone families. Clearly, this would be very tedious and there must be a way to take advantage of the many symmetries of isochrony to avoid doing all these calculations. The particular symmetry that will help us was already discussed in Chap. 7, but let us present it from scratch, for the reader's ease.

The radial action J generally depends on both constants of motion (ξ, Λ), as clearly expressed in (10.3). This dependence is twofold: direct, from the term $\xi - \Lambda^2/2r^2$ in the integrand; and indirect, from the boundary of the integrals r_P, r_A, that depend on (ξ, Λ) too. If one takes a partial derivative of (10.3) with respect to ξ (or Λ), the indirect dependence yields a vanishing contribution, since the integrand is precisely 0 when evaluated at r_P or r_A (recall Eq. (7.18)). Therefore, only the direct dependence contributes, and we readily find[3]

$$\frac{\partial J}{\partial \xi} = \frac{T}{2\pi} \quad \text{and} \quad \frac{\partial J}{\partial \Lambda} = -\frac{\Theta}{2\pi}\,, \tag{10.4}$$

as follows from comparing the result to the definitions of $T(\xi, \Lambda)$ (7.20) and $\Theta(\xi, \Lambda)$ (7.21). The identities (10.4) are true of any radial potential, not necessarily isochrone. However, in the case of an isochrone potential, we have by definition $T = T(\xi)$ and $\Theta(\Lambda)$. Therefore, the two PDE's in (10.4) are easily integrated and combined to give

$$J(\xi, \Lambda) = \frac{1}{2\pi} \int T(\xi)\, \mathrm{d}\xi - \frac{1}{2\pi} \int \Theta(\Lambda)\, \mathrm{d}\Lambda\,, \tag{10.5}$$

where any antiderivative can be considered at this stage. We emphasize that whereas (10.4) holds for any ψ, Eq. (10.5) only holds for isochrone ψ. To make more progress towards the explicit expression of H, we need to refer to Chap. 9, in particular Sect. 9.1 where we derived the generalized Kepler's third laws for T and Θ. Let us rewrite them here as

$$T^2 = -\frac{\pi^2}{4} \frac{\delta}{(a + b\xi)^3}\,, \quad \text{and} \quad \frac{\Theta^2}{\pi^2 \Lambda^2} = \frac{2b^2\Lambda^2 - d}{b^2\Lambda^4 - d\Lambda^2 + e} + \frac{2b}{\sqrt{b^2\Lambda^4 - d\Lambda^2 + e}}\,, \tag{10.6}$$

where we recall that $\delta \equiv ad - bc > 0$. These "Kepler laws" are valid for *any* particle of energy and angular momentum (ξ, Λ) orbiting in *any* isochrone potential,

---

[3] Formulae (10.4) are true in general, and explicit formulae such as the r.h.s of (10.3) is not necessary to derive (10.4) from $J = \frac{1}{2\pi} \oint p_r \mathrm{d}r$. Fundamentally, this can be understood from the fact that, locally around the equilibrium (circular orbit), the pair (H, t) itself defines symplectic coordinates (see [604] for more details).

parametrized by (a, b, c, d, e) as explained in Sect. 8.3.1. With the help of formulae (10.6), we can integrate explicitly equation in (10.5) and obtain (when $b \neq 0$)[4]

$$J(\xi, \Lambda) = \frac{1}{2b}\sqrt{\frac{-\delta}{a+b\xi}} - \frac{R(\Lambda)}{2b}, \tag{10.7}$$

where for convenience we introduced the function $R(\Lambda)$ independent of ξ and given by[5]

$$R(\Lambda) \equiv \sqrt{2b^2\Lambda^2 - d + 2b\sqrt{b^2\Lambda^4 - d\Lambda^2 + e}}\,. \tag{10.8}$$

It should be noted that while performing the integrals from (10.5) to (10.7), a constant of integration should be included in the latter expression. This constant can be shown to vanish by taking the Keplerian limit, for which on the one hand $(a, b, c, d, e) = (0, 1, -2\mu^2, 0, 0)$, and on the other $J_{\text{Ke}} = \mu/\sqrt{-2\xi} - \Lambda$, as a classical calculation readily shows. Equation (10.7) gives an exact formula for the radial action of all non-harmonic isochrone potentials. For the harmonic class ($b = 0$), the computation is given in Appendix B.2 (see Eq. (B.21) there).

Going back to the Hamiltonian H, for any pair (J, Λ) corresponding to a well-defined orbit, its numerical value is the energy ξ of the particle. Therefore, we may solve equation (10.7) for ξ in terms of (J, Λ), to obtain the expression of $H(J, \Lambda)$. This readily gives

$$H(J, \Lambda) = -\frac{a}{b} - \frac{\delta}{b\big(2bJ + R(\Lambda)\big)^2}, \tag{10.9}$$

with $R(\Lambda)$ given in (10.8). Equation (10.9) provides the general expression for the Hamiltonian of a particle in any non-harmonic isochrone potential, in action-angle variables (see Eq. (B.22) of Appendix B.2 for the harmonic class). In the Keplerian limit $(a, b, c, d, e) = (0, 1, -2\mu^2, 0, 0)$, we recover the Hamiltonian of the classical two-body problem in terms of the Delaunay variables (see, e.g., Eq. (E.1) of [591]). Equation (10.9) coincides with (and generalizes) the Hamiltonian of the Hénon potential as discussed in Sect. 3.5.2 of [591]. In action-angle variables, the equations of motion for the isochrone orbit are in their simplest form, given by the constancy of (J, Λ) and the linear-in-time evolution of the associated angles, namely

$$z_J(t) = \omega_J t + z_J(0) \quad \text{and} \quad z_\Lambda(t) = \omega_\Lambda t + z_\Lambda(0)\,. \tag{10.10}$$

where the Hamiltonian frequencies ω_i read, by definition,

---

[4] Although Eqs. (10.6) hold for any isochrone, including the harmonic family; starting from Eq. (10.7), most expressions differ in the harmonic case, because of the condition $b = 0$. We relegate them in Appendix B.2.

[5] In Eqs. (10.7) and (10.8), the quantities appearing in the square roots are all positive. For the former because of the isochrone Kepler's third law (9.7), and for the latter by geometrical considerations on parabolae.

$$\omega_J \equiv \frac{\partial H}{\partial J} \quad \text{and} \quad \omega_\Lambda \equiv \frac{\partial H}{\partial \Lambda}. \tag{10.11}$$

In action-angle variables, the four-dimensional phase space can be represented by embedding a torus of radii J, Λ in $\mathbb{R}^3$, allowing for a particularly nice representation, as depicted in Fig. 10.1. However, relating the angle variables (z_J, z_Λ) to the polar coordinates (r, θ) remains to be done. In the next section, doing so will enable us to derive a generalization of the Kepler equation and Kepler's third law, as well as true/eccentric anomaly relations. In the mean time, we can do a simple consistency check of our isochrone formulae. A straightforward computation from Eq. (10.9) reveals that the Hamiltonian frequencies read

$$\omega_J = \frac{4\delta}{\left(2bJ + R(\Lambda)\right)^3} \quad \text{and} \quad \omega_\Lambda = \frac{2\delta R'(\Lambda)}{\left(2bJ + R(\Lambda)\right)^3}, \tag{10.12}$$

where $R'(\Lambda) \equiv \mathrm{d}R/\mathrm{d}\Lambda$. Whenever the orbit is closed in real space, it should also be in phase space. Indeed, computing the ratio ω_Λ/ω_J using equations (10.12) and comparing the result to Eq. (10.6) readily gives

$$\frac{\omega_\Lambda}{\omega_J} = \frac{\Theta(\Lambda)}{2\pi}, \tag{10.13}$$

showing that all those isochrone formulae are consistent with one another. It is quite remarkable that all isochrone potentials admit a universal and closed-form expression for their Hamiltonian in action-angle variables. Coupled to the large variety of properties that these potentials and orbits therein offer (recall Sect. 9.3), this allows for a valuable pedagogical tool, e.g., illustrate Hamiltonian mechanics with applications beyond the harmonic oscillator and two-body problem. From a more practical point of view, it is very tempting to build toy-models in terms of isochrone potentials, as analytic expressions are rather rare for entire systems. As we mentioned in Chap. 7, availability of closed-form expressions has probably been the main reason for the success of the Hénon potential. The calculations performed in this section show that this feature holds for all isochrones.

10.1.2 Kepler Equation and Eccentric Anomaly

In Chap. 9, we showed that the equations of motion for a subclass of isochrone orbits (namely those associated with parabolae crossing the origin of the (x, y)-plane) could be integrated analytically in a parametric expression of the type $(r(s), \theta(s))$ for some parameter $s \in \mathbb{R}$. The parameter used in these expression was then a pure mathematical quantity, bearing, a priori, no physical meaning. In particular, these parametric equations were obtained by relating any isochrone orbit to a Keplerian one, through a linear transformation acting on their respective arcs of parabolae, in

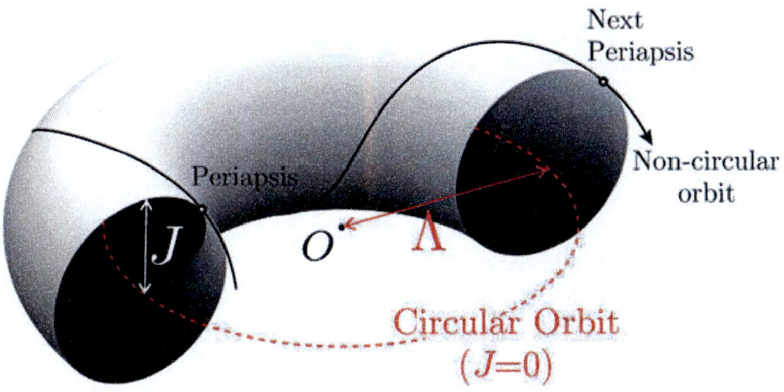

Fig. 10.1 One torus (J, Λ) in the phase space, depicting a generic (non-circular) orbit (black curve). The circular orbit ($J = 0$) with the same angular momentum Λ is depicted in red

the (x, y)-plane. In this section, we show that these formulae can be obtained (1) by direct integration, (2) without making any assumption as to the subclass of isochrone, (3) such that the parameter s admits a clear, physical interpretation.

We suppose that a particle of energy and angular momentum (ξ, Λ) orbits an isochrone potential $\psi(r)$. As argued before, $\psi(r)$ is in a one-to-one correspondence with a convex arc of parabola $y = Y(x)$. We start by deriving an explicit formula for the time t elapsed during orbit. Let T be the radial period and $t \in [0; T/2]$ be an instant between the initial-time periastron $r(0) = r_P$ and apoastron $r(T/2) = r_A$. By isolating $\mathrm{d}t$ in the radial equation of motion in the x variable, Eq. (8.4), and integrating, we readily obtain

$$t = \frac{1}{4} \int_{x_P}^{x} \left(a_0 + a_1 x + \sqrt{a_2 x + a_3}\right)^{-1/2} \mathrm{d}x \,, \tag{10.14}$$

where, for the sake of simplicity, we temporarily introduced the following coefficients that depend on the particle (ξ, Λ) and the potential (a, b, c, d, e):

$$(a_0, a_1, a_2, a_3) = \left(\frac{d}{2b^2} - \Lambda^2, \xi + \frac{a}{b}, \frac{\delta}{b^3}, \frac{d^2 - 4b^2 e}{4b^4} \right). \tag{10.15}$$

The change of variables $u = \sqrt{a_2 x + a_3}$ then turns the term in parenthesis in (10.14) into a pure quadratic, namely

$$t = \frac{\sqrt{u_0}}{\sqrt{2} a_2} \int_{u_P}^{u} \frac{u \, \mathrm{d}u}{\sqrt{v_0 - (u - u_0)^2}} \,, \tag{10.16}$$

where (u_0, v_0) are the coordinates of the apex of that quadratic, given by $u_0 = -a_2/2a_1$ and $v_0 = u_0^2 + 2a_0 u_0 + a_3$. Note that in the u variable, the periastron (lower bound of the integral (10.16)) corresponds to the smallest root of the quadratic in the denominator, namely $u_P = u_0 - \sqrt{v_0}$. To integrate equation (10.16), we start by turning the quadratic $v_0 - (u - u_0)^2$ in canonical form by performing the linear transformation $s = (u - u_0)/\sqrt{v_0}$, so that $s = -1$ when $u = u_P$. This turns (10.16) into

$$\Omega t = \int_{-1}^{s} \frac{1 + \varepsilon s}{\sqrt{1 - s^2}} \, \mathrm{d}s \,, \quad \text{where } \varepsilon = \frac{\sqrt{v_0}}{u_0} \,, \quad \Omega = \frac{\sqrt{2} a_2}{u_0^{3/2}} \,. \tag{10.17}$$

Notice that by construction, $0 < \varepsilon < 1$, since $u_P = u_0 - \sqrt{v_0} > 0$. The integral in (10.17) may then be finally integrated by defining an angle $E \in [0, \pi]$ such that $s = -\cos E$, with $s = 0$ at periastron ($t = 0 \Leftrightarrow s = -1$). Integrating in this fashion, we obtain the sought-after, expression for t, which takes the form of a *generalized Kepler equation*

$$\Omega t = E - \varepsilon \sin E \,. \tag{10.18}$$

Equation (10.18) looks exactly like the Kepler equation (B.19) found in the classical two-body problem. The expression of the constants (Ω, ε) can be given in terms of the generic Latin parameters (a, b, c, d, e) and (ξ, Λ) that characterize the potential and the particle, respectively. These expressions read:

$$\Omega^2 \equiv -16\delta^{-1}(a + b\xi)^3 \,, \tag{10.19a}$$
$$\varepsilon^2 \equiv 1 + 2\delta^{-1}(2b^2\Lambda^2 - d)(a + b\xi) + \delta^{-2}(d^2 - 4b^2 e)(a + b\xi)^2 \,, \tag{10.19b}$$

where $\delta = ad - bc$. Naturally, we may identify ε and E as the *isochrone eccentricity* and *isochrone eccentric anomaly*, respectively. The isochrone eccentricity verifies $0 \leq \varepsilon < 1$, vanishes only for circular orbits and coincides with the Keplerian eccentricity in the Keplerian limit (cf. Appendix B.2). The isochrone eccentric anomaly is a well-defined angle and coincides with its Keplerian counterpart as well (as we will see in the next subsection). It should be stressed that the frequency Ω also coincides with $2\pi/T$, as we see by comparing (10.19a) and (10.6). In other words, the left-hand side of the generalized Kepler equation (10.18) involves the frequency Ω of the radial motion $r(t)$. In particular, combining the angle coordinates (10.13) and the Kepler equation (10.18) provides

$$z_J = E - \varepsilon \sin E \,, \quad z_\Lambda = \frac{\Theta}{2\pi}(E - \varepsilon \sin E) \,, \tag{10.20}$$

where we have set $(z_J, z_\Lambda) = (0, 0)$ at $t = 0$. It is clear from (10.20) that z_J, the angle variable associated to the radial action J, generalizes in fact the Keplerian mean anomaly.

10.1.3 Parametric Polar Solution

Now that an eccentric anomaly E has been introduced via Kepler's equation (10.18), we derive its relation to the orbital radius r (or equivalently $x = 2r^2$) and the polar angle θ.

Radial Motion

For the radial part $r(E)$, we may simply go through the different changes of variables used to compute the integral (10.14) in the last subsection, but in reverse, i.e., $E \mapsto s \mapsto u \mapsto x$. After some easy algebra, we find that[6]

$$x(E) = \frac{4b^2e - d^2}{4b\delta} \pm \frac{\delta}{4|b|(a+b\xi)^2}(1 - \varepsilon \cos E)^2 . \tag{10.21}$$

where $\pm$ corresponds to the sign of b. We note that the first term on the right-hand side is actually x_v, the abscissa of the point with vertical tangent on the parabola introduced in Eq. (8.31). Whenever the potential is Keplerian, then $x_v = 0$ (and $b > 0$) and we recover the classical link $r = \alpha_{\mathrm{Ke}}(1 - \varepsilon \cos E)$, where α_{Ke} is the semi-major axis of the Keplerian ellipse. This motivates the following definition for an *isochrone semi-major axis* α such that the orbital radius $r(E)$ reads

$$r(E) = \sqrt{\frac{x_v}{2} \pm \alpha^2(1 - \varepsilon \cos E)^2} , \quad \text{where} \quad \alpha^2 \equiv \frac{\delta}{8|b|(a+b\xi)^2} . \tag{10.22}$$

This isochrone semi-major axis is, in general, not related to an ellipse axis, as isochrone orbits are not, in general, ellipses. However, it coincides with its Keplerian counterpart in the proper limit, and comparing equations (10.22) with (10.19) reveals the equality

$$\Omega^2\alpha^3 = \sqrt{\frac{\delta}{2|b|^3}} . \tag{10.23}$$

which we recognize as the generalization of the more traditional form of Kepler's third law of motion, relating the (square of) the orbital frequency to the (cube) of the semi-major axis. Indeed, the Keplerian limit $(a, b, c, d, e) = (0, 1, -2\mu^2, 0, 0)$ of Eq. (10.23) gives $\Omega^2\alpha^3 = \mu$, the well-known formulation of Kepler's third law [591]. In fact, the quantity appearing on the right-hand side is exactly the mass parameter μ introduced in the formulae for the isochrone potentials, cf. Eqs. (8.36)–(8.39).

---

[6] There is a subtlety in the case $a_2 < 0$, since then the function $u(x) = \sqrt{c_2 x + c_3}$ is decreasing. This is resolved by keeping track of $\mathrm{sign}(a_2) = \mathrm{sign}(b)$, which results in the $\pm|b|$ in (10.21).

Angular Motion

For the angular motion, we adapt the strategy developed in Sect. 9.2.4 and first construct an ODE to which $\theta(E)$ is a solution. We can do this with the Leibniz rule as follows

$$\frac{\mathrm{d}\theta}{\mathrm{d}E} = \frac{\mathrm{d}\theta}{\mathrm{d}t}\frac{\mathrm{d}t}{\mathrm{d}E} = \frac{\Lambda}{\Omega}\frac{1-\varepsilon\cos E}{r(E)^2}, \tag{10.24}$$

where in the second equality we used the angular equation of motion $\dot{\theta} = \Lambda/r^2$ and the Kepler equation (10.18). To obtain an expression $\theta(E)$ from (10.24), we simply need to inject the expression of $r(E)$ given in (10.22) and integrate the result. By doing so, we readily obtain

$$\theta(E) = \frac{2\Lambda}{\Omega}\int_0^E \frac{1-\varepsilon\cos\phi}{x_v + 2\alpha^2(1-\varepsilon\cos\phi)^2}\mathrm{d}\phi, \tag{10.25}$$

where we recall that α was defined in (10.22) and x_v in (8.31). When $x_v \leq 0$, which corresponds to the Hénon class of potentials, the integral can be easily integrated. Indeed, a partial fraction decomposition gives

$$\theta(E) = \frac{\Lambda}{2\Omega\alpha^2}\sum_{\pm}\frac{1}{1\pm\zeta}\int_0^E \frac{\mathrm{d}\phi}{1-\varepsilon_\pm\cos\phi}, \tag{10.26}$$

where $\varepsilon_\pm = \varepsilon/(1\pm\zeta)$ with $\zeta^2 = -x_v/2\alpha^2$ and is such that $0 \leq \zeta \leq \varepsilon < 1$, so that the integrals are well-defined. Equation (10.26) can be further simplified by computing explicitly the integral, and thus provides the final formula for θ in terms of E, namely

$$\theta(E) = \frac{\Lambda}{\Omega\alpha^2}\sum_{\pm}\frac{\varepsilon_\pm}{\sqrt{1-\varepsilon_\pm^2}}\arctan\left(\sqrt{\frac{1+\varepsilon_\pm}{1-\varepsilon_\pm}}\tan\frac{E}{2}\right). \tag{10.27}$$

The Keplerian limit of Eq. (10.27) consists in taking $\zeta \propto x_v = 0$ such that $\varepsilon_\pm = \varepsilon$, and thus provides a sum of two identical terms, recovering the well-known Keplerian result (cf. Eq. (B.18)). For consistency, one can check that when $E = \pi$, which should correspond to the apoastron of the orbit, Eq. (10.27) coincides with the general expression (10.6) of the apsidal angle Θ, which satisfies $\theta(\pi) = \Theta/2$ by definition.

As final word, let us mention that the derivation of formula (10.27) relies on the crucial assumption that $x_v \leq 0$, and thus only holds for the Hénon class of isochrone potentials (recall Fig. 8.7). When $x_v \geq 0$, the auxiliary quantity ζ, defined by $\zeta^2 = -x_v/2\alpha^2$, becomes imaginary and the $\varepsilon_\pm$ are now complex. It turns out that formula (10.27) also holds for this case. The reason is that, although both terms $+$ and $-$ in the $\sum_\pm$ sum are complex numbers, they are conjugate to one another. Their sum is therefore twice their real part, and is thus a real quantity. We provide the details of this in Appendix B.3.1. In particular, formula (10.27) for imaginary ζ is mathematically well-defined and the resulting $(r(E), \theta(E))$-orbit does coincide with the true isochrone dynamics. This "complex" number feature is reminiscent of the comment made below Eq. (9.41), regarding the apsidal angle in Hollowed potentials.

Once again, we end this section by a summary of the results. The equations of motion for a test particle in any isochrone potential can be solved analytically in the parametric form $(r(E), \theta(E))$ where E is a parameter that reduces to the Keplerien eccentric anomaly. These equations are given in (10.22) and (10.27). The polar coordinates (r, θ) along the orbit can also be related to orbital time t through a generalization of the Kepler equation (10.18), that holds for any isochrone orbit. Finally, we have shown that the radial action variable J is particularly well-adapted to the isochrone problem, as it (1) splits into a sum of ξ- and Λ-dependent terms and (2) can be used to derive the general Hamiltonian of the dynamics in action-angle variables (J, Λ).

10.2 Birkhoff Normal Forms and Invariants

The fundamental theorem of isochrony (8.22) is what allows one to derive all the analytical results for isochrone potentials and orbits therein, as we did in Sect. 10.1. As we mentioned earlier, this theorem was first proven by Michel Hénon in his seminal paper [583], although not without some (minor) mistakes. It was then discussed in [589] by borrowing techniques from complex analysis, and in Sect. 8.2.2 of the previous chapter we provided a proof with using Archimedean characterization. However, as we also mentioned there, Archimedes' original result was only an implication (In any case, at present, a self-consistent and natural proof of this central theorem relying only on classical mechanics, is nowhere to be found, to our knowledge. It is our goal, in the present and following sections to introduce and exploit a powerful tool of Hamiltonian mechanics: the *Birkhoff normal form*. In a nutshell, the Birkhoff normal form allows one to (quantitatively and rigorously) probe the neighborhood of equilibrium points in phase space, to obtain information on their stability, and thus on the integrability of the underlying Hamiltonian.

There exists a lot of specialized literature on this topic. Yet, introductory material on normal forms may be hard to find for non-specialists. Among the most accessible, we found that Arnold's classical textbook (Appendix 7 of [345]) and Hofer & Zehnder's lectures ([605], Sects. 1.7 and 1.8) are particularly relevant (see also Sect. (8.5) of [606, 607]). Other examples of accessible presentations (with applications) may be found in [602] (for the stability of the Lagrange points), and [608] (for solving PDE's). Other notable references, namely [609–611], present explicit computations of normal forms and use them to study the stability of the (restricted) N-body problem. The latter have largely motivated the present derivation.

Applications of our method go beyond the sole fundamental theorem of isochrony (8.22), as it allows us to prove the Bertrand theorem [345], as well as the generalized Kepler's third laws (10.6). We relegate these applications to Sect. 10.3, and only focus on the derivation of the normal form in the present section, which we organize as follows:

- in Sect. 10.2.1, we introduce the notion of Birkhoff normal form and Birkhoff invariants in a very simple case, sufficient for our purpose,
- in Sect. 10.2.2, we write the Birkhoff normal form N_1 for the Hamiltonian of a particle in a *generic* potential, which encodes information on the potential $Y(x)$,
- in Sect. 10.2.3, we write the Birkhoff normal form N_2 for the Hamiltonian of a particle in an *isochrone* potential, using the radial action (10.3) well-adapted to isochrony.

10.2.1 Birkhoff Normal Form and Invariants

Note: *The theory of Birkhoff normal forms is a rather technical, sub-branch of Hamiltonian dynamical systems. As a consequence, it is likely that the following presentation will (sometimes) lack mathematical rigor, but we chose a pedagogical style instead, and look at it from a physically-motivated point of view. For this reason, and as done several times already in this thesis, we shall start with a fundamentals: the pendulum.*

Pendular Motivation

Usually, the starting point of a Hamiltonian treatment of a mechanics problem is an expression written in terms of (phase-space) coordinates that are linked, somehow, to the physical coordinates (position in Euclidean space, for example). The point of using Hamiltonian mechanics, however, is that there exists coordinate transformations that leave the equations of motion invariant, the so-called canonical transformation. Let us consider, as an example, the classical pendulum. The Hamiltonian is the sum of the kinetic energy and the gravitational potential energy. Keeping the notations introduced in Chap. 7 alongside Galileo and Huygens, we have[7]

$$H(\theta, \omega) = \frac{1}{2}\omega^2 + (1 - \cos\theta)\,, \tag{10.28}$$

where $\omega \equiv \dot{\theta}$ is the momentum (angular velocity), conjugated to the position variable θ (angle). Around the $(\theta, \omega) = (0, 0)$ equilibrium position (pendulum at rest pointing down), we may expand the rightmost term of side of (10.28) to get

$$H(\theta, \omega) = \frac{1}{2}\omega^2 + \frac{1}{2}\theta^2 + o(\theta^2)\,. \tag{10.29}$$

[7] We chose our units of mass, time and length so that m, g and ℓ are equal to unity.

If we neglect the $o(\theta^2)$ term, then we are in the small-angle approximation. If we picture the phase space of the system described by (10.28), then Eq. (10.29) simply says that around the point $(\theta, \omega) = (0, 0)$, the level curves of the hamiltonian, say $H = \rho$ for some fixed $\rho \in \mathbb{R}$, are given by the equation $\omega^2 + \theta^2 = 2\rho$. If we consider (θ, ω) to be some kind of Cartesian coordinates, then this equation is that of a circle, with radius $\sqrt{2\rho}$. These circles are only approximate, however, owing to the $o(\theta^2)$ term. Nevertheless, this observation motivates the introduction of $\rho \equiv \frac{1}{2}(\theta^2 + \omega^2)$ as an independent variable, that labels the level curves of H according to the actual value of H along these curves. If we promote ρ to a new momentum variable, then let us chose the associated "position" φ so as to have a canonical transformation. It is straightforward, with the Poisson brackets for example, to show that the complete canonical transformation $(\theta, \omega) \mapsto (\rho, \varphi)$ reads

$$\theta = \sqrt{2\rho}\cos\varphi \quad \text{and} \quad \omega = -\sqrt{2\rho}\sin\varphi \tag{10.30}$$

We can now re-express the Hamiltonian in Eq. (10.29) in terms of these new variables, and since $\theta^2 = O(\rho)$, we readily find

$$H(\rho, \varphi) = \rho + o(\rho, \phi) \tag{10.31}$$

Now, notice that, by construction, up to very small terms hidden in the $o(\rho, \phi)$ this Hamiltonian does not depend explicitly on the variable ϕ. Consequently, in the small-oscillation approximation, (ρ, ϕ) are action-angle variables for the pendulum. This method of constructing approximate action-angle variables around an equilibrium point will be central to the following calculation, and is at the basis of the theory of Birkhoff invariants, which we will now introduce.

Definitions

The calculation performed in the previous section was made for a simple system: the classical pendulum. Given the Hamiltonian (10.28) and the equilibrium point (0, 0), we were able to construct approximate angle-action variables (ρ, φ) in the phase space, around that point. However, the method can be extended to a much wider class of Hamiltonian systems. For the sake of simplicity, we will only cover the very basics of Birkhoff normal forms and refer to the above literature for the details. In particular, we consider a 1-dimensional problem (2-dimensional phase space), but all can be generalized to any $2n$-dimensional phase space (see, e.g., Sect. 1.8 of [605]). Let $H(q, p)$ be a Hamiltonian defined in terms of some coordinates $(q, p) \in \mathbb{R}^2$. We assume, without any loss of generality, that the origin $(q, p) = (0, 0)$ is an elliptic[8] equilibrium of the system. A fundamental theorem of Hamiltonian systems, due

[8] We focus on elliptic equilibria since we want to describe the stable, periodic motion of the particle. Hyperbolic equilibria will in general be associated to semi-stable or unstable systems.

initially to Birkhoff[9] [613] and then refined/generalized since then (see [605] for references). This theorem essentially says that if the Hamiltonian behaves nicely around the equilibrium, *there exists a local, symplectic coordinate transformation* $(q, p) \mapsto (\rho, \varphi)$ *such that the Hamiltonian takes the form*

$$H(\rho, \varphi) = \mathfrak{l} + \mathfrak{b}\rho + \frac{1}{2}\mathfrak{B}\rho^2 + o(\rho^2)\,. \tag{10.32}$$

The real numbers $(\mathfrak{l}, \mathfrak{b}, \mathfrak{B})$ will be called *Birkhoff invariants* of zeroth, first and second order, respectively. In the general case where $(q, p) \in \mathbb{R}^n \times \mathbb{R}^n$, then $\rho \in \mathbb{R}^n$ and, accordingly, the Birkhoff invariants $\mathfrak{b}$ and $\mathfrak{B}$ are linear and bilinear forms, respectively. The Birkhoff invariants depend exclusively on H and not on the mapping $(q, p) \mapsto (\rho, \varphi)$, whence their name. They encode the information on the geometry of phase space around the equilibrium point. In whole generality, Birkhoff's theorem gives much stronger results than the result (10.32). In particular, it holds for any dimensions, explains how to extend (10.32) to any order in the powers of ρ, makes a distinction between resonant or non-resonant orbits, etc. However, quadratic order will be sufficient for our purposes, and we will refer the reader to Sect. (1.8) of [605] (and references therein) for a more detailed and rigorous discussion.

As we explained for the pendulum earlier, the main feature of the variables (ρ, φ) in (10.32) is that, up to $o(\rho^2)$ corrections, they are action-angle coordinates, as H does not depend on the angle φ. From now on, we call *Birkhoff normal form* (BNF)[10] of H, and denote by $N(\rho)$, the quadratic part of (10.32), namely

$$N(\rho) = \mathfrak{l} + \mathfrak{b}\rho + \frac{1}{2}\mathfrak{B}\rho^2\,. \tag{10.33}$$

Heuristically, the quantity $N(\rho)$ in (10.33) can be seen as a Hamiltonian that is (1) completely integrable and (2) describes the same dynamics as H in (10.32) in the $O(\rho^2)$-neighborhood of the equilibrium $(0, 0)$. Owing to the unicity of the Birkhoff invariants, the BNF (10.33) is itself unique, in the sense that any change of action-angle coordinates that leaves the equilibrium point at the origin must be the identity (see [614] for a detailed presentation as well as Appendix B.3.3). A natural method to construct a BNF is crystallized in Fig. 10.3, which shows the successive steps one may use to transform the geometry of the phase space (q, p) around $(0, 0)$, so as to introduce polar-symplectic coordinates (ρ, φ) (cf. Sect. 10.2.2). In fact, in Sect. 10.2.2 we explicitly provide a constructive example of such $(q, p) \mapsto (\rho, \varphi)$ mapping, that brings $H(q, p)$ into BNF (10.32).

The general picture to have in mind while working with a BNF is depicted in Fig. 10.2. It shows how, locally, the action-angle variables (ρ, φ) distort phase space so as to make the neighborhood of the equilibrium mimic that of the harmonic

[9] Obviously (cf. this chapter's main quote), some of Birkhoff's ideas on normal forms can be traced back to Poincaré's *New Methods*, as mentioned in [603, 612].

[10] BNF is also the acronym for the Bibliothèque Nationale de France (National French Library), where (almost) all of Poincaré's writings may be found and borrowed, cf. this webpage.

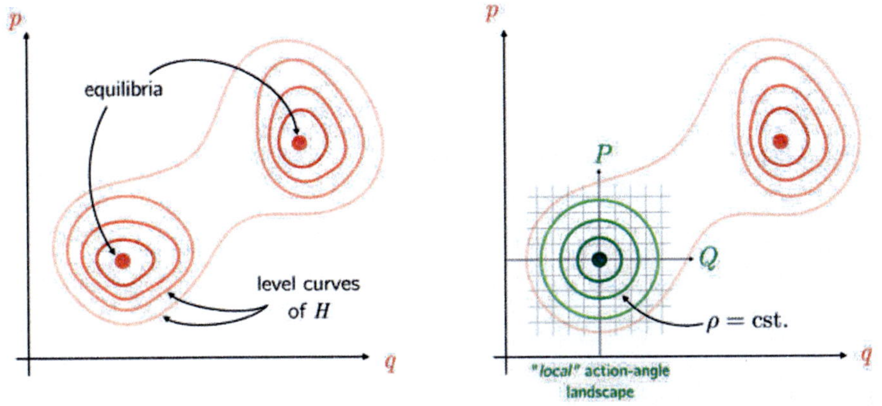

Fig. 10.2 *Left:* A phase space in arbitrary (p, q) with two equilibrium points. Periodic motion corresponds to closed curve that are level sets of the Hamiltonian H. *Left:* the construction of local action-angle variables around one of the two equilibria. The action $\rho = \frac{1}{2}(P^2 + Q^2)$ is defined from other coordinates (P, Q), and is related to the radius of the (approximate) circles, which are the level sets of H in the (P, Q) variables

oscillator. In the next section, we will construct these variables from scratch, in successive steps. The effect of each step will be to distort phase space more and more until we reach the BNF. These successive distortions are what is presented in Fig. 10.3, although this figure will be re-discussed in the following section.

10.2.2 BNF for a Generic Radial Potential

Let us start with the Hamiltonian of a particle in a radial potential $\psi(r)$, as given by (10.1), rewritten here for convenience as:

$$H(r, R, \theta, \Lambda) = \frac{R^2}{2} + \frac{\Lambda^2}{2r^2} + \psi(r)\,, \tag{10.34}$$

where (r, θ) are the coordinates and (R, Λ) their conjugated momenta. The complete, 4-dimensional phase space of the dynamics is a subset of $\mathbb{R}_+ \times \mathbb{R} \times [0; 2\pi[\times\mathbb{R}_+ \ni (r, R, \theta, \Lambda)$. However, since θ does not appear explicitly in (10.34) and its conjugated momentum Λ is then constant, (θ, Λ) is already a pair of angle-action coordinates. Therefore, it can be practical to think of (10.34) as a 1-dimensional family of Hamiltonians parametrized by Λ. In this way, we need just focus on the radial part (r, R) of the dynamics, and perform successive symplectic transformations to reach a BNF, the coefficients of which will thus be Λ-dependent. With this *2-dimensional phase space* point of view, we will write $H(r, R)$ instead of $H(r, R, \theta, \Lambda)$ to stick with the notations of Sect. 10.2.1, with no risk of confusion. Moreover, while performing

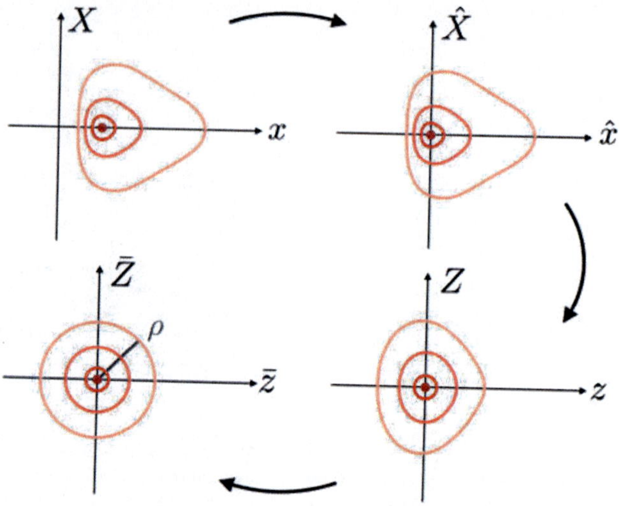

Fig. 10.3 The circular orbit (red point) and three non-circular orbits (red curves) in the 2-dimensional phase space under each transformation. The map $(x, X) \mapsto (\hat{x}, \hat{X})$ translate x_c at the origin, and $(\hat{x}, \hat{X}) \mapsto (z, Z)$ circularizes the orbits in the only in the close vicinity of the origin (the outer curves are not circular). Then $(z, Z) \mapsto (\bar{z}, \bar{Z})$ circularizes a larger neighborhood of the origin (all curves circular up to $O(\rho^2)$) allowing one to construct polar action-angle variables (ρ, φ) in that region

successive symplectic changes of coordinates on the phase space $(r, R) \in \mathbb{R}_+ \times \mathbb{R}$, we will keep the (lower case/upper case) notation for a coordinate $(r, x, z, \ldots)$ and its conjugated momentum $(R, X, Z, \ldots)$.

Although we have tried to be as pedagogical as possible (and we believe these computations are interesting in themselves), the following subsections are rather technical. On first reading (or for the reader in a hurry), it is possible to skip the following steps and just assume that there exists a pair of variables (ρ, φ), such that the Hamiltonian (10.34) admits a BNF $N_2(\rho)$, given by Eq. (10.49) below; before directly proceeding with Sect. 10.2.1.

Hénon Variable and Circular Orbits

Note: from this point on we denote by a lowercase subscript c any quantity related to the circular orbit, such as $(r_c, x_c, \xi_c, \ldots)$, and not an uppercase C like we did before.

The Hamiltonian (10.34) describes the same system as the Hamiltonian in (10.51) only if the potential is isochrone. Since the following computations are true for any radial potential (not just isochrones) we stay general and relegate the isochrone assumption to the next subsection. Still, as we have the isochrone theorem (8.22) in mind, we would prefer to speak in terms of $(x, Y(x))$ instead of $(r, \psi(r))$. Therefore, the first step is the change of variables $(r, R) \mapsto (x, X)$, where $x = 2r^2$ and X is the canonical momenta associated to x. The transformation is easily seen to be symplectic if and only if $R = \sqrt{8x}X$. In these variables, the Hamiltonian (10.34) now reads

$$H(x, X) = 4xX^2 + \frac{\Lambda^2}{x} + \frac{Y(x)}{x}. \tag{10.35}$$

where we recall that $Y(x) = x\psi(r(x))$. The derivation of a BNF starts with a choice of equilibrium point around which to write it. Using (10.35) for a given Λ, these points are simply given by $(x, X) = (x_c, 0)$, where $x_c = x_c(\Lambda)$ is a solution to the algebraic equation

$$x_c Y'(x_c) - Y(x_c) = \Lambda^2. \tag{10.36}$$

At this equilibrium, we have $X = 0 \propto \dot{r} = 0$, thus corresponding to circular orbits, the radius r_c of which is such that $x_c = 2r_c^2$ and x_c solves (10.36). In the complete 4-dimensional phase space, there is a family of such circular orbits, parametrized by Λ.

Moving the Equilibrium at the Origin

Now we are going to write the BNF of H around a given circular orbit $(x, X) = (x_c, 0)$. The main goal is to fix a Λ and to circularize the phase space around the equilibrium $(x_c, 0)$, in order to introduce symplectic polar coordinates, following the discussion in Sect. 10.2.1.

We start by translating the equilibrium $(x, X) = (x_c, 0)$ to the origin, by setting $(\hat{x}, \hat{X}) = (x - x_c, X)$ and then Taylor-expanding in the $\hat{x}$ variable, small by assumption. We obtain[11] the following expression

$$H(\hat{x}, \hat{X}) = Y_1 + 4x_c\hat{X}^2 + \frac{Y_2}{2x_c}\hat{x}^2 + 4\hat{X}^2\hat{x} + c_3\hat{x}^3 + c_4\hat{x}^4 + o(\hat{x}^4), \tag{10.37}$$

where we introduced the convenient notation $Y_n \equiv Y^{(n)}(x_c)$, and defined the following coefficients that depend on the derivatives of $Y(x)$ at $x = x_c$, namely

$$c_3 = \frac{x_c Y_3 - 3Y_2}{6x_c^2}, \quad \text{and} \quad c_4 = \frac{12Y_2 - 4x_c Y_3 + x_c^2 Y_4}{24x_c^3}. \tag{10.38}$$

In these variables, the circular orbits are at the origin $(\hat{x}, X) = (0, 0)$, and in the 4D phase space each coefficient depends on Λ through $x_c = x_c(\Lambda)$ (recall Eq. (10.36)). Notice that the energy of the circular orbit is $H(0, 0) = Y_1 = Y'(x_c(\Lambda))$. This is in

[11] In the 4D phase space, this change of variable is rendered symplectic by changing the angle accordingly. For example, the mapping $(x, \theta, X, \Lambda) \mapsto (x - x_c(\Lambda), \hat{\theta}, \hat{X}, \Lambda)$ is symplectic if we take $\hat{\theta} = \theta - x_c'(\Lambda)X$.

agreement with the way orbits are constructed in the Hénon variables, as we explained back in Chap. 8, see Fig. 8.1 there.

Lastly, and for the sake of completeness, let us deduce from (10.37) the nature of the equilibrium $(\hat{x}, \hat{X}) = (0, 0)$. Writing the Hamilton equations and linearizing around $(0, 0)$ readily gives

$$\frac{d\hat{x}}{dt} = \frac{\partial H}{\partial \hat{X}} = 8x_c\hat{X} + o(x, X)\,, \qquad \frac{d\hat{X}}{dt} = -\frac{\partial H}{\partial \hat{x}} = -\frac{Y_2}{x_c}\hat{x} + o(x, X)\,. \tag{10.39}$$

Now, since the potential $x \mapsto Y(x)$ must be convex, (again, recall the construction of an orbit, cf. Fig. 8.1), it is clear hat we must have $Y_2 \geq 0$. Consequently, the eigenvalues (ℓ_1, ℓ_2) of the linearized system (10.39) are

$$\ell_1 = \mathrm{i}\sqrt{8Y_2} \quad \text{and} \quad \ell_2 = -\mathrm{i}\sqrt{8Y_2}\,. \tag{10.40}$$

These eigenvalues are conjugate, imaginary numbers, allowing us to conclude that the equilibrium $(\hat{x}, \hat{X}) = (0, 0)$ is, indeed, elliptic, as our use of the BNF requires.

Circularizing the Equilibrium Neighborhood

Next, notice that the quadratic part of H in (10.37) describes ellipses in the $(\hat{x}, \hat{X})$-plane. As we aim, eventually, towards a polar-like system of action-angle coordinates, we would like to *circularize* these ellipses; that is, have the same coefficients in front of $\hat{x}^2$ and $\hat{X}^2$ in Eq. (10.37). This can be done easily by yet another change of variables. Explicitly, we set $(\hat{x}, \hat{X}) = (\eta z, \gamma Z)$ (a homothety for fixed Λ) and choose (η, γ) such that: (i) the transformation is symplectic, and (ii) the coefficients in front of z^2 and Z^2 are equal in the new variables. A calculation reveals that condition (i) holds if $\gamma = 1/\eta$, while condition (ii) holds if we set $\eta^4 = 8x_c^2/Y_2$. Expressing the Hamiltonian with the new (z, Z)-variables,[12] we find

$$H_\Lambda(z, Z) = Y_1 + \sqrt{2Y_2}\big(z^2 + Z^2 + c_0 z Z^2 + c_1 z^3 + c_2 z^4\big) + o(z^4)\,, \tag{10.41}$$

where we see that our phase-space ellipses have indeed been circularized, and where we defined new coefficients (c_0, c_1, c_2) by

$$c_0 = \frac{8^{1/4}}{Y_2^{1/4} x_c^{1/2}}\,, \quad c_1 = \frac{8^{1/4}}{3}\frac{x_c Y_3 - 3Y_2}{x_c^{1/2} Y_2^{5/4}} \quad \text{and} \quad c_2 = \frac{8^{1/2}}{12}\frac{12Y_2 - 4x_c Y_3 + x_c^2 Y_4}{x_c Y_2^{3/2}}\,. \tag{10.42}$$

One more time, we emphasize that, in the complete, 4-dimensional phase space, these coefficients all depend on Λ, through $x_c(\Lambda)$ and $Y_n(x_c(\Lambda))$. Next, we simplify the (z, Z)-dependent part in the parentheses of (10.41).

---

[12] We would also need to alter the angle $\hat{\theta} \mapsto \hat{\theta} + \frac{\eta'(\Lambda)}{\eta(\Lambda)}\hat{x}$, to ensure symplecticity in the 4D phase space.

Flowing Towards the BNF

For the moment, let us rewrite (10.41) in the form $H = Y_1 + \sqrt{2Y_2}\tilde{H} + o(z^4)$, where

$$\tilde{H}(z, Z) = z^2 + Z^2 + c_0 z Z^2 + c_1 z^3 + c_2 z^4 . \qquad (10.43)$$

As our final aim is to introduce polar-type coordinates, we would like $\tilde{H}$, a polynomial in (z, Z), to be written solely as powers of $\rho = z^2 + Z^2$, which would then correspond to the radial part of the polar-type coordinates. The best way to massage $\tilde{H}$ into this form is to make a transformation derived from the flow of another (polynomial) Hamiltonian Φ. Let us take a moment to explain this method more clearly.

Using the flow of a secondary Hamiltonian can be viewed as a very general procedure to produce symplectic transformations $(z, Z) \mapsto (\bar{z}, \bar{Z})$. Let $\Phi(z, Z)$ be some arbitrary Hamiltonian, and let ϕ_t be the *flow* associated to Φ, such that $\phi_t : (z, Z) \mapsto (\bar{z}, \bar{Z}) = (z(t), Z(t))$, where $(z(t), Z(t))$ is the solution to Hamilton's equations for Φ. For $t = 1$, the map $\phi \equiv \phi_1$ is appropriately called the *time-one flow*, as it sends a point $(z, Z) = (z(0), Z(0))$ (corresponding to some initial condition $t = 0$) to some other point $(\bar{z}, \bar{Z}) = (z(1), Z(1))$ (corresponding to its updated value at $t = 1$). Choosing Φ in the right way allows one to determine the dynamics between $t = 0$ and $t = 1$, and thus select the image $(\bar{z}, \bar{Z})$ of each point (z, Z). By construction, this mapping ϕ defines a symplectic transformation on the phase space, because it derives from a Hamiltonian system.

Returning to our problem, an explicit computation adapted from [609] (with a slight adjustment for the cross term zZ^2 in (10.43) absent there) shows that if $\tilde{H}(z, Z)$ is of the form (10.43), then the time-one flow ϕ of a well-chosen[13] $\Phi(z, Z)$ defines a set of coordinates $(\bar{z}, \bar{Z})$ precisely such that the polynomial Hamiltonian (10.43) now reads, in the "bar" variables:

$$\bar{H}(\bar{z}, \bar{Z}) = \bar{z}^2 + \bar{Z}^2 + C\,(\bar{z}^2 + \bar{Z}^2)^2 + O(5)\,, \qquad (10.44)$$

where $O(5)$ contains terms of order 5 or more in $(\bar{z}, \bar{Z})$, and C is expressed in terms of the constants appearing in (10.43), by

$$C = -\frac{3}{32}(5c_1^2 - 4c_2^2 + 2c_1 c_0 + c_0^2)\,. \qquad (10.45)$$

We can now go back to the original Hamiltonian (10.41) and express it in the new variables $(\bar{z}, \bar{Z})$. To this end, we replace $\tilde{H}(z, Z)$ in (10.41) (recall that $H = Y_1 + \sqrt{2Y_2}\tilde{H} + o(z^4)$) by $\bar{H}(\bar{z}, \bar{Z})$ as given in (10.44). We eventually find

$$H(\bar{z}, \bar{Z}) = Y_1 + \sqrt{2Y_2}(\bar{z}^2 + \bar{Z}^2) + C\sqrt{2Y_2}\,(\bar{z}^2 + \bar{Z}^2)^2 + O(5)\,. \qquad (10.46)$$

[13] Explicitly, $\Phi(z, Z) = b_1 Z z^2 + b_2 Z^3 + b_3 Z z^3 + b_4 z Z^3$, where (b_1, b_2, b_3, b_4) are combinations of (c_0, c_1, c_2), given by $b_1 = -\frac{1}{2}c_1$, $b_2 = -\frac{1}{6}(2c_1 + c_0)$, $b_3 = \frac{1}{64}(9c_1^2 - 20c_2 - 3c_0(2c_1 + c_0))$, $b_4 = \frac{1}{64}(15c_1^2 - 12c_2 - 5c_0(2c_1 + c_0))$.

where C is a function of x_c given by combining equations (10.45) and (10.42). Expression (10.46) is then directly amenable to a BNF, as we show in the next paragraph.

Polar Action-Angle Coordinates

The final step to extract the BNF of (10.46) is to promote $\bar{z}^2 + \bar{Z}^2$ to an action variable, just as we did for the classical pendulum in Sect. 10.2.1. A classical technique [345] is to think of $(\bar{z}, \bar{Z})$ as a kind of Cartesian-type coordinates and pass to (symplectic) polar coordinates, by setting

$$z = \sqrt{2\rho}\cos\varphi \quad \text{and} \quad Z = -\sqrt{2\rho}\sin\varphi\,, \tag{10.47}$$

where this expression is necessary to enforce symplecticity. Inserting into Eq. (10.46) the new coordinates (ρ, φ) gives us our final expression for the Hamiltonian

$$H(\rho,\varphi) = Y_1 + \sqrt{8Y_2}\rho + C\sqrt{32Y_2}\,\rho^2 + o(\rho^2)\,. \tag{10.48}$$

Now we can compute $C = C(\Lambda)$ in terms of x_c and the Y_n's from Eqs. (10.45) and (10.42). The BNF $N_1(\rho)$ of (10.48) is therefore

$$N_1(\rho) = Y_1 + \sqrt{8Y_2}\,\rho + \frac{1}{2}\left(\frac{4Y_3}{Y_2} + \frac{x_c}{3Y_2^2}(3Y_2Y_4 - 5Y_3^2)\right)\rho^2\,. \tag{10.49}$$

It should be noted that no assumption about the radial potential $Y(x)$ has been made to derive this BNF. In particular, $Y(x)$ is not required to be isochrone, and, much like in [609], this BNF is valid for any radial potential. We now turn to the derivation of the BNF for an isochrone potential.

10.2.3 BNF for an Isochrone Potential

In general, the strength and simplicity of a BNF is usually balanced by the (analytic) complexity involved in its derivation (see e.g. the BNF of the restricted N-body problem, [610, 611]). However, in the case of an isochrone potential, things are much simpler thanks to the symmetry at play. We explain, in this subsection, how to construct the BNF of the Hamiltonian in that particular, isochrone case. The simplicity of the argument should then be compared to the previous Sect. 10.2.2, where without the isochrone assumption the calculation was much more involved, and very close to that of [609].

We will follow the same steps used in Sect. 10.1.1. In particular, we start from the following result: if the potential is isochrone, the radial action J decomposes into a sum of two terms, one ξ-dependent and one Λ-dependent, as was shown in (10.5). For convenience, we rewrite this as

$$J(\xi, \Lambda) = F(\xi) - G(\Lambda), \tag{10.50}$$

where F, G are two functions[14] verifying $F'(\xi) = T(\xi)/2\pi$ and $G'(\Lambda) = \Theta(\Lambda)/2\pi$, since for any radial potential, (10.4) must hold. In Sect. 10.1.1 we had the explicit expressions of F and G (recall (10.7)) but these have been obtained by assuming what we are attempting to prove, namely the fundamental theorem of isochrony. As we shall see, these explicit forms are not required to make the computation.

Now let us fix a value of Λ, and solve equation (10.50) for the energy ξ in terms of the radial action J. Since that expression holds for any ξ, i.e. any numerical value of the Hamiltonian $H = \xi$, we have just obtained H expressed in terms of the action J, at fixed Λ. This expression reads

$$H(J, z_J) = F^{-1}(G(\Lambda) + J), \tag{10.51}$$

where we have re-introduced the dependence on the radial angle variable z_J associated to the radial action J, for completeness. Let us emphasize one more time that, at this stage, we do not know the expressions of F, G. They can only be computed once the isochrone theorem is demonstrated. When this is done Eqs. (10.50) and (10.51) will become (10.7) and (10.9), respectively.

As emphasized in the last section, for a given value of Λ, circular orbits are relative equilibria of H and correspond to $J = 0$. Let us then Taylor-expand (10.51) around $J = 0$ and set $H(0, z_J) \equiv \xi_c(\Lambda)$ as the energy of that circular orbit. We readily get

$$H(J, z_J) = \xi_c + \frac{1}{F'(\xi_c)} J + \frac{1}{2}\left(-\frac{F''(\xi_c)}{F'(\xi_c)^3}\right) J^2 + o(J^2). \tag{10.52}$$

We can now extract the BNF of the above Hamiltonian $H(J, z_j)$, which we denote by $N_1(J)$, such that $H(J, z_J) = N_1(J) + o(J^2)$. Using the property $F'(\xi) = T(\xi)/2\pi$ one more time, we find

$$N_2(J) = \xi_c + \frac{2\pi}{T(\xi_c)} J + \frac{1}{2}\left(-\frac{4\pi^2\, T'(\xi_c)}{T(\xi_c)^3}\right) J^2, \tag{10.53}$$

where $T(\xi_c)$ is understood as the limit of $T(\xi)$ when $\xi \to \xi_c(\Lambda)$ for fixed Λ, since the radial period of a circular orbit can be ambiguous to define. Equation (10.53) is, for a given Λ, a BNF for H, but we emphasize that it holds *only if the potential is isochrone*, otherwise Eq. (10.50) (from which (10.53) follows) does not hold in the first place. With this second BNF at hand, we can finally turn to the applications, in the next and last section.

---

[14] We assume that F, G behave nicely so they can be differentiated and inverted on their domain of definition. We know this will be the case as we know their explicit form, cf. Eq. (10.6).

10.3 Three Applications of the BNF

In this fourth and last section, we use the two BNFs (10.53) and (10.49) of the Hamiltonian describing a particle in an isochrone potential. By exploiting the equality between their respective Birkhoff invariants, we provide: (1) a proof of the fundamental theorem of isochrony (8.22), (2) a proof of the Bertrand theorem, and (3) a proof of the generalized Kepler's third law (10.6). These three items are presented in each of the three following subsections.

10.3.1 *Fundamental Theorem of Isochrony*

The two BNFs N_1 and N_2 derived in the previous section define two sets of three Birkhoff invariants (according to (10.33)), one for each BNF. They must be equal, by unicity of the BNF. From the first BNF (10.49), derived in the ρ action coordinate, their expression is

$$\mathfrak{l}_1 = Y_1 \,, \quad \mathfrak{b}_1 = \sqrt{8Y_2} \,, \quad \text{and} \quad \mathfrak{B}_1 = \frac{4Y_3}{Y_2} + \frac{x_c}{3Y_2^2}\left(3Y_2Y_4 - 5Y_3^2\right) , \tag{10.54}$$

where we emphasize that each of these invariants are Λ-dependent, through the derivatives of the potential $Y_n = Y^{(n)}(x_c(\Lambda))$ and (twice the square of) the radius of the circular orbit $x_c = x_c(\Lambda)$. The second BNF (10.53) then provides an alternative expression

$$\mathfrak{l}_2 = \xi_c(\Lambda) \,, \quad \mathfrak{b}_2 = \frac{2\pi}{T(\xi_c)} \,, \quad \text{and} \quad \mathfrak{B}_2 = -4\pi^2 \frac{T'(\xi_c)}{T(\xi_c)^3} \,, \tag{10.55}$$

where, once gain, they are Λ-dependent through the energy of the circular orbit $\xi_c = \xi_c(\Lambda)$. The invariants (10.55) are computed under the assumption that $Y(x)$ (or equivalently $\psi(r)$) is isochrone, while the invariants (10.54) are valid for any $Y(x)$ (not necessarily isochrone). However, if we assume $Y(x)$ isochrone, then (10.54) and (10.55) are the Birkhoff invariants of the same system (a particle of angular momentum Λ and energy $H = \xi$ in an isochrone potential). Therefore, from now on we assume that $Y(x)$ is isochrone, and derive the isochrone theorem (8.22) by exploring the consequences of the three equality's $(\mathfrak{l}_1, \mathfrak{b}_1, \mathfrak{B}_1) = (\mathfrak{l}_2, \mathfrak{b}_2, \mathfrak{B}_2)$ in three steps, one for each order of invariants.

Zeroth Order Invariant

The first equality $\mathfrak{l}_1 = \mathfrak{l}_2$ provides a link between the energy of the circular orbit of angular momentum Λ and the first derivative of Y, namely:

$$Y'(x_c(\Lambda)) = \xi_c(\Lambda)\,. \tag{10.56}$$

This equation is consistent with the construction of an orbit in the $x = 2r^2$ variable, as we explained in Fig. 8.1. Indeed, a circular orbit of energy ξ_c corresponds to the line $y = \xi_c x - \Lambda^2$ being tangent to the curve $Y(x)$. Therefore, their respective slope must be equal at the tangency point x_c, hence $\xi_c = Y'(x_c)$. The other consequence of that equation is how ξ_c varies with respect to Λ. Indeed, we have

$$\frac{\mathrm{d}\xi_c}{\mathrm{d}\Lambda} = \frac{\mathrm{d}Y_1}{\mathrm{d}\Lambda} = \frac{\mathrm{d}x_c}{\mathrm{d}\Lambda} Y_2\,, \tag{10.57}$$

where we note that the Leibniz rule must be used since Y_1 depends only on Lambda through $x_c(\Lambda)$ (recall that $Y_n \equiv Y^n(x_c)$ by definition). Equation (10.57) will be used below.

First Order Invariant

The second equality $\mathfrak{b}_1 = \mathfrak{b}_2$ implies a relation between the radial period and the second derivative of Y at x_c, namely

$$Y''(x_c(\Lambda)) = \frac{\pi^2}{2} \frac{1}{T(\xi_c(\Lambda))^2}\,. \tag{10.58}$$

Once we know $Y(x)$, this equation allows us to derive easily the generalization of the Kepler's third law of motion, which we saw back in (10.6). The other consequence of (10.58) is an equation for $\mathfrak{B}_2$. Indeed, differentiating (10.58) with respect to Λ readily gives

$$x_c' Y'''(x_c) = -\pi^2 \frac{\mathrm{d}\xi_c}{\mathrm{d}\Lambda} \frac{T'(\xi_c)}{T(\xi_c)^3}\,. \tag{10.59}$$

We see that (10.59) is very similar to the expression of $\mathfrak{B}_2$ in (10.55). In fact, inserting (10.57) in (10.59) and comparing the resulting with (10.55) readily gives the relation

$$\mathfrak{B}_2 = \frac{4Y_3}{Y_2}\,, \tag{10.60}$$

where we used the fact that $x_c'(\Lambda) \neq 0$, which follows by differentiating (10.36) with respect to Λ to obtain $x_c' x_c Y_2 = 2\Lambda$. With Eq. (10.60) at hand we may finally complete the proof of the fundamental theorem of isochrony.

Second-Order Invariant

Lastly, we insert in the equality $\mathfrak{B}_1 = \mathfrak{B}_2$ the expression (10.60) for $\mathfrak{B}_2$, and the expression (10.54) for $\mathfrak{B}_1$, to conclude that

$$\frac{x_c}{3Y_2^2}\left(3Y_2Y_4 - 5Y_3^2\right) = 0\,. \tag{10.61}$$

Since $x_c = 2r_c^2 \neq 0$, the parenthesis must vanish. Recalling the notation $Y_n = Y^{(n)}(x_c(\Lambda))$, and since (10.61) should hold for any Λ, we may now let Λ vary continuously. By continuity of $\Lambda \mapsto x_c(\Lambda)$, the equation $3Y_2Y_4 = 5Y_3^2$ is nothing but an ODE for the function $x_c \mapsto Y(x_c)$, i.e., the function Y. Therefore, at least on some open interval of $\mathbb{R}_+$, we must have

$$3Y^{(2)}Y^{(4)} = 5\left(Y^{(3)}\right)^2. \tag{10.62}$$

It turns out that Eq. (10.62) is *the universal differential equation for parabolae*, in the sense that its solutions cover all and only functions Y whose curve $y = Y(x)$ are parabolae in the (x, y) plane. A short proof of this statement is included in Appendix B.3.2. This concludes the proof of the isochrone theorem (8.22). Before going to the next paragraph, let us mention that (10.62) can be simply written as an ODE for the Λ-dependent Birkhoff invariants $(\mathfrak{l}, \mathfrak{b}, \mathfrak{B})$ themselves, namely

$$\mathfrak{B}\,\frac{\mathrm{d}\mathfrak{l}}{\mathrm{d}\Lambda} = \mathfrak{b}\,\frac{\mathrm{d}\mathfrak{b}}{\mathrm{d}\Lambda}\,. \tag{10.63}$$

In fact, we could have obtained (10.63) readily from the fact that $\mathfrak{B}_2\mathfrak{l}_2' = \mathfrak{b}_2\mathfrak{b}_2'$ (here a prime denotes $\mathrm{d}/\mathrm{d}\Lambda$), which can be seen easily from (10.55). That the isochrone theorem follows from such a simple differential relation between the Birkhoff invariants constitutes a very nice and fundamental characterization of isochrony. More insight on (10.63) is provided in Appendix B.3.3.

10.3.2 Bertrand Theorem

There is another fundamental result that we can derive from this formalism: the Bertrand theorem. As mentioned before, this was actually done in [609] and was the main motivation behind the present work. However, we would like to present it in the light of isochrony. Indeed: as we mentioned back in Chap. 7, the Bertrand theorem states that only the Harmonic and Kepler potentials generate closed and only closed orbits. But notice that both of these potentials are isochrone. Therefore, we expect the Bertrand theorem to be a corollary of the isochrone theorem (as was argued already in [589]). We prove the Bertrand theorem in two steps: first we show that a Bertrand potential $Y(x)$ must be a power law (up to a linear term); and second, that it must be isochrone.

Let us consider the BNF (10.49), which holds for any radial potential $Y(x)$, including isochrone and Bertrand potentials. Let us write the corresponding Hamiltonian $H(\rho, \phi, \Lambda, \vartheta)$ in the complete, 4D-phase space with the two pairs (ρ, ϕ), (Λ, ϑ) of action angle variables. We have seen that it reads

$$H(\rho, \Lambda) = \mathfrak{l}_1(\Lambda) + \mathfrak{b}_1(\Lambda)\rho + \frac{1}{2}\mathfrak{B}_1(\Lambda)\rho^2 + o(\rho^2) , \tag{10.64}$$

where $(\mathfrak{l}_1, \mathfrak{b}_1, \mathfrak{B}_1)$ are given in terms of $Y(x_c(\Lambda))$ in (10.49). Associated to the action variables (ρ, Λ), the corresponding frequencies $(\omega_\rho, \omega_\Lambda)$ of this Hamiltonian thus read

$$\omega_\Lambda \equiv \frac{\partial H}{\partial \Lambda} = \frac{\mathrm{d}\mathfrak{l}_1}{\mathrm{d}\Lambda} + o(1) , \quad \text{and} \quad \omega_\rho \equiv \frac{\partial H}{\partial \rho} = \mathfrak{b}_1(\Lambda) + o(1) . \tag{10.65}$$

If $Y(x)$ satisfies the Bertrand theorem, then all the orbits are closed in real space. In phase space, a closed orbit corresponds to a pair of actions (ρ, Λ) (recall Fig. 10.1) that defines a torus, on which the associated curve wraps around, but ultimately closes on itself. This is called a resonant orbit, i.e., an orbit for which there exists integers $(k_\Lambda, k_\rho) \in \mathbb{Z}$ such that $k_\Lambda\omega_\Lambda + k_\rho\omega_\rho = 0$. Now, since *each and every* orbit must be closed for a Bertrand potential, this means that these integers (k_Λ, k_ρ) are actually independent of the pair (ρ, Λ). In other words, there exists a $Q \in \mathbb{Q}$ such that for all (ρ, Λ),

$$\omega_\Lambda(\rho, \Lambda) = Q\, \omega_\rho(\rho, \Lambda) . \tag{10.66}$$

We emphasize that Eq. (10.66) should hold for any pair of actions (ρ, Λ). In particular, (10.66) should hold for a given Λ in the limit $\rho \to 0$ (quasi-circular orbits). According to (10.65), this means that

$$\frac{\mathrm{d}\mathfrak{l}_1}{\mathrm{d}\Lambda} = Q\, \mathfrak{b}_1 . \tag{10.67}$$

It is rather remarkable that the Bertrand theorem is equivalent to such a simple condition, namely a differential equation for the Birkhoff invariants. We can solve this equation easily. First we insert the definitions (10.54) of $\mathfrak{l}_1(\Lambda)$ and $\mathfrak{b}_1(\Lambda)$ in terms of Y_1 and Y_2. Then the calculation reads

$$(10.67) \;\Rightarrow\; x_c' Y_2 = Q\sqrt{8Y_2} \;\Rightarrow\; \Lambda^2 = 2x_c^2 Q^2 Y_2 \;\Rightarrow\; x_c Y_1 - Y = 2x_c^2 Q^2 Y_2 , \tag{10.68}$$

where in the first step we differentiated with the Leibniz rule (much like in (10.57)), in the second step we squared and used $x_c' x_c Y_2 = 2\Lambda$ which we obtain by differentiating (10.36) with respect to Λ, and in the last step we used (10.36) once more to remove

Λ. Much like Eq. (10.61) can be seen as an ODE for $x_c \mapsto Y(x_c)$, the rightmost equation in (10.68) is an ODE too, in which $Q \in \mathbb{Q}$ is a parameter. The solution to this ODE is simply found as

$$Y(x) = C_1 x + C_2 x^K, \quad \text{with} \quad K \equiv \frac{1}{2Q^2}, \tag{10.69}$$

with two integration constants $(C_1, C_2) \in \mathbb{R}^2$. The linear term $C_1 x$ corresponds to the addition of a constant in the potential $\psi(r)$ (recall $Y(2r^2) = 2r^2\psi(r)$). As it does not affect the dynamics, we leave it aside and set $C_1 = 0$.

On the one hand, we have shown that if $Y(x)$ is a Bertrand potential, then according to Eq. (10.69) it must be a power law. On the other hand, it is clear that a Bertrand potential must be isochrone: if all bounded orbits are closed, then the apsidal angle $\Theta(\xi, \Lambda)$ must be a constant, rational multiple of 2π. In particular, as a constant function it is also independent of the energy ξ of the particle. But this characterizes isochrony. Consequently, a Bertrand potential must, at once, have the form of a power law and that of a parabola. The only parabolae that verify this property are either the square root $Y \propto \sqrt{x}$ or the quadratic $Y \propto x^2$. In terms of the variable r, this means that either $\psi(r) \propto 1/r$ (the Kepler potential), or $\psi \propto r^2$ (the Harmonic potential). Moreover, according to (10.69), these two cases correspond to $K = 1/2$ and $K = 2$, i.e. to $Q = 1$ or $Q = 1/2$, respectively. In light of the link between the apsidal angle Θ and the ratio of Hamiltonian frequencies (10.12), we recover the classical formulae $\Theta = 2\pi$ for the Kepler ellipses, and $\Theta = \pi$ for the Harmonic ellipses.

10.3.3 Isochrone Kepler's Third Law

As a final application of the BNFs, let us consider once more the equality $\mathfrak{b}_2 = \mathfrak{b}_1$, which was written explicitly in terms of Y in (10.59). Re-arranging this equation provides, for any Λ,

$$T(\xi_c(\Lambda))^2 = \frac{\pi^2}{2} \frac{1}{Y''(x_c(\Lambda))}. \tag{10.70}$$

But now, recall the initial definition of isochrone potentials: T should be independent of Λ. Although here the equation holds for the circular orbit of energy ξ_c, there exist other, non-circular orbits with the same energy. Geometrically, they can be constructed by translating the line $y = \xi_c x - \Lambda^2$ upward on Fig. 8.1. By construction, all these orbits (defined by the translation) only see their angular momentum change, not their energy (a translation preserves the slope). Consequently, their radial period (squared) is numerically equal to (10.70). Summarizing, we can now write that an orbit of energy and angular momentum (ξ, Λ) has a radial period $T(\xi)$ given by

$$T(\xi) = \frac{\pi}{\sqrt{2}} \frac{1}{\sqrt{Y''(x_c(\xi))}}, \tag{10.71}$$

where now $x_c(\xi)$ denotes the abscissa of the circular orbit with energy ξ, obtained by a downward translation (cf. Fig. 8.1). Equation (10.71) is in complete agreement with formula the other form of Kepler's third laws given in Appendix C.2.1, where T was expressed in terms of the radius of curvature R_c of the parabola at the point of abscissa x_c. Recalling the link between curvature and the second derivative for explicit curves, equality between the two formulae follows. To obtain the general expression (10.6) in terms of ξ and the parabola parameters (a, b, c, d, e), one simply needs to compute the second derivative of a given parabola $Y(x)$, evaluate it at x_c and insert the result in (10.71). The result (10.6) follows immediately.

10.4 A Last Historical Note

We end this chapter with an historical note that involves both Hamiltonian mechanics and Huygens, closing the second part of this thesis just as it started, in Chap. 7. This note starts with a question. *How come that, in analytical mechanics, the conserved mechanical energy of a system, which is the most fundamental quantity of interest, is denoted H?* This is easy. It is because it refers to the *Hamiltonian* of the system. Another question then. *Who introduced that notation for the first time?* Again, a quick search reveals that it is Joseph Louis Lagrange, who used it first in the second edition of his masterwork "Mécanique Analytique", in 1811. A last question. *How old was Sir William Rowan Hamilton in 1811?* The answer is: *five years old.*

Although it is clear that Hamilton would have earned this privilege for his work on analytical mechanics (it is him who brought considerable novelty and valuable insight on Lagrange's work), the letter H was not dedicated to him. Did Lagrange choose H for lack of a better letter? No, either. In fact, a careful inspection of Lagrange's work reveals that he did dedicate this letter H to someone. That someone is none other than Christian Huygens. Indeed, as Lagrange mentions in his treatise, it his Huygens who proposed, for the first time, in his Horlogium Oscillatiorium (again!), to use what we now call the conservation of energy, to solve a mechanical problem. Lagrange knew Huygens' work very well, and makes many references to his brilliant idea throughout his writings.

This misuse of the letter H is not very well-known. It seems to have been first described by Jean-Marie Souriau in 1986 [551], and also by Patrick Iglésias in 1998 [615, 616]. These two accounts are based on Lagrange's printed work, and one may argue that we can never know for sure if Lagrange used that letter himself in his notes, and that the printer decided to choose the letter H. However, recently, a lost manuscript of Lagrange's second edition of Mécanique Analytique was found again, hidden in the library of the Ècole Nationale Supérieure des Techniques Avançées (ENSTA). In this particular copy, a leaflet can be found at the very end of the manuscript, in the binding of the book. On it, written from Lagrange's own hand, we can see a reference to the "vis-viva" constant of Huygens (the mechanical energy), just above the equation $T + V = H$, as depicted in Fig. 10.4. This book was given

Fig. 10.4 The leaflet added to Binet's copy of Lagrange's second edition of the *Mécanique Analytique*. We see the equation $T + V = H$ and a reference to Huygens' conservation of energy in the text above it. The gloved fingers belong to Jérôme Perez [617]

to the school by its owner, Jacques Binet,[15] who hand wrote on the cover page that "the handwritten leaflet at the end of the manuscript is from the author himself".

Of course, it is far too late to begin calling H the Huygensian of the system, and it will probably be called the Hamiltonian forever. Nevertheless, we see from this story a version of Arnold's theorem at work again: *if something is named after someone in science, then it is most probable that this someone did not discover nor invent it first*. Of course, this theorem from Arnold applies to itself as well.

Conclusions

To conclude this thesis, we come back to the main results derived in this work. We also discuss a number of research prospects associated to each of them and emphasize on potential applications.

---

[15] The same Binet who pointed out that the two-body problem is a harmonic oscillator when written in terms of the (now eponymous) variable $u \equiv 1/r$.

The First Law of Mechanics in General Relativity

The first chapter of this thesis contained a number of historical references. Although there is nothing there that cannot be found elsewhere with greater details, some of these historical references are perhaps not that well-known, and some figures may be of interest for pedagogical purposes. In any case, the point was to give some context to this work, and understand how our results fit into the bigger picture that is GW astronomy. Speaking of Chap. 1 and conclusions, I take this opportunity to reveal the "mystery" behind Fig. 1.3. Although this lithography depicts the properties and orbits of the (never to be found) planet Vulcan, it is also dated from 1846 (rectangular inset), namely 13 years before Le Verrier proposed it to the French Academy of Sciences for the first time [67]. Several experts on this topic and period of astronomy (including Guy Bertrand, David Aubin and Bill Sheehan) admitted never having seen this map and be somewhat baffled by it, having no convincing explanation as to the presence of Vulcan so early. References on this map can be found here, and any reader with the slightest piece of information is encouraged to contact me for further investigation. Solving the "mystery of the 1846 Vulcan map" would be an exciting side project which I intend to pursue with historians of physics at the Paris Observatory.

New material starts to appear in Chap. 2. Even though all approximation schemes discussed there are not new, we tried to add value to the discussion by comparing their main assumptions and outputs, and pointing out their unequal but overlapping range of applicability. In particular, emphasis was made on their complementarity: some of them take as an assumption the fact that we can describe an extended body by a point particle whereas others have this as an output; and some of them, based on SEM conservation, are a priori incompatible with the description of black holes; etc. In fact, we believe that an overall, thorough comparison analysis of all these relativistic approximation schemes would be of value, and could be an interesting line of research to explore. Regarding other prospects, we mention the possibility of extending Tuclzyjew's reduction scheme to higher multipolar orders. Indeed, we believe that the calculation at dipolar order (Appendix A.4) and quadrupolar order (cf. [344]) could be made systematic with the help of some computer algebra software. Currently, it seems that there is no obvious rule regarding the structure of higher order multipoles, derivatives and coefficients in the formula for a generic multipolar SEM tensor, contrary to the evolution equations, cf. [320]. This computer-assisted method could shed some light on this issue, and may show an interesting interplay between physics and combinatorics. Lastly, although straightforward to derive, the consequences of the evolution equations at any multipolar order (with a generic force and torque sourcing the equation of motion and precession), are given in Appendix A.4.4 and have not been written down before, to our knowledge. It would be interesting to examine these equations further to understand how each multipole contributes to the conservation laws.

In Chap. 3, we motivated, using leading-order post-Newtonian results for a two-body system, our need for the introduction of a helical Killing vector k^a. This strategy has been known and applied for a long time, in particular for co-rotating, extended-objects binaries and points particles. The main use of this helical Killing vector is the central identity

$$k^a|_\gamma = zu^a\,, \tag{10.72}$$

where u^a is the body's four-velocity, which mathematically encodes the circular motion of the body within the spacetime geometry at once. Our main result in this chapter is a covariant proof that the mere existence of a Killing vector enforces this equation automatically. In other words, given a helical spacetime and a multipolar point particle within it, it readily follows from a calculation that Eq. (10.72) holds. Additionally, we showed that this setup enforces a number of geometrical constraints on the multipoles equipping the particle, namely their Lie-dragging along k^a:

$$\mathcal{L}_k p^a = 0\,, \quad \mathcal{L}_k S^{ab} = 0 \quad \text{and} \quad \mathcal{L}_k J^{abcd} = 0\,, \tag{10.73}$$

with the four-momentum p^a, spin tensor S^{ab} and quadrupole tensor J^{abcd} of the particle. These results, which are expected to hold at any multipolar order, are of primary importance for subsequent chapters, but could also be used as a starting point to measure the effect of approximate isometries. In particular, given a spacetime that contains an "approximate" Killing vector (in a sense to be specified), our calculation could be repeated to see how the constraints (10.72) and (10.73) are modified. This could have potential applications in black hole perturbation theory, where the background isometries of the black hole are "slightly" broken by, say, a secondary object of any other type of field. Again, this could be of some interest to some groups, and we leave this as an interesting research prospect.

Chapter 4 discuses several of the plethora of forms that the first law of mechanics takes in the literature, depending on the context in which it is derived and/or used. Again, the added value here is to gather in a single place all these first laws and their associated references. We hope that it can be helpful as it is, and a more profound review could be of value for the community. However, the most important new result of Chap. 4 is found in Sect. 4.3, where we derive a general first law of mechanics valid for arbitrary, non-vacuum, helically isometric spacetime

$$\delta M - \Omega\,\delta J = \delta \int_{\mathcal{S}} \varepsilon_{abcd}\, T^{de} k_e - \frac{1}{2} \int_{\mathcal{S}} \varepsilon_{abcd}\, k^d\, T^{ef}\, \delta g_{ef}\,, \tag{10.74}$$

where T^{ab} is the SEM tensor of the fields in this spacetime and $\mathcal{S}$ any spacelike surface transverse to k^a. This first law generalizes that of Friedmann et al. [382] for perfect fluids. Because the formalism from which it is derived [470, 497] is very general, it should be straightforward to obtain a number of variations of this law, e.g., by adding other fields, or extend it to alternative (diffeomorphism-invariant) theories of gravity. The resulting first law should then include additional terms encoding deviations from GR, and could be used in the context of approximation schemes within alternative theories of gravity. Comparison to, say, post-Newtonian results within alternative theories of gravity could then be performed as a validity check for both results. Discussions on the matter have already been initiated.

In Chap. 5, we have combined all ingredients from previous chapters to derive a covariant, SSC-independent, first law of mechanics for dipolar particles. The coefficients appearing in the first law are shown to be related to the helical Killing field naturally. We also derived another, simpler formulation of that same law under the SSC assumption $u_a S^{ab} = 0$, and a further one only involving scalar quantities, valid up to linear order in the spin. These laws read

$$\delta M - \Omega\,\delta J = \sum_{\mathrm{i}} \Big(z_{\mathrm{i}}\,\delta m_{\mathrm{i}} - \tfrac{1}{2}\nabla_a k^b\,\delta S^a_{\mathrm{i}\,b} - k_a\,\delta \dot{D}^a_{\mathrm{i}} - \tfrac{1}{2}\dot{D}^a_{\mathrm{i}}\,\delta k_a \Big) \tag{10.75a}$$

$$\overset{\text{SSC}}{=} \sum_{\mathrm{i}} \Big(|k|_{\mathrm{i}}\,\delta m_{\mathrm{i}} - \tfrac{1}{2}(\nabla_a k^b)\,\delta S^a_{\mathrm{i}\,b} \Big) \tag{10.75b}$$

$$\overset{\text{SSC}}{=} \sum_{\mathrm{i}} \Big(|k|_{\mathrm{i}}\,\delta m_{\mathrm{i}} - |\nabla k|_{\mathrm{i}}\,\delta S_{\mathrm{i}} \Big) + O(S_{\mathrm{i}}^2)\,, \tag{10.75c}$$

respectively. We also showed that this law coincides (with the SSC and in the linear-in-spin limit) with the pre-existing first law for spinning particles derived in the Hamiltonian framework [406]. Although quadratic-in-spin corrections from quadrupolar contributions should be added for a self-consistent result, let us mention that our dipolar first law contains, by construction, all spin nonlinearities from the dipolar sector. Therefore, the first law for spinning particles accounting for all quadratic-in-spin effects only necessitates the computation of this quadrupolar contribution. Since non-linear spin effects are relatively well-understood in post-Newtonian theory, explicit verification of the first law is feasible. This PN verification is currently underway at dipolar order.

In that respect, we presented in Chap. 6 a number of preliminary but promising results regarding these extensions to quadrupolar order. In particular, we completed the derivation of one of the two integrals that enter the first law, namely the integral I in $\delta M - \Omega\delta J = -\delta I + \frac{1}{2}K$ (cf. Eq. (10.74)). Remarkably, this calculation does not require an explicit formula for the quadrupole J^{abcd} entering the description of the binary. Given the general mass-momentum-stress decomposition (6.2) of J^{abcd}, these contributions are given in Eqs. (6.30), (6.34) and (6.35), respectively. Although in the current state of our work this result needs to be double-checked, our confidence in it is rather high, owing to the fact that any non-covariant term (e.g., involving a dependence on the integration hypersurface $\mathcal{S}$ through the extrinsic curvature) present throughout the calculation vanishes identically at the end, based on very general symmetry arguments, and as should be expected. This integral, quadrupolar first law, when taken in the dipole limit, was compared to (and found to agree with) that found in the literature for spinning point particles [406]. Lastly, based on these results, we anticipated the final expression for the variational first law at quadrupolar order, in the particular case of a spin- and tidally-induced quadrupole:

$$\delta M - \Omega\delta J \overset{!!}{=} \sum_{\mathrm{i}} z_{\mathrm{i}} \Big(\delta m_{\mathrm{i}} - \omega_{\mathrm{i}}\delta S_{\mathrm{i}} + E^S_{\mathrm{i}}\,\delta(\kappa_{2,\mathrm{i}}\mathcal{Q}_{\mathrm{i}}) + \mathcal{E}_{\mathrm{i}}\,\delta(\mu_{2,\mathrm{i}}\mathcal{E}_{\mathrm{i}}) + \mathcal{B}_{\mathrm{i}}\,\delta(\sigma_{2,\mathrm{i}}\mathcal{B}_{\mathrm{i}}) \Big) \tag{10.76}$$

where the notations are defined in Chap. 6, Sect. 6.3. Regarding applications of this result, a first step would be to use the first law as a comparison tool, as we presented in Chap. 4. In particular, results for these spin and tidally-induced quadrupoles effects in the post-Newtonian expansions of quantities appearing in the first law are already available. For example, the work [618] contains the expressions for the binding energy and total angular momentum of a circular binary with spin-aligned spins, in terms of the polarizability constants (μ_2, σ_2), see Eqs. (6.5) and (6.6) there.[16] Regarding the spin-induced quadrupole, [320] contains similar PN expressions in terms of the constant κ_2 and could be used to verify our first law including our entire quadrupolar model. We note that these two references also contain octupolar (tidal and spin) contributions, which could be of interest if the first law were to be extended to this order. In practice, all these polarizability coefficients measure finite-size effects in binary systems of primary importance for the future of GW astronomy. Indeed, these "measurements" of the polarizability constants will put strong constraints on the equation of state of neutron stars in the supra-nuclear regime [619, 620], which cannot be tested in current laboratory experiment. These imprints left in the gravitational waveform of tidally-deformed objects have already been observed, in particular following the binary neutron-star merger event GW170817 [281, 621, 622].

Lastly, on more theoretical grounds, it would be very interesting to examine how the generalized Killing field formalism (touched upon in Chap. 2, Sect. 2.2.3) could be used to redo, and possibly simplify, our calculations. Indeed, this framework takes advantage of the existence of generalized Killing fields, and seems to be a natural choice in the context where one has a true Killing field, such as in our context. If our calculations are to be extended to octupolar order, the GKF formalism could be more adapted to handle these high (multipolar) order calculations, which are already intricate at quadrupolar order, at least in our formalism.

Isochrone Orbits in Newtonian Gravity

The last four chapters of this thesis presented our result regarding the solution of a problem posed by Michel Hénon: determining all isochrone potentials and studying orbits therein. In the first of these chapters, Chap. 7, we gave an overview of a related problem, solved by Huygens: the isochrone pendulum. As we pointed out several times, Huygens' solution contains a number of similarities with ours. This is expected, as both problems can be rephrased as, broadly speaking, "find a curve $y(x)$ such that a functional $T[y]$ is independent of one of the parameters of the problem". It would be very interesting to make a more complete dictionary between these two

[16] It was shown in that same reference that the first-law corollary $\partial_\Omega E - \Omega \partial_\Omega J = 0$ holds at the PN level considered in the paper, which includes both spin- and tidally-induced quadrupole considered in our work.

problems, and to study higher-dimensional versions of them. We will come back to these similarities below for this chapter.

In Chap. 8, we provided a complete classification of the set of isochrone potentials, which we split into five families, as summarized in Fig. 8.8. To obtain this classification, we first had to rewrite the integral definition of the radial period as the Hénon's formula

$$T(\xi) = \frac{\pi}{4} \frac{x_A - x_P}{\sqrt{\Lambda_C^2 - \Lambda^2}}, \tag{10.77}$$

which expresses the radial period as the ratio between two directly measurable (geometrically speaking) quantities built out of chords and tangents to the isochrone potential. With this geometrical rephrasing of isochrony, it was then possible to use Archimedes' characterization of parabolae to deduce the fundamental theorem of isochrony, namely

$$\psi \text{ is isochrone} \quad \Rightarrow \quad y = x\,\psi\big(\sqrt{x/2}\big) \text{ is a parabola}. \tag{10.78}$$

We then use this correspondence between parabolae in the plane (parameters (a, b, c, d, e)) and isochrone potentials to find and classify them, into five distinct families (parameters $(\epsilon, \lambda, \omega, \mu, \beta)$). Two of them are the most well-known, academic potentials (Harmonic and Kepler), the third is that of Hénon, the fourth one (Bounded), was found in [589], and the fifth one (Hollow) was put forward in this work. All these potentials exhibit a plethora of features and would be of great interest to act as toy-models for astrophysical systems or confined quantum systems, as pointed out in [3].

Having shed light on the nature of isochrone *potentials* in Chap. 8, we could then turn to the analysis of isochrone *orbits* in Chap. 9. Two of the most important results in that chapter were the generalization of Kepler's third law for both the radial period and the apsidal angle to all isochrone orbits:

$$T^2 = -\frac{\pi^2}{4} \frac{\delta}{(a + b\xi)^3}, \quad \text{and} \quad \frac{\Theta^2}{\pi^2 \Lambda^2} = \frac{2b^2\Lambda^2 - d}{b^2\Lambda^4 - d\Lambda^2 + e} + \frac{2b}{\sqrt{b^2\Lambda^4 - d\Lambda^2 + e}}, \tag{10.79}$$

which allows for a definitive check that these potentials are isochrone (T does not depend on Λ, nor Θ on ξ); and the explicit, closed-form expression of the polar coordinates (r, θ), given in Sect. 9.2.4. These formulae have been checked to agree with a direct, numerical integration of the equations of motion (cf. Appendix C.1.1). In particular, in Figs. 9.4 through 9.7, both the analytical and numerical orbits are plotted and indistinguishable from one another. To derive this analytical solution, we proposed a geometrical method that consisted in working in the Hénon plane, and mapping arcs of parabolae to one another. As this geometrical method does not rely on the isochrone nature of the potentials, it could be implemented for other types of dynamical systems, in the case where a family of potentials is in a one-to-one correspondence with a set of curves stable under an invertible group of transformations. This represents an interesting line of research to obtain analytical results for otherwise complicated equations of motion. This method could also be used to explore "distance" to isochrony, since any non-isochrone potential $Y(x)$ can always be written as $I(x) + f(x)$, where $I(x)$ is an isochrone potential (and thus an arc of parabola) and $f(x)$ an arbitrary function. When $f(x)$ is small (in some sense to be specified) then the curve $Y(x)$ is approximately a parabola and all results derived in this work could be appropriately modified. This could be helpful for a stability analysis of isochrone orbits, or to study other problems altogether (the geodesic motion of a particle in the Schwarzschild metric being one example of a "Kepler + perturbation" case.)

Lastly, we mention another exploration that did not make it into the thesis, but otherwise provided interesting results. Since all isochrone orbits are arcs of parabolae in the plane, constants of motion associated to this orbit should have an invariant formulation, i.e., an expression in terms of geometrical quantities whose numerical value does not change when going from a family of orbits to another. For instance, the Euclidean group maps parabolae arcs to parabolae arcs. Therefore, any physical quantity related to an orbit can be written in terms of the Euclidean invariants of the locus of points made of a parabola and a line. Such locus of points is a particular case of degree-three algebraic curve, and the theory of Euclidean invariants are rather involved. Thankfully, Euclidean invariants of parabolae are well-known (see the very nice article of MacDuffee [623]) and can easily be generalized to account for the "parabola+line" nature of this curve. Writing an expression for the physical, orbital invariants in terms of mathematical, Euclidean invariants of these curves does give very interesting (but preliminary) results, and these should be investigated further.

The last chapter on isochrony, Chap. 10, contains a revisit of the problem, with an Hamiltonian approach. This investigation needed to be done to fully understand the symmetries at play in isochrone mechanics. In particular, isochrone potentials are those for which the Hamiltonian of a test particle takes the form

$$H(J, \Lambda) = -\frac{a}{b} - \frac{\delta}{b\big(2bJ + R(\Lambda)\big)^2}, \tag{10.80}$$

in terms of the natural action-angle variables (J, Λ) of this (spherically symmetric) problem. The equations of motion derived from this Hamiltonian revealed that the notion of eccentric anomaly and eccentricity, usually defined for the Keplerian ellipse, both admit a generalization to all isochrone orbits. Although a direct, geometrical construction of this eccentric anomaly (like for the Keplerian case) could not be found, we have to mention the particular structure of the parametric solution (10.22) and (10.27), which take the (schematic) form $r(E) = \sqrt{r_+(E)r_-(E)}$ and $\theta(E) = \theta_+(E) + q\theta_-(E)$, where $q \in \mathbb{Q}$ and, remarkably, each pair $(r_\pm, \theta_\pm)$ depicts a Keplerian ellipse, in the Keplerian eccentric anomaly parametrization (the Keplerian case corresponds to $r_+ = r_-$ and $q = 0$.) This peculiarity would need to be investigated further, in particular to shed some light on the nature of this angle E in the case of isochrone, but non-Keplerian, orbits.

To conclude once and for all this last chapter, and perhaps most importantly regarding our work on isochrony, the Hamiltonian analysis of the isochrone problem culminated in two results. First, the derivation of the Kepler equation valid for all and any isochrone orbit, in the form

$$\Omega t = E - \epsilon \sin E \,, \tag{10.81}$$

with the parameter ϵ vanishing for circular orbits and satisfying $0 < \epsilon < 1$ for non-circular ones, and Ω being the radial frequency $2\pi/T$. These quantities are expressed simply in terms of the parameters (a, b, c, d, e) and (ξ, Λ) (cf. Eq. (10.19)) and give a complete solution to the problem of motion in any isochrone potential. Second, the explicit computation of the Birkhoff normal form of the system. The careful analysis of the (three first) Birkhoff invariants of this normal form revealed the profound link between parabolae and isochrony. Quite remarkably, the third Birkhoff invariant

$$\mathfrak{B} = \frac{4Y_3}{Y_2} + \frac{x_c}{3Y_2^2}\left(3Y_2Y_4 - 5Y_3^2\right), \tag{10.82}$$

encodes the universal ODE for parabolae in its second term. Since, as we showed, isochrony implies that $\mathfrak{B}_1 = 4Y_3/Y_2$, this second term vanishes, implying that $Y(x)$ satisfies this ODE, and hence is a parabola. As a direct corollary, we showed that the Birkhoff invariants also contain the Bertrand theorem, and the (generalized) Kepler's third law. The power of Hamiltonian mechanics revealed the deep connection between the symplectic nature of this problem of mechanics and its Euclidean counterparts involving parabolae and their chords.

The implications of the previous result (and in particular our method) are numerous. First, although we have not done it yet, we strongly suspect that the same treatment (Birkhoff normal form, Birkhoff invariants and an ODE) should give a very elegant solution to Huygens' isochrone pendulum. Indeed, much like parabolae, cycloids, which are the unique set of (symmetric) solution to the isochrone pendulum problem, admit a universal ODE, of the form $YY'' + Y = \text{const.}$. It should be straightforward to show that this ODE is encoded in the Birkhoff invariant of the pendulum Hamiltonian. Second, this Hamiltonian approach seems to be the most natural to generalize the isochrone problem to axially (and not merely spherically) symmetric potentials. This would induce another dimension to the orbit (not confined in a plane) but also an additional constant of motion, hence an integrable problem. Construction of angle-action variables are well-known in this case, and it should not be too difficult to generalize the normal form calculation. Third, although this was not included in the thesis, the expression for the Birkhoff invariants in an isochrone potential can be actually computed at any order in the normal form. We do not know if such analytic result, valid at any order in the normal form, is of any interest, but it may be one of the rare case where it is available analytically. We leave all these ideas for future explorations of the ever-expanding Land of Isochrony.

Appendix A
Lorentzian Geometry

A.1 Unicity of the Field Equations

In Chap. 1, Sect. 1.1.2, we presented, among other things, the Einstein field equation that characterizes GR, and Lovelock's theorem, which states the uniqueness of this equation under very natural requirements. In the literature, we find other names associated to Lovelock's theorem, among which Vermeil, Cartan and Weyl stand out. Below is a brief account on this priority (non-)issue.

- First of all, it is to be said that Einstein (probably thanks to Grossmann) and Hilbert both claim the unicity result as a motivation for the final form of the field equation, without giving a proof [624].
- Hermann Vermeil showed in 1917 that the Ricci scalar R is the only scalar field that is linear in the second derivative of g_{ab}, for manifolds of any dimensions [625]. It should be noted that this work by Vermeil was actually motivated by his mentor Felix Klein at the Göttingen university, directly following conversations between Felix Klein, Emmy Noether and David Hilbert about invariance and conservation laws [626]. At that time Hilbert had already derived the Einstein equation from an action principle, of which the Lagrangian was purely and simply the Ricci scalar; whence Vermeil and Klein's motivations.
- Hermann Weyl reproduced Vermeil's proof (in English) for the four-dimensional case in his treatise 1922 [627].
- The same year, Cartan [628], working at the level of the field equations, showed that the only tensor that is symmetric and linear in the second derivatives of the metric tensor is a linear combination of g_{ab}, Rg_{ab} and the Ricci tensor R_{ab}. He also showed that enforcing the divergence-free condition imposes that g_{ab} and Rg_{ab} must combine in the Einstein tensor.[1]

[1] Joël Merker provides a nice revisit of Cartan's work on the Einstein equations, with additional mathematical details and physical applications in [628].

P. Ramond, *The First Law of Mechanics in General Relativity & Isochrone Orbits in Newtonian Gravity*, Springer Theses,
https://doi.org/10.1007/978-3-031-17964-8

- Finally, David Lovelock showed in a four-papers program [40, 629–631] the strongest unicity result, relaxing most of Cartan's assumptions. Lovelock aknowledges Cartan's theorem and points out the differences between their work. Lovelock proved, in particular, that symmetry is not required and is a consequence of the four-dimensionality of spacetime. He also showed that asking for a mere dependence in second-order derivatives is sufficient, as the result is then linear in these. For a revisited, easier-to-read proof of Lovelock's work, see [632]. Note that Weyl also mentions that linearity is a consequence, not an assumption, at the very end of Appendix II of his treatise [627], but no proof nor reference is given there.

A.2 Killing Fields and Bitensors

We provide a number of important mathematical results used especially in Chap. 3 when working with a Killing vector field.

A.2.1 *Kostant Formula*

Combining the defining property of the Riemann tensor with Killing's equation (3.8) yields

$$\nabla_{ab}\xi_c + \nabla_{bc}\xi_a = R_{abc}{}^{d}\xi_d\,. \tag{A.1}$$

Performing a cyclic permutation on the indices a, b, c, and considering the linear combination $(abc) + (cab) - (bca)$, we readily obtain

$$2\nabla_{ab}\xi_c = \left(R_{abc}{}^{d} + R_{cab}{}^{d} - R_{bca}{}^{d}\right)\xi_d = -2R_{bca}{}^{d}\xi_d\,, \tag{A.2}$$

where the algebraic symmetry property $R_{[abc]}{}^{d} = 0$ was used in the last equality. Finally with $R_{bca}{}^{d} = -R_{cba}{}^{d}$ we obtain the Kostant formula

$$\nabla_{ab}\xi_c = R_{cba}{}^{d}\xi_d\,, \tag{A.3}$$

or equivalently $\nabla_{ab}\xi^c = R_{dab}{}^{c}\xi^d$. Equation (A.3) implies that the 2-form $\nabla_a\xi_b = \nabla_{[a}\xi_{b]}$, as well as its norm $|\nabla\xi|$, are conserved along the integral curves of ξ^a. Indeed, by virtue of the antisymmetry of the Riemann tensor with respect to its last two indices,

$$\xi^c\nabla_c\nabla_a\xi_b = -R_{abcd}\xi^c\xi^d = 0\,. \tag{A.4}$$

A.2.2 Commutation of the Covariant and Lie Derivatives

In this subsection we prove that, for any tensor field, the Lie derivative operator $\mathcal{L}_\xi$, such that $\mathcal{L}_\xi g_{ab} = 0$, commutes with the metric-compatible covariant derivative operator ∇_c, such that $\nabla_c g_{ab} = 0$. First, consider a tensor field T^N of type $(n, 0)$, where $N \equiv c_1 \dots c_n$ denotes an abstract multi-index with n indices. By definition of the Lie derivative operator in terms of the Levi-Civita connection, we have

$$\mathcal{L}_\xi T^N = \xi^e \nabla_e T^N - \sum_{i=1}^{n} T^{N_e} \nabla_e \xi^{c_i} , \tag{A.5a}$$

$$\mathcal{L}_\xi \nabla_a T^N = \xi^e \nabla_{ea} T^N - (\nabla^e T^N) \nabla_e \xi_a - \sum_{i=1}^{n} (\nabla_a T^{N_e}) \nabla_e \xi^{c_i} , \tag{A.5b}$$

where we used the shorthand $N_e \equiv c_1 \dots e \dots c_n$, with the abstract index e at the ith slot. Taking the covariant derivative of the first equation yields a formula that will be shown to be identical to Eq. (A.5b). Indeed,

$$\nabla_a \mathcal{L}_\xi T^N = (\nabla_a \xi^e) \nabla_e T^N + \xi^e \nabla_{ae} T^N - \sum_{i=1}^{n} (\nabla_a T^{N_e}) \nabla_e \xi^{c_i} - \sum_{i=1}^{n} T^{N_e} \nabla_{ae} \xi^{c_i} . \tag{A.6}$$

By Killing's equation, the first term of (A.6) is identical to the second term of (A.5b). Since (A.6) and (A.5b) share the same third term, we get the following expression for their difference:

$$(\nabla_a \mathcal{L}_\xi - \mathcal{L}_\xi \nabla_a) T^N = 2 \xi^e \nabla_{[ae]} T^N - \sum_{i=1}^{n} T^{N_e} \nabla_{ae} \xi^{c_i} . \tag{A.7}$$

Then, we apply the defining property of the curvature tensor to the first term on the right-hand side of (A.7), we use $R_{aeb}{}^{c_i} = -R_{eab}{}^{c_i}$, and rename some indices to obtain

$$(\nabla_a \mathcal{L}_\xi - \mathcal{L}_\xi \nabla_a) T^{c_1 \dots c_n} = \sum_{i=1}^{n} T^{c_1 \dots e \dots c_n} (R_{bae}{}^{c_i} \xi^b - \nabla_{ae} \xi^{c_i}) = 0 , \tag{A.8}$$

where the last equality follows by noticing that each term in parenthesis vanishes, by virtue of Kostant's formula (A.3). Finally, since the metric satisfies $\mathcal{L}_\xi g_{ab} = 0$ and $\nabla_c g_{ab} = 0$, it can be used to "lower" indices in (A.8), such that the result holds for a tensor field of any type. In summary, we have proven that for any Killing vector field ξ^a and for any tensor field χ,

$$\nabla (\mathcal{L}_\xi \chi) = \mathcal{L}_\xi (\nabla \chi) . \tag{A.9}$$

A.2.3 Bitensors and Their Lie Derivatives

In this appendix we shall briefly review the concepts of bitensor, coincidence limit, parallel propagator, invariant Dirac functional and Lie derivative operator. We shall then prove that the Lie derivative of the invariant Dirac distribution along the flow of a Killing field vanishes identically. The reader is referred to, e.g., Refs. [197, 633] for more details on those notions.

Bitensors and coincidence limit

Just like a tensor field is a multilinear map on the points x of a spacetime manifold $\mathscr{E}$, a *bitensor* field is a multilinear map on two points $(x, x') \in \mathscr{E} \times \mathscr{E}$. A generic bitensor will then be denoted as

$$\Omega^{ab\ldots}{}_{a'b'\ldots}(x, x')\,, \tag{A.10}$$

where the abstract indices $a, b, c, \ldots$ and $a', b', c', \ldots$ refer to the points x and x', respectively. Two examples of bitensors are used in this paper: the parallel propagator $g^{a}{}_{a'}(x, x')$ and the invariant Dirac functional $\delta_4(x, x')$, both defined below.

An important operation for bitensors is the *coincidence limit*, which consists in evaluating a bitensor at the same point. It is defined as

$$\left[\Omega^{ab\ldots}{}_{a'b'\ldots}\right](x) \equiv \lim_{x'\to x} \Omega^{ab\ldots}{}_{a'b'\ldots}(x, x')\,. \tag{A.11}$$

The coincidence limit of a bitensor is thus an ordinary tensor field. We will assume that this coincidence limit always exist and is independent of the direction in which x' approaches x. For more details regarding bitensors and the coincidence limit, see e.g. Ref. [197].

Parallel propagator

An important example of bitensor is the parallel propagator. If the points x and x' are "close enough," i.e., if x' is in a normal neighborhood of x, then there is a unique geodesic segment λ that joins them. On this geodesic segment, we introduce an orthonormal tetrad (e^{a}_{A}) that is parallel-transported on λ, where the subscript $A \in \{0, 1, 2, 3\}$ labels the vectors of the basis. By definition, this tetrad obeys the orthonormality and completion relations

$$g_{ab} e^{a}_{A} e^{b}_{B} = \eta_{AB} \quad \text{and} \quad g_{ab} = \eta_{AB} e^{A}_{a} e^{B}_{b}\,, \tag{A.12}$$

where the Minkowski metric $\eta_{AB} = \mathrm{diag}\,(-1, 1, 1, 1)$ is used to lower the Greek indices, and its inverse η^{AB} to raise them. The 1-form e^{A}_{a} is defined by metric duality as $e^{A}_{a} \equiv \eta^{AB} g_{ab} e^{b}_{B}$.

Next, we introduce a generic vector field v^a defined on λ. At any point $z \in \lambda$, this vector field can be expanded with respect to the tetrad (e^{a}_{A}), according to

$$v^a(z) = v^A(z) e^{a}_{A}(z)\,, \quad \text{where} \quad v^A \equiv v^a e^{A}_{a}\,. \tag{A.13}$$

Now we make the following remark: if v^a is parallely transported along λ, then it is clear from (A.13) that the tetrad components v^A remain constant along λ, and thus have the same numerical value at $z = x$ and at $z = x'$. By substituting the definition given in the right-hand side of Eq. (A.13) in each side of the equality $v^A(x) = v^A(x')$, and by using the orthogonal properties of the tetrad (A.12), we obtain

$$v^a(x) = g^a{}_{a'}(x, x') v^{a'}(x') \,, \quad \text{where} \quad g^a{}_{a'}(x, x') \equiv e^a_A(x) e^A_{a'}(x') \,. \tag{A.14}$$

The bitensor $g^a{}_{a'}(x, x')$ is the so-called *parallel propagator* from x' to x. It takes the vector v^a at the point x' and extends it by parallel transport to the point x. As long as the underlying geodesic is unique, this extension is unique as well. The formula (A.14) can be generalized to a generic tensor field of arbitrary rank. The parallel propagator (A.14) is used in Sect. 3.2.2 to extend the multipoles off the worldline of each quadrupolar particle.

Invariant Dirac distribution

The gravitational skeleton model reviewed in Chap. 2 relies crucially on a 4-dimensional, covariant generalization of the ordinary, noncovariant Dirac distribution. In four spacetime dimensions, the *invariant* Dirac functional $\delta_4(x, x')$ is the distributional biscalar defined by the relations [197]

$$\int_{\mathscr{V}} f(x)\, \delta_4(x, x')\, \mathrm{d}V = f(x') \quad \text{and} \quad \int_{\mathscr{V}'} f(x')\, \delta_4(x, x')\, \mathrm{d}V' = f(x) \,, \tag{A.15}$$

where f is a smooth scalar field (a test function), $\mathscr{V}$ and $\mathscr{V}'$ any four-dimensional regions of spacetime that contain the points x' and x, respectively, and $\mathrm{d}V = \sqrt{-g}\, \mathrm{d}^4x$ is the invariant volume element, with g the determinant of the metric tensor g_{ab} in a given coordinate basis. The definition (A.15) ensures that δ_4 is symmetric in its arguments,

$$\delta_4(x, x') = \delta_4(x', x) \,, \tag{A.16}$$

such that it depends necessarily on the *difference* of the events' coordinates. More precisely, given a coordinate system (x^α), one can easily show that [197]

$$\delta_4(x, x') = \prod_{\alpha=0}^{3} \frac{\delta(x^\alpha - x'^\alpha)}{\sqrt{-g}} \,, \tag{A.17}$$

where δ is the ordinary, noncovariant Dirac distribution, such that $\int_{\mathbb{R}} \phi(t) \delta(t)\, \mathrm{d}t = \phi(0)$ for any test function ϕ. Together with the consequence $\nabla_a\, g = 0$ of metric compatibility, the explicit formula (A.17) implies the property

$$(\nabla_a + \nabla_{a'})\, \delta_4(x, x') = 0 \,. \tag{A.18}$$

Finally, by recalling the notation (A.11) for the coincidence limit where $x' \to x$, an important distributional identity satisfied by δ_4, valid for any bitensor $\Omega^{ab\ldots}{}_{a'b'\ldots}$, is

$$\Omega^{ab\ldots}{}_{a'b'\ldots}(x, x')\,\delta_4(x, x') = \left[\Omega^{ab\ldots}{}_{a'b'\ldots}\right]\delta_4(x, x')\,. \tag{A.19}$$

Lie derivative operator

Our derivation of the relations (3.30), (3.36), (3.46), (3.48) and (3.49) relies on the invariance (3.13) of the quadrupolar SEM tensors (2.58)–(2.59) and (3.17) along the integral curves of the helical Killing vector (3.7). Since these tensor fields involve the (distributional) bitensor $\delta_4(x, x')$, we require a generalization to bitensor fields of the ordinary definition of the Lie derivative of a smooth tensor field.

Let ϕ_ϵ denote a one-parameter group of diffeomorphism generated by a vector field $\xi^a(x)$. The "push-forward" ϕ^*_ϵ can then be used to carry any smooth bitensor field $\Omega^{ab\ldots}{}_{a'b'\ldots}(x, x')$ along the flow of ξ^a, by acting independently on both spacetime points x and x'. By analogy with the definition of the Lie derivative of a smooth tensor field, we define the Lie derivative $\mathcal{L}_\xi$ along ξ^a of a smooth bitensor field as [6]

$$\mathcal{L}_\xi \Omega^{ab\ldots}{}_{a'b'\ldots} \equiv \lim_{\epsilon\to 0} \frac{1}{\epsilon}\left[\phi^*_{-\epsilon}\Omega^{ab\ldots}{}_{a'b'\ldots} - \Omega^{ab\ldots}{}_{a'b'\ldots}\right], \tag{A.20}$$

where all bitensor appearing in (A.20) are evaluated at the same combination (x, x') of points. For a generic biscalar $S(x, x')$, this general definition reduces to

$$\mathcal{L}_\xi S(x, x') = \xi^a \nabla_a S(x, x') + \xi^{a'} \nabla_{a'} S(x, x')\,. \tag{A.21}$$

This definition coincides with that used by Harte [346], who defines the Lie derivative of any bitensor as acting independently and linearly on each spacetime point.

Lie derivative of the invariant Dirac functional

Finally, we wish to establish a formula for the Lie derivative $\mathcal{L}_\xi \delta_4$ along a vector field ξ^a of the invariant Dirac distribution δ_4. By applying the definition (A.21) of the Lie derivative to the distributional biscalar (A.17), and by using the property (A.18), the Lie derivative of the invariant Dirac distribution along a smooth vector field ξ^a reads

$$\mathcal{L}_\xi \delta_4(x, x') = (\xi^a - \xi^{a'})\nabla_a \delta_4(x, x')\,. \tag{A.22}$$

This form can be further simplified by integrating $\mathcal{L}_\xi \delta_4$ against an arbitrary "test function." Indeed, for any smooth scalar field f with compact support, the formula (A.22) implies

$$\int_{\mathscr{V}} f(x)\,\mathcal{L}_\xi \delta_4(x, x')\,\mathrm{d}V = \int_{\mathscr{V}} \nabla_a\big(f(\xi^a - \xi^{a'})\delta_4\big)\,\mathrm{d}V - \int_{\mathscr{V}} \left(\nabla_a f(\xi^a - \xi^{a'}) + f\nabla_a \xi^a\right)\delta_4\,\mathrm{d}V\,, \tag{A.23}$$

where we integrated by parts and used $\nabla_a \xi^{a'} = 0$. The first integral in the right-hand side can be converted into a surface integral by applying Stokes' theorem, and easily shown to vanish thanks to the distributional identity (A.19) and the coincidence limit $[\xi^a - \xi^{a'}] = 0$:

$$\int_{\mathcal{V}} \nabla_a\big(f(\xi^a - \xi^{a'})\delta_4\big)\,\mathrm{d}V = \oint_{\partial\mathcal{V}} [f(\xi^a - \xi^{a'})]\delta_4\,\mathrm{d}\Sigma_a = 0\,. \tag{A.24}$$

Moreover, by using the distributional identity (A.19), the coincidence limit $[\xi^a - \xi^{a'}] = 0$ and the defining property (A.15) of the invariant distribution δ_4, the second term in the right-hand side of Eq. (A.23) simply reads

$$\int_{\mathcal{V}} \Big([\nabla_a f(\xi^a - \xi^{a'})] + f\nabla_a\xi^a\Big)(x)\,\delta_4(x,x')\,\mathrm{d}V = \big(f\nabla_a\xi^a\big)(x')\,. \tag{A.25}$$

Hence, by substituting (A.24) and (A.25) into (A.23), while using the formula $\nabla_a\xi^a = \frac{1}{2}g^{ab}\mathcal{L}_\xi g_{ab}$, we obtain the distributional identity

$$\mathcal{L}_\xi\delta_4(x,x') = -\frac{1}{2}\,\delta_4(x,x')\,g^{ab}(x)\mathcal{L}_\xi g_{ab}(x)\,. \tag{A.26}$$

This agrees with Eq. (136) in [347], where the same definition of the Lie derivative acting on bitensors was introduced.

In the case where ξ^a is a Killing vector field (see Appendix A.2 below), Eq. (A.26) shows that the Dirac functional $\delta_4(x,x')$ is invariant along the integral curves of a Killing field. In particular, for the helical Killing field (3.7) considered in this work, the distributional identity (A.26) implies that

$$\mathcal{L}_k\delta_4(x,y) = 0 \tag{A.27}$$

for any point $y \in \gamma$. This result was used in Sect. 3.2.1 to establish the helical constraint (3.30), and in Sect. 3.2.2 to derive the Lie-dragging along k^a of the velocity u^a, momentum p^a, spin S^{ab} and quadrupole J^{abcd} of each quadrupolar particle, Eqs. (3.36), (3.46), (3.48) and (3.49).

A.3 Tulczyjew's Theorems

In this appendix, we shall review Tulczyjew's two theorems [338], which play a central in any work that relies on the gravitational skeleton formalism. The first theorem ensures the existence and unicity of the normal form of a tensor expressed as a distributional multipolar expansion. The second theorem gives a necessary and sufficient condition for such a distributional multipolar expansion to vanish: that the multipoles of its normal form all vanish identically. Finally, we give a proof of the proposition (3.29), which was used to derive the key helical constraint (3.30). In what follows, the notation $[\![p, q]\!]$ to denotes the set of integers between any two given integers $(p, q) \in \mathbb{N} \times \mathbb{N}$ with $p < q$.

A.3.1 *Tulczyjew's First Theorem*

First we introduce some notation. Let $Y^M \equiv Y^{a_1 \dots a_m}$ denote a contravariant tensor field of rank $m \in \mathbb{N}$. We assume that its support is restricted to a worldline γ with proper time τ and unit tangent u^a, and that it can be written as a distributional multipolar expansion of order $n \in \mathbb{N}$. Therefore, at any point $x \in \mathscr{E}$ we have

$$Y^M(x) = \sum_{k=0}^{n} \nabla_K \int_\gamma \mathcal{Y}^{MK}(y)\, \delta_4(x, y)\, \mathrm{d}\tau \,, \tag{A.28}$$

where $y \in \gamma$ and $(\mathcal{Y}^{MK})_{k \in [\![0,n]\!]}$ is a collection of $n+1$ multipoles, i.e., contravariant tensors of rank $m+k$ defined along γ. We introduced the notations $\nabla_K \equiv \nabla_{c_1 \dots c_k}$ and $\mathcal{Y}^{MK} \equiv \mathcal{Y}^{Mc_1 \dots c_k}$ for $k \geqslant 1$, while $\nabla_K = \mathrm{id}$ and $\mathcal{Y}^{MK} = \mathcal{Y}^M$ for $k = 0$. We may now state the first theorem.

Theorem 1 *For any given $(m, n) \in \mathbb{N} \times \mathbb{N}$, let Y^M be defined as in Eq. (A.28). Then there exists a collection of multipoles $(\mathscr{Y}^{MK})_{k \in [\![0,n]\!]}$ that are (i) symmetric with respect to any pair of indices of the multi-index K, (ii) orthogonal to u^a with respect to any index of K, and (iii) such that*

$$Y^M(x) = \sum_{k=0}^{n} \nabla_K \int_\gamma \mathscr{Y}^{MK}(y)\, \delta_4(x, y)\, \mathrm{d}\tau \,. \tag{A.29}$$

Moreover, the multipolar expansion (A.29) is unique and the multipoles $(\mathscr{Y}^{MK})_{k \in [\![0,n]\!]}$ can be written explicitly in terms of the multipoles $(\mathcal{Y}^{MK})_{k \in [\![0,n]\!]}$ of (A.28). Equation (A.29) is referred to as the normal form of Y^M.

Unicity of the normal form is straightforward once we have Theorem 2 below. For the existence, we construct in Appendix A.4 below the explicit normal form associated with the quadrupolar ($n = 2$) gravitational skeleton of the generic tensor Y^M. In particular, the multipoles $\mathscr{Y}^{MK}$ of the quadrupolar normal form are given in terms of the multipoles $\mathcal{Y}^{MK}$ in Eqs. (A.50). For the existence of the normal form when $n > 2$, see e.g. Ref. [344] and references therein.

A.3.2 *Extension of Tulczyjew's Second Theorem*

Again, we first introduce some notation. Let $p \in \mathbb{N}^\star$ and let $(Y_\mathrm{i}^M)_{\mathrm{i} \in [\![1,p]\!]}$ denote a collection of p contravariant tensor fields of rank $m \in \mathbb{N}$. We assume that the support of each Y_i^M is restricted to a worldline γ_i with proper time τ_i and unit tangent u_i^a, and that it can be expressed as a distributional multipolar expansion of order $n \in \mathbb{N}$. Therefore, at any point $x \in \mathscr{E}$, we have

$$Y_{\rm i}^{M}(x) = \sum_{k=0}^{n} \nabla_K \int_{\gamma_{\rm i}} \mathscr{Y}_{\rm i}^{MK}(y_{\rm i})\, \delta_4(x, y_{\rm i})\, \mathrm{d}\tau_{\rm i}\,, \tag{A.30}$$

where, for each $\mathrm{i} \in [\![1, p]\!]$, $(\mathscr{Y}_{\rm i}^{MK})_{k\in[\![0,n]\!]}$ is a collection of $n+1$ multipoles, i.e., contravariant tensors of rank $m+k$ defined along γ. We may now state the second theorem.

Theorem 2 *For any given $(m, n, p) \in \mathbb{N} \times \mathbb{N} \times \mathbb{N}^*$, let $(Y_{\rm i}^{M})_{\mathrm{i}\in[\![1,p]\!]}$ be a collection of p tensors defined as in Eq. (A.30), and let $Y^M \equiv \sum_{\rm i} Y_{\rm i}^M$ denote their sum. The following result holds. If for all $(\mathrm{i}, k) \in [\![1, p]\!] \times [\![0, n]\!]$, $\mathscr{Y}_{\rm i}^{MK}$ is symmetric with respect to any pair of indices of the multi-index K and is orthogonal to $u_{\rm i}^a$ with respect to any index of K, then*

$$Y^M = 0 \iff \forall\, (\mathrm{i}, k) \in [\![1, p]\!] \times [\![0, n]\!]\,,\ \mathscr{Y}_{\rm i}^{MK} = 0\,. \tag{A.31}$$

Clearly, if all the multipoles $\mathscr{Y}_{\rm i}^{MK}$ vanish, then $Y_{\rm i}^M = 0$ by (A.30), and the sum $Y^M = \sum_{\rm i} Y_{\rm i}^M$ vanishes as well. The heart of the proof therefore resides in showing that if $Y_{\rm i}^M$ is in normal form, then $Y^M = 0$ implies $\mathscr{Y}_{\rm i}^{MK} = 0$ for all $(\mathrm{i}, k) \in [\![1, p]\!] \times [\![0, n]\!]$. See Refs. [311, 338, 344] and references therein for more details.

Importantly, in the literature this result is proven for a single multipolar particle, whereas Theorem 2 is stated for an arbitrary number $p \in \mathbb{N}^*$ of particles. However, we now argue that the multi-particle case can easily be reduced to the single-particle case. Indeed, the general idea behind the proof for a single particle is the following: take an arbitrary rank-m tensor Z_M, whose compact support $\mathscr{V}$ intersects the worldline γ of the particle. Contract Z_M with Y^M, given as a multipolar expansion in normal form (A.29), and integrate over $\mathscr{V}$. The goal is then to show that this integral vanishes for every Z_M only if $\mathscr{Y}^{MK} = 0$ for all $k \in [\![0, n]\!]$. Since this holds for any compact support $\mathscr{V}$, the vanishing of $\mathscr{Y}^{MK}$ must hold for any portion of γ, and thus on all of γ. Now, if there are $p \in \mathbb{N}^*$ multipolar particles, one may choose the volume $\mathscr{V}$ such that it intersects *only one* of the p worldlines, say γ, and proceed with the single-particle proof, as summarized above.

A.3.3 Proof of Proposition (3.29)

We now give a proof of the proposition (3.29), which was used in Sect. 3.2.2 to derive the colinearity (3.30) of k^a and $u_{\rm i}^a$ along the worldline $\gamma_{\rm i}$ of the ith particle. For clarity's sake we will drop the subscript i, as the proof holds for any of the two particles of the binary system. The proposition (3.29) is an implication which is most easily proven by contraposition.

First, we introduce Fermi coordinates (t, x^i) in a neighborhood $\mathscr{N} \subset \mathscr{E}$ of the worldline γ. Using Fermi coordinates, γ is parameterized by the proper time $\tau \in \mathbb{R}$ according to $(t, x^i) = (\tau, 0, 0, 0)$. Let $I \subset \mathbb{R}$ be a finite interval and γ_I be the finite portion of γ parameterized by $\tau \in I$. We also set $\epsilon > 0$ and let $\mathscr{V}$ denote

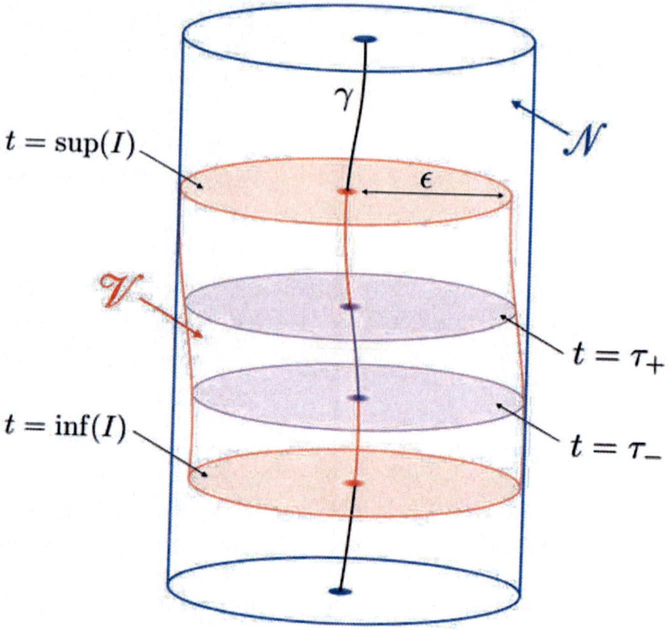

Fig. A.1 The geometrical setup used in Appendix A.3.3 to prove the proposition (3.29)

the 3-cylinder of Fermi coordinate radius ϵ that surrounds γ_I. We assume that ϵ is small enough such that $\mathscr{V} \subseteq \mathscr{N}$. This geometric setup is depicted on Fig. A.1. Finally, for given tensor fields $(\tilde{\mathscr{T}}^{ab}, \tilde{\mathscr{T}}^{abc}, \tilde{\mathscr{T}}^{abcd})$ defined in Eq. (3.19), we let $\mathbb{F} \equiv \{\tilde{\mathscr{T}}^{ab} f_{ab} + \tilde{\mathscr{T}}^{abc}\nabla_c f_{ab} + \tilde{\mathscr{T}}^{abcd}\nabla_{cd} f_{ab}\,,\, f_{ab} \in C^\infty_{\mathscr{V}}\}$, with $C^\infty_{\mathscr{V}}$ the set of tensor fields with compact support $\mathscr{V}$ that are smooth on the interior $\mathscr{V}^\circ$.

Proof that z is constant along γ

Since all the scalar fields are evaluated along γ in the following integrals, we view them as functions of the proper time τ. We first prove by contraposition the part of the proposition (3.29) that implies that $\dot{z} = 0$ along γ, i.e., we show that the following implication is true:

$$\exists\, \tau_0 \in I\,,\ \dot{z}(\tau_0) \neq 0 \implies \exists\, f \in \mathbb{F}\,,\ \int_{\mathbb{R}} \big(z\dot{f} + w^a \nabla_a f\big)\, \mathrm{d}\tau \neq 0\,. \tag{A.32}$$

Let $\tau_0 \in I$ be as in Eq. (A.32). Since $\tau \mapsto \dot{z}(\tau)$ is continuous, there exists a neighborhood of τ_0, say $]\tau_-, \tau_+[\, \subset I$, such that $\forall \tau \in\,]\tau_-, \tau_+[\,, \dot{z}(\tau) \neq 0$ and is of constant sign. Now consider the scalar field f defined on $\mathscr{V}^\circ$ by

$$f(t, x^i) \equiv \begin{cases} \exp\left([(t - \tau_-)(t - \tau_+)]^{-1}\right) & \text{if } t \in\,]\tau_-, \tau_+[\,, \\ 0 & \text{if } t \notin\,]\tau_-, \tau_+[\,, \end{cases} \tag{A.33}$$

and $f \equiv 0$ elsewhere. We claim that such f verifies the right-hand side of (A.32).

Indeed, in the Fermi coordinate system, the 4-velocity u^a has components $u^\alpha = (1, 0, 0, 0)$, so that $w^a u_a = 0$ implies that w^a has components $w^\alpha = (0, w^i)$. Hence $w^a \nabla_a f = w^i \partial_i f = 0$ since f does not depend on x^i. Moreover, the function $\tau \mapsto f(\tau)$ is smooth on $\mathbb{R}$ and vanishes for $\tau \notin]\tau_-, \tau_+[$. Consequently $\int_\gamma (z\dot{f} + w^a \nabla_a f)\,\mathrm{d}\tau = \int_{\tau_-}^{\tau_+} z(\tau)\dot{f}(\tau)\,\mathrm{d}\tau$, and an integration by parts gives

$$\int_\gamma \left(z\dot{f} + w^a \nabla_a f\right) \mathrm{d}\tau = -\int_{\tau_-}^{\tau_+} \dot{z}(\tau) f(\tau)\,\mathrm{d}\tau\,. \tag{A.34}$$

But the integral on the right-hand side of (A.34) cannot vanish, as $\dot{z}$ is nonzero with constant sign over $]\tau_-, \tau_+[$ by assumption, and $f(\tau) > 0$ for all $\tau \in]\tau_-, \tau_+[$. Therefore, f as defined in Eq. (A.33) verifies the proposition (A.32), provided that it belongs to the set $\mathbb{F}$.

To establish that $f \in \mathbb{F}$, consider the tensor $f_{ab} \equiv \phi g_{ab}$, where the scalar field ϕ is defined over $\mathscr{V}^\circ$ by

$$\phi(t, x^i) \equiv \begin{cases} \exp\left([(t-\tau_-)(t-\tau_+)]^{-1}\right) \Phi(t)^{-1} & \text{if } t \in]\tau_-, \tau_+[\,, \\ 0 & \text{if } t \notin]\tau_-, \tau_+[\,, \end{cases} \tag{A.35}$$

and $\phi \equiv 0$ elsewhere, where $\Phi(t) \equiv g_{ab} \tilde{\mathscr{T}}^{ab}$, the latter being evaluated at the point $(t, 0, 0, 0)$. With $f_{ab} = \phi g_{ab}$ and ϕ given in (A.35), one can readily check that $f = \tilde{\mathscr{T}}^{ab} f_{ab} - \tilde{\mathscr{T}}^{abc} \nabla_c f_{ab} + \tilde{\mathscr{T}}^{abcd} \nabla_{cd} f_{ab}$, for f given in Eq. (A.33). The computation involves (i) the metric compatibility $\nabla_c g_{ab} = 0$, (ii) the independence of ϕ with respect to x^i, (iii) the fact that the Christoffel symbols $\Gamma^t{}_{ij}|_\gamma$ vanish in Fermi coordinates, and (iv) the normal form of the tensors $\tilde{\mathscr{T}}^{ab}$, $\tilde{\mathscr{T}}^{abc}$ and $\tilde{\mathscr{T}}^{abcd}$.

Proof that $w^a = 0$ along γ

Having proven that z is constant along γ, it is clear that for any compactly supported f, we have $\int_\gamma z\dot{f}\,\mathrm{d}\tau = 0$. Consequently, we will now establish that $w^a = 0$ along γ by proving the following proposition:

$$\exists\, \tau_0 \in I\,,\ w^a(\tau_0) \neq 0 \implies \exists\, f \in \mathbb{F}\,,\ \int_\gamma w^a \nabla_a f\,\mathrm{d}\tau \neq 0\,. \tag{A.36}$$

As noted before, in the Fermi coordinate system we have $w^\alpha = (0, w^i)$ so that $w^a \nabla_a f = w^i \partial_i f$. Because w^i is continuous and $w^i(\tau_0) \neq 0$, there exists a neighborhood $]\tau_-, \tau_+[\subset I$ of τ_0 such that *at least* one component of w^i, say w^1, is nonzero and of constant sign over $]\tau_-, \tau_+[$. Now consider the following scalar field defined on $\mathscr{V}^\circ$:

$$f(t, x^i) \equiv \begin{cases} x^1 \exp\left([(t-\tau_-)(t-\tau_+)]^{-1}\right) & \text{if } t \in]\tau_-, \tau_+[\,, \\ 0 & \text{if } t \notin]\tau_-, \tau_+[\,, \end{cases} \tag{A.37}$$

and $f \equiv 0$ elsewhere. Because f does not depend on x^2 and x^3, the integral in the right-hand side of Eq. (A.36) is simply

$$\int_\gamma w^a \nabla_a f \, \mathrm{d}\tau = \int_{\tau_-}^{\tau_+} w^1(\tau) \exp\left([(\tau - \tau_-)(\tau - \tau_+)]^{-1} \right) \mathrm{d}\tau \,. \qquad \text{(A.38)}$$

As earlier this integral does not vanish since, by assumption, $w^1(\tau)$ is nonzero and of constant sign over $]\tau_-, \tau_+[$. Therefore, the scalar field (A.37) verifies Eq. (A.36) provided that it belongs to $\mathbb{F}$. Once again, let us consider the tensor field $f_{ab} \equiv \phi g_{ab}$, with the scalar field ϕ now defined on $\mathscr{V}^\circ$ by

$$\phi(t, x^i) \equiv \begin{cases} \frac{1}{6}(x^1)^3 \exp\left([(t - \tau_-)(t - \tau_+)]^{-1} \right) \Phi(t)^{-1} & \text{if } t \in \,]\tau_-, \tau_+[\,, \\ 0 & \text{if } t \notin \,]\tau_-, \tau_+[\,, \end{cases} \qquad \text{(A.39)}$$

and $\phi \equiv 0$ elsewhere, where this time $\Phi(t) \equiv g_{\alpha\beta}\left(\frac{1}{6}(x^1)^2 \tilde{\mathscr{T}}^{\alpha\beta} - \frac{1}{2}x^1 \tilde{\mathscr{T}}^{\alpha\beta 1} + \tilde{\mathscr{T}}^{\alpha\beta 11}\right)$, the latter being evaluated at the point $(t, 0, 0, 0)$. With $f_{ab} = \phi g_{ab}$ and ϕ given in Eq. (A.39), one can readily check that $f = \tilde{\mathscr{T}}^{ab} f_{ab} - \tilde{\mathscr{T}}^{abc} \nabla_c f_{ab} + \tilde{\mathscr{T}}^{abcd} \nabla_{cd} f_{ab}$, for f given in (A.37). This time, the computation involves (i) the metric compatibility $\nabla_c g_{ab} = 0$, (ii) the independence of ϕ with respect to the coordinates x^2 and x^3, (iii) the fact that the Christoffel symbols $\Gamma^\alpha{}_{ij}|_\gamma$ vanish in Fermi coordinates, and (iv) the normal form of the tensors $\tilde{\mathscr{T}}^{ab}$, $\tilde{\mathscr{T}}^{abc}$ and $\tilde{\mathscr{T}}^{abcd}$.

A.4 Normal Form of a Quadrupolar Skeleton

In this appendix, we shall detail the computations that lead to the unique normal form associated with the quadrupolar gravitational skeleton of a generic tensor field. This normal form can for instance be used to derive the equations of evolution for the momentum and spin of a *dipolar* particle, i.e. Eq. (2.57) with $J^{abcd} = 0$, or to obtain the Lie-dragging constraints (3.22) for a *quadrupolar* particle.

A.4.1 A Useful Formula

Before deriving this normal form, we first prove a simple formula that will turn out crucial in order to carry out the following computations. Let χ denote a generic tensor field defined along the worldline γ with unit tangent u^a. Then we have

$$\begin{aligned}\nabla_a \int_\gamma \chi(y')\, u^{a'}(y')\, \delta_4(x, y')\, \mathrm{d}\tau &= \int_\gamma \chi(y')\, u^{a'}(y') \nabla_a \delta_4(x, y')\, \mathrm{d}\tau \\ &= -\int_\gamma \chi(y')\, u^{a'}(y') \nabla_{a'} \delta_4(x, y')\, \mathrm{d}\tau \\ &= -\int_\gamma \left[\chi(y')\, \delta_4(x, y')\right]^{\cdot} \mathrm{d}\tau + \int_\gamma \dot{\chi}(y')\, \delta_4(x, y')\, \mathrm{d}\tau\,, \end{aligned} \tag{A.40}$$

where we used the fact that the covariant derivative ∇_a acts on points $x \in \mathscr{E}$ but not on points $y' \in \gamma$ in the first equality, the property (A.18) of the invariant Dirac functional in the second equality, and we integrated by parts in the third and last equality. Assuming that χ vanishes as $\tau \to \pm\infty$ to discard the boundary terms, we conclude that for any tensor field χ defined along γ,

$$\nabla_a \int_\gamma \chi\, u^a\, \delta_4\, \mathrm{d}\tau = \int_\gamma \dot{\chi}\, \delta_4\, \mathrm{d}\tau\,. \tag{A.41}$$

A.4.2 *Derivation of the Normal Form*

Note: to fit equations on a single line, in what follows we use the shorthand $\mathrm{d}\!\!\mathrm{d}\tau \equiv \delta_4 \mathrm{d}\tau$.

We now turn to the derivation of the normal form at quadrupolar order. Let Y^M denote a generic tensor field of rank $m \in \mathbb{N}$, expressed as a gravitational skeleton at quadrupolar order, i.e., Eq. (A.28) with $n = 2$, such that

$$Y^M = \int_\gamma \mathcal{Y}^M \mathrm{d}\!\!\mathrm{d}\tau + \nabla_a \int_\gamma \mathcal{Y}^{Ma} \mathrm{d}\!\!\mathrm{d}\tau + \nabla_{ab} \int_\gamma \mathcal{Y}^{Mab} \mathrm{d}\!\!\mathrm{d}\tau\,, \tag{A.42}$$

with $\mathcal{Y}^M$, $\mathcal{Y}^{Ma}$ and $\mathcal{Y}^{Mab}$ the monopole, dipole and quadrupole of Y^M, respectively. From Theorem 1 the first term in (A.42) is already in normal form. For the second term, we perform an orthogonal decomposition of $\mathcal{Y}^{Ma}$ with respect to the index a by means of the projector (2.62) orthogonal to the 4-velocity u^a, namely $\mathcal{Y}^{Ma} = \mathcal{Y}^{M\hat{a}} - \mathcal{Y}^{Mu} u^a$. (Recall the notations introduced below Eqs. (3.18).) Using the formula (A.41) then gives

$$\nabla_a \int_\gamma \mathcal{Y}^{Ma} \mathrm{d}\!\!\mathrm{d}\tau = \nabla_a \int_\gamma \mathcal{Y}^{M\hat{a}} \mathrm{d}\!\!\mathrm{d}\tau - \int_\gamma \left(\mathcal{Y}^{Mu}\right)^{\cdot} \mathrm{d}\!\!\mathrm{d}\tau\,. \tag{A.43}$$

Regarding the third term on the right-hand side of Eq. (A.42), we start again by performing an orthogonal decomposition of the integrand, yielding $\mathcal{Y}^{Mab} = \mathcal{Y}^{M\hat{a}\hat{b}} - \mathcal{Y}^{Mu\hat{b}} u^a - \mathcal{Y}^{Mau} u^b$. Substituting this decomposition into the integral and using the formula (A.41), we obtain

$$\nabla_{ab} \int_\gamma \mathcal{Y}^{Mab} \mathrm{d}\tau = \nabla_{ab} \int_\gamma \mathcal{Y}^{M\hat{a}\hat{b}} \mathrm{d}\tau - \nabla_{ab} \int_\gamma \mathcal{Y}^{Mu\hat{b}} u^a \mathrm{d}\tau - \nabla_a \int_\gamma (\mathcal{Y}^{Mau})^{\cdot} \mathrm{d}\tau \,. \tag{A.44}$$

We shall now consider those three terms successively.

We begin with the first term of (A.44). We split the second covariant derivative into its symmetric and antisymmetric part, $\nabla_{ab} \int_\gamma \mathcal{Y}^{M\hat{a}\hat{b}} \mathrm{d}\tau = \nabla_{ab} \int_\gamma \mathcal{Y}^{M(\hat{a}\hat{b})} \mathrm{d}\tau + \nabla_{[ab]} \int_\gamma \mathcal{Y}^{M\hat{a}\hat{b}} \mathrm{d}\tau$, the first term of which being already in normal form (integrand symmetric with respect to a and b and orthogonal to u^a). For the second term we use the definition of the Riemann tensor and its algebraic symmetries to get

$$\nabla_{[ab]} \int_\gamma \mathcal{Y}^{M\hat{a}\hat{b}} \mathrm{d}\tau = -\frac{1}{2} \sum_{j=1}^{m} \int_\gamma R_{abe}{}^{c_j} \mathcal{Y}^{M_e \hat{a}\hat{b}} \mathrm{d}\tau \,, \tag{A.45}$$

where M_e is the multi-index M with e at the jth slot. This term is in normal form since it does not involve any derivative, just like the first term on the right-hand side of Eq. (A.42).

Next, for the second term of Eq. (A.44), we commute the two covariant derivatives and use once again the definition of the Riemann tensor. Using the formula (A.41) we obtain

$$\nabla_{ab} \int_\gamma \mathcal{Y}^{Mu\hat{b}} u^a \mathrm{d}\tau = \nabla_b \int_\gamma (\mathcal{Y}^{Mu\hat{b}})^{\cdot} \mathrm{d}\tau - \sum_{j=1}^{m} \int_\gamma R_{abe}{}^{c_j} \mathcal{Y}^{M_e u\hat{b}} u^a \mathrm{d}\tau \,, \tag{A.46}$$

where the rightmost term is in normal form. However the first term is not, because it needs not be orthogonal to u_b. But it can be handled simply by writing the integrand $(\mathcal{Y}^{Mu\hat{b}})^{\cdot}$ as $(\mathcal{Y}^{Muc})^{\cdot} h^b{}_c + \mathcal{Y}^{Muc} \dot{h}^b{}_c$. The Leibniz rule and metric compatibility imply $\dot{h}^b{}_c = \dot{u}^b u_c + u^b \dot{u}_c$. We combine these formulas and use the formula (A.41) one last time to get

$$\nabla_b \int_\gamma (\mathcal{Y}^{Mu\hat{b}})^{\cdot} \mathrm{d}\tau = \nabla_b \int_\gamma \big[(\mathcal{Y}^{Muc})^{\cdot} h^b{}_c + \mathcal{Y}^{Muu} \dot{u}^b \big] \mathrm{d}\tau + \int_\gamma (\mathcal{Y}^{Muc} \dot{u}_c)^{\cdot} \mathrm{d}\tau \,. \tag{A.47}$$

Finally, for the third and last term of (A.44), we write, again, an orthonormal decomposition with respect to the abstract index a, namely $\mathcal{Y}^{Mau} = \mathcal{Y}^{M\hat{a}u} - \mathcal{Y}^{Muu} u^a$. Taking the covariant derivative along u^a and using the Leibniz rule, along with the formula (A.41), then gives

$$\nabla_a \int_\gamma (\mathcal{Y}^{Mau})^{\cdot} \mathrm{d}\tau = \nabla_a \int_\gamma (\mathcal{Y}^{M\hat{a}u})^{\cdot} \mathrm{d}\tau - \nabla_a \int_\gamma \mathcal{Y}^{Muu} \dot{u}^a \mathrm{d}\tau - \int_\gamma (\mathcal{Y}^{Muu})^{\cdot\cdot} \mathrm{d}\tau \,. \tag{A.48}$$

The second to last term is in normal form since $\dot{u}^a$ is orthogonal to u_a, and the last one is in normal form too. Finally, the first term in the right-hand side of Eq. (A.48) can be brought into normal form by following the steps that yielded Eq. (A.47).

To conclude, we can combine Eqs. (A.45)–(A.48) to write the normal form of (A.44). Combining the latter with (A.43) gives, at last, the normal form of the quadrupolar expansion (A.42) of Y^M according to

$$Y^M = \int_\gamma \mathscr{Y}^M \mathrm{d}\tau + \nabla_a \int_\gamma \mathscr{Y}^{Ma} \mathrm{d}\tau + \nabla_{ab} \int_\gamma \mathscr{Y}^{Mab} \mathrm{d}\tau \,, \tag{A.49}$$

where $\mathscr{Y}^M$, $\mathscr{Y}^{Ma}$ and $\mathscr{Y}^{Mab}$ are given explicitly in terms of $\mathcal{Y}^M$, $\mathcal{Y}^{Ma}$ and $\mathcal{Y}^{Mab}$ by

$$\mathscr{Y}^M \equiv \mathcal{Y}^M - (\mathcal{Y}^{Mu} - (\mathcal{Y}^{Muu})^{\cdot} + 2\mathcal{Y}^{M(cu)}\dot{u}_c)^{\cdot} + \sum_{j=1}^{m} R_{abe}{}^{c_j}\, \mathcal{Z}^{M_e ab} \,, \tag{A.50a}$$

$$\mathscr{Y}^{Ma} \equiv \mathcal{Y}^{M\hat{a}} - 2(\mathcal{Y}^{M(cu)})^{\cdot} h^a{}_c - \mathcal{Y}^{Muu}\dot{u}^a \,, \tag{A.50b}$$

$$\mathscr{Y}^{Mab} \equiv \mathcal{Y}^{M(\hat{a}\hat{b})} \,, \tag{A.50c}$$

where $\mathcal{Z}^{M_e ab} \equiv \mathcal{Y}^{M_e u\hat{b}} u^a - \frac{1}{2}\mathcal{Y}^{M_e \hat{a}\hat{b}}$ in the first term. By construction, we have $\mathscr{Y}^{Ma} u_a = 0$, $\mathscr{Y}^{M[ab]} = 0$ and $\mathscr{Y}^{Mab} u_b = 0$. Consequently, Eq. (A.49) is the normal form of (A.42). This normal form was used in Chap. 4, Sect. 3.2.1 to go from Eq. (3.20) to the associated normal form (3.21a)–(3.22). As a final note, we should mention that the calculations performed above are not sufficient by themselves to derive the reduced form (2.58)–(2.59) of the SEM tensor of a quadrupolar particle, nor the associated equations of evolution (2.57), which were achieved in Ref. [344]. Indeed, while imposing the SEM conservation $\nabla_a T^{ab} = 0$ to the generic quadrupolar SEM tensor (2.58)–(2.59), one must in particular put into normal form the quadrupolar contribution $\nabla_{bcd} \int_\gamma \mathcal{T}^{abcd} \mathrm{d}\tau$, which involves a *triple* covariant derivative. Such (rather tedious) calculation was made is present in [344].

A.4.3 Tulczyjew's Reduction Process for a Dipolar Particle

We now derive the evolution equations and SEM tensor at dipolar order. To this end, we apply the normal form (A.49)–(A.50) to the tensor $Y^e \equiv \nabla_a T^{ae}$, thus taking the multi-index $M = e$ of length 1. Our starting point is therefore the generic dipolar Ansatz given in Eq. (2.25), to which we apply the covariant derivative ∇_b. In the notations of Eq. (A.49) our starting coefficients $\mathcal{Y}$ are therefore just

$$\mathcal{Y}^e \equiv 0 \,, \quad \mathcal{Y}^{ea} = \mathcal{T}^{ea} \quad \text{and} \quad \mathcal{Y}^{eab} = \mathcal{T}^{eab} \,. \tag{A.51}$$

The normal form of $\nabla_a T^{ae}$ and can therefore be written, as claimed in Eq. (2.27) of Chap. 2, namely

$$\nabla_b T^{ab} = \int_\gamma \mathscr{X}^a \delta_4 \mathrm{d}\tau + \nabla_b \int_\gamma \mathscr{X}^{ab} \delta_4 \mathrm{d}\tau + \nabla_b \nabla_c \int_\gamma \mathscr{X}^{abc} \delta_4 \mathrm{d}\tau \,. \tag{A.52}$$

where

$$\mathscr{X}^e \equiv \left((\mathcal{T}^{euu})^{\cdot} - \mathcal{T}^{eu} - 2\mathcal{T}^{e(cu)}\dot{u}_c\right)^{\cdot} - \tfrac{1}{2}R_{abc}{}^e\left(\mathcal{T}^{c\hat{a}\hat{b}} + 2\mathcal{T}^{cu\hat{a}}u^b\right), \quad \text{(A.53a)}$$

$$\mathscr{X}^{ea} \equiv \mathcal{T}^{e\hat{a}} - 2\left(\mathcal{T}^{e(cu)}\right)^{\cdot} h^a{}_c - \mathcal{T}^{euu}\dot{u}^a, \quad \text{(A.53b)}$$

$$\mathscr{X}^{eab} \equiv \mathcal{T}^{e(\hat{a}\hat{b})}. \quad \text{(A.53c)}$$

By Tulczyjew's second theorem all the $\mathscr{X}$ vanish, and we obtain three equations. Let us explore their consequences, by introducing the 3+1 expansions of the quantities $\mathcal{T}$ discussed in Sect. 2.2.2, which we rewrite here for convenience

$$\mathcal{T}^{ab} = \mathscr{m}u^a u^b + 2\mathscr{m}^{(a}u^{b)} + \mathscr{m}^{ab} \quad \text{(A.54a)}$$

$$\mathcal{T}^{abc} = u^a u^b \mathscr{n}^c + 2u^{(a}\mathscr{n}^{b)c} + \mathscr{n}^{abc} + \mathscr{o}^{ab}u^c, \quad \text{(A.54b)}$$

Equation $\mathscr{X}^{eab} = 0$

In terms of the quantities appearing in Eqs. (A.54), the equation $\mathscr{X}^{eab} = 0$ implies that $u^a\mathscr{n}^{(bc)} + \mathscr{n}^{a(bc)} = 0$. Contracting this with u_a yields $\mathscr{n}^{(bc)} = 0$ and $\mathscr{n}^{a(bc)} = 0$. Hence $\mathscr{n}^{ab}$ is antisymmetric and $\mathscr{n}^{abc}$ vanishes identically because it also satisfies $\mathscr{n}^{[ab]c} = 0$ by construction.[2] These results simplify the two other equations as follows

Equation $\mathscr{X}^{ea} = 0$

Now Eq. (A.53b) reads, thanks to the previous results,

$$\mathscr{m}^b u^a + \mathscr{m}^{ab} = -h^b{}_c(\mathscr{o}^{ac} + \mathscr{n}^c u^a + \mathscr{n}^{ac})^{\cdot},. \quad \text{(A.55)}$$

Taking its space-space component with the help of the orthogonal projector, antisymmetrizing the result and using $\mathscr{m}^{[ab]=0}$ leads to $h^a{}_b h^c{}_d\dot{\sigma}^{cd} = 0$, where $\sigma^{ab} \equiv u^{[a}\mathscr{n}^{b]} + \mathscr{n}^{[ab]}$. Re-expanding the projector as $h^a{}_b = \delta^a{}_b + u^a u_b$ and using the orthogonalities then yields

$$\dot{\sigma}^{ab} = -\dot{\mathscr{n}}^{[a}u^{b]} + 2u_c\dot{\mathscr{n}}^{c[a}u^{b]}. \quad \text{(A.56)}$$

This will eventually turn into the precession equation for the spin tensor S^{ab}, that we will define at the end, i.e., once the last equation has been massaged, as follows.

Equation $\mathscr{X}^e = 0$

Equation (A.53c), with the help of the previously established results and the quantity σ^{ab} defined above, can readily be transformed into

$$\left(\mathscr{m}u^a + \mathscr{m}^a + (\mathscr{o}^{ab} + u^a\mathscr{n}^b + \mathscr{n}^{ab})\dot{u}_b\right)^{\cdot} = -\frac{1}{2}R_{cbd}{}^a\left(2\sigma^{dc}u^b + \sigma^{bc}u^d\right), \quad \text{(A.57)}$$

[2] To show this, one can keep applying the antisymmetry and symmetry property alternatively until one reaches $\mathscr{n}^{abc} = -\mathscr{n}^{abc}$.

where we simply inserted the projections (A.54) into Eq. (A.53a). Let us denote the quantity that is being differentiated on the left-hand side of this equation by p^a (thus giving the definition in Eq. (2.28)). On the one hand, a n explicit calculation shows that the antisymmetric tensorial product between u^a and p^a, and Eq. (A.56), gives

$$p^{[a}u^{b]} = \dot{\sigma}^{ab} . \tag{A.58}$$

On the other hand, using the antisymmetry of σ^{ab} and the algebraic symmetries of the Riemann tensor allows one to simplify the right-hand side of Eq. (A.57), to obtain

$$\dot{p}^a = -2R_{bcd}{}^a u^b \sigma^{cd} . \tag{A.59}$$

Equations (A.59) and (A.58) are the dipolar equations of motion and precession respectively, and the spin tensor S^{ab} is simply twice the quantity σ^{ab} (thus giving the definition in Eq. (2.28)). It is then straightforward to rewrite the $\mathcal{T}$'s of the SEM Ansatz in terms of p^a, S^{ab} and derivatives of other terms. Thanks to the magic formula (A.41), these cancel one another and one obtains the pole-dipole SEM tensor (2.29). This concludes the derivation.

A.4.4 Equations of Motion of a Multipolar Particle

Here we give the consequences of the dipolar evolution equations given in Sect. 2.3.2, generalized to any multipolar order. In particular, starting from

$$\dot{p}^a = \frac{1}{2}R_{bcd}{}^a S^{bc} u^d + F^a \quad \text{and} \quad \dot{S}^{ab} = 2p^{[a}u^{b]} + M^{ab} , \tag{A.60}$$

with a generic force vector F^a and torque tensor M^{ab}, we can perform the same analysis as that provided in that section to include the multipolar contributions to the Eqs. (2.66)–(2.76). They read

$$\dot{S} = \frac{1}{2S} M^{ab} S_{ab} , \tag{A.61a}$$

$$\dot{m} = -F^u - M^{ua}\ddot{u}_a , \tag{A.61b}$$

$$\dot{S}_a = -\frac{1}{2}\varepsilon_{abcd}\dot{u}^b S^{cd} - \tfrac{1}{2}\varepsilon_{abcd}u^b M^{cd} , \tag{A.61c}$$

$$p^a = mu^a - \varepsilon^a{}_{bcd}u^b \dot{u}^c S^d + M^{ab}u_b , \tag{A.61d}$$

$$m\dot{u}^a = B^{ab}S_b - S^{ab}\ddot{u}_b - \dot{M}^{ab}u_b - 2M^{\hat{a}b}\dot{u}_b - F^{\hat{a}} . \tag{A.61e}$$

We stress that these equations hold for any multipolar order, but depend on our choice of SSC $u_a S^{ab} = 0$.

Appendix B
Symplectic Geometry

B.1 Newtonian First Law of Mechanics

In this section, we derive the quasi-Newtonian first law of point-particle mechanics, for circular orbits, as touched upon. It is only quasi-Newtonian because one needs to add the mass energy mc^2 of the particles to write the result in a form similar to the first law derived in Chap. 5. For completeness, we start from scratch with the classical reduction of the newtonian two-body problem.

B.1.1 Reduction of the Two-Body Problem

Let us consider two test particles labeled by $\mathrm{i} \in \{1; 2\}$, of mass m_i and located in the Euclidean three-dimensional space of Newtonian mechanics. The Lagrangian of this system is

$$\mathscr{L}(r_1, r_2, \dot{r}_1, \dot{r}_2) \equiv \frac{1}{2} m_1 |\dot{r}_1|^2 + \frac{1}{2} m_2 |\dot{r}_2|^2 + \frac{G m_1 m_2}{|r_1 - r_2|}, \tag{B.1}$$

where r_i are the position vectors and G the gravitational constant. The conjugated variables read $p_\mathrm{i} \equiv \partial_{\dot{r}_\mathrm{i}} \mathscr{L} = m_\mathrm{i} \dot{r}_\mathrm{i}$ and allows us to write the Hamiltonian of the system $\mathscr{H} \equiv \dot{r}_1 p_1 + \dot{r}_2 p_2 - \mathscr{L}$ as

$$\mathscr{H}(r_1, r_2, p_1, p_2) \equiv \frac{|p_1|^2}{2m_1} + \frac{|p_2|^2}{2m_2} - \frac{G m_1 m_2}{|r_1 - r_2|}. \tag{B.2}$$

It is clear from that equation that $\mathscr{H}$ depends on (r_1, r_2, p_1, p_2) only through the norms of the momenta and of the distance between the two particles. It is therefore natural to perform a first change of coordinates (or physically a change of reference frame) in order to reduce the complexity of the problem. To this end, let u be the vector pointing from the origin of the old coordinates to the centre of mass of the

P. Ramond, *The First Law of Mechanics in General Relativity & Isochrone Orbits in Newtonian Gravity*, Springer Theses,
https://doi.org/10.1007/978-3-031-17964-8

system. Then,

$$u \equiv \beta_1 r_1 + \beta_2 r_2 \,, \quad \text{where} \quad \beta_{\rm i} \equiv \frac{m_{\rm i}}{m} \quad \text{and} \quad m \equiv m_1 + m_2 \,. \tag{B.3}$$

Then, let $r \equiv r_1 - r_2$ be the separation vector between the two particles. The new variables (u, r) are related to the old ones (r_1, r_2) by a linear transformation, which we may represent as

$$\begin{pmatrix} u \\ r \end{pmatrix} = \begin{pmatrix} \beta_1 \mathbb{I} & \beta_2 \mathbb{I} \\ -\mathbb{I} & \mathbb{I} \end{pmatrix} \begin{pmatrix} r_1 \\ r_2 \end{pmatrix} \,. \tag{B.4}$$

where $\mathbb{I}$ is the 3×3 identity matrix. Having performed a linear transformation of the coordinates constrains us to change the momenta linearly too, if we want the system to remain hamiltonian. Indeed, if (q, p) are canonical variables and $\mathcal{A}$ is an invertible matrix, then the mapping $(q, p) \mapsto (Q, P)$ with $Q = \mathcal{A}q$ is canonical if and only if $p = \mathcal{A}^\top P$, where $\mathcal{A}^\top$ denotes the transpose of $\mathcal{A}$. Applying this to the present case, we need for the new momenta $P \equiv (v, p)$ to satisfy

$$v \equiv p_1 + p_2 \quad \text{and} \quad p \equiv \beta_1 p_2 - \beta_2 p_2 \,. \tag{B.5}$$

We note that v is the total linear momentum of the system. In the new variables (u, r, v, p), the Hamiltonian reads

$$\mathscr{H}(u, r, v, p) \equiv \frac{|p|^2}{2\mu} + \frac{|v|^2}{2m} - \frac{Gm\mu}{|r|} \,, \tag{B.6}$$

where we have introduced the reduced mass $\mu \equiv m_1 m_2 / m$ to further simplify the expression. By comparing equations (B.2) and (B.6), we see that, numerically, all we have done is replace (p_1, p_2, m_1, m_2) by (p, v, μ, m), respectively. In particular, the hamiltonian of the initial (m_1, m_2)-system is equivalent to that of (μ, m). The main advantage of this new expression is that, now, $\mathscr{H}$ does not depend explicitly on the coordinate u associated to the total momentum. This is reminiscent of the momentum conservation. In particular, since v is constant, we may as well set it equal to zero, in which case $p_1 = -p_2$, and therefore $p = p_1 = -p_2$. Leaving aside the couple (u, v) all together, we have reduced the initial problem to the dynamics of the Hamiltonian

$$\mathscr{H}(r, p) \equiv \frac{|p|^2}{2\mu} - \frac{Gm\mu}{|r|} \,. \tag{B.7}$$

But now we see that the Hamiltonian is invariant under a fixed rotation of the separation vector r as well as its associated momentum p. Intuitively, this should reduce the problem from the phase space $\mathbb{R}^3 \times \mathbb{R}^3 \ni (r, p)$ to $\mathbb{R} \times \mathbb{R}$. To do this properly, the usual spherical coordinates (ρ, θ, ϕ) should be particularly useful. But to do things self-consistently, we would have to compute the new momenta $(p_\rho, p_\theta, p_\phi)$ associated to (ρ, θ, ϕ). To do this, we can use a brute force, Poisson brackets-like calculation. I prefer to take a less straightforward route, with much less calculations. First, one of the two Hamilton equations in the Cartesian variables (x, y, z), which I denote by $r = (r_1, r_2, r_3)$, gives us $\dot{r}_i = \partial_{p_i}\mathscr{H} = p_i/\mu$. Consequently, we have $|p|^2 = \mu^2|\dot{r}|^2$, and we can now perform the inverse Legendre transform $\mathscr{L}(r, \dot{r}) = \mu|\dot{r}|^2 - \mathscr{H}(r, \mu\dot{r})$, so that

$$\mathscr{L}(r, \dot{r}) = \frac{1}{2}\mu|\dot{r}|^2 + \frac{Gm\mu}{r}. \tag{B.8}$$

The advantage now is that we can write the separation vector r in any coordinates, spherical being the one we are after. In the natural, spherical basis $(e_\rho, e_\theta, e_\phi)$ we simply have $r = \rho e_\rho$ and therefore $|r| = \rho$ and $|\dot{r}|^2 = \dot{\rho}^2 + \rho^2\dot{\theta}^2 + \rho^2\sin^2\theta\dot{\phi}^2$. Inserting this in the previous expression of the Lagrangian and computing the momenta gives

$$p_\rho \equiv \frac{\partial\mathscr{L}}{\partial\dot{\rho}} = \mu\dot{\rho}\,, \quad p_\theta \equiv \frac{\partial\mathscr{L}}{\partial\dot{\theta}} = \mu\rho^2\dot{\theta} \quad \text{and} \quad p_\phi \equiv \frac{\partial\mathscr{L}}{\partial\dot{\phi}} = \mu\rho^2\sin^2\theta\dot{\phi}\,. \tag{B.9}$$

One last time, a Legendre transform allows us to write the Hamiltonian in the new, spherical-type canonical coordinates $(\rho, \theta, \phi, p_\rho, p_\theta, p_\phi)$, with respect to the center of mass of the system. It reads

$$\mathscr{H} = \frac{p_\rho^2}{2\mu} + \frac{p_\theta^2}{2\mu\rho^2} + \frac{p_\phi^2}{2\mu\rho^2\sin^2\theta} - \frac{Gm\mu}{\rho}. \tag{B.10}$$

Several remarks are in order. First, the fact that, in these coordinates, $\mathscr{H}$ is independent of (ϕ), implying that p_ϕ is constant. Without loss of generality, we can set $(p_\phi, \phi = (0, \pi/2)$, effectively defining an orbital plane, on which the coordinate system (r, θ) are the usual polar coordinates. By doing so, $\mathscr{H}$ becomes independent of θ, and therefore p_θ is constant as well. This is the conservation of angular momentum within the orbital plane.

In the end, we see that the classical two-body problem, which started with a 12-dimensional phase-space can be reduced to the analysis of a hamiltonian on $\mathbb{R}_+ \times \mathbb{R} \ni (r, p_r)$. This is the "reduction" of the two-body problem, a remarkable consequence of the underlying symmetries of the Keplerian potential (and Euclidean nature of Newtonian mechanics in general).

B.1.2 First Law of Mechanics

Let us consider the hamiltonian of the system as a function of both the coordinates/momenta as well as the masses of the particles:

$$H(r, p, L, m_1, m_2) = \frac{p^2}{2\mu} + \frac{L^2}{2\mu r^2} - \frac{Gm\mu}{r}, \tag{B.11}$$

where the reduced and total masses are functions of the individual masses through $m = m_1 + m_2$ and $\mu = m_1 m_2/m$. With the help of Hamilton's equations, the differential $\delta\mathscr{H}$ of this hamiltonian can be simply written as

$$\delta\mathscr{H} = -\dot{p}\delta r + \dot{r}\delta p + \dot{\theta}\delta L + \sum_{\rm i} z_{\rm i}\delta m_{\rm i}, \tag{B.12}$$

where for any i $\in \{1; 2\}$, the coefficients $z_{\rm i}$ are defined by

$$z_{\rm i} \equiv -\frac{1}{m_{\rm i}}\left(\frac{p^2}{2m_{\rm i}} + \frac{L^2}{2m_{\rm i}r^2} - \frac{Gm}{r}\right), \tag{B.13}$$

In the case of a circular orbit of radius r_o, we have $\dot{p} = \dot{r} = 0$ and the true anomaly θ evolves linearly with time, at frequency Ω (the orbital frequency). Thanks to Kepler's third law $\Omega^2 r_o^2 = Gm$ and the expression of the angular momentum $L^2 = Gmr_o\mu^2$, Eq. (B.12) simply reads

$$\delta\mathscr{H} = \Omega\delta L - \frac{Gm_2}{r_o}\left(1 + \frac{m_2}{2m}\right) - \frac{Gm_1}{r_o}\left(1 + \frac{m_1}{2m}\right). \tag{B.14}$$

How can we compare this Newtonian first law to the relativistic case? In GR, the hamiltonian has the numerical value of the ADM mass M, which encompasses all energy within the spacetime. The link between the total energy H at play here and this ADM mass is simply $Mc^2 = H + mc^2$. Let us add the total mass energy mc^2 to the Hamiltonian and regroup all terms to get

$$\delta Mc^2 - \Omega\delta L = \left(1 - \frac{Gm_2}{r_o c^2}\left(1 + \frac{m_2}{2m}\right)\right)\delta m_1 c^2 + (1 \leftrightarrow 2). \tag{B.15}$$

This is precisely the Newtonian first law of mechanics derived by the authors of [387]. Indeed, in Eq. (1.1) there, we read the same equation with $z_{\rm i}$ given in Eq. (2.37) as

$$z_1 = 1 + \left(-\frac{3}{4} - \frac{3}{4}\Delta + \frac{\nu}{2}\right)x + O(x^2), \tag{B.16}$$

where $\nu \equiv m_1 m_2/m^2$, $\Delta \equiv (m_2 - m_1)/m$ and $x = (m\Omega)^3/2$. Using Kepler's third law, we find agreement between the two results.

B.2 Keplerian and Harmonic Dynamics

B.2.1 Keplerian Dynamics

The dynamics in the Kepler potential $\psi(r) = -\mu/r$ is most easily found by writing an ODE for the function $u(\theta)$ where $u = 1/r$ is the Binet variable (see [345]). This ODE is linear in the case of the Kepler potential and reads $u'' + u = \mu/\Lambda^2$, with $u' = \mathrm{d}u/\mathrm{d}\theta$. The solution for $r(\theta)$ is then easily obtained, and using $(r(0), \theta(0)) = (r_p, 0)$ as initial conditions gives

$$r(\theta) = \frac{p}{1+\varepsilon\cos\theta}, \quad \text{where} \quad p = \frac{\Lambda^2}{\mu}, \quad \varepsilon = \sqrt{1 + \frac{2\xi\Lambda^2}{\mu^2}}, \tag{B.17}$$

where $p > 0$ is the semi-latus rectum and $\varepsilon \in [0; 1[$ the eccentricity. In celestial mechanics, the angle θ is called the true anomaly. One can also parametrize the orbit with the so-called eccentric anomaly E, such that

$$r(E) = \alpha(1 - \varepsilon\cos E), \quad \text{where} \quad \tan\frac{E}{2} = \sqrt{\frac{1+\varepsilon}{1-\varepsilon}}\tan\frac{\theta}{2}, \tag{B.18}$$

where $\alpha = p/(1-\varepsilon^2) = -\mu/2\xi$ is the semi-major axis of the elliptic orbit. Contrary to the true anomaly, the eccentric anomaly E can be linked analytically to the orbital time t elapsed along the orbit, as encoded in the Kepler equation

$$\frac{\sqrt{\mu}}{\alpha^{3/2}}t = E - \varepsilon\sin E, \tag{B.19}$$

where it is assumed that $E = \theta = 0$ at $t = 0$, as our initial conditions require. Finally, Kepler's third law of motion comes as a corollary of (B.19). By definition, every Keplerian orbit is an ellipse, thus the radial period T coincides with the orbital period, and corresponds to $E = 2\pi$. Consequently, (B.19) implies Eq. (B.19) since the semi-major axis α is linked to the energy ξ by $\alpha = -\mu/2\xi$, as follows from (B.17).

B.2.2 Harmonic Dynamics

In an harmonic potential, the orbit is an ellipse, except that the origin of coordinates is at the center of the ellipse, not at one of its focii (like the Keplerian case). The easiest way to solve the equations of motion is through Cartesian coordinates. This is well-known and derived in most classical textbooks (e.g., Sect. 3.1.(a) of [591]). It was mentioned in Chap. 8 that an isochrone potential ψ always belongs to one of the five families ($\psi_{\rm Ha}$, $\psi_{\rm He}$, $\psi_{\rm Ke}\psi_{\rm Bo}$, $\psi_{\rm Ho}$,) up to a gauge-term of the form $\varepsilon + \frac{\lambda}{2r^2}$, where $(\varepsilon, \lambda) \in \mathbb{R}^2$. The analytical results derived in this paper are general enough to take into account this gauge-liberty, for all non-harmonic potentials. For the (non-gauged) harmonic potential, the results of the last section is very classical and applies. There only remains the case of gauged-harmonic potentials, of the form (8.28). Let us solve the dynamics in this general potential.

Hamiltonian in action-angle variables

The computation of the radial action J for the harmonic class is easily found by setting $b = 0$ in the expressions (10.6) for T and Θ. They are given by

$$T = \frac{\pi}{2}\frac{\sqrt{-d}}{a} \quad \text{and} \quad \Theta = \frac{\pi\Lambda}{\sqrt{\Lambda^2 - e/d}}, \tag{B.20}$$

where we recall that $d < 0$ for an orbit to exist (otherwise the parabola is not convex) and $a > 0$ since by assumption $\delta = ad > 0$. From the above formulae, the expression (10.5) of J as a sum of integrals over T and Θ can be easily turned into

$$J(\xi, \Lambda) = J_o + \frac{\sqrt{-d}}{4a}\xi - \frac{1}{2}\sqrt{\Lambda^2 - \frac{e}{d}}, \tag{B.21}$$

where J_o is an integration constant that depends on (a, b, c, d, e). Since we must have $J = 0$ in the case of a circular orbit, a calculation (see e.g. Sect. IV.A.*1* of [3]) gives $J_o = -\frac{c}{4a\sqrt{-d}}$. Now, as in Sect. 10.1.1, we get the expression of the Hamiltonian $H(J, \Lambda)$ by simply solved the above formula for the energy ξ, giving

$$H(J, \Lambda) = -\frac{c}{d} + \frac{4a}{\sqrt{-d}} J + \frac{2a}{\sqrt{-d}}\sqrt{\Lambda^2 - \frac{e}{d}}. \tag{B.22}$$

The Hamiltonian of the isochrone class is therefore always linear in J, and linear in Λ only in the case $e = 0$, i.e., when the parabola crosses the origin of the (x, y)-plane (cf. Eq. (8.27)). Naturally, we recover the gauge term $-c/d$ (energy shift) of the harmonic potential (8.28). The Hamiltonian frequencies $(\omega_J, \omega_\Lambda)$ read in the present, harmonic case

$$\omega_J := \frac{\partial H}{\partial J} = \frac{4a}{\sqrt{-d}} \quad \text{and} \quad \omega_\Lambda := \frac{\partial H}{\partial \Lambda} = \frac{2a}{\sqrt{-d}}\frac{\Lambda}{\sqrt{\Lambda^2 - e/d}}. \tag{B.23}$$

Comparing these frequencies with (B.20) shows that the ratio ω_Λ/ω_J coincides with $\Theta(\Lambda)/2\pi$, as discussed in Sect. 10.1.1 in the non-harmonic isochrone case.

Explicit polar solution

First, let us re-write the equation of motion (8.4) with the explicit form (8.28). We have

$$\frac{1}{16}\left(\frac{\mathrm{d}x}{\mathrm{d}t}\right)^2 = (\xi + \frac{c}{d})x + \frac{e}{d} - \Lambda^2 + \frac{a^2}{d}x^2 \,, \tag{B.24}$$

and we suppose as always that $x(t = 0) = x_p$. The right-hand side of (B.24) vanishes at $t = 0$ when $x = x_p$ (periastron) and at $t = T/2$ when $x = x_a$ (apoastron). Both x_p, x_a are easily found as the quadratic roots of the right-hand side. Between the two, i.e., for $t \in]0; T/2[$, it is strictly positive. Therefore, for we can factorize the right-hand side of (B.24) and rearrange the result into[3]

$$\frac{\mathrm{d}x}{\sqrt{(x - x_p)(x_a - x)}} = 4\sqrt{\frac{a^2}{-d}}\mathrm{d}t \,. \tag{B.25}$$

Now we integrate this equation using the Euler substitution $x \mapsto \chi$ defined by $\chi^2 = \frac{x - x_p}{x_a - x}$. Integrating in the χ variable and going back to x then gives the simple formula

$$x(t) = x_a + (x_p - x_a)\cos(\Omega t)^2 \,, \quad \text{where} \quad \Omega = \sqrt{\frac{4a^2}{-d}} = \frac{2\pi}{T} \,, \tag{B.26}$$

in which we fixed the integration constant by requiring $x(0) = x_p$. As a verification, we see that $x(t = T/2) = x_a$, in agreement with Kepler's generalized third law (10.6) for $b = 0$ (harmonic class). Equation (B.26) gives the solution for the radial part of the dynamics $r(t)$ via $x = 2r^2$. For the angular part, we may use the angular equation of motion $\dot{\theta} = \Lambda/r^2$ and integrate with respect to t. We readily obtain:

$$\theta(t) = 2\Lambda \int_0^t \frac{\mathrm{d}\tau}{x_a + (x_p - x_a)\cos(\Omega\tau)^2} \,. \tag{B.27}$$

Let us now set $\phi = \Omega\tau$ and $\varepsilon^2 = 1 - x_p/x_a$ (such that $0 < \varepsilon < 1$) in (B.27), and integrate with respect to ϕ using a partial fraction decomposition (same technique as around Eq. (4.55)). We then find the explicit expression:

$$\theta(t) = \frac{2\Lambda}{\Omega x_p x_a} \sum_{\pm} \arctan\left(\sqrt{\frac{1 + \varepsilon_\pm}{1 - \varepsilon_\pm}} \tan\frac{\Omega t}{2}\right) \,, \quad E = \Omega t \,. \tag{B.28}$$

where we used the identity $x_a^2(1 - \varepsilon^2) = x_p x_a$ and set $\varepsilon_\pm = \mp\varepsilon$. Equations (B.26) and (B.28) are true of any isochrone potential of the harmonic class. They should be compared to their non-harmonic equivalent (10.21) and (10.27), respectively. Based on the fact that the Kepler equation (B.19) reduces to $\Omega t = E$ in the limit $b \to 0$, there probably exists a way to gather all these isochrone results (both harmonic and non-harmonic) under the same formulation. We have not managed to find such

[3] For the harmonic case $b = 0$, we necessarily have $d < 0$, since $y = Y(x)$ as given by (8.28) must be convex for the physical orbit to be well-defined.

formulae, but encourage the interested reader to give it a try. It would provide a strong proof of universality to the isochrone paradigm. As a perspective, we mention the following formulae for arctangent which is seldom found in the literature and which may be of some help. Consider the classical trigonometric identity: $\tan(x + y) = \frac{\tan x + \tan y}{1 - \tan x \tan y} + \kappa\pi$ with $\kappa \in \{-1, 0, 1\}$ depending on where $\tan(x + y)$ lies in $[-\pi; \pi]$. Setting $(X, Y) = (\tan x, \tan y)$ we readily get the other well-known identity (assuming $\kappa = 0$)

$$\arctan \frac{X + Y}{1 - XY} = \arctan X + \arctan Y \,. \tag{B.29}$$

Now set $2U := X + Y$ and $2V := 1 - XY$, so that we can express X and Y in terms of (U, V) by solving a quadratic equation. Noticing that the left-hand side of (B.29) is $\arctan U/V$, we obtain a formula for the arctangent of a quotient

$$\arctan \frac{U}{V} = \sum_{\pm} \arctan(U \pm \sqrt{U^2 + 2V - 1}) \tag{B.30}$$

Noticing that each formula for $\theta(E)$ ((10.27) non-harmonic and (B.28) for harmonic) always involves terms of the form (B.30), it may be a possible starting point to find the common point between the harmonic and non-harmonic results. We leave this for future work.

B.3 Details on the Hamiltonian Treatment

B.3.1 Analytic Continuations

The explicit formula (10.27) for $\theta(E)$ has been obtained for all (non-harmonic) isochrone potentials with $x_v < 0$. In particular, the derivation does not hold a priori for the Bounded and Hollowed class, for which $x_v > 0$. Indeed, if $x_v > 0$, then $\zeta^2 = -x_v/2\alpha^2$ is negative, and thus ζ is imaginary (see between (10.25) and (4.55)). However, note that the final formula (10.27) is a sum of two terms, one with $+\zeta$ and another with $-\zeta$ (recall that $\varepsilon_\pm = \varepsilon/(1 \pm \zeta)$ there). Since $\zeta \in \mathbb{I} \subset \mathbb{C}$, this means that Eq. (10.27) reads $\theta \propto F(\zeta) + F(\bar{\zeta})$ where the function F is simply

$$F : z \in \mathbb{C} \mapsto \frac{\varepsilon}{\sqrt{(1+z)^2 - \varepsilon^2}} \arctan\left(\sqrt{\frac{1 + z + \varepsilon}{1 + z - \varepsilon}} \tan \frac{E}{2}\right), \tag{B.31}$$

with $(\varepsilon, E) \in [0; 1] \times [0 : \pi]$ seen as fixed parameters here. Now if F is holomorphic around II (and since its restriction to real z is real-valued) then we automatically have $F(z) + F(\bar{z}) = F(z) + \bar{F(z)}$ from standard results of complex analysis. But up to multiplicative positive constants F can be written as $F(z) = \sqrt{f(z)} \arctan(\sqrt{g(z)})$,

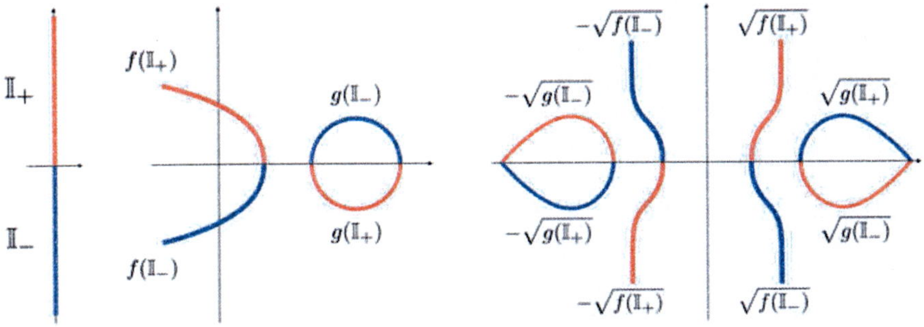

Fig. B.1 Successive images of the imaginary axis $\mathbb{I} = \mathbb{I}_+ \cup \mathbb{I}_-$ under f, g and then $\sqrt{f}$, $\sqrt{g}$. Images for both branches $\pm\sqrt{\ }$ of the (complex) square root are represented

where

$$f : z \mapsto (1+z)^2 - \varepsilon^2 \quad \text{and} \quad g : z \mapsto \frac{1+z+\varepsilon}{1+z-\varepsilon}, \tag{B.32}$$

Now let $\mathbb{I} = \mathrm{i}\mathbb{R}$ be the imaginary axis and $\mathbb{I}_\pm = \mathrm{i}\mathbb{R}_\pm$ be the lower (−) and upper (+) part of the imaginary axis. By a direct calculation, the image of $\mathbb{I}_+$ (resp. $\mathbb{I}_-$) under f is the upper (resp. lower) part of a parabola, and under g, it is the upper (resp. lower) part of a circle (Möbius transformation). Under $z \mapsto \sqrt{z}$, and irrespective of the chosen principal value, the parabola $f(\mathbb{I})$ is then mapped to a set of two disconnected curves that are complex conjugate to one-another (corresponding to the two $f(\mathbb{I}_+)$ and $f(\mathbb{I}_-)$ parts). The same is true for the circle $g(\mathbb{I})$. In particular, $\sqrt{g(\mathbb{I})}$ is a closed curve on the right-half plane (for the $+\sqrt{\ }$ branch) that does not intersect the imaginary axis, such that its (complex) arctangent is well-defined and holomorphic (square roots and inverse trig functions can all be defined in terms of the complex logarithm, with which it is easy to check that all is well-defined and holomoprhic). A summary of all this is depicted on Fig. B.1. The conclusion is that the function F defined in (B.31) is holomorphic on $\mathbb{I}$, and therefore $F(z) + F(\bar{z}) = 2\mathrm{Re}(F(z))$, making the formula for $\theta(E)$ also real-valued and well-defined, even in the case $x_v > 0$, i.e., for Bounded and Hollowed potentials.

B.3.2 Universal ODE for Parabolae

In this appendix, we solve the so-called universal ODE for parabolae $3Y''Y'''' = 5(Y''')^2$. This ODE was already used in appendix B of [589] to characterize parabolae. We start by the case where $Y''' = 0$ which clearly is a solution. Then Y'' is a constant function and, therefore, $Y(x)$ is a quadratic polynomial. This corresponds to the harmonic class of parabolae (8.34). If $Y''' \neq 0$, then re-arranging the equation yields $Y''''/Y''' = \frac{5}{3}Y'''/Y''$, which can be readily integrated as $Y'''(Y'')^{-5/3} = C$ where $C \in \mathbb{R}$. From this which we directly get $-\frac{3}{2}(Y'')^{-2/3} = Cx + D$ with $D \in \mathbb{R}$. This implies that $Y''(x) \propto (Cx + D)^{-3/2}$, and therefore Y is of the form (8.30), encompassing all non-harmonic types of parabolae. Reciprocally, each parabola is a solution of the ODE, which finishes the proof.

B.3.3 Action-Angle Transformation and Birkhoff Invariants

In Sects. 10.2 and 10.3, we have used the Birkhoff normal form for a 1-dimensional system (2-dimensional phase space), and thus worked with scalar Birkhoff invariants. In this section, we would like to provide an alternative way of looking at the relation between these invariants. In particular, we consider the 2-dimensional point of view of the problem (4-dimensional phase space), and consider other, more general, types of invariants. In particular, this will allow us to understand more deeply the unicity of the Birkhoff invariants.

Let $\boldsymbol{\omega}$ be the frequency vector made of the two natural frequencies associated with some Hamiltonian $\mathcal{H}(A, B)$, given in terms of some action variables (A, B) (we are not interested in their respective angles here). It is given by

$$\boldsymbol{\omega} = \left(\frac{\partial \mathcal{H}}{\partial A}, \frac{\partial \mathcal{H}}{\partial B} \right) . \tag{B.33}$$

Let us know construct simple quantities using $\boldsymbol{\omega}$ whose value remain unchanged under a transformation from one set of action-angle variables to another. We will use the fact that transformation between sets of action-angles is not arbitrary. Indeed, when going from a set (A, B) to another, say (A', B'), symplecticity imposes that the old and new actions must be related by a matrix $M \in \mathrm{SL}(2, \mathbb{Z})$. These are 2×2 matrices with determinant 1 and coefficients in $\mathbb{Z}$. Roughly speaking, this is because angles must be transformed so that they remain angles, i.e. make $\mathbb{Z}$-linear combinations of them and not mix them with actions. Then symplecticity imposes that the actions be transformed similarly. For more on these action-angle transformations, we refer to the very clear discussion in [614], and to the book [634] for more technical details (see around proposition (6.5.3) there). Summarizing, we must have

$$\begin{pmatrix} A' \\ B' \end{pmatrix} = \begin{pmatrix} m & p \\ n & q \end{pmatrix} \begin{pmatrix} A \\ B \end{pmatrix} , \quad \text{where} \quad \begin{cases} (m, n, p, q) \in \mathbb{Z}^4 , \\ mq - np = 1 . \end{cases} \tag{B.34}$$

The change of actions (B.34) induces a change in the Hamiltonian such that the frequency vector (B.33) is transformed as $\boldsymbol{\omega} \mapsto M^{\top} \boldsymbol{\omega}$, where $M^{\top}$ is the transpose of $M = \left(\begin{smallmatrix} m & p \\ n & q \end{smallmatrix}\right)$. From then, it is easy to construct invariants by taking advantage of the fact that $\det M = 1$. For example, consider the following scalar quantities

$$\mathcal{J} := \frac{\partial \boldsymbol{\omega}}{\partial A} \wedge \boldsymbol{\omega} , \quad \mathcal{G} := \boldsymbol{\omega} \wedge \frac{\partial \boldsymbol{\omega}}{\partial B} \quad \text{and} \quad \mathcal{T} := \frac{\partial \boldsymbol{\omega}}{\partial A} \wedge \frac{\partial \boldsymbol{\omega}}{\partial B} , \tag{B.35}$$

where $\wedge$ denotes the usual determinant between two vectors. Then the transformation (B.34) leaves $\mathcal{J}, \mathcal{G}, \mathcal{T}$ invariant. Indeed, consider $\mathcal{J}'$, the expression of $\mathcal{J}$ in the new action variables. Then

$$\mathcal{J}' = \frac{\partial(M\boldsymbol{\omega})}{\partial A} \wedge (M\boldsymbol{\omega}) = (\det M)^2 \frac{\partial \boldsymbol{\omega}}{\partial A} \wedge \boldsymbol{\omega} = \mathcal{J}\,, \tag{B.36}$$

where we used in the first equality $\boldsymbol{\omega}' = M\boldsymbol{\omega}$, in the second the fact that M has (constant) coefficients in $\mathbb{Z}$ and in the third $\det M = 1$. A similar computation holds for both $\mathcal{G}$ and $\mathcal{T}$. Since $\boldsymbol{\omega} \wedge \boldsymbol{\omega} = 0$, the three quantities (B.35) are the most simple scalars built out of $\boldsymbol{\omega}$ that are invariant under (B.34). The link between $(\mathcal{J}, \mathcal{G}, \mathcal{T})$ (functions of (A, B)) and the Birkhoff invariants used in Sects. 10.2 and 10.3 is easily obtained as follows. Setting $(A, B) = (I, \Lambda)$ where Λ is the angular momentum action and $I = \rho$ or J. Then if the Hamiltonian is in a normal form of the type $N(I, \Lambda) = \mathfrak{l}(\Lambda) + \mathfrak{b}(\Lambda) I + \frac{1}{2}\mathfrak{B}(\Lambda) I^2$, the quantities $(\mathcal{J}, \mathcal{G}, \mathcal{T})$ are easily found to be

$$\mathcal{J} = \mathfrak{l}'\mathfrak{B} - \mathfrak{b}\mathfrak{b}'\,, \quad \mathcal{G} = \mathfrak{l}'\mathfrak{b}' - \mathfrak{b}\mathfrak{l}'' \quad \text{and} \quad \mathcal{T} = \mathfrak{l}''\mathfrak{B} - \mathfrak{b}'\mathfrak{b}'\,. \tag{B.37}$$

The quantity $\mathcal{T}(A, B)$ in (B.35) is the torsion of the torus (A, B). The vanishing of $\mathcal{T}$ and $\mathcal{G}$ is a necessary and sufficient condition for the Bertrand theorem to hold, as explained in [609]. We see from Eq. (10.63) that, in fact, the vanishing of $\mathcal{J}$ is a necessary and sufficient condition for the isochrone theorem to hold. What's more, it is clear from their definition (B.35) that if both $\mathcal{T}$ and $\mathcal{G}$ vanish, then so does $\mathcal{J}$, since the $\mathcal{T} = 0 = \mathcal{G}$ implies that $\partial_B \boldsymbol{\omega}$ is parallel to both $\partial_A \boldsymbol{\omega}$ and $\boldsymbol{\omega}$, consequently $\partial_A \boldsymbol{\omega}$ is parallel to $\boldsymbol{\omega}$ and thus $\mathcal{J} = 0$. Physically, this parallelism relations between the frequency vectors encodes the remarkable fact that Bertrand potentials are necessarily isochrone.

Appendix C
Euclidean Geometry

C.1 Details on Radial Potentials

Finite mass and attractive nature

Before going further, we would like to discuss some physical properties of the isochrone potentials associated with the five families $(\mathscr{P}_i)$. We start with a (non-necessarily isochrone) central potential ψ and the Poisson equation (7.15), from which we can easily infer the mass contained within a (spherical) shell surrounding the origin. We choose the units so that $G = 1$ in order to simplify the equations.

Let $\epsilon > 0$ be the inner radius of such a shell, and $R > \epsilon$ be its outer radius, so that $\psi(R)$ is well-defined, and we let $M_\epsilon(R)$ be the mass contained within this shell $[\epsilon, R]$. By definition, $M_\epsilon(R)$ is given by $\int_\varepsilon^R \rho(r) 4\pi r^2 \mathrm{d}r$. Multiplying Eq. (7.15) by r^2 and integrating over the shell $[\epsilon, R]$ readily give

$$M_\epsilon(R) = R^2 \psi'(R) - \epsilon^2 \psi'(\epsilon) \,. \tag{C.1}$$

From this equation, it is clear that the total mass $M(R)$ contained within the sphere or radius $r = R$ is simply given by the $\epsilon \to 0$ limit of $M_\epsilon(R)$. Therefore, for any radius R, $M(R)$ is finite if and only if the rightmost term $\epsilon^2 \psi'(\epsilon)$ in Eq. (C.1) remains bounded as $\epsilon \to 0$. If this limit is infinite, the potential is sourced by an infinite amount of mass at the physical origin.

With the Hénon variables, it is very simple to see geometrically if $M(R)$ is infinite or not. Indeed, if we differentiate $Y(x) = 2r^2 \psi(r)$ with respect to r we obtain $r^2 \psi'(r) = (xY'(x) - Y(x))/r$. Evaluating this at $r = \epsilon$ and Taylor-expanding around $\epsilon = 0$ give easily

$$\epsilon^2 \psi'(\epsilon) = -\frac{Y(0)}{\epsilon} + o(\epsilon) \,. \tag{C.2}$$

It is clear from Eqs. (C.1) and (C.2) that the mass $M(R)$ is finite if and only if $Y(0) = 0$; a result true for any central potential ψ. In other words, we have the

P. Ramond, *The First Law of Mechanics in General Relativity & Isochrone Orbits in Newtonian Gravity*, Springer Theses,
https://doi.org/10.1007/978-3-031-17964-8

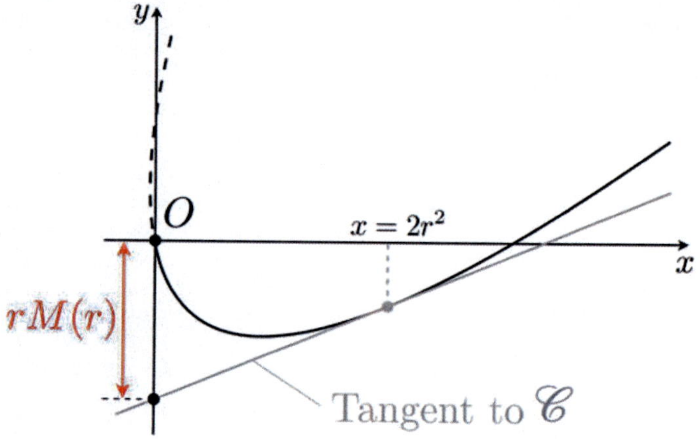

Fig. C.1 The curve $\mathscr{C}$ is in solid black, and the rest of the parabola in dashed black. The curve passes through the origin; therefore, the mass $M(r)$ inside any sphere of radius r is finite. It can be read off as the y-intercept of the tangent at the point of abscissa x (in red). Note that this construction for the mass holds for any central potential

following geometrical result: *A potential ψ is sourced by a finite mass at the origin if and only if its curve $\mathscr{C}$ in Hénon's variables passes through the origin.* In the isochrone context, this means that any isochrone parabola whose convex branch does not cross the origin is associated with an infinite mass at the origin. Moreover, we see on Eq. (C.2) that if $Y(0) > 0$, i.e., the convex branch crosses the y-axis above the origin, then the central mass is infinite and positive.[4]

Now if we focus on a potential satisfying $Y(0) = 0$, the mass $M(r)$ is finite within any sphere, and it can be read off of the curve $\mathscr{C}$ as follows. Plugging $r^2\psi'(r) = (xY'(x) - Y(x))/r$ into Eq. (C.1) allows us to write $M(r)$ in the following evocative form

$$M(r) = -\frac{Y'(x)(0 - x) + Y(x)}{r}. \tag{C.3}$$

Notice that the numerator in Eq. (C.3) is nothing but the y-intercept of the tangent of $\mathscr{C}$ at the point of abscissa x. Therefore, given a central potential, the mass contained within a sphere of radius r can be measured simply by reading this y-intercept. In particular, for the isochrone potentials the mass within a given sphere is also something that can be geometrically read off the parabola, as depicted in Fig. C.1.

[4] We shall see this feature at play in the classification of isochrone orbits, later in Sect. 9.3.

C.1.1 Dynamical System

In order to draw the orbit, we write the equations of motion as a three-dimensional dynamical system. Although, in general, a generic three-dimensional motion in classical mechanics involves 6 degrees of freedom, namely the three coordinates and their associated momenta, the spherical symmetry here at play reduces this number to three. Moreover, the radial motion is decoupled from the polar one. To see this, differentiate Eq. (7.16) with respect to r to obtain a second-order ODE for $r(t)$, or equivalently, a two-dimensional dynamical system for the radial motion in $(r, \dot{r})$. To get the polar motion, and thus the full orbit $(r(t), \theta(t))$, one may simply use the definition of the angular momentum $\Lambda = r^2\dot{\theta}$, which gives $\theta(t)$ directly from $r(t)$. These three pieces together give the following three-dimensional dynamical system in $(r, \dot{r}, \theta)$

$$\frac{\mathrm{d}r}{\mathrm{d}t} = \dot{r}\,, \quad \frac{\mathrm{d}\dot{r}}{\mathrm{d}t} = \frac{\Lambda^2}{r^3} - \psi'(r) \quad \text{and} \quad \frac{\mathrm{d}\theta}{\mathrm{d}t} = \frac{\Lambda}{r^2}\,. \tag{C.4}$$

The system (C.4) is sufficient to compute the trajectory of any particle in any central potential $\psi(r)$. In particular, once $\psi(r)$ is plugged into Eqs. (C.4) and some initial conditions $(r(0), \dot{r}(0), \theta(0))$ are provided, the motion can be solved using, e.g., a classical Runge–Kutta numerical method. Since we are interested in periodic, bounded orbits, we must, however, choose the initial conditions carefully. In order to find these orbits more easily, we choose to express $(r(0), \dot{r}(0))$ in terms of the two constants of motion ξ and Λ, and take $\theta(0) = 0$, as the latter does not change the periodic nature of an orbit. Since the set of (ξ, Λ) that produces periodic orbits is precisely the one we found in Sect. 8.3.3 depicted in Figs. 8.9, 8.10 and 8.11, this procedure allows for an easy picking of initial conditions and allows us to draw any periodic orbit in any isochrone potential. This has been used to draw the orbits in Figs. 9.4–9.6, and to check the validity of all our analytic isochrone formulae.

C.2 Details on Isochrone Potentials

C.2.1 Alternative Form of the Third Laws

We have seen that the Hénon's formula (8.13) gives the period $T(\xi)$ of an orbit (ξ, Λ) in any isochrone potential. For any value of ξ, there exists a unique value Λ_C such that the orbit is circular, corresponding to the line $\mathscr{L}_C : y = \xi x - \Lambda_C^2$ being tangent to the isochrone parabola. Using the notations h and $L(h)$ introduced in Sect. 8.2.2, this circular limit corresponds to $h \to 0$. Taking this well-defined limit in Eq. (8.18) gives

$$T^2 = \frac{\pi^2}{16(1+\xi^2)^{3/2}} \lim_{h\to 0} \frac{L(h)^2}{h}\,. \tag{C.5}$$

As it can be intuited from the discussion of Sect. 8.2.2, it turns out that the limit on the right-hand side of Eq. (C.5) is independent of the global aspect of the curve. In fact, this limit is simply eight times the radius of curvature R_C at the point C corresponding to the circular orbit.[5] In other words, we have

$$T^2 = \frac{\pi^2}{2} \frac{R_C}{(1+\xi^2)^{3/2}} \,. \tag{C.6}$$

Equation (C.6) provides a geometrical way to find the period of any given orbit in an isochrone potential, without any algebraic reference to the parabola itself. First take a line $\mathscr{L}$ intersecting an isochrone parabola $\mathscr{P}$ at P and A, and then, perform a translation of this line to construct $\mathscr{L}_C$, tangent to $\mathscr{P}$ at C. The curvature radius r_C of the parabola at the tangency point C gives the period, via Eq. (C.6). This is the local version of the result given in Eq. (8.13).

In a similar fashion, the law for the apsidal angle can also be written in terms of curvature, albeit for the effective potential. If $\Psi_e(u) = \psi_e(r)$ with $u = 1/r$, we have

$$\Theta^2 = \frac{4\pi^2\Lambda^2}{\Psi_e''(u_C)} \,. \tag{C.7}$$

This law provides a way to compute the apsidal angle in the effective potential Ψ_e in the Binet variable $u = 1/r$, or in the real effective potential ψ_e, using $\Psi_e''(u_C) = r_C^4 \psi_e''(r_C)$.

C.2.2 Hénon's Formula for Θ

We detail the computation of the integral (7.21) for Θ, with the method used to derive the Hénon formula (8.13) for T. According to the dictionary in Table. 8.1, this time we use the Binet variable $u := 1/r$ and define a potential $\Psi_e(u)$ by $\psi_e(r) = \Psi_e(u)$. Inserting these notations in (7.21) readily gives

$$\Theta = \sqrt{2}\Lambda \int_{u_A}^{u_P} \frac{du}{\sqrt{D_\Theta(u)}}\,, \quad \text{with} \quad D_\Theta(u) := \xi - \Psi_e(u)\,. \tag{C.8}$$

This is the equivalent for Θ, of Eq. (8.6) for T. The bounds of the integral are $u_A := 1/r_A$ and $u_P = 1/r_P \geq u_A$. In the (u, y) plane, the quantity $D_\Theta(u) := \xi - \Psi_e(u)$ appearing in Eq. (C.8) is the vertical distance between the curve $\mathscr{C} : y = \Psi_e(u)$ and the line $\mathscr{L} : y = \xi$. Once again, the fact that $D_\Theta(u) \geq 0$ follows from the requirement $\xi - \psi_e(r) \geq 0$. Next we rewrite the distance D_Θ as $D_\Theta(u) = \varepsilon^2 - z(u)^2$, with $\varepsilon^2 :=$

[5] The intuition comes from the following remark: The information on the period should be encoded somewhere on the curve, but be independent of Λ and thus of the height of the line $\mathscr{L}$. By varying Λ we see that the only place that is not altered by this translation is the point C. In particular, the slope of the tangent encodes ξ, and the curvature at that point encodes T.

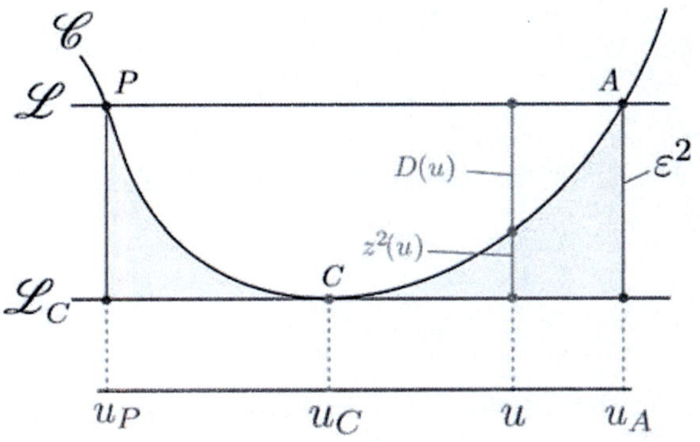

Fig. C.2 Illustration of the geometrical analysis involved in the computation of Θ, according to the dictionary of Table. 8.1. D_Θ is the vertical distance between the straight line $\mathscr{L} : y = \xi$ and a generic curve $\mathscr{C} : y = \Psi_e(u)$. The line $\mathscr{L}_C : y = \xi_C$ is the unique line both parallel to $\mathscr{L}$ and tangent to $\mathscr{C}$

$\xi - \xi_C$ the vertical distance between the two lines $\mathscr{L}$ and $\mathscr{L}_C$, and $z(u)^2 := \Psi_e(u) - \xi_C$, as depicted in Fig. C.2. As we did for T, we may conveniently choose $z(u)$ to be negative on $[u_A, u_C]$ and positive on $[u_C, u_P]$. The formula for Θ becomes

$$\Theta = \sqrt{2}\Lambda \int_{u_A}^{u_P} \frac{du}{\sqrt{\varepsilon^2 - z(u)^2}} . \tag{C.9}$$

As we argued for T, the function $u \mapsto z(u)$ is by construction monotonically increasing so that we can perform the change of variables $u \to z(u)$ and introduce f such that $u = f(z)$. Since $z(u_A) = -\varepsilon$ and $z(u_A) = \varepsilon$, the integral becomes

$$\Theta = \sqrt{2}\Lambda \int_{u_A}^{u_P} \frac{f'(z)\mathrm{d}z}{\sqrt{\varepsilon^2 - z^2}} = \sqrt{2}\Lambda \int_{-\pi/2}^{\pi/2} f'(\varepsilon \sin\phi)\mathrm{d}\phi\,, \tag{C.10}$$

where the last equality follows from the change of variables $z \to \varepsilon \sin\phi$, with ϕ varying between $-\pi/2$ and $\pi/2$ when $z \in [-\varepsilon, \varepsilon]$. Now we assume for f' a Taylor expansion around 0 of the form $f(z) = a_0 + \sum_{n\geq 1} a_n z^n$, and integrate term by term to get

$$\Theta = \sqrt{2}\pi\Lambda a_0 + 2\sqrt{2}\Lambda \sum_{n\geq 1} a_{2n} W_{2n} \varepsilon^{2n} , \tag{C.11}$$

with W_n the Wallis integral as given in Eq. (8.10), and the odd terms vanishing by integration over the symmetric interval $[-\varepsilon, \varepsilon]$. Now if Θ is to be independent of ξ, then it must also be independent of ε, since ξ_C depends only on Λ. Therefore, we must have $a_{2n} = 0$ for all $n \geq 1$. We thus obtain the formula $\Theta = \sqrt{2}\pi\Lambda a_0$ and the Taylor expansion of f' therefore writes

$$f'(z) = \frac{\Theta}{\sqrt{2}\pi\Lambda} + \sum_{n\geq 1} a_{2n+1} z^{2n+1} . \tag{C.12}$$

Integrating this equation over the interval $[z(u_A), z(u_P)] = [-\varepsilon, \varepsilon]$, we can make the same remarks as we did in the paragraph below Eq. (8.12), except that in this case $f(z_A) = u_A$ and $f(z_P) = u_P$. In the end, we obtain Eq. (8.15), which is the equivalent to Eq. (8.13) for T. The right-hand side of that equation is independent of ξ, even though the quantities u_P, u_A depend explicitly on ξ.

C.2.3 *Analysis of* $\Theta(\Lambda)$

Let ψ_i be a potential in one of the four families $i \in \{1, 2, 3, 4\}$ as defined in Sect. 8.2.2. In this appendix, we study the properties of the function $\Theta_i(\Lambda)$ defined in Eq. (9.11). These formulae are used in Sect. 9.3 to classify the orbits in each isochrone potential ψ_i. The claims of Sect. 9.3 regarding each function $\Lambda \mapsto \Theta_i(\Lambda)$ follow easily from the mathematical analysis detailed here, with the Latin parameters (a, b, c, d, e) replaced by the Greek ones $(\omega, \varepsilon, \lambda, \mu, \beta)$.

Harmonic and Kepler family

For the Harmonic potentials ψ_1, the analysis of $\Theta_1(\Lambda) = \pi\Lambda(\Lambda^2+\lambda)^{-1/2}$ is straightforward. For $\lambda < 0$, it is strictly decreasing and Θ varies in $]\pi, +\infty[$ when $\Lambda \in]\sqrt{-\lambda}, +\infty[$. For $\lambda = 0$, $\Theta = \pi$ for all $\Lambda \in \mathbb{R}$. For $\lambda > 0$, it is strictly increasing and Θ varies in $]0, \pi[$ when $\Lambda \in]0, +\infty[$. For the Kepler family ψ_4 the analysis is also straightforward since we have for any Λ and λ the identity $\Theta_4(\Lambda) = 2\Theta_1(\Lambda)$ by direct examination of Eq. (9.11) when $b = 0$ (harmonic) and $d^2 = 4b^2e$ (Kepler).

Bounded family

For Bounded potentials ψ_2, the analysis is more involved. For any $\mu > 0$ and $\beta > 0$ we write $\alpha := \lambda/(\lambda + 4\mu\beta)$. We also define a function $f(\Lambda, \lambda)$ of the real variables Λ, λ by the formula

$$f(\Lambda, \lambda) := \frac{\Lambda}{(\Lambda^2+\lambda)^{1/2}} - \frac{\Lambda}{(\Lambda^2+\lambda+4\mu\beta)^{1/2}}\,. \tag{C.13}$$

With these notations, we have $\Theta_2(\Lambda) = \pi f(\Lambda, \lambda)$ (cf. Eq. (9.37)). We want to study the three cases $\lambda > 0$, $\lambda = 0$ and $\lambda < 0$, used to classify the orbits in Sect. 9.3.

• Case $\lambda = 0$. In this case, we simply plug $\lambda = 0$ in Eq. (C.13) and we see that $f(\Lambda, 0) \in [0, 1]$. Moreover, we have easily $\partial_\Lambda f < 0$. Therefore, $\Theta(\Lambda)$ is strictly decreasing and varies $[0, \pi]$.

• Case $\lambda > 0$. In this case, $0 < \alpha < 1$ and for a fixed λ, we have $\partial_\Lambda f(\Lambda, \lambda) = \lambda(\Lambda^2+\lambda)^{-3/2} - (\lambda+4\mu\beta)(\Lambda^2+\lambda+4\mu\beta)^{-3/2}$. Then, a few algebraic manipulation show that $\partial_\Lambda f(\Lambda, \lambda)$ vanishes for a value Λ_o given by

$$\Lambda_o^2 = \lambda\frac{\alpha^{1/3}-1}{\alpha-\alpha^{1/3}} \quad \Rightarrow \quad f(\Lambda_o, \lambda) = \frac{\Lambda_o(1-\alpha^{1/3})}{(\Lambda_o^2+\lambda)^{1/2}}\,. \tag{C.14}$$

Since $0 < \alpha < 1$ and $0 < \Lambda_o < (\Lambda_o + 4\mu\beta)^{1/2}$, we have readily $0 < f(\Lambda_o, \lambda) < 1$. Now, for any fixed $\lambda > 0$, $\Lambda \mapsto f(\Lambda, \lambda)$ is continuous, $\partial_\Lambda f$ vanishes only once at Λ_o and furthermore $0 < f(\Lambda_o, \lambda) < 1$. Furthermore, it is clear that $f(\Lambda, \lambda)$ goes to 0 as $\Lambda \to 0$ and $\Lambda \to +\infty$. With all these results, the general shape of the curve $\Lambda \mapsto f(\Lambda, \lambda)$ can be easily inferred.

• Case $\lambda < 0$. In this case, $\Lambda \mapsto f(\Lambda; \lambda)$ is defined only when $\Lambda^2 > -\lambda$. First subcase: $\lambda < 0$ and $\lambda + 4\mu\beta < 0$. Then, this is the same as in the $\lambda > 0$ case, where we saw that $\partial_\Lambda f(\Lambda, \lambda) > 0$. Second subcase: $\lambda < 0$ but $\lambda + 4\mu\beta \geq 0$, then setting $g(\Lambda, \lambda) := \lambda(\Lambda^2 + \lambda)^{-3/2}$, we have for any $\Lambda^2 > -\lambda$

$$\frac{\partial g}{\partial \lambda}(\Lambda, \lambda) = \frac{2\Lambda^2 - \lambda}{2(\Lambda^2 + \lambda)^{5/2}} \,. \tag{C.15}$$

Now, since $\Lambda^2 > -\lambda$, the right-hand side of Eq. (C.15) is strictly positive, and therefore g is an increasing function of λ. In particular, we have $\lambda + 4\mu\beta > \lambda \Rightarrow g(\Lambda, \lambda + 4\mu\beta) > g(\Lambda, \lambda)$ and by definition of g, the latter is exactly $\partial_\Lambda f(\Lambda, \lambda) > 0$. To conclude, in the $\lambda < 0$ case, $\Lambda \mapsto f(\Lambda, \lambda)$ is strictly decreasing. Furthermore, it is clear that $f(\Lambda, \lambda)$ goes to $+\infty$ as $\Lambda \to (-\lambda)^{1/2}$, and to 0 as $\Lambda \to +\infty$. With all these results, the general shape of the curve $\Lambda \mapsto f(\Lambda, \lambda)$ can be easily inferred.

Hénon family

For Hénon potentials ψ_3, the analysis is similarly more involved. As for the Bounded potentials we fix $\mu > 0$ and $\beta > 0$ and write $\alpha := \lambda/(\lambda + 4\mu\beta)$. This time we define a function $h(\Lambda, \lambda)$ of the real variables Λ, λ by the formula

$$h(\Lambda, \lambda) := \frac{\Lambda}{(\Lambda^2 + \lambda)^{1/2}} + \frac{\Lambda}{(\Lambda^2 + \lambda + 4\mu\beta)^{1/2}} \,. \tag{C.16}$$

With these notations, we have $\Theta_3(\Lambda) = \pi h(\Lambda, \lambda - 2\mu\beta)$ (cf. Eq. (9.39)). The analysis follows the same lines as what was done for $\Theta_2(\Lambda)$. We want to study the three cases $\lambda > 0$, $\lambda = 0$ and $\lambda < 0$, used to classify the orbits in Sect. 9.3.

• Case $\lambda > 0$. Then we have $0 < \alpha < 1$ and there is no problem in showing that $\Lambda \mapsto h(\Lambda, \lambda)$ is strictly increasing and that $0 < h(\Lambda, \lambda) < 1$.

• Case $\lambda = 0$. Once again, there is no problem in showing that $\Lambda \mapsto h(\Lambda, \lambda)$ is strictly increasing and that $0 < h(\Lambda, \lambda) < 2$.

• Case $\lambda < 0$. In this case, h is only defined when $\Lambda^2 > -\lambda$. For any such (Λ, λ), we have

$$\frac{\partial h}{\partial \Lambda}(\Lambda, \lambda) = \frac{\lambda}{(\Lambda^2 + \lambda)^{3/2}} + \frac{\lambda + 4\mu\beta}{(\Lambda^2 + \lambda + 4\mu\beta)^{3/2}} \,. \tag{C.17}$$

There are two subcases. First subcase: $\lambda < 0$ and $\lambda + 4\mu\beta < 0$. Then from Eq. (C.17), $\partial_\Lambda h(\Lambda, \lambda) < 0$. Furthermore, $h(\Lambda, \lambda)$ goes to $+\infty$ as $\Lambda \to (-\lambda)^{1/2}$, and to 2 as $\Lambda \to +\infty$. Second subcase: $\lambda < 0$ and $\lambda + 4\mu\beta < 0$. If $|\alpha| < 1$, then there is a value Λ_o that makes $\partial_\Lambda h(\Lambda, \lambda)$ vanish. It is given by

$$\Lambda_o^2 = \lambda \frac{|\alpha|^{1/3}+1}{|\alpha|-|\alpha|^{1/3}} \quad \Rightarrow \quad f(\Lambda_o,\lambda) = \frac{\Lambda_o(1+|\alpha|^{1/3})}{(\Lambda_o^2+\lambda)^{1/2}} . \tag{C.18}$$

In this case, the function $\Lambda \mapsto h(\Lambda, \lambda)$ decreases on $[(-\lambda)^{1/2}, \Lambda_o]$ and increases on $[\Lambda_o, +\infty[$. The value $h(\Lambda_o, \lambda)$ is always strictly between 1 and 2. If $|\alpha| \geq 1$, then the function f is strictly decreasing. (It can be seen as the limit $\Lambda_o \to +\infty$.) The value $h(\Lambda_o, \lambda)$ is in this case always above 2.

C.2.4 *Proof that* $c + d\xi < 0$ *for Isochrone Orbits Around Finite Central Mass*

In Sect. 9.2 we used the fact that $a + b\xi < 0$ and $c + d\xi < 0$ for isochrone orbits in order to prove that our formulae in Sect. 9.2 covers all isochrone orbits. The former identity follows from the generalized Kepler's third law, and here, we prove the latter identity. By assumption, we have a particle (ξ, Λ) on an isochrone orbit in a potential with finite central mass whose parabola $\mathscr{P} : (ax+by)^2 + cx + dy = 0$ verifies all hypotheses (H_i) of Sect. 8.3.2. First we can check easily that $cx + dy = 0$ is an equation for the tangent to $\mathscr{P}$ at the origin. geometrically, since two intersections exist between $\mathscr{L}$ and $\mathscr{P}$, the slope of $\mathscr{L}$ must be bigger than that of this tangent, i.e., we must have $\xi > -c/d$. We just have to show that $d \leq 0$ and the result will follow. First, if $b = 0$ (harmonic case), then we necessarily have $d < 0$ (top-oriented parabola). Second, if $b \neq 0$, then since $\lambda = 0$ ($\mathscr{P}$ crosses the origin) we have by Eq. (8.32) the equality $-d = \sqrt{d^2}$, which implies $d \leq 0$. Therefore, we always have $d \leq 0$ and thus $c + d\xi < 0$.

C.2.5 *Peaks of Orbits in Bounded Potentials*

Let an arbitrary orbit be given by a polar equation $r(\theta)$, and let us compute the value of $|\mathrm{d}r/\mathrm{d}\theta|$. The latter is a measure of the change of $\mathrm{d}r$ when moving from θ to $\theta + \mathrm{d}\theta$. It vanishes for circles $r =$ cst and is *infinite* for straight lines $\theta =$ cst. With the help of Eq. (7.16) and $\Lambda = r^2\dot{\theta}$, we obtain easily $|\mathrm{d}r/\mathrm{d}\theta|^2 = 2r^4(\xi - \psi_e(r))/\Lambda^2$. Using a Taylor expansion of $\psi(r)$ and Eq. (7.18), we can linearize this equation around the apoapsis r_A. We then obtain

$$\left|\frac{\mathrm{d}r}{\mathrm{d}\theta}\right|^2 = \frac{2r_A^4}{\Lambda^2}\left(\psi'(r_A) - \frac{\Lambda^2}{r_A^3}\right)(r_A - r) + o(r_A - r) . \tag{C.19}$$

Examining Eq. (C.19), we see that as $r \to r_A$ the right-hand side goes to zero as every term is finite in front of $(r_A - r)$. The orbit is therefore smooth and differentiable around the apoapsis. However, the quantity $\psi'(r_A)$ turns out to be very large for the Bounded family, in general. This is because the slope of a Bounded potential increases to infinity as r grows toward β from below, as can be seen readily on Eq. (9.36). Therefore, a line $\mathscr{L}$ can intersect $\mathscr{C}$ such that r_A is very close to β, and it is clear from Eq. (9.36) that $\psi_2'(r) \to \infty$ as $r \to \beta$. As a conclusion, before the apoapsis, the term $(r_A - r)$ does not yet compensate the $\psi'(r_A)$ which is large for the Bounded potential, making $\frac{\mathrm{d}r}{\mathrm{d}\theta}$ large and the curve resembles a $\theta = \text{cst}$ line. This is why we see such abrupt and pointy turns in Fig. 9.5.

Appendix D
The Abel–Ruffini Theorem

This appendix contains an article published in The American Mathematical Monthly, on the Abel–Ruffini theorem. Using only elementary knowledge of complex numbers, we sketch a proof of the celebrated Abel–Ruffini theorem, which states that the general solution to an algebraic equation of degree five or more cannot be written using radicals, that is, using its coefficients and arithmetic operations $+, -, \times, \div$, and $\sqrt{\ }$. The present article is written purposely with concise and pedagogical terms and dedicated to students and researchers not familiar with Galois theory, or even group theory in general, which are the usual tools used to prove this remarkable theorem. In particular, the proof is self-contained and gives some insight as to why formulae exist for equations of degree four or less (and how they are constructed), and why they do not for degree five or more.

Introduction

Historical background

Finding a general expression for the solutions of an algebraic equation has been one of the oldest and most fruitful problems in mathematics. The history behind what was once called the "theory of equations" [635], is almost as rich and old as the history of mathematics itself. For example, methods for solving linear and quadratic equations have been known for at least four millennia [636], in independent places in the world. The quadratic formula taught today in school, with modern notation, was first written down by R. Descartes in 1637 [637]. The introduction of the definitive $\sqrt{\ }$ notation (with the horizontal overbar called the "vinculum") was only introduced in 1525 [638]. Regarding cubic and quartic equations, they

P. Ramond, *The First Law of Mechanics in General Relativity & Isochrone Orbits in Newtonian Gravity*, Springer Theses,
https://doi.org/10.1007/978-3-031-17964-8

too had to wait until the sixteenth century to be finally solved. By then, a group of rivaling Italian mathematicians, including S. del Ferro, N. Tartaglia, G. Cardano, and L. Ferrari, made the serendipitous discovery of complex numbers while solving the general cubic equation. In 1545, a few years before their quarrels settled in a public mathematical contest [635], L. Ferrari solved the quartic equation by reducing it to a cubic one. The quintic equation, however, would still keep these mathematicians (and all others) in check, while the idea of it being unsolvable slowly started to emerge.

Unsolvable equations

The idea of examining permutations of the solutions to study the (un-)solvability of algebraic equations dates back to the pioneering works of J.-L. Lagrange in 1771 [639]. Lagrange's ideas matured, and were finally extended to the quintic equation by P. Ruffini in early 1800 [640]. For twenty years, Ruffini tried to convince the mathematical community of the importance of his results, without success. It is only in 1821, with the help of L.-A. Cauchy, that Ruffini's work was recognized as a stepping stone in the theory of algebraic equations. Although it turned out that Ruffini did not prove the theorem that now bears his name *per se*, his results were strong enough to place serious doubt about the possibility of finding a solution to the general quintic equation. The wait was finally over in 1824 when N.-H. Abel wrote the first complete proof of the theorem (a short proof published in 1824 [641] at his own expense, and a longer, more detailed version two years later [642]). His work still remained unworthy of interest to the eyes of most mathematicians, including Gauss and Cauchy themselves. Abel died aged 26 in 1829, just before his work on the unsolvability of the quintic finally received all the appreciation it deserved. He received posthumously the *Grand Prix de l'Académie des Sciences de Paris* in 1830, in recognition of his work. The same year also marks the publication of É. Galois's first paper on these topics [643], in which he gives the premises of (now) Galois theory, a novel and elegant extension of all previous results. He, too, died young (aged 20 in 1832) and his work also took several decades to be fully published and recognized as revolutionary.

This short historical account lacks many interesting stories about these mathematicians, such as conflict of interests, encrypted communications, fatal duels, long-lost and recovered memoirs, etc. The interested reader could start with Sesiano's [636] and Stillwell's [635] books and references therein for well-written and thorough presentations of these fascinating pieces of history.

Aim and content

The aim of this article is to sketch an accessible and self-contained proof of the *Abel–Ruffini theorem*:

> *No formula exists for the solution to the general equation of degree five or more, using only the operations* $+, -, \times, \div,$ *and* $\sqrt{\ }$.

The word *general* is important: it emphasizes that a formula that *holds for any coefficients* cannot be found. However, the theorem does not prevent *some* equations to have a solution that can be written in terms of $+, -, \times, \div, \sqrt{\ }$. In most textbooks, the proof of this remarkable theorem relies on a powerful subbranch of mathematics called *Galois theory*, developed quasi-exclusively by the French mathematician É. Galois at the beginning of the nineteenth century. Galois theory solved all "unsolvability problems" once and for all, as well as other millennia-long problems [635]. However, it is also rather advanced, usually taught in the second/third years of specialized, university-level mathematics. The first complete proof (by Abel) of the Abel–Ruffini theorem is a few years older than the birth of Galois theory. Moreover, the works of Galois took several decades to be broadly known to other mathematicians. In other words, neither Ruffini nor Abel used the methods developed by Galois to prove that some equations were unsolvable. Because it usually relies on advanced mathematics, few people in the scientific community are aware of this theorem and its underlying principles. But because Abel did not prove it this way, there must be another, perhaps simpler, way of understanding the reason why the general quintic equation does not have a solution in terms of radicals. In particular, Galois's, Abel's, and Ruffini's ideas all rely on a unique, fundamental, common point: *the symmetry of an algebraic equation under the permutation of its solutions*. Based solely on this fundamental symmetry, we propose to sketch a proof of Abel–Ruffini's theorem using only elementary knowledge about complex numbers. Familiarity with complex numbers and a pen (and paper!) to draw appropriate figures are the only prerequisites to get a grasp of how the proof works. Everything else is elementary mathematics and useful notations that help present the ideas more clearly.

Motivation

The proof given here cannot be considered new. It is the result of several adaptations and simplifications of ideas that we feel compelled to attribute to the theoretical physicist B. Katz. His ideas are presented concisely in an online video [644], which can be used as complementary material with dynamic illustrations. Katz's inspiration for making this video comes from a series of lectures given by physicist and mathematician V. Arnold, which were nicely crystallized in a problems-and-solutions book published by V. B. Alekseev, who was Arnold's student at the time of these lectures. This book, although very well written and complete, is, however, not elementary in any sense. While Katz's video does a very good job at explaining the general idea of the proof, we found that some gaps could be filled, and some arguments could be made much simpler, especially when we get to the end of the proof. Other references dealing with the present ideas are rather scarce in the literature (academic or not). A non-exhaustive selection is located at the conclusion of the article, and can serve as complementary material to deepen one's understanding of the proof.

Outline

The remainder of this article is organized as follows. After some prerequisites and reminders regarding complex numbers are introduced, we spend some time on the quadratic equation, explaining why a quadratic formula cannot be built out of only the four basic arithmetic operations (our first impossibility result). Similar ideas are then extended successively to the cubic and quartic equation, giving stronger impossibility results at each step. By the time we get to the quintic equation, the reader should be comfortable enough with the strategy (hopefully) to see how the quadratic, cubic, and quartic cases foreshadow the proof of the Abel–Ruffini theorem. Along the way, we also derive the cubic and quartic formulae, scarcely presented in the nonspecialized literature. Although the derivation of these formulae is interesting enough to justify their presence, they will be especially useful in light of our temporary results, and will naturally guide us step by step to Abel–Ruffini's theorem. Finally, we note that animated versions of Figs. D.2, D.4, D.5 and D.6 are available as supplementary material for a better understanding.

Prerequisites

In this article, we are dealing with algebraic equations of degree $n \geq 2$. These equations are always of the form

$$z^n + c_{n-1}z^{n-1} + \cdots + c_1 z + c_0 = 0\,, \tag{D.1}$$

where $z \in \mathbb{C}$ is the *unknown* and the n complex numbers $(c_0, \ldots, c_{n-1})$ are the *coefficients*. It is a remarkable fact, often cited as the fundamental theorem of algebra, that Eq. (D.1) always has exactly n complex solutions. (We use *solutions*, instead of *roots* of polynomials, to avoid confusion with the "root" operation $\sqrt{\ }$ later on.) These solutions will always be denoted $(s_1, \ldots, s_n)$, and we use s as a placeholder for *any* of the solutions.

Permutations

Our strategy will be based on picturing the solutions $(s_1, \ldots, s_n)$ in the complex plane and make them move around to exchange their positions, i.e., *permute* them. We will need two kinds of permutation:

- *transpositions*, denoted (ij), exchanging the position of *two* solutions, i.e., $s_i \leftrightarrow s_j$. The transposition (12) is depicted on the left in Fig. D.1,
- *cycles*, denoted (ijk), exchanging the position of *three* solutions cyclically, i.e., $s_i \to s_j, s_j \to s_k$, and $s_k \to s_i$. The cycle (123) is depicted on the right in Fig. D.1.

Two permutations next to each other are to be performed successively, from *left to right*. For example, (12)(23) consists in exchanging s_1 and s_2, *then* s_2 with s_3.

Fig. D.1 The paths-induced *transposition* (12) and *cycle* (123) on the solutions (s_1, s_2, s_3) of some algebraic equation of degree $n \geq 3$. *See supplementary material for animated version*

Notice that the result is equivalent to the cycle (132), hence there is no unique way of writing permutations. However, permutations *do not commute* in general. Indeed, $(12)(23) = (132)$ and $(23)(12) = (123)$; therefore $(12)(23) \neq (23)(12)$.

Loops

One way of visualizing permutations of $(s_1, \ldots, s_n)$ is to locate them in the complex plane and make them travel along some *paths*. Paths in the complex plane are just continuous curves that connect two points (we assume that they do not self-intersect, otherwise things get unnecessarily complicated). A path that *closes*, i.e., connects a point to itself, is called a *loop* and denoted γ, whereas a path that connects two *distinct* points is simply called an *unclosed path*, denoted ω. These paths will be represented by arrows in our figures, and will be used to induce permutations on $(s_1, \ldots, s_n)$. For example, in Fig. D.1 are depicted the transposition (12) on the left, and the cycle (123) on the right. Notice that to induce (12), s_3 follows a loop γ so that only s_1 and s_2 swap places by following the unclosed paths ω_1, ω_2. When speaking of permutations of solutions, we will *always* imagine them traveling on these paths.

Roots

Now let us examine how roots of complex numbers move around the complex plane. Fixing some complex number z, a *root* of z is some number $\zeta \in \mathbb{C}$ such that $\zeta^k = z$ for some $k \in \mathbb{N}$. Such a ζ is then called a kth root of z; and z admits exactly k such kth roots (this follows from the fundamental theorem of algebra). We will deliberately use the ambiguous notation $\sqrt[k]{\ }$ as a multivariable notation, i.e., for a given k, $\sqrt[k]{z}$ means *any* of the kth roots of z. Fixing $k \in \mathbb{N}$ and assuming that z itself follows a loop γ, let us examine what kind of path $\sqrt[k]{z}$ follows. To this end, we use the exponential form of z, i.e., $z = r\mathrm{e}^{\mathrm{i}\theta}$ with $r = |z|$ and $\theta = \arg z$, from which we find that all kth roots $(\zeta_1, \ldots, \zeta_k)$ can be written explicitly as

$$\zeta_\ell = r^{1/k}\mathrm{e}^{\mathrm{i}(\theta+2\ell\pi)/k}, \quad \ell \in \{1, \ldots, k\}. \tag{D.2}$$

From Eq. (D.2), one can already tell that all kth roots of z have the same modulus : $r^{1/k}$. Geometrically, this means that they lie on the same circle (of radius $r^{1/k}$) in the complex plane. Moreover, we readily see from Eq. (D.2) that

$$\arg \zeta_\ell = \frac{\theta}{k} + \ell\frac{2\pi}{k}, \tag{D.3}$$

which means that all roots are equally spaced on this circle, at angle $2\pi/k$ apart. Now suppose that z goes on a journey exploring the complex plane, by traveling on a loop γ winding once (say) around the origin, in the counterclockwise direction (in red, on the left in Fig. D.2). As z travels along γ, its kth roots also move around, and their position can be tracked from Eq. (D.2) (see the red paths on the right in Fig. D.2). Since γ is a loop, the radius r comes back to its original (i.e., pre-loop) value, and so does $r^{1/k}$. In other words: the roots remain on their circle after the path γ (see the gray, dashed-circle on the right of Fig. D.2). However, $\arg z$ went from

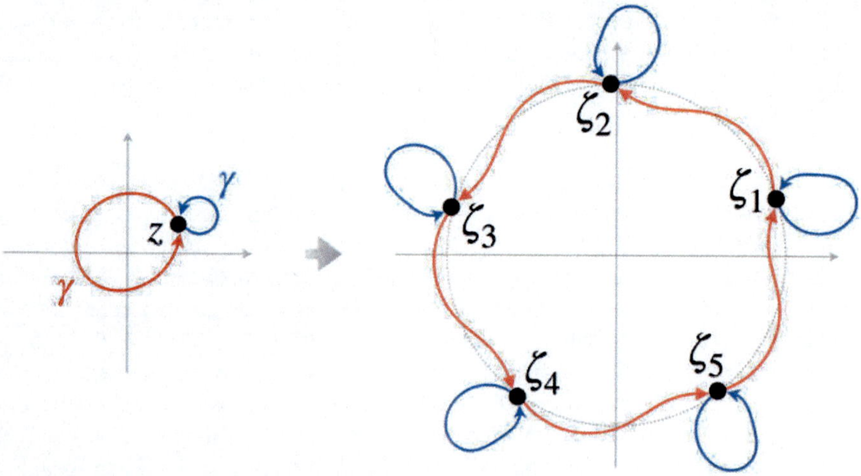

Fig. D.2 When z follows a loop γ that does not wind around the origin, its roots ζ_i follow loops as well (right). However, for the loop γ that winds once, the roots then follow the red, unclosed paths. *See supplementary material for animated version*

θ to $\theta + 2\pi$ (one counterclockwise turn). Therefore, from Eq. (D.3), each kth root ζ_ℓ has moved to its closest, counterclockwise neighbor, $\zeta_{\ell+1}$. In particular, the roots have followed an unclosed path. Had z not wound around the origin (in blue, on the left in Fig. D.2), its argument θ would have seen no net change after the loop γ, and the roots would have followed their own loops (in blue on the right in Fig. D.2).

We have seen two example of loops followed by z and the result is not the same for its roots $\sqrt[k]{z}$: in one case the roots follow a loop (blue part of Fig. D.2), in the other they *do not* (red part of Fig. D.2). Consequently, we conclude that *when z follows a loop, $\sqrt[k]{z}$ does not always follow a loop.* This conclusion holds for any type of root (i.e., any k in $\sqrt[k]{z}$). Since we will not need to differentiate between all these roots, we will denote by $\sqrt{z}$ *any* root of z (that is, *any* kth root, whatever the value $k \in \mathbb{N}$). With this notation, the takeaway result of this paragraph is simply:

When z follows a loop, $\sqrt{z}$ does not always follow a loop.

Formula ingredients

In this article we question the existence of a general *formula* for the solutions of the general algebraic equation of degree n, Eq. (D.1). By *formula*, we mean some equality

$$s = \Phi(c_0, \ldots, c_{n-1}), \tag{D.4}$$

where s is a solution of equation (D.1) and Φ is some function of its coefficients $(c_0, \ldots, c_{n-1})$. The Abel–Ruffini theorem states that for $n \geq 5$, no formula *in terms of radicals* exists. "In terms of radicals" simply mean that the function Φ in Eq. (D.4) can be constructed solely in terms of (1) the coefficients $(c_0, \ldots, c_{n-1})$, and (2) the elementary operations $+, -, \times, \div$ and $\sqrt{\ }$.

Leaving $\sqrt{\ }$ aside, if we constrain ourselves to a formula combining the coefficients $(c_0, \ldots, c_n)$ and the four operations $+, -, \times, \div$, we obtain what we will call an F-formula, or simply an *F-function*. Examples of such F-functions are

$$F = 1\,, \quad F = -\frac{c_6}{2}\,, \quad F = c_8^2 - 7c_2\,. \tag{D.5}$$

They are the elementary building blocks for constructing formulae. In particular, they encompass integers, the coefficients themselves, as well as polynomials and rational functions of the coefficients. Clearly, if two coefficients each follow a loop simultaneously, then their sum, difference, product, and quotient also follow a loop. As they are built with only these four operations, F-functions enjoy the same property. In other words:

When $(c_0, \ldots, c_{n-1})$ follow a loop, F-functions also follow a loop.

This property of F-functions is not shared by $\sqrt{F}$-functions, i.e., expressions that are roots of F-functions, e.g., $\sqrt[7]{c_0}$ or $\sqrt[2]{1 - 3c_2}$. (recall the notation in the subsection "*Loops*"). In particular, if we denote by G-function a combination of F- and $\sqrt{F}$-functions together with $+, -, \times, \div$, then we have the following:

When $(c_0, \ldots, c_{n-1})$ follow a loop, G-functions do not always follow a loop.

A G-function is a new type of ingredient as it may include expressions with one level of roots, such as

$$G = -\frac{c_5}{2} + \frac{1}{2}\sqrt[2]{c_4^2 - 4c_1}\,. \tag{D.6}$$

We can keep going like this to construct formulae with higher number of *nested roots*, i.e., roots in roots. For example, we can combine G-functions and $\sqrt{G}$-functions with $+, -, \times, \div$ to make H-functions. These may contain up to two levels of nested roots, such as

$$H = c_4 - \sqrt[3]{7c_2} + \sqrt[5]{-\frac{c_0}{2} + \sqrt[5]{c_1^2 - 4c_6}}\,, \tag{D.7}$$

and so on, as summarized in Fig. D.3. With this nomenclature, we can make arbitrarily complex expressions involving $+, -, \times, \div$ and $\sqrt{\ }$, and at the same time keep track of the number of nested roots appearing in the formula. Conversely, any formula constructed with $+, -, \times, \div, \sqrt{\ }$ can be built using this procedure, provided that we look high enough in the "..." of the list of ingredients $(F, G, H, \ldots)$.

We have now covered all the tools necessary: permutations of $(s_1, \ldots, s_n)$, loops, and F, G, H-functions. Let us now apply all these concepts to the degree n equation, starting with $n = 2$, to understand the Abel–Ruffini theorem when $n = 5$.

The quadratic equation

Our journey toward the Abel–Ruffini theorem starts with considerations of the much more familiar quadratic equation. In particular, considering only the case $n = 2$, we

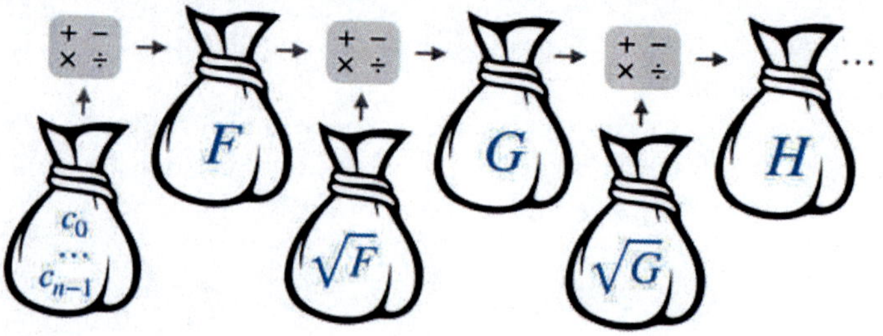

Fig. D.3 Ingredients used to build a formula. Combining coefficients $(c_0, \ldots, c_{n-1})$ with $+, -, \times, \div$ defines an F-function. Combining F-functions and their roots $\sqrt{F}$ with $+, -, \times, \div$ defines a G-function, etc. *See supplementary material for animated version*

will prove a first impossibility result, actually valid for $n \geq 2$. The ideas developed here are rather simple but also at the heart of the proof of the Abel–Ruffini theorem.

Vieta's formulae

Let us consider the general quadratic equation

$$z^2 + c_1 z + c_0 = 0 \,. \tag{D.8}$$

As mentioned previously, the fundamental theorem of algebra informs us that this equation admits exactly two complex solutions s_1 and s_2. Let us then write it in the factored form $(z - s_1)(z - s_2) = 0$ and expand this product, ordering the terms by power of z. We find a new expansion that can be compared to Eq. (D.8). By identification, we obtain the so-called Vieta's formulae:

$$c_1 = -(s_1 + s_2) \quad \text{and} \quad c_0 = s_1 s_2 \,. \tag{D.9}$$

This kind of relation between the coefficients and the solutions can be established for any degree $n \geq 2$. For example, Eq. (D.9) generalizes nicely to $c_{n-1} = -\sum_i s_i$ and $c_0 = (-1)^n \prod_i s_i$, for any $n \geq 2$. In any case, these formulae always reveal the same fundamental property:

Coefficients $(c_0, \ldots, c_{n-1})$ are symmetric functions of the solutions $(s_1, \ldots, s_n)$.

In particular, for the $n = 2$ case here at hand, if one permutes s_1 and s_2 by moving them continuously in the complex plane (using, for example, the transposition (12) depicted in Fig. D.1), then the coefficients (c_0, c_1) will each move on some path, but eventually they must come back to their original location as they are symmetric in (s_1, s_2). In other words, they will follow a *loop*, as depicted in Fig. D.4.

A first impossibility result

Because it is the central idea at play, let us rephrase the symmetry in Vieta's formulae geometrically:

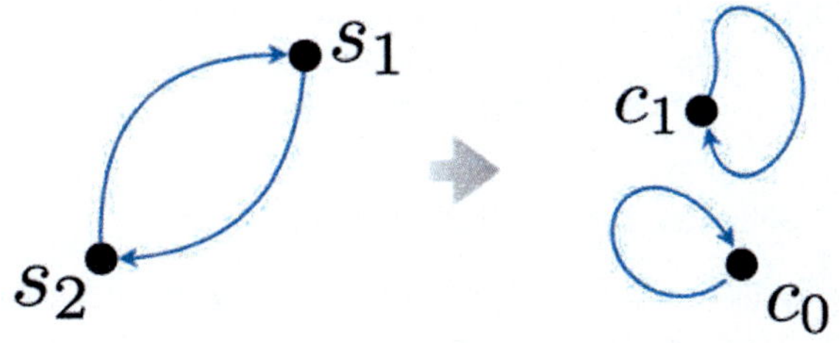

Fig. D.4 The transposition (12) of the solutions (s_1, s_2) induces a loop on the coefficients (c_0, c_1). *See supplementary material for animated version*

When $(s_1, \ldots, s_n)$ undergo a permutation, $(c_0, \ldots, c_{n-1})$ each follow a loop.

This remarkable fact can be used to obtain a first impossibility result, as follows. Suppose that the solutions s_1 and s_2 of the quadratic equation are given by two formulae of the type

$$s_1 = F_1(c_0, c_1) \quad \text{and} \quad s_2 = F_2(c_0, c_1), \tag{D.10}$$

with F_1, F_2 two F-functions (i.e., expressions involving (c_0, c_1) and the symbols $+, -, \times, \div$). Now, picture (s_1, s_2) and (c_0, c_1) in the complex plane, and study the following process:

- connect s_1 and s_2 with paths inducing the transposition (12), and make them move along these paths (see Fig. D.4);
- as s_1 and s_2 move around, c_0 and c_1 each travel on a loop, as seen previously (see Fig. D.4),
- while c_0 and c_1 follow their own loop, the two F-functions F_1 and F_2 will also follow a loop, as argued earlier (see Fig. D.5).

At the end of this process, s_1 and s_2 have permuted, yet both F_1 and F_2 have followed a loop. Consequently, F_1 and F_2 cannot equal s_2 and s_2 respectively, and no formula such as in (D.10) exists. This impossibility result actually holds for an equation of any degree $n \geq 2$. Indeed, it suffices to pick two of the n solutions to the degree n equation, name them s_1 and s_2, and apply the above recipe. The conclusion is thus:

Using only F-functions, no formula solving
the general Eq. (D.1) can be found for $n \geq 2$.

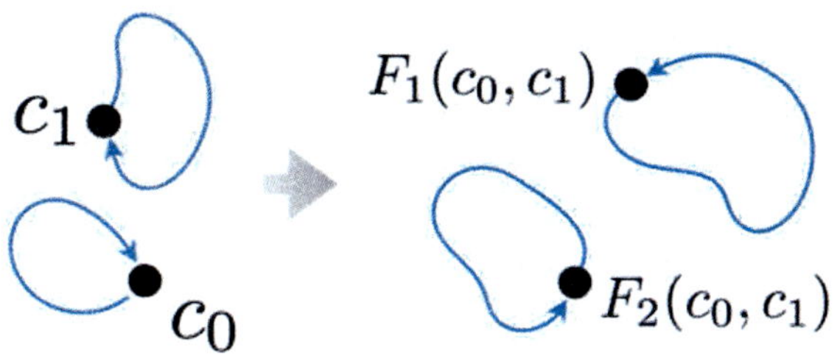

Fig. D.5 A loop followed by (c_0, c_1) also induced a loop on the F-functions $F_1(c_0, c_1)$ and $F_2(c_0, c_1)$. *See supplementary material for animated version*

This is our *first impossibility result*. In particular, it means that we have no chance of finding a formula for the cubic equation using only F-functions either. To see which extra ingredients are needed, let us examine closely the well-known quadratic formula.

Discussion: the quadratic formula

The quadratic formula is derived most easily by "completing the square" in Eq. (D.8) to get $(z + \frac{c_1}{2})^2 = \frac{1}{4}c_1^2 - c_0$. Using our notation $\sqrt[2]{\ }$ for any of the two square roots, we easily obtain a formula for the general solution s of Eq. (D.8) as

$$s = -\frac{c_1}{2} + \frac{1}{2}\sqrt[2]{c_1^2 - 4c_0}\,. \tag{D.11}$$

This formula alone corresponds to two solutions, one for each of the two square roots on the right-hand side. Moreover, notice how this root indeed points to the same direction as our impossibility result: we need to add $\sqrt{F}$-function to the list of ingredients. One last note: just as the Abel–Ruffini theorem, the impossibility result just derived tells something about the *general* quadratic equation. However, there exists *some* quadratic equations with given, explicit coefficient that admit a formula in terms of $+, -, \times, \div$.

The cubic equation

Let us now try to construct a formula for the solutions of the general cubic equation. The equation reads

$$z^3 + c_2 z^2 + c_1 z + c_0 = 0\,. \tag{D.12}$$

Let (s_1, s_2, s_3) be its three complex solutions. Learning from our previous findings, we now add $\sqrt{F}$-functions to the list of ingredients. Therefore, we assume that there exists some formulae of the type

$$s_{\mathrm{i}} = G_{\mathrm{i}}(c_0, c_1, c_2)\,, \quad \text{for i} \in \{1, \ldots, 3\}\,, \tag{D.13}$$

with G_1, G_2, G_3 three G-functions (combinations of F and $\sqrt{F}$ with $+, -, \times, \div$). Our second impossibility result will consist in showing that such a formula cannot exist. Our previous method is not guaranteed to work: *yes*, the coefficients still follow loops as solutions permute, but *no*, G-functions do not generally follow loops in these circumstances, as we have already seen. We need to modify our plan.

Introducing commutators

Consider the transposition (12) that induces a loop γ_1 on F and thus an unclosed path ω_1 on $\sqrt{F}$. Consider also (23), inducing a loop γ_2 on F and a path ω_2 on $\sqrt{F}$. Now perform the following sequence of transpositions, called the *commutator* of (12) and (23), and denoted

$$[(12), (23)] = (12)(23)(12)^{-1}(23)^{-1}\,. \tag{D.14}$$

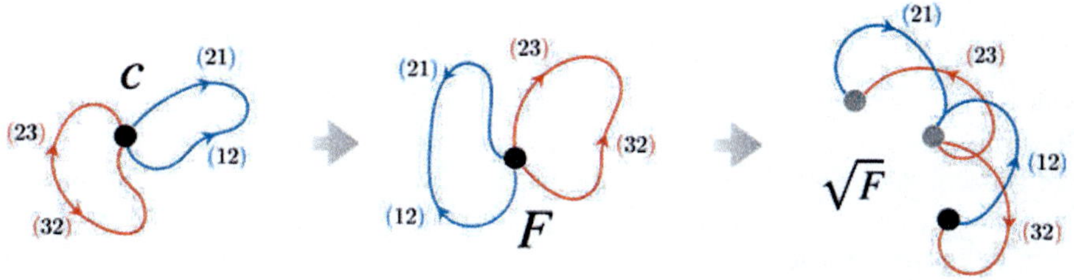

Fig. D.6 Effect of the commutator [(12), (23)] on a coefficient c (left), on an F-function (center) and on $\sqrt{F}$-function (right). After the process, both F and $\sqrt{F}$ have followed a loop. Notice the loop followed by $\sqrt{F}$ consisting in four unclosed paths. *See supplementary material for animated version*

Since $(12)^{-1}$ is simply (21), and $(23)^{-1} = (32)$, it turns out that [(12), (23)] is simply the cycle (123). In fact, this is true with any pair of transposition, i.e.,

$$[(ij), (jk)] = (ijk)\,. \tag{D.15}$$

Therefore, [(12), (23)] does permute the three solutions (s_1, s_2, s_3). But what is its effect on numbers like F and $\sqrt{F}$? Clearly, F follows a sequence of loops $\gamma_1\gamma_2\gamma_1^{-1}\gamma_2^{-1}$, which is itself a loop. The number $\sqrt{F}$, however, follows a sequence of unclosed paths $\omega_1\omega_2\omega_1^{-1}\omega_2^{-1}$ (visiting other roots) but closes on itself by construction; see Fig. D.6.

Conclusion

With the permutation (123) written as the commutator [(12), (23)], we reach the same conclusion as in the quadratic case: while (s_1, s_2, s_3) undergoes the permutation (123), both F and $\sqrt{F}$ follow a loop (and thus any G-function). Consequently, there cannot be equalities given by (D.13). Again, this holds for the general equation of degree $n \geq 3$, too, as it suffices to pick up three solutions out of the $n \geq 3$, label them s_1, s_2, s_3, and apply the above recipe. Therefore, we conclude:

Using only G-functions, no formula solving
the general Eq. (D.1) *can be found for* $n \geq 3$.

This is our *second impossibility result*. We must emphasize that it works only if we apply the cycle (123) as a commutator such as in Eq. (D.14). Had we just applied the cycle (123) directly (i.e., without writing it as a commutator), there would have been no guarantee that $\sqrt{F}$ followed a loop. It is the commutator that allows us to *discard* one level of roots, and thus $\sqrt{F}$, from the list of ingredients. Let us now put this new impossibility result to the test, by solving explicitly the cubic equation.

Discussion: the cubic formula

We follow the classical method found by Italian mathematicians of the sixteenth century. First, perform the change of variables $Z = z + c_2/3$, which "removes" the z^2 term in Eq. (D.12), transforming it into

$$Z^3 + 3PZ + 2Q = 0\,, \tag{D.16}$$

where $P = \frac{c_1}{3} - \frac{c_2^2}{9}$ and $Q = \frac{c_0}{2} + \frac{c_2^3}{27} - \frac{c_1 c_2}{6}$. Notice that both P and Q are F-functions of (c_0, c_1, c_2). To solve Eq. (D.16), one then writes $Z = v + w$, where v, w are two complex numbers to be chosen freely later on. Then, Eq. (D.16) becomes $v^3 + w^3 + 3(vw + P)(v + w) + 2Q = 0$, from which we can remove the second term by imposing that v, w satisfy $vw = -P$. By cubing the latter, we then obtain two equations for two unknowns, namely

$$v^3 + w^3 = -2Q \quad \text{and} \quad v^3 w^3 = -P^3\,. \tag{D.17}$$

These equations can be solved simultaneously for v^3 and w^3, since they explicitly give their sum and product, respectively. (These are nothing but Vieta's formulae for $n = 2$; See Eq. (D.9).) Using the quadratic formula, one obtains v^3 and w^3 in terms of P and Q, takes their cube root and adds the result to obtain $v + w = Z$. Going back to the original unknown $z = Z - c_2/3$ gives the famous "cubic formula"

$$s = -\frac{c_2}{3} + \sqrt[3]{-Q + \sqrt{Q^2 + P^3}} + \sqrt[3]{-Q - \sqrt{Q^2 + P^3}}\,. \tag{D.18}$$

This formula gives three solutions (s_1, s_2, s_3), one for each cube root. It is clear that this expression involves more than F and $\sqrt{F}$ functions: indeed, the two cube roots are actually $\sqrt{G}$-functions. In a sense, the cubic formula above contains "two levels" of roots, whereas G-functions can only contain one, by definition. This kind of expression is thus called a *nested root*. Our "commutator trick" was only able to remove *one* level of roots. Perhaps two levels of commutators will remove two? If so, then it looks like a pattern is emerging...

The quartic equation

We now turn to the quartic equation

$$z^4 + c_3 z^3 + c_2 z^2 + c_1 z + c_0 = 0\,. \tag{D.19}$$

For the cubic, we saw that G-functions are not enough to construct a formula, as we also needed $\sqrt{G}$ functions. Therefore, we start by assuming the existence of some formula for the four solutions

$$s_\text{i} = H_\text{i}(c_0, c_1, c_2, c_3)\,, \quad \text{for i} \in \{1, \ldots, 4\}\,. \tag{D.20}$$

As before, the four functions H_i are assumed to be H-functions, i.e., G- and $\sqrt{G}$-functions combined with $+, -, \times, \div$. As should be clear by now, it turns out that even with the extra ingredient $\sqrt{G}$, no general quartic formula can be constructed. Once again we will prove this by constructing an appropriate permutation of (s_1, s_2, s_3, s_4).

A brief checkpoint

Once again, just as the first method did not work for cubic equations, the method used for the cubic case is not guaranteed to work for quartic equations either. Indeed, the commutator of transpositions induced a loop on F and $\sqrt{F}$ (and thus on G). But a loop on G generally does not induce a loop on $\sqrt{G}$, as we have seen many times. A summary of these previous methods is given in Table D.1.

But now a natural solution presents itself: what if we take the commutator of, say, (123) and (234), written as commutators themselves, using equation (D.15)? Let us examine this in detail.

Commutators, yet again

First we need to check that the commutator of (123) and (234) does indeed permute the four solution (s_1, s_2, s_3, s_4). Fortunately it does, as a quick check reveals that

$$[(123), (234)] = (14)(23)\,, \tag{D.21}$$

which is a particular case of the more general formula $[(ijk), (jk\ell)] = (i\ell)(jk)$. Therefore, our commutator [(123), (234)] does indeed permute (s_1, s_2, s_3, s_4). Now, let us examine how it affects G- and $\sqrt{G}$-functions, one step at a time:

- First, we apply the cycles (123) = [(12), (23)] then (234) = [(23), (34)]. Since they are commutators, G-functions will follow two loops γ_1, γ_2 successively, coming back to their original positions. However, quantities like $\sqrt{G}$ will move along two (generally unclosed) paths ω_1 and ω_2. All this is exactly as in the cubic case.
- Second, we apply these two paths backwards, in reverse i.e., (432) = [(43), (32)] and then (321) = [(32), (21)]. During these two, G-functions will follow $\gamma_2^{-1}\gamma_1^{-1}$, i.e. the previous loops backwards. Similarly, $\sqrt{G}$-functions will travel along $\omega_2^{-1}\omega_1^{-1}$.

What just happened is exactly the same as in the cubic case, except with G-functions in place of F-functions. In particular, G-functions follow the loop $\gamma = \gamma_1\gamma_2\gamma_1^{-1}\gamma_2^{-1}$; and $\sqrt{G}$-functions a sequence of unclosed paths $\omega_1\omega_2\omega_1^{-1}\omega_2^{-1}$, which closes on itself by construction. In other words, both G- and $\sqrt{G}$-functions followed

Table D.1 Summary of the methods used to prove the first two impossibility results

Ingredient	F-functions	G-functions
Nested roots	0	1
Discarded by	transpositions	commutator of transpositions
With the path	(12)	[(12), (23)] = (123)
For degree	$n \geq 2$	$n \geq 3$

a loop and thus any H-function will, too. Our conclusion has therefore been reached: while (s_1, s_2, s_3, s_4) undergoes the permutation $(14)(23)$ written as a commutator of commutators, any H-function follows a loop. Consequently, no formula (D.20) can exist. This result extends to any equation of degree $n \geq 4$, as before, and constitutes our *third impossibility result:*

Using only H-functions, no formula solving
the general Eq. (D.1) can be found for $n \geq 4$.

In particular, we can extend Table D.1 with an additional column for the new ingredient, H-functions.

Discussion: the quartic formula

As for the cubic case, our impossibility result does not imply that there is no quartic (nor quintic) formula. It just means that to construct one, one needs at least three levels of nested roots, and H-functions contain only two. It turns out that the quartic equation can be solved as follows and, indeed, it involves three levels of nested roots. As for the cubic case, we start by removing the z^3 term by the change of variables $Z = z + c_3/4$. This brings Eq. (D.19) into the form

$$Z^4 + PZ^2 + QZ + R = 0\,, \tag{D.22}$$

where P, Q, R are three F-functions of (c_0, c_1, c_2, c_3), whose expressions are long, but easily obtained. The next step is to transform Eq. (D.22) into one that is quadratic in Z^2. For now, nothing guarantees that $PZ^2 + QZ + R$ is a perfect square, but if it were, then Eq. (D.22) could be factored into two equations quadratic in Z^2. One way is to write Z^4 in the equivalent form $Z^4 = (Z^2 + A)^2 - 2AZ^2 - A^2$, for some complex number A to be chosen freely later on. Inserting this in Eq. (D.22) gives

$$(Z^2 + A)^2 + (P - 2A)Z^2 + QZ + R - A^2 = 0\,. \tag{D.23}$$

Now we can choose A in Eq. (D.23) such that the quadratic part $(P - 2A)Z^2 + QZ + R - A^2$ has the form of a perfect square. This will be the case if its discriminant $Q^2 + 4(P - 2A)A^2$ vanishes. The latter amounts to

$$8A^3 - 4PA^2 - Q^2 = 0\,, \tag{D.24}$$

which is a cubic equation A. It can be solved using the cubic formula, giving a value of A in terms of P and Q that is an H-function (recall the cubic formula (D.18)). Once A takes this special value, Eq. (D.23) becomes $(Z^2 + A)^2 + (P - 2A)(Z - A)^2 = 0$, which can be factored easily into two quadratic polynomials in Z. The latter equations are solved easily using the quadratic formula. Since A is an H-function, the solution for Z will necessarily involve some $\sqrt{H}$ quantities, something which we did not include in Eq. (D.20). This confirms our impossibility result, once again.

Table D.2 Extension of Table D.1 to H-functions, for degree $n \geq 4$.

Ingredient	H-functions
Nested roots	2
Discarded by	commutator of commutator of transpositions
With the path	$[[(12), (23)], [(23), (34)]] = (14)(23)$
For degree	$n \geq 4$

The quintic equation

It seems at this point that things are becoming repetitive, and that a clear pattern emerges. For $n = 2, 3, 4$, commutators could be used to reject formulae with too few nested roots in their expressions. However, we were still able to solve the equation simply by allowing more levels of roots. But at $n = 5$, this all breaks down, and this is why the quintic equation is a very special case. The goal of this section is to apply our methods to the case of degree $n = 5$ and understand why it allows, not only to discard 4 levels of nested roots (i.e., one more than the quartic case), but actually *any* number of roots.

Let us pretend that we found a quintic formula, e.g.,

$$s_{\rm i} = \Phi_{\rm i}(c_0, \ldots, c_4), \quad \text{for i} \in \{1, \ldots, 5\}, \tag{D.25}$$

with the five functions $\Phi_{\rm i}$ built out of $H-$ and $\sqrt{H}$–functions. If we follow the previous methods, summarized in Tables D.1 and D.2, it should be clear that (1) all H-functions will follow a loop from a commutator of commutators of the solutions (as in the quartic case), but (2) we will need one more level of commutators for the $\sqrt{H}$ terms.

As we now have five solutions to play with, let us consider for example the permutations (123) and (345) to construct a first commutator $[(123), (345)]$. An easy check shows that the latter is equal to (235), and this commutator therefore permutes three of our solutions. In general, the following result holds at $n = 5$:

$$[(ijk), (k\ell m)] = (jkm)\,. \tag{D.26}$$

But now, contrary to the previous cases, we have something rather remarkable with Eq. (D.26). It shows that any cycle (jkm) can be written as a commutator of two other cycles, namely $[(ijk), (k\ell m)]$. But notice that this is true for *any* cycle (jkm), including (ijk) and $(k\ell m)$ on the left-hand side of equation (D.26) itself. In other words, this formula can be applied to itself, again and again, allowing us to write (jkm) as a commutator of as many commutators as needed. Since a number $N \in \mathbb{N}$

of commutators allows us to discard precisely N levels of roots in a formula (see Tables D.1 and D.2), we can actually discard *any* number of roots in any candidate quintic formula. The Abel–Ruffini theorem follows immediately from this remark, but let us give a more detailed explanation.

Suppose that, in the quintic formula, Eq. (D.25), we use a Φ-function made of $+, -, \times, \div$, along with N levels of roots, for some $N \in \mathbb{N}$. To construct this Φ-function, we have at our disposal several ingredients: F-functions (no roots), G-functions (one level of root), H-functions (two levels of root), and so on. As always, we start by choosing a permutation of the solutions, say (123), that discards any F-functions (no roots). Next, using equation (D.26), we write (123) as a commutator, for example:

$$(123) = [(412), (253)]\,. \tag{D.27}$$

When applied to $(s_1, \ldots, s_5)$, this commutator discards the G-functions from the list of ingredients (one level of roots). Now we keep going: we write the cycles (412), (253) appearing in Eq. (D.27) as commutators themselves, again using equation (D.26). We obtain (123) expressed with two commutators:

$$(123) = [[(341), (152)], [(425), (513)]]\,, \tag{D.28}$$

which removes H-functions (expression with two levels of roots). By iterating equation (D.26) $N - 2$ more times, we end up writing (123) as a combination of N commutators. When the latter is applied to $(s_1, \ldots, s_5)$, the solutions s_1, s_2, s_3 will permute; and yet any expression of the coefficients with N or fewer roots will follow a loop. Since a Φ-function is made up of all these ingredients, Φ_1, Φ_2, Φ_3 go back to their original position. Clearly this contradicts Eq. (D.25). This result generalizes to an equation of any degree $n \geq 5$ by picking five of its solutions, as before. Moreover, since N is arbitrary, we conclude that no number of roots will be sufficient to write a formula. Our conclusion is therefore:

No formula exists for the solution to the general equation of degree five or more, using only the operations $+, -, \times, \div$, *and* $\sqrt{\ }$.

i.e., the Abel–Ruffini theorem itself. A last remark is in order. Why the fifth degree, and not the fourth or sixth? This all boils down to the possibility of writing a permutation of at least two solutions as a sequence of commutators. A formula such as Eq. (D.26) can only be iterated indefinitely when it involves five or more elements. For four or fewer elements, any sequence of commutators of transposition and/or cycles will necessarily end, i.e., end up being the trivial permutation that "does nothing." The reader familiar with group theory will here recognize the notions of perfect or solvable group.

Conclusions

To conclude this article, we would like to first make some comments on the various advantages and disadvantages of this proof, compared to the usual proof using Galois theory. First, the present proof does not say that no equations of degree five or higher can be solved; but only that a general formula (valid for the general equation) cannot be written using only $+$, $-$, $\times$, $\div$, and $\sqrt{\ }$. Indeed, some equations of degree $n \geq 5$ can actually be solved explicitly (see [645] for a nice and thorough discussion on quintics of the form $z^5 + az + b = 0$ that are solvable by radicals.) Galois theory, on the other hand, is perfectly able to say whether a given equation is solvable or not. On the other hand, the present proof can be extended to also account for continuous (and single-valued) functions of the coefficients (such as exp, sin, ...) in the list of ingredients. Indeed, just like $+$, $-$, $\times$, $\div$, these functions follow a loop when the coefficients do. Galois theory is unable to provide for this, as it only accounts for expressions in terms of radicals.

We hope that the present proof will be seen not only as a simplified and elementary demonstration of the Abel–Ruffini's theorem, but also as a complementary result, as it helps to explain why the $n = 5$ case is so special, and why the quadratic, cubic, and quartic formulae have such "nested roots" structures. It is also a good and instructive exercise to complete the present proof sketch with rigorous arguments (we encourage students to give it a try!). For more insight on this topic, one should definitely watch Katz's video [644] and read Goldmakher's paper [646], which have both inspired this article. We end this paper by providing more references that should help the interested reader to get started with topics that are based on (and broadly extend) the ideas presented here: (1) an interactive blog article by Akalin [647]; (2) an article by Zoladek [648] that deals with similar but more advanced ideas; (3) the original book of Alekseev mentioned in the introduction [649] on the Abel–Ruffini theorem. Devised as a problems-and-solutions book, it discusses many advanced concepts in a very pedagogical and extremely well-written manner.

References

1. Ramond P, Le Tiec A (2020) Multipolar particles in helically symmetric spacetimes. Class Quantum Gravit 38(13). https://doi.org/10.1088/1361-6382/abebef, arXiv:2005.00602 [gr-qc]
2. Ramond P, Le Tiec A (2022) First law of mechanics for extended compact binaries: quadrupolar order. Phys Rev D. To appear. arXiv:2202.09345 [gr-qc]
3. Ramond P, Perez J (2020) The geometry of isochrone orbits: from archimedes' parabolae to kepler's third law. Celest Mech Dyn Astron 132(4):22. https://doi.org/10.1007/s10569-020-09960-w, arXiv:2003.13456
4. Ramond P, Perez J (2021) New methods of isochrone mechanics. J Math Phys 62(11):112704. https://doi.org/10.1063/5.0056957, arXiv:2104.05643
5. Ramond P (2022) The Abel–Ruffini theorem: complex but not complicated. Am Math Mon 129(3):231–245. https://doi.org/10.1080/00029890.2022.2010494, arXiv:2011.05162
6. Wald RM (1984) General relativity. University of Chicago Press, Chicago
7. Friedman M (2014) Foundations of space-time theories: relativistic physics and philosophy of science, vol 113. Princeton University Press
8. Heisenberg W, Northrop FSC (1958) Physics and philosophy: the revolution in modern science. Harper, New York
9. Girard P (1981) Histoire de la Relativité Générale d'Einstein: Développement Conceptuel de la Théorie. PhD thesis, University of Wisconsin-Madison USA. https://hal.archives-ouvertes.fr/tel-02321688/document
10. Sauer T (2005) Albert Einstein, review paper on general relativity theory (1916) (Chap. 63). In: Grattan-Guinness I, Cooke R, Corry L, Crépel P, Guicciardini N (eds) Landmark writings in western mathematics 1640-1940. Elsevier Science, Amsterdam, pp. 802–822. https://doi.org/10.1016/B978-044450871-3/50144-3, arXiv:0405066 [physics.hist-ph]
11. Schilpp PA (1951) Albert Einstein: philosopher. Scientist
12. Robinson A (2015) Einstein: a hundred years of relativity. Princeton University Press
13. Aichelburg PC, Sexl RU (2012) Albert Einstein: his influence on physics, philosophy and politics. Springer Science & Business Media
14. Pais A, Goldberg S (1984) "Subtle is the Lord...": the science and the life of Albert Einstein
15. Heller M (1991) The happiest thought of Einstein's life. Astron Q 8(3):177–180
16. Janssen M, Renn J (2015) History: Einstein was no lone genius. Nat News 527(7578):298

P. Ramond, *The First Law of Mechanics in General Relativity & Isochrone Orbits in Newtonian Gravity*, Springer Theses,
https://doi.org/10.1007/978-3-031-17964-8

17. Klein É (2016) Le pays qu'habitait Albert Einstein. Éditions Actes Sud
18. Ni W-T (2016) Genesis of general relativity - a concise exposition. Int J Mod Phys D 25(14):1630004. https://doi.org/10.1142/s0218271816300044, arXiv:1612.09498 [physics.hist-ph]
19. https://einsteinpapers.press.princeton.edu/
20. https://www.einstein.caltech.edu/
21. Einstein A (1907) On the relativity principle and the conclusions drawn from it. Jahrb Radioaktivitat Elektronik 4:411–462
22. Gourgoulhon E (2013) Special relativity in general frames. Graduate texts in physics. Springer, New York
23. Riemann B (1954) Über die hypothesen, welche der geometrie zu grunde liegen. Physikalische Blätter 10(7):296–306
24. Klein F (1893) Vergleichende betrachtungen über neuere geometrische forschungen. Math Ann 43(1):63–100
25. Bianchi L (1894) Lezioni di geometria differenziale
26. Levi-Civita T (1925) Lezioni di calcolo differenziale assoluto. Zanichelli
27. Fré PG (2018) A conceptual history of space and symmetry. Springer
28. Einstein A, Grossmann M (1913) Entwurf einer verallgemeinerten Relativitätstheorie und einer Theorie der Gravitation. BG Teubner
29. Einstein A (1914) The formal foundation of the general theory of relativity. Sitzungsber Preuss Akad Wiss Berlin (Math Phys) 1914:1030–1085
30. Einstein A (1915) Grundgedanken der allgemeinen relativitätstheorie und anwendung dieser theorie in der astronomie. Preussische Akademie der Wissenschaften Sitzungsberichte 315:778–786
31. Einstein A (1915) Zur allgemeinen relativitätstheorie. Preussische Akademie der Wissenschaften, Sitzungsberichte
32. Einstein A (1915) Erklärung der perihelbewegung des merkur aus der allgemeinen relativitätstheorie. Preussische Akademie der Wissenschaften, Sitzungsberichte 1932:831–839
33. Einstein A (1915) Die feldgleichungen der gravitation. Sitzung der physikalische-mathematischen Klasse 25:844–847
34. Wheeler JA (1990) A journey into gravity and spacetime. Scientific American Library
35. Einstein A (1916) Näherungsweise integration der feldgleichungen der gravitation. Sitzber. Preuss. Akad. Wiss., p 688
36. Einstein A (1917) Kosmologische betrachtungen zur allgemeinen relativitätstheorie. Sitzber. Preuss. Akad. Wiss., p 142
37. Orlin B (2018) Change is the only constant: the wisdom of calculus in a madcap world. Black Dog and Leventhal
38. Jetzer P, Sereno M (2006) Two-body problem with the cosmological constant and observational constraints. Phys Rev D 73(4):044015
39. Mana PP (2021) Dimensional analysis in relativity and in differential geometry. Eur J Phys. arXiv:2007.14217 [gr-qc]
40. Lovelock D (1969) The uniqueness of the Einstein field equations in a four-dimensional space. Arch Ration Mech Anal 33:54
41. Van Aelst K (2016) Spinors and cosmology in Einstein-Cartan theory. PhD thesis, Imperial College London
42. Provost J-P, A brief history of the energy-momentum tensor 1912-1915: Einstein physicist's logic in compromising gravitation with relativity, pp 3348–3353. https://doi.org/10.1142/9789813226609_0429
43. Frankel T (2011) The geometry of physics: an introduction. Cambridge University Press
44. Choquet-Bruhat Y (2008) General relativity and the Einstein equations. OUP Oxford
45. Harte AI, Vines J (2016) Generating exact solutions to Einstein's equation using linearized approximations. Phys Rev D 94(8). https://doi.org/10.1103/physrevd.94.084009
46. Harte AI (2014) Taming the nonlinearity of the Einstein equation. Phys Rev Lett 113:261103. arXiv:1409.4674 [gr-qc]

47. Einstein A (1916) Die Grundlage der allgemeinen Relativitätstheorie. Ann der Phys 354(7):769–822. https://doi.org/10.1002/andp.19163540702
48. Schild A (1960) Equivalence principle and red-shift measurements. Am J Phys 28(9):778–780
49. Pound RV, Rebka GA Jr (1960) Apparent weight of photons. Phys Rev Lett 4(7):337
50. Pound RV, Snider JL (1965) Effect of gravity on gamma radiation. Phys Rev 140(3B):B788
51. Do T, Hees A, Ghez A, Martinez GD, Chu DS, Jia S, Sakai S, Lu JR, Gautam AK, O'neil KK, et al (2019) Relativistic redshift of the star s0-2 orbiting the galactic center supermassive black hole. Science 365(6454):664–668. arXiv:1907.10731 [astro-ph.GA]
52. Hernández JG, Rebolo R, Pasquini L, Curto GL, Molaro P, Caffau E, Ludwig H-G, Steffen M, Esposito M, Mascareño AS et al (2020) The solar gravitational redshift from HARPS-LFC moon spectra-a test of the general theory of relativity. Astron & Astrophys 643:A146
53. Deruelle N, Uzan JP (2014) Théories de la relativité. Belin, Paris
54. Will CM (1988) Henry cavendish, johann von soldner, and the deflection of light. Am J Phys 56(5):413–415
55. Einstein A (1911) Über den einfluß der schwerkraft auf die ausbreitung des lichtes. Ann der Phys 340(10):898–908
56. Kennefick D (2019) No shadow of a doubt: the 1919 eclipse that confirmed Einstein's theory of relativity. Princeton University Press
57. Dyson FW, Eddington AS, Davidson C (1920) IX. A determination of the deflection of light by the sun's gravitational field, from observations made at the total eclipse of may 29, 1919. Philos Trans R Soc A: Math Phys Eng Sci 220(571–581):291–333
58. Crelinsten J (1983) William wallace campbell and the "Einstein problem": an observational astronomer confronts the theory of relativity. Hist Stud Phys Sci 14(1):1–91
59. Cervantes-Cota JL, Galindo-Uribarri S, Smoot GF (2020) The legacy of Einstein's eclipse, gravitational lensing. Universe 6(1):9
60. Lemon CA, Auger MW, McMahon RG, Ostrovski F (2018) Gravitationally lensed quasars in Gaia-II. Discovery of 24 lensed quasars. Mon Not R Astron Soc 479(4):5060–5074
61. Aghanim N, Akrami Y, Ashdown M, Aumont J, Baccigalupi C, Ballardini M, Banday AJ, Barreiro R, Bartolo N, Basak S et al (2020) Planck 2018 results-VIII. Gravitational lensing. Astron & Astrophys 641:A8
62. Campbell WW, Trumpler R (1923) Observations on the deflection of light in passing through the sun's gravitational field, made during the total solar eclipse of September 21. Publ Astron Soc Pac 35(205):158–163
63. Johnson TL, Sharon K, Bayliss MB, Gladders MD, Coe D, Ebeling H (2014) Lens models and magnification maps of the six hubble frontier fields clusters. Astrophys J 797(1):48
64. Richard J, Kneib J-P, Limousin M, Edge A, Jullo E (2010) Abell 370 revisited: refurbished hubble imaging of the first strong lensing cluster. Mon Not R Astron Soc: Lett 402(1):L44–L48
65. https://hubblesite.org/contents/media/images/2017/20/4024-Image?Topic=105-galaxies&keyword=abell%20370
66. Le Verrier UJ (1859) Theorie du mouvement de mercure. Ann de l'Observatoire de Paris 5
67. Laskar J (2017) Des premiers travaux de Le Verrier à la découverte de Neptune. Comptes Rendus Physique 18(9–10):504–519. https://doi.org/10.1016/j.crhy.2017.10.011, arXiv:1710.03688 [physics.hist-ph]
68. Roseveare NT (1982) Mercury's perihelion, from Le Verrier to Einstein. Oxford University Press, New York; Clarendon Press, Oxford
69. Fienga A, Laskar J, Exertier P, Manche H, Gastineau M (2015) Numerical estimation of the sensitivity of INPOP planetary ephemerides to general relativity parameters. Celest Mech Dyn Astron 123(3):325–349
70. Nordtvedt K Jr (1968) Equivalence principle for massive bodies. II. Theory. Phys Rev 169(5):1017
71. Will CM, Nordtvedt K Jr (1972) Conservation laws and preferred frames in relativistic gravity. I. Preferred-frame theories and an extended PPN formalism. Astrophys J 177:757
72. Will CM (1993) Theory and experiment in gravitational physics. Cambridge University Press, Cambridge

73. Bertotti B, Iess L, Tortora P (2003) A test of general relativity using radio links with the Cassini spacecraft. Nature 425(6956):374–376
74. Fienga A, Laskar J, Kuchynka P, Manche H, Desvignes G, Gastineau M, Cognard I, Theureau G (2011) The inpop10a planetary ephemeris and its applications in fundamental physics. Celest Mech Dyn Astron 111(3):363–385
75. Ni W-T (2017) One hundred years of general relativity: from genesis and empirical foundations to gravitational waves, cosmology and quantum gravity, vol 2. World Scientific
76. Will CM (2014) The confrontation between general relativity and experiment. Living Rev Relat 17:4. arXiv:1403.7377 [gr-qc]
77. Montgomery C, Orchiston W, Whittingham I (2009) Michell, Laplace and the origin of the black hole concept. J Astron Hist Herit 12:90–96
78. Eisenstaedt J (1982) Histoire et singularités de la solution de schwarzschild (1915–1923). Arch Hist Exact Sci 27(2):157–198
79. Schwarzschild K (1916) On the gravitational field of a mass point according to Einstein's theory. Sitzber. Preuss. Akad. Wiss. 189. arXiv:physics/9905030. Translated in English by S. Antoci and reprinted in Gen Rel Gravit 35:951 (2003)
80. Schwarzschild K (1916) On the gravitational field of a sphere of incompressible fluid according to Einstein's theory, Sitzber. Preuss. Akad. Wiss. 424. arXiv:physics/9912033. Translated in English by S. Antoci
81. Droste J (2002) Physics-"the field of a single center in Einstein's theory of gravitation and the motion of a particle in the field". Gen Rel Gravit 34:1545–1563. https://doi.org/10.1023/A:1020747322668
82. Gourgoulhon É (2021) Geometry and physics of black holes. https://luth.obspm.fr/~luthier/gourgoulhon/bh16
83. Hagihara Y (1930) Theory of the relativistic trajectories in a gravitational field of Schwarzschild. Jpn J Astron Geophys 8:67
84. Oppenheimer JR, Snyder H (1939) On continued gravitational contraction. Phys Rev 56:455
85. Giacconi R (2003) Nobel lecture: the dawn of x-ray astronomy. Rev Mod Phys 75(3):995
86. Matthews TA, Sandage AR (1963) Optical identification of 3C 48, 3C 196, and 3C 286 with stellar objects. Astrophys J 138:30. https://doi.org/10.1086/147615
87. Hewish A, Bell S, Pilkington J, Scott P, Collins R (1969) Observation of a rapidly pulsating radio source. Nature 224(5218):472 (Reprinted from nature, february 24, 1968)
88. Kerr RP (1963) Gravitational field of a spinning mass as an example of algebraically special metrics. Phys Rev Lett 11:237
89. Teukolsky SA (2015) The Kerr metric. Class Quantum Gravit 32(12):124006. https://doi.org/10.1088/0264-9381/32/12/124006
90. Penrose R (1965) Gravitational collapse and space-time singularities. Phys Rev Lett 14(3):57
91. Hawking SW (1967) The occurrence of singularities in cosmology. III. Causality and singularities. Proc R Soc Lond Ser A Math Phys Sci 300(1461):187–201
92. Penrose R (1969) Gravitational collapse: the role of general relativity. Riv Nuovo Cim 1:242
93. Hawking SW, Penrose R (1970) The singularities of gravitational collapse and cosmology. Proc R Soc Lond A Math Phys Sci 314(1519):529–548
94. Regge T, Wheeler JA (1957) Stability of a Schwarzschild singularity. Phys Rev 108(4):1063
95. Teukolsky SA, Press WH (1974) Perturbations of a rotating black hole. III. Interaction of the hole with gravitational and electromagnetic radiation. Astrophys J 193:443
96. Finster F (2018) Lectures on linear stability of rotating black holes. In: Domoschool-International Alpine School of Mathematics and Physics. Springer, pp 61–91
97. Barack L, Cardoso V, Nissanke S, Sotiriou TP, Askar A, Belczynski C, Bertone G, Bon E, Blas D, Brito R et al (2019) Black holes, gravitational waves and fundamental physics: a roadmap. Class Quantum Gravit 36(14):143001. arXiv:1806.05195 [gr-qc]
98. Doeleman S, E. Collaboration et al (2019) Focus on the first event horizon telescope results. Astrophys J Lett
99. Baade W, Zwicky F (1934) Remarks on super-novae and cosmic rays. Phys Rev 46(1):76
100. Oppenheimer JR, Volkoff GM (1939) On massive neutron cores. Phys Rev 55(4):374

101. Özel F, Freire P (2016) Masses, radii, and the equation of state of neutron stars. Ann Rev Astron Astrophys 54:401–440. https://doi.org/10.1146/annurev-astro-081915-023322, arXiv:1603.02698 [astro-ph.HE]
102. Primack JR (2005) Precision cosmology. New Astron Rev 49(2–6):25–34
103. Cabanac R, Valls-Gabaud D, Jaunsen A, Lidman C, Jerjen H (2005) Discovery of a high-redshift Einstein ring. Astron & Astrophys 436(2):L21–L25. https://doi.org/10.1051/0004-6361:200500115
104. Belokurov V, Evans N, Hewett P, Moiseev A, McMahon R, Sanchez S, King L (2009) Two new large-separation gravitational lenses from SDSS. Mon Not R Astron Soc 392(1):104–112. https://doi.org/10.1111/j.1365-2966.2008.14075.x
105. Hennawi JF, Gladders MD, Oguri M, Dalal N, Koester B, Natarajan P, Strauss MA, Inada N, Kayo I, Lin H et al (2008) A new survey for giant arcs. Astron J 135(2):664. https://doi.org/10.1088/0004-6256/135/2/664
106. Bartelmann M, Schneider P (2001) Weak gravitational lensing. Phys Rep Ser 340(4–5):291–472
107. Kilbinger M (2015) Cosmology with cosmic shear observations: a review. Rep Prog Phys 78(8):086901. https://doi.org/10.1088/0034-4885/78/8/086901, arXiv:1411.0115 [gr-qc]
108. Aghanim N et al (Planck Collaboration) (2020) Planck 2018 results. VI. Cosmological parameters. Astron Astrophys 641:A6. arXiv:1807.06209 [astro-ph.CO]
109. Ade PA, Aghanim N, Arnaud M, Ashdown M, Aumont J, Baccigalupi C, Banday A, Barreiro R, Bartolo N, Battaner E et al (2016) Planck 2015 results-XIV. Dark energy and modified gravity. Astron & Astrophys 594:A14
110. Salvatelli V, Piazza F, Marinoni C (2016) Constraints on modified gravity from Planck 2015: when the health of your theory makes the difference. J Cosmol Astropart Phys 09:027. arXiv:1602.08283 [astro-ph.CO]
111. Ishak M (2019) Testing general relativity in cosmology. Living Rev Relat 22(1):1–204
112. Einstein A (1918) Gravitationswellen. Sitzber. Preuss. Akad. Wiss. p 154
113. Eddington AS (1922) The propagation of gravitational waves. Proc R Soc of Lond Ser A Contain Pap Math Phys Character 102(716):268–282
114. Eddington AS (1923) Cxvii. the spontaneous loss of energy of a spinning rod according to the relativity theory. Lond Edinb Dublin Philos Mag J Sci 46(276):1112–1117
115. Kennefick DJ (2007) Traveling at the speed of thought: Einstein and the quest for gravitational waves. Princeton University Press, Princeton
116. Bondi H (1957) Plane gravitational waves in general relativity. Nature 179:1072
117. Pirani FAE (1957) Invariant formulation of gravitational radiation theory. Phys Rev 105:1089
118. Bondi H, Pirani FA, Robinson I (1959) Gravitational waves in general relativity III. Exact plane waves. Proc R Soc Lond Ser A Math Phys Sci 251(1267):519–533
119. Pirani F (1957) Measurement of classical gravitation fields. In: The role of gravitation in physics. Proceedings of the 1957 Chapel Hill conference, Max Planck Research Library for the History and Development of Knowledge
120. Einstein A, Infeld L, Hoffmann B (1938) Gravitational equations and the problem of motion. Ann Math 39:65
121. Lorentz HA, Droste J (1917) De beweging van een stelsel lichamen onder den invloed van hunne onderlinge aantrekking, behandeld volgens de theorie van Einstein I, II. Versl K Akad Wet Amst 26:392 & 649
122. de Sitter W (1916) On Einstein's theory of gravitation and its astronomical consequences. Mon Not R Astron Soc 77:155
123. Landau LD, Lifshitz EM (1975) The classical theory of fields, 4th edn. Butterworth-Heinemann, Oxford
124. Ehlers J, Rosenblum A, Goldberg JN, Havas P (1976) Comments on gravitational radiation damping and energy loss in binary systems. Astrophys J 208(2):L77–L81
125. Chandrasekhar S, Esposito FP (1970) The 5/2-post-Newtonian equations of hydrodynamics and radiation reaction in general relativity. Astrophys J 160:153

126. Burke WL, Thorne KS (1970) Gravitational radiation damping. In: Carmeli M, Fickler SI, Witten L (eds) Relativity. Springer, US, Boston, MA, pp 209–228
127. Bel L, Damour T, Deruelle N, Ibanez J, Martin J (1981) Poincaré-invariant gravitational field and equations of motion of two pointlike objects: the postlinear approximation of general relativity. Gen Relat Gravit 13(10):963–1004
128. Damour T, Deruelle N (1981) Radiation reaction and angular momentum loss in small angle gravitational scattering. Phys Lett A 87(3):81–84
129. Damour T, Deruelle N (1981) Lagrangien généralisé du système de deux masses ponctuelles, à l'approximation post-post-newtonienne de la relativité générale. CR Acad Sci Ser II 293:537–540
130. Damour T (1982) Problème des deux corps et freinage de rayonnement en relativité générale. C R Acad Sci Paris 294:1355
131. Hobbs G, Manchester R, Teoh A, Hobbs M (2004) The ATNF pulsar catalog. In: Young neutron stars and their environments, vol 218, p 139
132. Atnf pulsar catalogue. https://www.atnf.csiro.au/research/pulsar/psrcat/
133. Antoniadis J (2020) Gaia pulsars and where to find them in EDR3. Res Notes Am Astron Soc 4(12):223. https://doi.org/10.3847/2515-5172/abd189, arXiv:2012.06335 [astro-ph.HE]
134. Antoniadis J (2020) Gaia pulsars and where to find them. Mon Not R Astron Soc 501(1):1116–1126. https://doi.org/10.1093/mnras/staa3595
135. Wex N, Kramer M (2020) Gravity tests with radio pulsars. Universe 6(9). https://doi.org/10.3390/universe6090156, https://www.mdpi.com/2218-1997/6/9/156
136. Maggiore M (2008) Gravitational waves: theory and experiments, vol 1. Oxford University Press, Oxford
137. Hulse RA, Taylor JH (1975) Discovery of a pulsar in a binary system. Astrophys J 195:L51
138. Altschuler DR, Salter CJ (2013) The Arecibo observatory. Phys Today 66(11):43. https://doi.org/10.1063/PT.3.2179
139. Peters PC, Mathews J (1963) Gravitational radiation from point masses in a Keplerian orbit. Phys Rev 131:435
140. Peters PC (1964) Gravitational radiation and the motion of two point masses. Phys Rev 136:B1224
141. Poisson E, Will CM (2014) Gravity: Newtonian, post-Newtonian, relativistic. Cambridge University Press, Cambridge
142. Damour T (2015) 1974: the discovery of the first binary pulsar. Class Quantum Gravit 32(12):124009. arXiv:1411.3930 [gr-qc]
143. Weisberg JM, Huang Y (2016) Relativistic measurements from timing the binary pulsar PSR B1913+16. Astrophys J 829:55. arXiv:1606.02744 [astro-ph.HE]
144. Gair JR, Porter EK (2013) Observing EMRIs with eLISA/NGO. In: Auger G, Binétruy P, Plagnol E (eds) 9th LISA symposium. ASP conference series, vol 467. Astronomical Society of the Pacific, San Fransisco, p 173. arXiv:1210.8066 [gr-qc]
145. Blanchet L (2014) Gravitational radiation from post-Newtonian sources and inspiralling compact binaries. Living Rev Relat 17:2. arXiv:1310.1528 [gr-qc]
146. Jackson JD (1998) Classical electrodynamics, 3rd edn. Wiley, New York
147. Feynman R (2018) Feynman lectures on gravitation. CRC Press
148. Futamase T, Itoh Y (2007) The post-Newtonian approximation for relativistic compact binaries. Living Rev Relat 10:2
149. Stewart J (1991) Advanced general relativity. Cambridge University Press, Cambridge
150. Straumann N (2013) General relativity, 2nd edn. Springer, New York
151. Jackson JD, Okun LB (2001) Historical roots of gauge invariance. Rev Mod Phys 73(3):663
152. Bonetti M, Barausse E, Faye G, Haardt F, Sesana A (2017) About gravitational-wave generation by a three-body system. Class Quantum Gravit 34(21):215004. https://doi.org/10.1088/1361-6382/aa8da5, arXiv:1707.04902 [gr-qc]
153. Caprini C, Figueroa DG (2018) Cosmological backgrounds of gravitational waves. Class Quantum Gravit 35(16):163001. arXiv:1801.04268 [gr-qc]

154. Maggiore M (2000) Gravitational wave experiments and early Universe cosmology. Phys Rep 331:283. arXiv:gr-qc/9909001
155. Thorne KS (1980) Multipole expansions of gravitational radiation. Rev Mod Phys 52:299
156. Blanchet L (1998) On the multipole expansion of the gravitational field. Class Quantum Gravit 15:1971. arXiv:gr-qc/9801101
157. Maselli A et al (2017) Dark stars: gravitational and electromagnetic observables. Phys Rev D 96(2):023005. https://doi.org/10.1103/PhysRevD.96.023005, arXiv:1704.07286 [astro-ph.HE]
158. Sennett N et al (2017) Distinguishing boson stars from black holes and neutron stars from tidal interactions in inspiraling binary systems. Phys Rev D 96(2):024002. https://doi.org/10.1103/PhysRevD.96.024002, arXiv:1704.08651 [gr-qc]
159. Lisa mission l3 proposal. https://www.elisascience.org/files/publications/LISA_L3_20170120.pdf
160. Abbott BP (LIGO Scientific Collaboration and Virgo Collaboration) et al (2016) Observation of gravitational waves from a binary black hole merger. Phys Rev Lett 116(6):061102. https://doi.org/10.1103/PhysRevLett.116.061102, arXiv:1602.03837 [gr-qc]
161. Rees MJ (1997) Gravitational waves from galactic centres? Class Quantum Gravit 14:1411
162. Milosavljević M, Merritt D (2003) Long-term evolution of massive black hole binaries. Astrophys J 596:860. arXiv:astro-ph/0212459
163. Khan FM, Capelo PR, Mayer L, Berczik P (2018) Dynamical evolution and merger timescales of Lisa massive black hole binaries in disk galaxy mergers. Astrophys J 868(2):97
164. Campanelli M, Lousto C, Zlochower Y, Merritt D (2007) Large merger recoils and spin flips from generic black hole binaries. Astrophys J Lett 659(1):L5–L8. https://doi.org/10.1086/516712, arXiv:gr-qc/0701164 [gr-qc]
165. Ryan FD (1995) Gravitational waves from the inspiral of a compact object into a massive, axisymmetric body with arbitrary multipole moments. Phys Rev D 52:5707
166. Ryan FD (1996) Effect of gravitational radiation reaction on nonequatorial orbits around a Kerr black hole. Phys Rev D 53:3064 arXiv:gr-qc/9511062
167. Gair JR, Vallisneri M, Larson SL, Baker JG (2013) Testing general relativity with low-frequency, space-based gravitational-wave detectors. Living Rev Relat 16(1):1–109 arXiv:1212.5575 [gr-qc]
168. Drasco S, Hughes SA (2006) Gravitational wave snapshots of generic extreme mass ratio inspirals. Phys Rev D 73:024027. arXiv:gr-qc/0509101. Errata: Phys Rev D 88:109905(E) (2013) & Phys Rev D 90:109905(E) (2014)
169. Amaro-Seoane P et al (2013) eLISA: astrophysics and cosmology in the millihertz regime. GW Notes 6:4–110 arXiv:1201.3621 [astro-ph.CO]
170. Mihaylov DP, Gair JR (2017) Transition of EMRIs through resonance: corrections to higher order in the on-resonance flux modification. J Math Phys 58(11):112501. https://doi.org/10.1063/1.5006336, arXiv:1706.06639 [gr-qc]
171. Cutler C, Thorne KS (2002) An overview of gravitational-wave sources. In: Bishop NT, Maharaj SD (eds) General relativity and gravitation: proceedings of the 16th international conference. World Scientific, Singapore, p 72. arXiv:gr-qc/0204090
172. Finn LS (1991) Detectability of gravitational radiation from stellar-core collapse. In: Buchler JR, Detweiler S, Ipser JR (eds) Nonlinear problems in relativity and cosmology. Annals of the New York Academy of Sciences, vol 631. New York Academy of Sciences, New York, p 156
173. Ott CD (2009) The gravitational-wave signature of core-collapse supernovae. Class Quantum Gravit 26:063001. arXiv:0809.0695 [astro-ph]
174. Ott C, Abdikamalov E, O'Connor E, Reisswig C, Haas R, Kalmus P, Drasco S, Burrows A, Schnetter E (2012) Correlated gravitational wave and neutrino signals from general-relativistic rapidly rotating iron core collapse. Phys Rev D 86(2):024026
175. Fryer CL, New KC (2011) Gravitational waves from gravitational collapse. Living Rev Relat 14(1):1–93

176. Ott CD, Abdikamalov E, Mösta P, Haas R, Drasco S, O'Connor EP, Reisswig C, Meakin CA, Schnetter E (2013) General-relativistic simulations of three-dimensional core-collapse supernovae. Astrophys J 768(2):115. https://doi.org/10.1088/0004-637X/768/2/115, arXiv:1210.6674 [astro-ph.HE]
177. Necula V, Klimenko S, Mitselmakher G (2012) Transient analysis with fast Wilson-Daubechies time-frequency transform. J Phys Conf Ser 363:012032. https://doi.org/10.1088/1742-6596/363/1/012032
178. Portilla ML, Di Palma I, Drago M, Cerda-Duran P, Ricci F (2020) Deep learning for multi-messenger core-collapse supernova detection. arXiv:2011.13733
179. Klimenko S, Vedovato G, Drago M, Salemi F, Tiwari V, Prodi G, Lazzaro C, Ackley K, Tiwari S, Da Silva C et al (2016) Method for detection and reconstruction of gravitational wave transients with networks of advanced detectors. Phys Rev D 93(4):042004
180. Torres-Forné A, Cerdá-Durán P, Passamonti A, Font JA (2018) Towards asteroseismology of core-collapse supernovae with gravitational-wave observations-I. Cowling approximation. Mon Not R Astron Soc 474(4):5272–5286
181. Torres-Forné A, Cerdá-Durán P, Passamonti A, Obergaulinger M, Font JA (2019) Towards asteroseismology of core-collapse supernovae with gravitational wave observations-II. Inclusion of space-time perturbations. Mon Not R Astron Soc 482(3):3967–3988
182. Rosado PA (2011) Gravitational wave background from binary systems. Phys Rev D 84(8):084004. https://doi.org/10.1103/PhysRevD.84.084004, arXiv:1106.5795 [gr-qc]
183. Nissanke S, Vallisneri M, Nelemans G, Prince TA (2012) Gravitational-wave emission from compact galactic binaries. Astrophys J 758(2):131. https://doi.org/10.1088/0004-637X/758/2/131, arXiv:1201.4613 [astro-ph.GA]
184. Zhong S-Q, Dai Z-G, Li X-D (2019) Gravitational waves from newborn accreting millisecond magnetars. Phys Rev D 100(12):123014
185. Gittins F, Andersson N (2019) Population synthesis of accreting neutron stars emitting gravitational waves. Mon Not R Astron Soc 488(1):99–110
186. Arca-Sedda M, Gualandris A (2018) Gravitational wave sources from inspiralling globular clusters in the galactic centre and similar environments. Mon Not R Astron Soc 477(4):4423–4442
187. Sathyaprakash BS, Schutz BF (2009) Physics, astrophysics and cosmology with gravitational waves. Living Rev Relat 12:2. arXiv:0903.0338 [gr-qc]
188. Lasky PD et al (2016) Gravitational-wave cosmology across 29 decades in frequency. Phys Rev X 6(1):011035. https://doi.org/10.1103/PhysRevX.6.011035, arXiv:1511.05994 [astro-ph.CO]
189. Riles K (2013) Gravitational waves: sources, detectors and searches. Prog Part Nucl Phys 68:1–54. https://doi.org/10.1016/j.ppnp.2012.08.001, arXiv:1209.0667 [astro-ph.GA]
190. Einstein A (1905) Ist die trägheit eines körpers von seinem energieinhalt abhängig? Ann der Phys 323(13):639–641
191. Gourgoulhon É (2014) Relativité générale. luth.obspm.fr
192. Blanchet L (2011) Post-Newtonian theory and the two-body problem. Fund Theor Phys 162:125 arXiv:0907.3596 [gr-qc]
193. Schäfer G (2011) Post-Newtonian methods: analytic results on the binary problem 162:167–210. https://doi.org/10.1007/978-90-481-3015-3_6
194. Schäfer G, Jaranowski P (2018) Hamiltonian formulation of general relativity and post-Newtonian dynamics of compact binaries. Living Rev Relat 21(1):7. https://doi.org/10.1007/s41114-018-0016-5, arXiv:1805.07240 [gr-qc]
195. Levi M (2020) Effective field theories of post-Newtonian gravity: a comprehensive review. Rep Prog Phys Ser 83:075901. arXiv:1807.01699 [hep-th]
196. Marchand T (2018) Studying gravitational waves of compact binary systems using post-Newtonian theory. PhD thesis, Sorbonne université
197. Poisson E, Pound A, Vega I (2011) The motion of point particles in curved spacetime. Living Rev Relat 14:7 arXiv:1102.0529 [gr-qc]

198. Barack L (2014) Gravitational self-force: orbital mechanics beyond geodesic motion. Fund Theor Phys 177:147
199. Barack L, Pound A (2018) Self-force and radiation reaction in general relativity. Rep Prog Phys 82:016904. arXiv:1805.10385 [gr-qc]
200. Pound A, Wardell B (2021) Black hole perturbation theory and gravitational self-force. In: Bambi C, Katsanevas S, Kokkotas K (eds) Handbook of gravitational wave astronomy. Springer, Singapore
201. Kokkotas KD, Schmidt B (1999) Quasi-normal modes of stars and black holes. Living Rev Relat 2:2. arXiv:gr-qc/9909058
202. Ferrari V, Gualtieri L (2008) Quasi-normal modes and gravitational wave astronomy. Gen Relat Gravit 40(5):945–970. arXiv:0709.0657 [gr-qc]
203. Pani P (2013) Advanced methods in black-hole perturbation theory. Int J Mod Phys A 28(22n23):1340018. arXiv:1305.6759 [gr-qc]
204. Le Tiec A et al (2011) Periastron advance in black-hole binaries. Phys Rev Lett 107:141101. arXiv:1106.3278 [gr-qc]
205. Le Tiec A (2014) The overlap of numerical relativity, perturbation theory and post-Newtonian theory in the binary black hole problem. Int J Mod Phys D 23:1430022. arXiv:1408.5505 [gr-qc]
206. Le Tiec A, Grandclément P (2018) Horizon surface gravity in corotating black hole binaries. Class Quantum Gravit 35:144002. arXiv:1710.03673 [gr-qc]
207. van de Meent M, Pfeiffer H, Intermediate mass-ratio black hole binaries: applicability of small mass-ratio perturbation theory. Phys Rev Lett 125(20):181101. arXiv:2006.12036 [gr-qc]
208. Warburton N, Pound A, Wardell B, Miller J, Durkan L (2021) Gravitational-wave energy flux for compact binaries through second order in the mass ratio. arXiv:2107.01298
209. Pretorius F (2005) Evolution of binary black-hole spacetimes. Phys Rev Lett 95:121101. arXiv:gr-qc/0507014
210. Baker JG, Centrella J, Choi D-I, Koppitz M, van Meter J (2006) Gravitational-wave extraction from an inspiraling configuration of merging black holes. Phys Rev Lett 96:111102. arXiv:gr-qc/0511103
211. Campanelli M, Lousto CO, Marronetti P, Zlochower Y (2006) Accurate evolutions of orbiting black-hole binaries without excision. Phys Rev Lett 96:111101. arXiv:gr-qc/0511048
212. Mroué AH et al (2013) Catalog of 174 binary black hole simulations for gravitational wave astronomy. Phys Rev Lett 111(24):241104. https://doi.org/10.1103/PhysRevLett.111.241104, arXiv:1304.6077 [gr-qc]
213. Jani K, Healy J, Clark JA, London L, Laguna P, Shoemaker D (2016) Georgia tech catalog of gravitational waveforms. Class Quantum Gravit 33(20):204001. https://doi.org/10.1088/0264-9381/33/20/204001, arXiv:1605.03204 [gr-qc]
214. Healy J, Lousto CO, Zlochower Y, Campanelli M (2017) The RIT binary black hole simulations catalog. Class Quantum Gravit 34(22):224001. arXiv:1703.03423 [gr-qc]
215. Damour T, Nagar A (2016) The effective-one-body approach to the general relativistic two body problem. In: Astrophysical black holes. Springer, pp 273–312
216. Bohé A, Shao L, Taracchini A, Buonanno A, Babak S, Harry IW, Hinder I, Ossokine S, Pürrer M, Raymond V et al (2017) Improved effective-one-body model of spinning, nonprecessing binary black holes for the era of gravitational-wave astrophysics with advanced detectors. Phys Rev D 95(4):044028
217. Ajith P et al (2007) A phenomenological template family for black-hole coalescence waveforms. Class Quantum Gravit 24(19):S689–S699. https://doi.org/10.1088/0264-9381/24/19/S31, arXiv:0704.3764 [gr-qc]
218. Buonanno A, Damour T (1999) Effective one-body approach to general relativistic two-body dynamics. Phys Rev D 59:084006. arXiv:gr-qc/9811091
219. Buonanno A, Damour T (2000) Transition from inspiral to plunge in binary black hole coalescences. Phys Rev D 62:064015. arXiv:gr-qc/0001013
220. Abbott BP, Abbott R, Abbott TD, Abernathy M, Acernese F, Ackley K, Adams C, Adams T, Addesso P, Adhikari RX et al (2017) Effects of waveform model systematics on the interpretation of GW150914. Class Quantum Gravit 34(10):104002. arXiv:gr-qc/1611.07531

221. Abbott B, Abbott R, Abbott T, Abraham S, Acernese F, Ackley K, Adams C, Adhikari RX, Adya V, Affeldt C et al (2019) Tests of general relativity with the binary black hole signals from the ligo-virgo catalog GWTC-1. Phys Rev D 100(10):104036. arXiv:gr-qc/1903.04467
222. Buonanno A, Sathyaprakash BS (2014) Sources of gravitational waves: theory and observations. arXiv:1410.7832 [gr-qc]
223. Hinderer T et al (2013) Periastron advance in spinning black hole binaries: comparing effective-one-body and numerical relativity. Phys Rev D 88:084005. arXiv:1309.0544 [gr-qc]
224. Le Tiec A et al (2013) Periastron advance in spinning black hole binaries: gravitational self-force from numerical relativity. Phys Rev D 88:124027. arXiv:1309.0541 [gr-qc]
225. Jaranowski P, Królak A (2012) Gravitational-wave data analysis. formalism and sample applications: the gaussian case. Living Rev Relat 15(1):1–47
226. Abbott BP et al (LIGO Scientific Collaboration and Virgo Collaboration) (2016) GW151226: observation of gravitational waves from a 22-solar-mass binary black hole coalescence. Phys Rev Lett 116:241103. arXiv:1606.04855 [gr-qc]
227. Fritschel P, Evans M, Frolov V (2014) Balanced homodyne readout for quantum limited gravitational wave detectors. Opt Express 22(4):4224–4234
228. Pitkin M, Reid S, Rowan S, Hough J (2011) Gravitational wave detection by interferometry (ground and space). Living Rev Relat 14(1):1–75
229. McIver J, Shoemaker D (2021) Discovering gravitational waves with advanced ligo
230. L. Collaboration (2015) Advanced LIGO. Class Quantum Gravit 32(7):074001. https://doi.org/10.1088/0264-9381/32/7/074001
231. V. Collaboration (2014) Advanced virgo: a second-generation interferometric gravitational wave detector. Class Quantum Gravit 32(2):024001. https://doi.org/10.1088/0264-9381/32/2/024001
232. Coughlin MW (LIGO Scientific Collaboration and Virgo Collaboration) (2011) Identification of long-duration noise transients in LIGO and Virgo. Class Quantum Gravit 28(23):235008. https://doi.org/10.1088/0264-9381/28/23/235008, arXiv:1108.1521 [gr-qc]
233. Abbott BP et al (LIGO Scientific Collaboration and Virgo Collaboration) (2016) Binary black hole mergers in the first Advanced LIGO observing run. Phys Rev X 6:041015. arXiv:1606.04856 [gr-qc]
234. Weber J (1969) Evidence for discovery of gravitational radiation. Phys Rev Lett 22(24):1320–1324. https://doi.org/10.1103/PhysRevLett.22.1320
235. Weber J (1960) Detection and generation of gravitational waves. Phys Rev 117:306
236. Aguiar OD (2011) Past, present and future of the Resonant-mass gravitational wave detectors. Res Astron Astrophys 11(1):1–42. https://doi.org/10.1088/1674-4527/11/1/001, arXiv:1009.1138 [astro-ph.IM]
237. Levin J (2016) Black hole blues and other songs from outer space. Anchor
238. Oliveira NF, Aguiar OD (2016) The Mario Schenberg gravitational wave antenna. Braz J Phys 46(5):596–603. https://doi.org/10.1007/s13538-016-0436-1
239. De Waard A et al (2005) Minigrail progress report 2004. Class Quantum Grav 22(10):S215
240. de Waard A et al (2006) Preparing for science run 1 of MiniGRAIL. Class Quantum Gravit 23(8):S79–S84. https://doi.org/10.1088/0264-9381/23/8/S11
241. De Waard A et al (2004) Cooling down minigrail to milli-kelvin temperatures. Class Quantum Grav 21(5):S465
242. Sazhin MV (1978) Opportunities for detecting ultralong gravitational waves. Sov Astron 22:36–38
243. Detweiler S (1979) Pulsar timing measurements and the search for gravitational waves. Astrophys J 234:1100–1104
244. Kramer M, Champion DJ (2013) The European pulsar timing array and the large European array for pulsars. Class Quantum Gravit 30(22):22400. https://doi.org/10.1088/0264-9381/30/22/2240099
245. Hobbs G (2015) The Parkes pulsar timing array project. Publ Korean Astron Soc 30(2):577–581. https://doi.org/10.5303/PKAS.2015.30.2.577

246. Ransom S et al (2019) The NANOGrav program for gravitational waves and fundamental physics. Bull Am Astron Soc 51:195. arXiv:1908.05356 [astro-ph.IM]
247. Indian pulsar timing array website. https://inpta.gitlab.io/profile/index.html
248. Manchester RN, IPTA (2013) The international pulsar timing array. Class Quantum Gravit 30(22):224010. https://doi.org/10.1088/0264-9381/30/22/224010, arXiv:1309.7392 [astro-ph.IM]
249. Susobhanan A et al (2021) pinta: the uGMRT data processing pipeline for the Indian pulsar timing array. Publ Astron Soc Aust 38
250. Kerr M et al (2020) The parkes pulsar timing array project: second data release. Publ Astron Soc Aust 37:e020. https://doi.org/10.1017/pasa.2020.11, arXiv:2003.09780 [astro-ph.IM]
251. Perera BBP et al (2019) The international pulsar timing array: second data release. Mon Not R Astron Soc 490(4):4666–4687. https://doi.org/10.1093/mnras/stz2857, arXiv:1909.04534 [astro-ph.HE]
252. Demorest P (2019) The NANOGrav observing program and 12.5-year data release. Am Astron Soc Meet Abstr 233:149.17
253. Carilli C, Rawlings S (2004) Science with the square kilometer array: motivation, key science projects, standards and assumptions. New Astron Rev 48(11–12):979–984. https://doi.org/10.1016/j.newar.2004.09.001
254. Wang Y, Mohanty SD (2017) Pulsar timing array based search for supermassive black hole binaries in the square kilometer array era. Phys Rev Lett 118(15):151104. arXiv:1611.09440 [gr-qc]
255. Abbott BP et al (LIGO Scientific Collaboration and Virgo Collaboration) (2020) Gwtc-2: compact binary coalescences observed by ligo and virgo during the first half of the third observing run. arXiv:2010.14527
256. Abbott BP et al (LIGO Scientific Collaboration and Virgo Collaboration) (2019) GWTC-1: a gravitational-wave transient catalog of compact binary mergers observed by LIGO and Virgo during the first and second observing runs. Phys Rev X 9:031040. arXiv:1811.12907 [astro-ph.HE]
257. https://www.ligo.caltech.edu/page/detection-companion-papers
258. Abbott BP et al (LIGO Scientific Collaboration and Virgo Collaboration) (2017) The basic physics of the binary black hole merger GW150914. Ann der Phys 529(1–2):1600209
259. L. S. Collaboration, V. Collaboration (2017) GW170814: a three-detector observation of gravitational waves from a binary black hole coalescence. Phys Rev Lett 119:141101. arXiv:1709.09660 [gr-qc]
260. Abbott BP et al (LIGO Scientific Collaboration and Virgo Collaboration) (2017) GW170817: observation of gravitational waves from a binary neutron star inspiral. Phys Rev Lett 119:161101. arXiv:1710.05832 [gr-qc]
261. Côté B, Fryer CL, Belczynski K, Korobkin O, Chruślińska M, Vassh N, Mumpower MR, Lippuner J, Sprouse TM, Surman R et al (2018) The origin of r-process elements in the milky way. Astrophys J 855(2):99
262. Abbott BP et al (LIGO Scientific Collaboration and Virgo Collaboration) (2017) Multi-messenger observations of a binary neutron star merger. Astrophys J 848(2):L12. https://doi.org/10.3847/2041-8213/aa91c9, arXiv:1710.05833 [astro-ph.HE]
263. LIGO Scientific Collaboration and Virgo Collaboration (2017) Gravitational waves and gamma-rays from a binary neutron star merger: GW170817 and GRB 170817a. Astrophys J Lett 848(2):L13. arXiv:1710.05834 [gr-qc]
264. Abbott R, Abbott T, Abraham S, Acernese F, Ackley K, Adams C, Adhikari R, Adya V, Affeldt C, Agathos M et al (2020) Gw190521: a binary black hole merger with a total mass of 150 solar mass. Phys Rev Lett 125(10):101102. arXiv:2009.01075 [gr-qc]
265. Abbott R, Abbott T, Abraham S, Acernese F, Ackley K, Adams C, Adhikari R, Adya V, Affeldt C, Agathos M et al (2020) Properties and astrophysical implications of the 150 solar mass binary black hole merger GW190521. Astrophys J Lett 900(1):L13. arXiv:2009.01190 [gr-qc]

266. Abbott R, Abbott T, Abraham S, Acernese F, Ackley K, Adams C, Adhikari R, Adya V, Affeldt C, Agathos M et al (2020) GW190814: gravitational waves from the coalescence of a 23 solar mass black hole with a 2.6 solar mass compact object. Astrophys J Lett 896(2):L44. arXiv:2006.12611 [gr-qc]
267. Mills C, Fairhurst S, Measuring gravitational-wave higher-order modes. arXiv:2007.04313 [gr-qc]
268. Abbott R, Abbott T, Abraham S, Acernese F, Ackley K, Adams C, Adhikari R, Adya V, Affeldt C, Agathos M et al (2020) Gw190412: observation of a binary-black-hole coalescence with asymmetric masses. Phys Rev D 102(4):043015. arXiv:2004.08342 [gr-qc]
269. Abbott R, Abbott T, Abraham S, Acernese F, Ackley K, Adams A, Adams C, Adhikari R, Adya V, Affeldt C et al (2021) Observation of gravitational waves from two neutron star-black hole coalescences. Astrophys J Lett 915(1):L5. arXiv:2106.15163 [astro-ph.HE]
270. Abbott R, Abbott T, Abraham S, Acernese F, Ackley K, Adams A, Adams C, Adhikari R, Adya V, Affeldt C et al (2020) Population properties of compact objects from the second ligo-virgo gravitational-wave transient catalog. arXiv:2010.14533
271. Margalit B, Metzger BD (2017) Constraining the maximum mass of neutron stars from multi-messenger observations of GW170817. Astrophys J Lett 850(2):L19. arXiv:1710.05938 [astro-ph.HE]
272. Troja E, Piro L, Van Eerten H, Wollaeger R, Im M, Fox O, Butler N, Cenko S, Sakamoto T, Fryer C et al (2017) The x-ray counterpart to the gravitational-wave event GW170817. Nature 551(7678):71–74. arXiv:1710.05433 [astro-ph.CO]
273. Evans PA et al (2017) Swift and nustar observations of GW170817: detection of a blue kilonova. Science 358:1565–1570. arXiv:astro-ph/1710.05437
274. Arcavi I et al (2017) Optical emission from a kilonova following a gravitational-wave-detected neutron-star merger. Nature 551(7678):64–66. https://doi.org/10.1038/nature24291, arXiv:1710.05843 [astro-ph.HE]
275. Lattimer J (2019) Impact of GW170817 for the nuclear physics of the eos and the r-process. Ann Phys 411:167963
276. Lazzati D, Perna R, Morsony BJ, Lopez-Camara D, Cantiello M, Ciolfi R, Giacomazzo B, Workman JC (2018) Late time afterglow observations reveal a collimated relativistic jet in the ejecta of the binary neutron star merger GW170817. Phys Rev Lett 120(24):241103. arXiv:1712.03237 [astro-ph.HE]
277. Ghirlanda G, Salafia OS, Paragi Z, Giroletti M, Yang J, Marcote B, Blanchard J, Agudo I, An T, Bernardini MG et al (2019) Compact radio emission indicates a structured jet was produced by a binary neutron star merger. Science 363(6430):968–971
278. Ezquiaga JM, Zumalacarregui M (2017) Dark energy after GW170817: dead ends and the road ahead. Phys Rev Lett 119:251304. https://doi.org/10.1103/PhysRevLett.119.251304, arXiv:1710.05901 [astro-ph.CO]
279. Toubiana A, Marsat S, Babak S, Barausse E, Baker J (2020) Tests of general relativity with stellar-mass black hole binaries observed by LISA. Phys Rev D 101(10):104038. https://doi.org/10.1103/PhysRevD.101.104038, arXiv:2004.03626 [gr-qc]
280. Maggiore M et al (2020) (2020) Science case for the Einstein telescope. J Cosmol Astropart Phys 3:050. https://doi.org/10.1088/1475-7516/2020/03/050, arXiv:1912.02622 [astro-ph.CO]
281. De S et al (2018) Tidal deformabilities and radii of neutron stars from the observation of GW170817. Phys Rev Lett 121:091102. arXiv:1804.08583 [astro-ph.HE]
282. Bauswein A, Just O, Janka H-T, Stergioulas N (2017) Neutron-star radius constraints from GW170817 and future detections. Astrophys J Lett 850(2):L34. arXiv:1710.06843 [astro-ph.HE]
283. Radice D, Perego A, Zappa F, Bernuzzi S (2018) Gw170817: joint constraint on the neutron star equation of state from multimessenger observations. Astrophys J Lett 852(2):L29. arXiv:1711.03647 [astro-ph.HE]
284. Bird S et al (2016) Did LIGO detect dark matter? Phys Rev Lett 116(20):201301. https://doi.org/10.1103/PhysRevLett.116.201301, arXiv:1603.00464 [astro-ph.CO]

285. Clesse S, García-Bellido J (2017) The clustering of massive primordial black holes as dark matter: measuring their mass distribution with advanced LIGO. Phys Dark Univ 15:142–147. https://doi.org/10.1016/j.dark.2016.10.002, arXiv:1603.05234 [astro-ph.CO]
286. Belgacem E et al (2019) (2019) Cosmology and dark energy from joint gravitational wave-GRB observations. J Cosmol Astropart Phys 8:015. https://doi.org/10.1088/1475-7516/2019/08/015, arXiv:1907.01487 [astro-ph.CO]
287. Riess AG et al (2016) A 2.4% Determination of the local value of the hubble constant. Astrophys J 826(1):56. https://doi.org/10.3847/0004-637X/826/1/56, arXiv:1604.01424 [astro-ph.CO]
288. Verde L, Treu T, Riess AG (2019) Tensions between the early and late Universe. Nat Astron 3:891–895. https://doi.org/10.1038/s41550-019-0902-0, arXiv:1907.10625 [astro-ph.CO]
289. Di Valentino E et al (2021) In the Realm of the Hubble tension – a review of solutions. arXiv:2103.01183 [astro-ph.CO]
290. Armano M et al (2016) Sub-femto-g free fall for space-based gravitational wave observatories: Lisa pathfinder results. Phys Rev Lett 116(23):231101
291. Klein A, Barausse E, Sesana A, Petiteau A, Berti E, Babak S, Gair J, Aoudia S, Hinder I, Ohme F et al (2016) Science with the space-based interferometer eLISA: supermassive black hole binaries. Phys Rev D 93(2):024003. arXiv:1511.05581 [gr-qc]
292. Sesana A, Volonteri M, Haardt F (2007) The imprint of massive black hole formation models on the LISA data stream. Mon Not R Astron Soc 377:1711. arXiv:astro-ph/0701556
293. Sesana A (2016) Prospects for multiband gravitational-wave astronomy after GW150914. Phys Rev Lett 116(23):231102. arXiv:astro-ph/1602.06951
294. Cornish N, Robson T (2017) Galactic binary science with the new LISA design. J Phys Conf Ser 840:012024. https://doi.org/10.1088/1742-6596/840/1/012024, arXiv:1703.09858 [astro-ph.IM]
295. Korol V, Toonen S, Klein A, Belokurov V, Vincenzo F, Buscicchio R, Gerosa D, Moore C, Roebber E, Rossi E et al (2020) Populations of double white dwarfs in milky way satellites and their detectability with LISA. Astron & Astrophys 638:A153. arXiv:2002.10462 [gr-qc]
296. Babak S, Gair J, Sesana A, Barausse E, Sopuerta CF, Berry CP, Berti E, Amaro-Seoane P, Petiteau A, Klein A (2017) Science with the space-based interferometer LISA. V. Extreme mass-ratio inspirals. Phys Rev D95(10):103012. arXiv:1703.09722 [gr-qc]
297. Caprini C, Figueroa DG, Flauger R, Nardini G, Peloso M, Pieroni M, Ricciardone A, Tasinato G (2019) Reconstructing the spectral shape of a stochastic gravitational wave background with LISA. J Cosmol Astropart Phys 11:017. https://doi.org/10.1088/1475-7516/2019/11/017, arXiv:1906.09244 [astro-ph.CO]
298. Cornish N, Robson T (2017) Galactic binary science with the new LISA design. J Phys Conf 840:012024. https://doi.org/10.1088/1742-6596/840/1/012024, arXiv:1703.09858
299. Mangiagli A, Klein A, Bonetti M, Katz M, Sesana A, Volonteri M, Colpi M, Marsat S, Babak S (2020) On the inspiral of coalescing massive black hole binaries with LISA in the era of multi-messenger astrophysics. arXiv:2006.12513
300. Barausse E et al (2020) Prospects for fundamental physics with LISA. Gen Relat Gravit 52:81 arXiv:2001.09793 [gr-qc]
301. Bailes M, Berger B, Brady P, Branchesi M, Danzmann K, Evans M, Holley-Bockelmann K, Iyer B, Kajita T, Katsanevas S et al (2021) Gravitational-wave physics and astronomy in the 2020s and 2030s. Nat Rev Phys 1–23
302. Sato S et al (2017) The status of decigo. J Phys: Conf Ser 840:012010. IOP Publishing
303. Isoyama S, Nakano H, Nakamura T (2018) Multiband gravitational-wave astronomy: observing binary inspirals with a decihertz detector, B-DECIGO. Prog Theor Exp Phys 2018(7):073E01
304. Akutsu T et al (2018) Kagra: 2.5 generation interferometric gravitational wave detector. arXiv:1811.08079
305. Unnikrishnan C (2013) Indigo and ligo-India: scope and plans for gravitational wave research and precision metrology in India. Int J Mod Phys D 22(01):1341010

306. Punturo M, Abernathy M, Acernese F, Allen B, Andersson N, Arun K, Barone F, Barr B, Barsuglia M, Beker M et al (2010) The Einstein telescope: a third-generation gravitational wave observatory. Class Quantum Gravit 27(19):194002
307. Abbott BP et al (2017) Exploring the sensitivity of next generation gravitational wave detectors. Class Quantum Gravit 34(4):044001. arXiv:1607.08697 [gr-qc]
308. Sedda MA et al (2021) The missing link in gravitational-wave astronomy. Exp Astron 1–14. https://doi.org/10.1007/s10686-021-09713-z, arXiv:2104.14583pdf [gr-qc]
309. Connes A, Damour T, Fayet P (1997) Aspherical gravitational monopoles. Nucl Phys B 490(1–2):391–431. arXiv:9611051 [gr-qc]
310. Steinhoff J (2015) Spin and quadrupole contributions to the motion of astrophysical binaries. Fund Theor Phys 179:615. arXiv:1412.3251 [gr-qc]
311. Trautman A (2002) Lectures on general relativity. Gen Relat Gravit 34:721. Reprinted from lectures delivered in 1958
312. Blanchet L, Faye G (2000) Hadamard regularization. J Math Phys 41:7675. arXiv:gr-qc/0004008
313. Damour T, Jaranowski P, Schäfer G (2001) Dimensional regularization of the gravitational interaction of point masses. Phys Lett B 513:147. arXiv:gr-qc/0105038
314. Barack L, Golbourn DA, Sago N (2007) m-mode regularization scheme for the self-force in Kerr spacetime. Phys Rev D 76(12):124036. https://doi.org/10.1103/PhysRevD.76.124036, arXiv:0709.4588 [gr-qc]
315. Vega I, Detweiler S (2008) Regularization of fields for self-force problems in curved spacetime: foundations and a time-domain application. Phys Rev D 77(8):084008. https://doi.org/10.1103/PhysRevD.77.084008, arXiv:0712.4405 [gr-qc]
316. Bailey I, Israel W (1975) Lagrangian dynamics of spinning particles and polarized media in general relativity. Commun Math Phys 42:65
317. Hanson AJ, Regge T (1974) The relativistic spherical top. Ann Phys 87:498
318. Steinhoff J (2011) Canonical formulation of spin in general relativity. Ann Phys 523:296. arXiv:1106.4203 [gr-qc]
319. Marsat S, Bohé A, Faye G, Blanchet L (2013) Next-to-next-to-leading order spin-orbit effects in the equations of motion of compact binary systems. Class Quantum Gravit 30:055007. arXiv:1210.4143 [gr-qc]
320. Marsat S (2015) Cubic-order spin effects in the dynamics and gravitational wave energy flux of compact object binaries. Class Quantum Gravit 32:085008. arXiv:1411.4118 [gr-qc]
321. Hartle JB (2003) Gravity: an introduction to Einstein's general relativity. Addison Wesley, San Fransisco
322. Dixon WG (1974) Dynamics of extended bodies in general relativity III. Equations of motion. Philos Trans R Soc Lond A 277:59–119
323. Goldberger WD, Rothstein IZ (2006) Effective field theory of gravity for extended objects. Phys Rev D 73:104029. arXiv:hep-th/0409156
324. Foffa S, Sturani R (2014) Effective field theory methods to model compact binaries. Class Quantum Gravit 31:043001. arXiv:1309.3474 [gr-qc]
325. Porto RA (2016) The effective field theorist's approach to gravitational dynamics. Phys Rep 663:1. arXiv:1601.04914 [gr-qc]
326. Cardoso V, Kimura M, Maselli A, Senatore L (2018) Black holes in an effective field theory extension of general relativity. Phys Rev Lett 121:251105. arXiv:1808.08962 [gr-qc]
327. Dixon WG (2015) The new mechanics of Myron Mathisson and its subsequent development. In: Equations of motion in relativistic gravity. Springer, pp 1–66
328. Mathisson M (1937) Neue mechanik materieller systeme. Acta Phys Polon 6:163–200. Translated in English by A. Ehlers and reprinted in Gen Relat Gravit 42:1011 (2010)
329. Mathisson M (1937) Das zitternde elektron und seine dynamik. Acta Phys Polon 6:218–227
330. Bielecki A, Mathisson M, Weyssenhoff J (1939) Sur un théorème concernant une transformation d'intégrales quadruples en intégrales curvilignes dans l'espace de Riemann. Polska Akademia Umiejętności

331. Mathisson M (1940) The variational equation of relativistic dynamics. In: Mathematical proceedings of the Cambridge philosophical society, vol 36. Cambridge University Press, pp 331–350
332. Sauer T, Trautman A (2008) Myron Mathisson: what little we know of his life. arXiv:0802.2971 [physics.hist-ph]
333. Mathisson M (1942) Relativistic dynamics of a spinning magnetic particle. In: Mathematical proceedings of the Cambridge philosophical society, vol 38. Cambridge University Press, pp 40–60
334. Hadamard J (1942) The problem of diffusion of waves. Ann Math 510–522
335. Hönl H, Papapetrou A (1940) Über die innere bewegung des elektrons. III. Z. für Phys 116(3–4):153–183
336. Papapetrou A (1951) Spinning test-particles in general relativity. I. Proc. R. Soc. Lond. A 209:248
337. Corinaldesi E, Papapetrou A (1951) Spinning test-particles in general relativity. II. Proc. R. Soc. Lond. A 209:259
338. Tulczyjew W (1959) Motion of multiple particles in general relativity theory. Acta Phys Polon 18:393
339. Tulczyjew B, Tulczyjew W (1962) On multipole formalism in general relativity, p 465
340. Dixon WG (1964) A covariant multipole formalism for extended test bodies in general relativity. Il Nuovo Cimento 34:317
341. Dixon WG (1967) Description of extended bodies by multipole moments in special relativity. J Math Phys 8(8):1591–1605
342. Dixon WG (1970) Dynamics of extended bodies in general relativity I. Momentum and angular momentum. Proc R Soc Lond A 314:499–527
343. Dixon WG (1970) Dynamics of extended bodies in general relativity II. Moments of the charge-current vector. Proc R Soc Lond A 319:509–547
344. Steinhoff J, Puetzfeld D (2010) Multipolar equations of motion for extended test bodies in general relativity. Phys Rev D 81:044019. arXiv:0909.3756 [gr-qc]
345. Arnold VI (1995) Mathematical methods of classical mechanics. Springer, New York. https://doi.org/10.1007/978-1-4757-2063-1
346. Harte AI (2012) Mechanics of extended masses in general relativity. Class Quantum Gravit 29:055012. arXiv:1103.0543 [gr-qc]
347. Harte AI (2015) Motion in classical field theories and the foundations of the self-force problem. Fund Theor Phys 179:327. arXiv:1405.5077 [gr-qc]
348. Landau LD, Lifshitz EM (1987) Fluid mechanics, 2nd edn. Butterworth-Heinemann, Oxford
349. Dixon WG (1979) Extended bodies in general relativity: their description and motion. In: Ehlers J (ed) Isolated gravitating systems in general relativity. Proceedings of the international school of physics Enrico Fermi, vol 67. North-Holland, Amsterdam, p 156
350. Harte AI (2021) Private communication
351. Pound A (2010) The self-consistent gravitational self-force. Phys Rev D 81:024023. arXiv:0907.5197 [gr-qc]
352. Bender CM, Orszag SA (2013) Advanced mathematical methods for scientists and engineers I: asymptotic methods and perturbation theory. Springer Science & Business Media
353. Geroch RP (1970) Multipole moments. II. Curved space. J Math Phys 11:2580
354. Hansen RO (1974) Multipole moments of stationary space-times. J Math Phys 15:46
355. Poisson E (2004) A relativist's toolkit. Cambridge University Press, Cambridge
356. Pound A (2012) Second-order gravitational self-force. Phys Rev Lett 109:051101. arXiv:1201.5089 [gr-qc]
357. Pound A (2017) Nonlinear gravitational self-force. second-order equation of motion. Phys Rev D 95:104056. arXiv:1703.02836 [gr-qc]
358. Mino Y, Sasaki M, Tanaka T (1997) Gravitational radiation reaction to a particle motion. Phys Rev D 55:3457. arXiv:gr-qc/9606018
359. Quinn TC, Wald RM (1997) Axiomatic approach to electromagnetic and gravitational radiation reaction of particles in curved spacetime. Phys Rev D 56:3381. arXiv:gr-qc/9610053

360. Gralla SE, Wald RM (2008) A rigorous derivation of gravitational self-force. Class Quantum Gravit 25:205009. arXiv:0806.3293 [gr-qc]
361. Gralla SE (2012) Second order gravitational self force. Phys Rev D 85:124011. arXiv:1203.3189 [gr-qc]
362. Upton SD, Pound A (2021) Second-order gravitational self-force in a highly regular gauge. Phys Rev D 103(12):124016. arXiv:2101.11409 [gr-qc]
363. Detweiler S, Whiting BF (2003) Self-force via a Green's function decomposition. Phys Rev D 67:024025. arXiv:gr-qc/0202086
364. Pound A, Miller J (2014) Practical, covariant puncture for second-order self-force calculations. Phys Rev D 89:104020. arXiv:1403.1843 [gr-qc]
365. Akcay S, Dolan SR, Kavanagh C, Moxon J, Warburton N, Wardell B (2020) Dissipation in extreme mass-ratio binaries with a spinning secondary. Phys Rev D 102(6):064013. arXiv:1912.09461 [gr-qc]
366. Detweiler S (2012) Gravitational radiation reaction and second order perturbation theory. Phys Rev D 85:044048. arXiv:1107.2098 [gr-qc]
367. Barausse E, Racine E, Buonanno A (2009) Hamiltonian of a spinning test particle in curved spacetime. Phys Rev D 80:104025. arXiv:0907.4745 [gr-qc]. Erratum: Phys. Rev. D 85:069904(E) (2012)
368. Harte AI (2020) Extended-body motion in black hole spacetimes: what is possible? Phys Rev D 102(12):124075. https://doi.org/10.1103/PhysRevD.102.124075, arXiv:2011.00110 [gr-qc]
369. Pound A, Wardell B, Warburton N, Miller J (2020) Second-order self-force calculation of gravitational binding energy in compact binaries. Phys Rev Lett 124(2):021101. arXiv:1908.07419 [gr-qc]
370. Bohé A, Faye G, Marsat S, Porter EK (2015) Quadratic-in-spin effects in the orbital dynamics and gravitational-wave energy flux of compact binaries at the 3PN order. Class Quantum Gravit 32:195010. arXiv:1501.01529 [gr-qc]
371. Ehlers J, Rudolph E (1977) Dynamics of extended bodies in general relativity: center-of-mass description and quasirigidity. Gen Rel Gravit 8:197
372. Gralla SE, Harte AI, Wald RM (2010) Bobbing and kicks in electromagnetism and gravity. Phys Rev D 81:104012. arXiv:1004.0679 [gr-qc]
373. Costa LFO, Natário J (2015) Center of mass, spin supplementary conditions, and the momentum of spinning particles. Fund Theor Phys 179:215. arXiv:1410.6443 [gr-qc]
374. Kyrian K, Semerák O (2007) Spinning test particles in a Kerr field - II. Mon Not R Astron Soc 382:1922
375. Costa LF, Herdeiro C, Natário J, Zilhão M (2012) Mathisson's helical motions for a spinning particle: are they unphysical? Phys Rev D 85:024001. arXiv:1109.1019 [gr-qc]
376. Costa LFO, Lukes-Gerakopoulos G, Semerák O (2018) Spinning particles in general relativity: momentum-velocity relation for the Mathisson-Pirani spin condition. Phys Rev D 97:084023. arXiv:1712.07281 [gr-qc]
377. Burgay M et al (2003) An increased estimate of the merger rate of double neutron stars from observations of a highly relativistic system. Nature 426(6966):531–533. https://doi.org/10.1038/nature02124, arXiv:astro-ph/0312071 [astro-ph]
378. Hinderer T, Flanagan ÉÉ (2008) Two-timescale analysis of extreme mass ratio inspirals in Kerr spacetime: orbital motion. Phys Rev D 78:064028. arXiv:0805.3337 [gr-qc]
379. Bonazzola S, Gourgoulhon E, Marck J-A (1999) Numerical models of irrotational binary neutron stars in general relativity. Phys Rev Lett 82(5):892. arXiv:9810072 [gr-qc]
380. Gourgoulhon E, Grandclément P, Bonazzola S (2002) Binary black holes in circular orbits. I. A global spacetime approach. Phys Rev D 65:044020. arXiv:gr-qc/0106015
381. Detweiler S (1989) Kepler's third law in general relativity. In: Evans CR, Finn LS, Hobill DW (eds) Frontiers in numerical relativity. Cambridge University Press, Cambridge, p 43
382. Friedman JL, Uryū K, Shibata M (2002) Thermodynamics of binary black holes and neutron stars. Phys Rev D 65:064035. arXiv:gr-qc/0108070. Erratum: Phys Rev D 70:129904(E) (2004)

383. Stephani H, Kramer D, MacCallum M, Hoenselaers C, Herlt E (2003) Exact solutions of Einstein's field equations. Cambridge University Press, Cambridge
384. Klein C (2004) Binary black hole spacetimes with a helical Killing vector. Phys Rev D 70:124026. arXiv:gr-qc/0410095
385. Gibbons GW, Stewart JM (1984) Absence of asymptotically flat solutions of Einstein's equations which are periodic and empty near infinity. In: Bonnor WB, Islam JN, MacCallum MAH (eds) Classical general relativity. Cambridge University Press, Cambridge, p 77
386. Shibata M, Uryū K, Friedman JL (2004) Deriving formulations for numerical computation of binary neutron stars in quasicircular orbits. Phys Rev D 70:044044. arXiv:gr-qc/0407036. Erratum: Phys Rev D 70:129901(E) (2004)
387. Le Tiec A, Blanchet L, Whiting BF (2012) First law of binary black hole mechanics in general relativity and post-newtonian theory. Phys Rev D 85:064039. arXiv:1111.5378 [gr-qc]
388. Gralla SE, Le Tiec A (2013) Thermodynamics of a black hole with a moon. Phys Rev D 88:044021. arXiv:1210.8444 [gr-qc]
389. Isenberg J, Nester J (1980) Canonical gravity. In: Held A (ed) General relativity and gravitation: one hundred years after the birth of Albert Einstein, vol 1. Plenum Press, New York, p 23
390. Wilson JR, Mathews GJ (1989) Relativistic hydrodynamics. In: Evans CR, Finn LS, Hobill DW (eds) Frontiers in numerical relativity. Cambridge University Press, Cambridge, p 306
391. Isenberg JA (2008) Waveless approximation theories of gravity. Int J Mod Phys D 17:265. arXiv:gr-qc/0702113
392. Detweiler S (1994) Periodic solutions of the Einstein equations for binary systems. Phys Rev D 50:4929. arXiv:gr-qc/9312016
393. Geroch R, Traschen J (1987) Strings and other distributional sources in general relativity. Phys Rev D 36:1017
394. Steinbauer R, Vickers JA (2009) On the geroch-traschen class of metrics. Class Quantum Gravit 26(6):065001. arXiv:0811.1376 [gr-qc]
395. Damour T, Jaranowski P, Schäfer G (2016) Conservative dynamics of two-body systems at the fourth post-Newtonian approximation of general relativity. Phys Rev D 93(8):084014. https://doi.org/10.1103/PhysRevD.93.084014, arXiv:1601.01283 [gr-qc]
396. Marchand T, Bernard L, Blanchet L, Faye G (2018) Ambiguity-free completion of the equations of motion of compact binary systems at the fourth post-Newtonian order. Phys Rev D 97:044023. arXiv:1707.09289 [gr-qc]
397. 't Hooft G, Veltman M (1972) Regularization and renormalization of gauge fields. Nucl Phys B 44:189
398. Bollini CG, Giambiagi JJ (1972) Lowest order divergent graphs in ν-dimensional space. Phys Lett B 40:566
399. Grandclément P, Gourgoulhon E, Bonazzola S (2002) Binary black holes in circular orbits. ii. Numerical methods and first results. Phys Rev D 65:044021. arXiv:gr-qc/0106016
400. Detweiler S (2008) Consequence of the gravitational self-force for circular orbits of the Schwarzschild geometry. Phys Rev D 77:124026. arXiv:0804.3529 [gr-qc]
401. Blanchet L, Detweiler S, Le Tiec A, Whiting BF (2010) Post-Newtonian and numerical calculations of the gravitational self-force for circular orbits in the Schwarzschild geometry. Phys Rev D 81:064004. arXiv:0910.0207 [gr-qc]
402. Blanchet L, Detweiler S, Le Tiec A, Whiting BF (2010) High-order post-Newtonian fit of the gravitational self-force for circular orbits in the Schwarzschild geometry. Phys Rev D 81:084033. arXiv:1002.0726 [gr-qc]
403. Blanchet L, Detweiler S, Le Tiec A, Whiting BF (2011) High-accuracy comparison between the post-Newtonian and self-force dynamics of black-hole binaries. Fund Theor Phys 162:415. arXiv:1007.2614 [gr-qc]
404. Blanchet L, Faye G, Whiting BF (2014) Half-integral conservative post-Newtonian approximations in the redshift factor of black hole binaries. Phys Rev D 89:064026. arXiv:1312.2975 [gr-qc]

405. Blanchet L, Faye G, Whiting BF (2014) High-order half-integral conservative post-Newtonian coefficients in the redshift factor of black hole binaries. Phys Rev D 90:044017. arXiv:1405.5151 [gr-qc]
406. Blanchet L, Buonanno A, Le Tiec A (2013) First law of mechanics for black hole binaries with spins. Phys Rev D 87:024030. arXiv:1211.1060 [gr-qc]
407. Fujita R, Isoyama S, Le Tiec A, Nakano H, Sago N, Tanaka T (2017) Hamiltonian formulation of the conservative self-force dynamics in the Kerr geometry. Class Quantum Gravit 34:134001. arXiv:1612.02504 [gr-qc]
408. Le Tiec A, Barausse E, Buonanno A (2012) Gravitational self-force correction to the binding energy of compact binary systems. Phys Rev Lett 108:131103. arXiv:1111.5609 [gr-qc]
409. Zimmerman A, Lewis AGM, Pfeiffer HP (2016) Redshift factor and the first law of binary black hole mechanics in numerical simulations. Phys Rev Lett 117:191101. arXiv:1606.08056 [gr-qc]
410. Mano S, Suzuki H, Takasugi E (1996) Analytic solutions of the Teukolsky equation and their low frequency expansions. Prog Theor Phys 95:1079. arXiv:gr-qc/9603020
411. Mano S, Takasugi E (1997) Analytic solutions of the Teukolsky equation and their properties. Prog Theor Phys 97:213. arXiv:gr-qc/9611014
412. Bini D, Damour T (2014) Analytic determination of the eight-and-a-half post-Newtonian self-force contributions to the two-body gravitational interaction potential. Phys Rev D 89:104047. arXiv:1403.2366 [gr-qc]
413. Bini D, Damour T (2015) Detweiler's gauge-invariant redshift variable: analytic determination of the nine and nine-and-a-half post-Newtonian self-force contributions. Phys Rev D 91:064050. arXiv:1502.02450 [gr-qc]
414. Kavanagh C, Ottewill AC, Wardell B (2015) Analytical high-order post-Newtonian expansions for extreme mass ratio binaries. Phys Rev D 92:084025. arXiv:1503.02334 [gr-qc]
415. Kavanagh C, Ottewill AC, Wardell B (2016) Analytical high-order post-Newtonian expansions for spinning extreme mass ratio binaries. Phys Rev D 93:124038. arXiv:1601.03394 [gr-qc]
416. Bini D, Damour T, Geralico A, Kavanagh C (2018) Detweiler's redshift invariant for spinning particles along circular orbits on a Schwarzschild background. Phys Rev D 97:104022. arXiv:1801.09616 [gr-qc]
417. Bini D, Geralico A, Steinhoff J (2020) Detweiler's redshift invariant for extended bodies orbiting a Schwarzschild black hole. Phys Rev D 102:024091. arXiv:2003.12887 [gr-qc]
418. Shah AG, Friedman JL, Whiting BF (2014) Finding high-order analytic post-Newtonian parameters from a high-precision numerical self-force calculation. Phys Rev D 89:064042. arXiv:1312.1952 [gr-qc]
419. Johnson-McDaniel NK, Shah AG, Whiting BF (2015) Experimental mathematics meets gravitational self-force. Phys Rev D 92:044007. arXiv:1503.02638 [gr-qc]
420. Damour T, Jaranowski P, Schäfer G (2016) Conservative dynamics of two-body systems at the fourth post-Newtonian approximation of general relativity. Phys Rev D 93:084014. arXiv:1601.01283 [gr-qc]
421. Barausse E, Buonanno A, Le Tiec A (2012) Complete nonspinning effective-one-body metric at linear order in the mass ratio. Phys Rev D 85:064010. arXiv:1111.5610 [gr-qc]
422. Akcay S, Barack L, Damour T, Sago N (2012) Gravitational self-force and the effective-one-body formalism between the innermost stable circular orbit and the light ring. Phys Rev D 86:104041. arXiv:1209.0964 [gr-qc]
423. Bini D, Damour T (2016) Conservative second-order gravitational self-force on circular orbits and the effective one-body formalism. Phys Rev D 93:104040. arXiv:1603.09175 [gr-qc]
424. Barack L, Sago N (2011) Beyond the geodesic approximation: conservative effects of the gravitational self-force in eccentric orbits around a Schwarzschild black hole. Phys Rev D 83:084023. arXiv:1101.3331 [gr-qc]
425. Akcay S, Le Tiec A, Barack L, Sago N, Warburton N (2015) Comparison between self-force and post-Newtonian dynamics: beyond circular orbits. Phys Rev D 91:124014. arXiv:1503.01374 [gr-qc]

426. Le Tiec A (2015) First law of mechanics for compact binaries on eccentric orbits. Phys Rev D 92:084021. arXiv:1506.05648 [gr-qc]
427. Blanchet L, Le Tiec A (2017) First law of compact binary mechanics with gravitational-wave tails. Class Quantum Gravit 34:164001. arXiv:1702.06839 [gr-qc]
428. Bini D, Damour T, Geralico A (2016) Confirming and improving post-Newtonian and effective-one-body results from self-force computations along eccentric orbits around a Schwarzschild black hole. Phys Rev D 93:064023. arXiv:1511.04533 [gr-qc]
429. Akcay S, van de Meent M (2016) Numerical computation of the effective-one-body potential q using self-force results. Phys Rev D 93:064063. arXiv:1512.03392 [gr-qc]
430. Sago N, Barack L, Detweiler S (2008) Two approaches for the gravitational self-force in black hole spacetime: comparison of numerical results. Phys Rev D 78:124024. arXiv:0810.2530 [gr-qc]
431. Shah AG, Friedman JL, Keidl TS (2012) Extreme-mass-ratio inspiral corrections to the angular velocity and redshift factor of a mass in circular orbit about a Kerr black hole. Phys Rev D Ser 86:084059. arXiv:1207.5595 [gr-qc]
432. Keidl TS, Shah AG, Friedman JL, Kim D-H, Price LR (2010) Gravitational self-force in a radiation gauge. Phys Rev D 82:124012. arXiv:1004.2276 [gr-qc]
433. Pound A (2014) Conservative effect of the second-order gravitational self-force on quasicircular orbits in Schwarzschild spacetime. Phys Rev D 90:084039. arXiv:1404.1543 [gr-qc]
434. Schattner R, Streubel M (1981) Properties of extended bodies in spacetimes admitting isometries. Ann Inst H Poincaré Sect A 34:117
435. Harte AI, Flanagan ÉÉ, Taylor P (2016) Self-forces on static bodies in arbitrary dimensions. Phys Rev D 93:124054. arXiv:1603.00052 [gr-qc]
436. Dolan SR, Warburton N, Harte AI, Le Tiec A, Wardell B, Barack L (2014) Gravitational self-torque and spin precession in compact binaries. Phys Rev D 89:064011. arXiv:1312.0775 [gr-qc]
437. Bini D, Damour T (2014) Two-body gravitational spin-orbit interaction at linear order in the mass ratio. Phys Rev D 90:024039. arXiv:1404.2747 [gr-qc]
438. Bini D, Damour T (2015) Analytic determination of high-order post-Newtonian self-force contributions to gravitational spin precession. Phys Rev D 91:064064. arXiv:1503.01272 [gr-qc]
439. Akcay S, Dempsey D, Dolan SR (2017) Spin-orbit precession for eccentric black hole binaries at first order in the mass ratio. Class Quantum Gravit 34:084001. arXiv:1608.04811 [gr-qc]
440. Akcay S (2017) Self-force correction to geodetic spin precession in Kerr spacetime. Phys Rev D 96:044024. arXiv:1705.03282 [gr-qc]
441. Bini D, Damour T, Geralico A, Kavanagh C, van de Meent M (2018) Gravitational self-force corrections to gyroscope precession along circular orbits in the Kerr spacetime. Phys Rev D Ser 98:104062. arXiv:1809.02516 [gr-qc]
442. Kavanagh C, Bini D, Damour T, Hopper S, Ottewill AC, Wardell B (2017) Spin-orbit precession along eccentric orbits for extreme mass ratio black hole binaries and its effective-one-body transcription. Phys Rev D 96:064012. arXiv:1706.00459 [gr-qc]
443. Cohen MI, Kaplan JD, Scheel MA (2012) Toroidal horizons in binary black hole inspirals. Phys Rev D 85(2):024031
444. Bardeen JM, Carter B, Hawking SW (1973) The four laws of black hole mechanics. Commun Math Phys 31:161
445. Hawking SW (1975) Particle creation by black holes. Commun Math Phys 43:199. Erratum: Commun Math Phys 46:206 (1976)
446. Bekenstein JD (1973) Black holes and entropy. Phys Rev D 7:2333
447. Wald RM (2001) The thermodynamics of black holes. Living Rev Relat 4:6. https://doi.org/10.12942/lrr-2001-6, arXiv:gr-qc/9912119
448. Wald RM (1993) Black hole entropy is the noether charge. Phys Rev D 48(8):R3427
449. Isi M, Farr WM, Giesler M, Scheel MA, Teukolsky SA (2021) Testing the black-hole area law with gw150914. Phys Rev Lett 127(1):011103. arXiv:gr-qc/2012.04486
450. Kochanek CS (1992) Coalescing binary neutron stars. Astrophys J 398:234

451. Bildsten L, Cutler C (1992) Tidal interactions of inspiraling compact binaries. Astrophys J Ser 400:175–180
452. Baumgarte TW, Shapiro SL (2003) Numerical relativity and compact binaries. Phys Rep 376(2):41–131. arXiv:0211028 [gr-qc]
453. Uryū K, Limousin F, Friedman JL, Gourgoulhon E, Shibata M (2006) Binary neutron stars: equilibrium models beyond spatial conformal flatness. Phys Rev Lett 97(17):71101. arXiv:0511136 [gr-qc]
454. Uryū K, Limousin F, Friedman JL, Gourgoulhon E, Shibata M (2009) Nonconformally flat initial data for binary compact objects. Phys Rev D 80(12):124004. arXiv:0908.0579 [gr-qc]
455. Tsokaros A, Uryū K, Rezzolla L (2015) New code for quasiequilibrium initial data of binary neutron stars: corotating, irrotational, and slowly spinning systems. Phys Rev D 91(10):104030. arXiv:1502.05674 [gr-qc]
456. Gourgoulhon E, Grandclement P, Taniguchi K, Marck J-A, Bonazzola S (2001) Quasiequilibrium sequences of synchronized and irrotational binary neutron stars in general relativity: method and tests. Phys Rev D 63(6):064029. arXiv:0007028 [gr-qc]
457. Taniguchi K, Gourgoulhon E, Bonazzola S (2001) Quasiequilibrium sequences of synchronized and irrotational binary neutron stars in general relativity. II. Newtonian limits. Phys Rev D 64(6):064012. arXiv:0103041 [gr-qc]
458. Taniguchi K, Gourgoulhon E (2002) Quasiequilibrium sequences of synchronized and irrotational binary neutron stars in general relativity. III. Identical and different mass stars with $\gamma = 2$. Phys Rev D 66(10):104019. arXiv:0207098 [gr-qc]
459. Uryū K, Gourgoulhon E, Markakis C (2010) Thermodynamics of magnetized binary compact objects. Phys Rev D 82:104054. arXiv:1010.4409 [gr-qc]
460. Faber JA, Rasio FA (2012) Binary neutron star mergers. Living Rev Relat 15(1):1–83. https://doi.org/10.12942/lrr-2012-8
461. Uryū K, Tsokaros A (2011) New code for equilibriums and quasiequilibrium initial data of compact objects. Phys Rev D 85:064014. arXiv:1108.3065 [gr-qc]
462. Damour T, Jaranowski P, Schäfer G (2014) Nonlocal-in-time action for the fourth post-Newtonian conservative dynamics of two-body systems. Phys Rev D 89:064058. arXiv:1401.4548 [gr-qc]
463. Jaranowski P, Schäfer G (2015) Derivation of local-in-time fourth post-Newtonian ADM Hamiltonian for spinless compact binaries. Phys Rev D 92:124043. arXiv:1508.01016 [gr-qc]
464. Gourgoulhon E (2012) 3+1 formalism in general relativity: bases of numerical relativity, vol 846. Lecture notes in physics. Springer, New York
465. Marsat S (2013) Approche post-Newtonienne pour les effets de spin dans les binaires d'objets compacts et modèles alternatifs pour la masse manquante en cosmologie. PhD thesis. Thèse de doctorat dirigée par Blanchet, Luc Physique Théorique Paris, 6. http://www.theses.fr/2013PA066406
466. Antonelli A, Kavanagh C, Khalil M, Steinhoff J, Vines J (2020) Gravitational spin-orbit and aligned spin1-spin2 couplings through third-subleading post-Newtonian orders. Phys Rev D 102:124024. arXiv:2010.02018 [gr-qc]
467. Le Tiec A (2014) A note on celestial mechanics in Kerr spacetime. Class Quantum Gravit 31:097001. arXiv:1311.3836 [gr-qc]
468. Carter B (1968) Global structure of the Kerr family of gravitational fields. Phys Rev 174:1559
469. Le Tiec A (2021) Black hole physics and relativistic celestial mechanics
470. Iyer V, Wald RM (1994) Some properties of the Noether charge and a proposal for dynamical black hole entropy. Phys. Rev. D 50:846. arXiv:gr-qc/9403028
471. Jaramillo J, Gourgoulhon E (2009) Mass and angular momentum in general relativity. Springer
472. Wald RM, Zoupas A (2000) General definition of conserved quantities" in general relativity and other theories of gravity. Phys Rev D 61:084027. arXiv:gr-qc/9911095
473. Damour T, Jaranowski P, Schäfer G (2000) Dynamical invariants for general relativistic two-body systems at the third post-Newtonian approximation. Phys Rev D 62:044024. arXiv:gr-qc/9912092

474. Damour T, Jaranowski P, Schäfer G (2000) Determination of the last stable orbit for circular general relativistic binaries at the third post-Newtonian approximation. Phys Rev D 62:084011. arXiv:gr-qc/0005034
475. Damour T, Nagar A, Pollney D, Reisswig C (2012) Energy versus angular momentum in black hole binaries. Phys Rev Lett 108:131101. arXiv:1110.2938 [gr-qc]
476. Barack L, Sago N (2009) Gravitational self-force correction to the innermost stable circular orbit of a Schwarzschild black hole. Phys Rev Lett 102:191101. arXiv:0902.0573 [gr-qc]
477. Damour T. (2010) Gravitational self-force in a Schwarzschild background and the effective one-body formalism. Phys Rev D 81:024017. arXiv:0910.5533 [gr-qc]
478. Barack L, Sago N (2010) Gravitational self-force on a particle in eccentric orbit around a Schwarzschild black hole. Phys Rev D 81:084021. arXiv:1002.2386 [gr-qc]
479. Isoyama S et al (2014) Gravitational self-force correction to the innermost stable circular equatorial orbit of a Kerr black hole. Phys Rev Lett 113:161101. arXiv:1404.6133 [gr-qc]
480. van de Meent M (2017) Self-force corrections to the periapsis advance around a spinning black hole. Phys Rev Lett 118:011101. arXiv:1610.03497 [gr-qc]
481. Barack L, Colleoni M, Damour T, Isoyama S, Sago N (2019) Self-force effects on the marginally bound zoom-whirl orbit in schwarzschild spacetime. Phys Rev D 100:124015. arXiv:1909.06103 [gr-qc]
482. Colleoni M, Barack L (2015) Overspinning a Kerr black hole: the effect of the self-force. Phys Rev D 91:104024. arXiv:1501.07330 [gr-qc]
483. Colleoni M, Barack L, Shah AG, van de Meent M (2015) Self-force as a cosmic censor in the kerr overspinning problem. Phys Rev D 92:084044. arXiv:1508.04031 [gr-qc]
484. Sorce J, Wald RM (2017) Gedanken experiments to destroy a black hole. II. Kerr-Newman black holes cannot be overcharged or overspun. Phys Rev D 96(10). https://doi.org/10.1103/PhysRevD.96.104014
485. Jaranowski P, Schäfer G (2012) Towards the fourth post-Newtonian Hamiltonian for two-point-mass systems. Phys Rev D 86:061503(R). arXiv:1207.5448 [gr-qc]
486. Jaranowski P, Schäfer G (2013) Dimensional regularization of local singularities in the fourth post-Newtonian two-point-mass Hamiltonian. Phys Rev D 87:081503(R). arXiv:1303.3225 [gr-qc]
487. Bernard L, Blanchet L, Bohé A, Faye G, Marsat S (2016) Fokker action of nonspinning compact binaries at the fourth post-Newtonian approximation. Phys Rev D 93:084037. arXiv:1512.02876 [gr-qc]
488. Bernard L, Blanchet L, Bohé A, Faye G, Marsat S (2017) Energy and periastron advance of compact binaries on circular orbits at the fourth post-Newtonian order. Phys Rev D 95:044026. arXiv:1610.07934 [gr-qc]
489. Bernard L, Blanchet L, Faye G, Marchand T (2018) Center-of-mass equations of motion and conserved integrals of compact binary systems at the fourth post-Newtonian order. Phys Rev D 97:044037. arXiv:1711.00283 [gr-qc]
490. Barausse E et al (2012) Modeling multipolar gravitational-wave emission from small mass-ratio mergers. Phys Rev D 85:024046. arXiv:1110.3081 [gr-qc]
491. Bini D, Damour T, Geralico A (2015) Spin-dependent two-body interactions from gravitational self-force computations. Phys Rev D 92:124058. arXiv:1510.06230 [gr-qc]. Erratum: Phys Rev D 93:109902 (2016)
492. Bini D, Damour T, Geralico A (2016) New gravitational self-force analytical results for eccentric orbits around a Schwarzschild black hole. Phys Rev D 93:104017. arXiv:1601.02988 [gr-qc]
493. Bini D, Damour T, Geralico A (2019) Novel approach to binary dynamics: application to the fifth post-Newtonian level. Phys Rev Lett 123(23):231104. https://doi.org/10.1103/PhysRevLett.123.231104, arXiv:1909.02375 [gr-qc]
494. Bini D, Damour T, Geralico A (2020) Binary dynamics at the fifth and fifth-and-a-half post-Newtonian orders. Phys Rev D 102(2):024062. https://doi.org/10.1103/PhysRevD.102.024062, arXiv:2003.11891 [gr-qc]

495. Wald RM (1993) Black hole entropy is the Noether charge. Phys Rev D 48:3427. arXiv:gr-qc/9307038
496. Gao S, Wald RM (2001) "Physical process version" of the first law and the generalized second law for charged and rotating black holes. Phys Rev D 64:084020. arXiv:gr-qc/0106071
497. Iyer V (1997) Lagrangian perfect fluids and black hole mechanics. Phys Rev D 55:3411. arXiv:gr-qc/9610025
498. Lee J, Wald RM (1990) Local symmetries and constraints. J Math Phys 31:725
499. Cartan É (1945) Les systèmes differentiels extérieurs et leur applications géométriques
500. Iyer V, Wald RM (1995) Comparison of the Noether charge and Euclidean methods for computing the entropy of stationary black holes. Phys Rev D 52:4430 arXiv:gr-qc/9503052
501. Rossi L (2020) The first law of black hole mechanics. arXiv:2012.04593 [gr-qc]
502. Belitsky AV, Korchemsky GP, Müller D (2006) Integrability of two-loop dilatation operator in gauge theories. Nucl Phys B 735(1):17–83. https://doi.org/10.1016/j.nuclphysb.2005.11.015, arXiv:hep-th/0509121 [hep-th]
503. Bonga B, Grant AM, Prabhu K (2020) Angular momentum at null infinity in Einstein-Maxwell theory. Phys Rev D 101(4):044013. https://doi.org/10.1103/PhysRevD.101.044013, arXiv:1911.04514 [gr-qc]
504. Beig R (1978) Arnowitt-Deser-Misner energy and g_{00}. Phys Lett A 69:153
505. Ashtekar A, Magnon-Ashtekar A (1979) On conserved quantities in general relativity. J Math Phys 20:793
506. Gourgoulhon E, Bonazzola S (1994) A formulation of the virial theorem in general relativity. Class Quantum Gravit 11:443
507. Cook GB, Pfeiffer HP (2004) Excision boundary conditions for black-hole initial data. Phys Rev D 70:104016. arXiv:gr-qc/0407078
508. Ansorg M (2005) Double-domain spectral method for black hole excision data. Phys Rev D 72:024018. arXiv:gr-qc/0505059
509. Caudill M, Cook GB, Grigsby JD, Pfeiffer HP (2006) Circular orbits and spin in black-hole initial data. Phys Rev D 74:064011. arXiv:gr-qc/0605053
510. Ansorg M (2007) A multi-domain spectral method for initial data of arbitrary binaries in general relativity. Class Quantum Gravit 24:S1. arXiv:gr-qc/0612081
511. Katz J (1985) A note on Komar's anomalous factor. Class Quantum Gravit 2:423
512. Gourgoulhon E (2007) 3+1 formalism and bases of numerical relativity. Lecture notes
513. Rüdiger R (1981) Conserved quantities of spinning test particles in general relativity. I. Proc R Soc Lond A Math Phys Sci 375(1761):185–193
514. Rüdiger R (1983) Conserved quantities of spinning test particles in general relativity. II. Proc R Soc Lond A Math Phys Sci 385(1788):229–239
515. Dolan SR, Nolan P, Ottewill AC, Warburton N, Wardell B (2015) Tidal invariants for compact binaries on quasicircular orbits. Phys Rev D 91:023009. arXiv:1406.4890 [gr-qc]
516. Damour T, Jaranowski P, Schäfer G (2008) Hamiltonian of two spinning compact bodies with next-to-leading order gravitational spin-orbit coupling. Phys Rev D 77:064032. arXiv:0711.1048 [gr-qc]
517. Damour T, Jaranowski P, Schäfer G (2008) Effective one body approach to the dynamics of two spinning black holes with next-to-leading order spin-orbit coupling. Phys Rev D Ser 78:024009. arXiv:0803.0915 [gr-qc]
518. Bohé A, Marsat S, Faye G, Blanchet L (2013) Next-to-next-to-leading order spin-orbit effects in the near-zone metric and precession equations of compact binaries. Class Quantum Gravit 30:075017. arXiv:1212.5520 [gr-qc]
519. Harte AI (2007) Extended-body effects in cosmological spacetimes. Class Quantum Gravit 24(20):5161. arXiv:0706.2909 [gr-qc]
520. Dixon WG (1973) The definition of multipole moments for extended bodies. Gen Relat Gravit Ser 4:199
521. Porto RA (2006) Post-Newtonian corrections to the motion of spinning bodies in nonrelativistic general relativity. Phys Rev D 73:104031. arXiv:gr-qc/0511061

522. Porto RA, Rothstein IZ (2008) Next to leading order spin(1)spin(1) effects in the motion of inspiralling compact binaries. Phys Rev D 78:044013. arXiv:0804.0260 [gr-qc]. Erratum: Phys Rev D 81:029905(E) (2010)
523. Buonanno A, Faye G, Hinderer T (2013) Spin effects on gravitational waves from inspiraling compact binaries at second post-Newtonian order. Phys Rev D 87:044009. arXiv:1209.6349 [gr-qc]
524. Levi M, Steinhoff J (2015) Leading order finite size effects with spins for inspiralling compact binaries. JHEP 1506:059. arXiv:1410.2601 [gr-qc]
525. Levi M, Steinhoff J (2015) Spinning gravitating objects in the effective field theory in the post-Newtonian scheme. JCAP 1509:219. arXiv:1501.04956 [gr-qc]
526. Poisson E (1998) Gravitational waves from inspiraling compact binaries: the quadrupole-moment term. Phys Rev D 57:5287. arXiv:gr-qc/9709032
527. Laarakkers WG, Poisson E (1999) Quadrupole moments of rotating neutron stars. Astrophys J 512:282. arXiv:gr-qc/9709033
528. Pappas G, Apostolatos TA (2012) Revising the multipole moments of numerical spacetimes and its consequences. Phys Rev Lett 108:231104. arXiv:1201.6067 [gr-qc]
529. Goldberger WD, Rothstein IZ (2006) Dissipative effects in the worldline approach to black hole dynamics. Phys Rev D 73:104030. arXiv:hep-th/0511133
530. Damour T, Nagar A (2010) Effective one body description of tidal effects in inspiralling compact binaries. Phys Rev D 81:084016. arXiv:0911.5041 [gr-qc]
531. Bini D, Damour T, Faye G (2012) Effective action approach to higher-order relativistic tidal interactions in binary systems and their effective one-body description. Phys Rev D Ser 85:124034. arXiv:1202.3565 [gr-qc]
532. Steinhoff J, Puetzfeld D (2012) Influence of internal structure on the motion of test bodies in extreme mass ratio situations. Phys Rev D 86:044033. arXiv:1205.3926 [gr-qc]
533. Damour T, Nagar A (2009) Relativistic tidal properties of neutron stars. Phys Rev D 80:084035. arXiv:0906.0096 [gr-qc]
534. Binnington T, Poisson E (2009) Relativistic theory of tidal Love numbers. Phys Rev D 80:084018. arXiv:0906.1366 [gr-qc]
535. Kol B, Smolkin M (2012) Black hole stereotyping: induced gravito-static polarization. JHEP 1202:010. arXiv:1110.3764 [hep-th]
536. Gürlebeck N (2015) No-hair theorem for black holes in astrophysical environments. Phys Rev Lett 114:151102. arXiv:1503.03240 [gr-qc]
537. Le Tiec A, Casals M, Franzin E (2020) Tidal Love numbers of kerr black holes. Phys Rev D 103:084021. arXiv:2010.15795 [gr-qc]
538. Le Tiec A, Casals M (2021) Spinning black holes fall in Love. Phys Rev Lett 126:131102. arXiv:2007.00214 [gr-qc]
539. Landry P, Poisson E (2015) Tidal deformation of a slowly rotating material body: external metric. Phys Rev D 91:104018. arXiv:1503.07366 [gr-qc]
540. Landry P, Poisson E (2015) Gravitomagnetic response of an irrotational body to an applied tidal field. Phys Rev D 91:104026. arXiv:1504.06606 [gr-qc]
541. Pani P, Gualtieri L, Maselli A, Ferrari V (2015) Tidal deformations of a spinning compact object. Phys Rev D 92:024010. arXiv:1503.07365 [gr-qc]
542. Pani P, Gualtieri L, Ferrari V (2015) Tidal Love numbers of a slowly spinning neutron star. Phys Rev D 92:124003. arXiv:1509.02171 [gr-qc]
543. Landry P (2017) Tidal deformation of a slowly rotating material body: Interior metric and Love numbers. Phys Rev D 95:124058. arXiv:1703.08168 [gr-qc]
544. Chakrabarti S, Delsate T, Steinhoff J (2013) New perspectives on neutron star and black hole spectroscopy and dynamic tides. arXiv:1304.2228 [gr-qc]
545. Steinhoff J, Hinderer T, Buonanno A, Taracchini A (2016) Dynamical tides in general relativity: Effective action and effective-one-body Hamiltonian. Phys Rev D 94:104028. arXiv:1608.01907 [gr-qc]
546. Katz J (1972) A note on energy extremal properties for rotating stars in general relativity. J Phys A: Gen Phys 5(6):781

547. Bini D, Damour T (2014) Gravitational self-force corrections to two-body tidal interactions and the effective one-body formalism. Phys Rev D 90:124037. arXiv:1409.6933 [gr-qc]
548. Shah AG, Pound A (2015) Linear-in-mass-ratio contribution to spin precession and tidal invariants in Schwarzschild spacetime at very high post-Newtonian order. Phys Rev D 91:124022. arXiv:1503.02414 [gr-qc]
549. Einstein A (1921) Geometry and experience. Springer, Berlin
550. Weinstein A (1977) Lectures on symplectic manifolds, vol 29. American Mathematical Society
551. Souriau J-M (1986) La structure symplectique de la mécanique décrite par Lagrange en 1811. Mathématiques et sciences humaines 94:45–54
552. Howe R (1988) The classical groups and invariants of binary forms. Math Herit Hermann Weyl 48:133–166
553. Arnold V (2005) Lobachevsky triangle altitudes theorem as the Jacobi identity in the lie algebra of quadratic forms on symplectic plane. J Geom Phys 53(4):421–427
554. Ivanov NV (2011) Arnol'd, the jacobi identity, and orthocenters. Am Math Mon 118(1):41–65. https://doi.org/10.4169/amer.math.monthly.118.01.041
555. Braginsky VB, Polnarev AG, Thorne KS (1984) Foucault pendulum at the south pole: proposal for an experiment to detect the earth's general relativistic gravitomagnetic field. Phys Rev Lett 53(9):863. https://doi.org/10.1103/PhysRevLett.53.863
556. Krivoruchenko MI (2009) Rotation of the swing plane of Foucault's pendulum and thomas spin precession: two sides of one coin. Phys-Uspekhi 52(8):821. https://doi.org/10.3367/UFNe.0179.200908e.0873, arXiv:0805.1136 [nucl-th]
557. Sommeria J (2017) Foucault and the rotation of the earth. Comptes Rendus Physique 18(9–10):520–525. https://doi.org/10.1016/j.crhy.2017.11.003
558. Defossez L (1946) Les savants du XVIIe siècle et la mesure du temps. Lausanne
559. Vardi I, Henein S, À la recherche du temps précis: la découverte de l'oscillateur
560. Mohazzabi P, Shankar SP et al (2017) Damping of a simple pendulum due to drag on its string. J Appl Math Phys 5(01):122. https://doi.org/10.4236/jamp.2017.51013
561. Drake S (1973) Galileo's discovery of the law of free fall. Sci Am 228(5):84–92. https://doi.org/10.1038/scientificamerican0573-84
562. Ariotti P (1968) Galileo on the isochrony of the pendulum. ISIS 59(4):414–426
563. Galileo G (1639) Nouvelles pensées de Galilée, mathématicien et ingénieur du Duc de Florence (Translated and commented in French by M. Mersenne). Paris
564. Palmieri P (2009) A phenomenology of Galileo's experiments with pendulums. Br J Hist Sci 42(4):479–513
565. Gould RT (1989) The marine chronometer: its history and development. Woodbridge
566. Nelson RA, Olsson M (1986) The pendulum-rich physics from a simple system. Am J Phys 54(2):112–121. https://doi.org/10.1119/1.14703
567. Carvalhaes CG, Suppes P (2008) Approximations for the period of the simple pendulum based on the arithmetic-geometric mean. Am J Phys 76(12):1150–1154
568. Yoder JG (2004) Unrolling time: Christiaan Huygens and the mathematization of nature. Cambridge University Press
569. Taton R (1970) Pascal Blaise. Dict Sci Biogr 10:330–342
570. Mahoney MS (2000) Huygens and the pendulum: from device to mathematical relation. In: The growth of mathematical knowledge. Springer, pp 17–39
571. Terra P, de Melo e Souza R, Farina C (2016) Is the tautochrone curve unique? Am J Phys 84(12):917–923. arXiv:1610.01006 [physics.class-ph]
572. Martin J (2010) The Helen of geometry. Coll Math J 41(1):17–28
573. Hénon M (1976) A two-dimensional mapping with a strange attractor. In: The theory of chaotic attractors. Springer, pp 94–102
574. Alimi J-M, Mohayaee R, Perez J (2016) Une vie dédié aux systèmes dynamiques. In: Michel Hénon memoriam. Hermann
575. Herschel W (1789) XX. Catalogue of a second thousand of new nebulæ and clusters of stars; with a few introductory remarks on the construction of the heavens. Philos Trans R Soc Lond 79:212–255

576. Kulkarni S, Hut P, McMillan SJ (1993) Stellar black holes in globular clusters. Nature 364(6436):421–423
577. Coleman Miller M, Hamilton DP (2002) Production of intermediate-mass black holes in globular clusters. Mon Not R Astron Soc 330(1):232–240
578. Morscher M, Pattabiraman B, Rodriguez C, Rasio FA, Umbreit S (2015) The dynamical evolution of stellar black holes in globular clusters. Astrophys J Ser 800(1):9
579. Harris GL, Poole GB, Harris WE (2014) Globular clusters and supermassive black holes in galaxies: further analysis and a larger sample. Mon Not R Astron Soc Ser 438(3):2117–2130
580. Pancino E, Bellazzini M, Giuffrida G, Marinoni S (2017) Globular clusters with Gaia. Mon Not R Astron Soc 467(1):412–427
581. Baumgardt H, Hilker M, Sollima A, Bellini A (2019) Mean proper motions, space orbits, and velocity dispersion profiles of galactic globular clusters derived from Gaia DR2 data. Mon Not R Astron Soc 482(4):5138–5155
582. Vasiliev E (2019) Proper motions and dynamics of the milky way globular cluster system from Gaia DR2. Mon Not R Astron Soc 484(2):2832–2850
583. Hénon M (1959) L'amas isochrone: I. Ann d'Astrophysique 22:126–139
584. Hénon M (1959) L'amas isochrone: II. Le calcul des orbites. Ann d'Astrophysique 22:491–498
585. Hénon M (1960) L'amas isochrone: III. Fonction de distribution. Ann d'Astrophysique 23:474–477
586. Binney J (2014) Hénon's isochrone model. arXiv:1411.4937
587. McGill C, Binney J (1990) Torus construction in general gravitational potentials. Mon Not R Astron Soc 244:634–645
588. Simon-Petit A, Perez J, Plum G (2019) The status of isochrony in the formation and evolution of self-gravitating systems. Mon Not R Astron Soc 484(4):4963–4971. arXiv:1902.01095
589. Simon-Petit A, Perez J, Duval G (2018) Isochrony in 3D radial potentials. From Michel Hénon's ideas to isochrone relativity: classification, interpretation and applications. Commun Math Phys 363:605–653. https://doi.org/10.1007/s00220-018-3212-y, arXiv:1804.11282
590. Lynden-Bell D (2014) Variations on the theme of michel Hénon's isochrone. arXiv:1411.4926
591. Binney J, Tremaine S (2008) Galactic dynamics, Second edn. Princeton University Press, Princeton. https://doi.org/10.1515/9781400828722
592. Evans N, De Zeeuw P, Lynden-Bell D (1990) The flattened isochrone. Mon Not R Astron Soc 244:111–129
593. Binney J (2014) Self-consistent flattened isochrones. Mon Not R Astron Soc 440(1):787–798
594. Chandrasekhar S (1943) Stochastic problems in physics and astronomy. Rev Mod Phys 15(1):1
595. Santos F, Soares V, Tort A (2009) Determination of the apsidal angles and Bertrand's theorem. Phys Rev E Ser 79(3):036605. arXiv:0809.2069
596. Heath T (2002) The works of Archimedes. Dover Publications Inc., New-York
597. Stein S (1999) Archimedes: what did he do beside cry eureka? The Mathematical Association of America, Washington
598. Bényi À, Szeptycki P, Vleck FV (2003) A generalized Archimedean property. Real Anal Exch 29(2):881–889
599. Poincaré H (1892–99) Les méthodes nouvelles de la mécanique céleste, Tome 1-3. Gauthiers-Vilard, Paris
600. Chenciner A (2012) Poincaré and the three-body problem. Séminaire Poincaré XV I:45–133
601. Borisovich SM (2003) The classical KAM theory at the dawn of the twenty-first century. Moscow Math J 3(3):1113–1144. https://doi.org/10.17323/1609-4514-2003-3-3-1113-1144
602. Boccaletti D, Pucacco G (2003) Theory of orbits 1: integrable systems and non perturbative methods. Springer - Astronomy & Astrophysics Library, Heidelberg. https://doi.org/10.1007/978-3-662-03319-7
603. Chenciner A (2015) Poincaré and the three-body problem. In: Henri Poincaré, 1912–2012. Springer, pp 51–149
604. Féjoz J (2013) On action-angle coordinates and the Poincaré coordinates. Regul Chaotic Dyn 18(6):703–718. https://doi.org/10.1134/S1560354713060105

605. Hofer H, Zehnder E (2012) Symplectic invariants and Hamiltonian dynamics. Birkhäuser
606. Boccaletti D, Pucacco G (2004) Theory of orbits: volume 2: perturbative and geometrical methods. Springer - Astronomy & Astrophysics Library, Heidelberg. https://doi.org/10.1007/978-3-662-09240-8
607. Pinzari G (2013) Aspects of the planetary Birkhoff normal form. Regul Chaotic Dyn 18(6):860–906. https://doi.org/10.1134/s1560354713060178, arXiv:1310.0181
608. Grébert B (2007) Birkhoff normal form and Hamiltonian PDEs
609. Féjoz J, Kaczmarek L (2004) Sur le théorème de Bertrand (d'après Michael Herman). Ergod Theory Dyn Syst 24(5):1583–1589. https://doi.org/10.1017/S0143385704000434
610. Féjoz J (2004) Démonstration du théorème d'Arnold sur la stabilité du système planétaire (d'après Herman). Ergod Theory Dyn Syst 24(5):1521–1582. https://doi.org/10.1017/S0143385704000410
611. Chierchia L, Pinzari G (2011) Planetary Birkhoff normal forms. J Mod Dyn 5(4):623–664. https://doi.org/10.3934/jmd.2011.5.623
612. Chenciner A (1984) La dynamique au voisinage d'un point fixe elliptique conservatif: de Poincaré et Birkhoff à Aubry et Mather. Éditeur inconnu
613. Birkhoff GD (1927) Stability and the equations of dynamics. Am J Math 49(1):1–38. https://doi.org/10.2307/2370769, http://www.jstor.org/stable/2370769
614. Antonowicz M (1981) On the freedom of choice of the action-angle variables for Hamiltonian systems. J Phys A: Math Gen 14(5):1099–1106. https://doi.org/10.1088/0305-4470/14/5/029
615. Iglesias P (1998) Les origines du calcul symplectique chez Lagrange. Enseignement Mathématique 44:257–278
616. Iglesias P, Le mystère de la lettre H. http://math.huji.ac.il/~piz/documents/LMLH.pdf
617. Perez J (2014) La clé du mystère de la lettre H? Images des Mathématiques. http://images.math.cnrs.fr/La-cle-du-mystere-de-la-lettre-H.html
618. Henry Q, Faye G, Blanchet L (2020) Tidal effects in the equations of motion of compact binary systems to next-to-next-to-leading post-Newtonian order. Phys Rev D 101(6):064047. https://doi.org/10.1103/PhysRevD.101.064047, arXiv:1912.01920 [gr-qc]
619. Chatziioannou K (2020) Neutron-star tidal deformability and equation-of-state constraints. Gen Relat Gravit 52(11):109. https://doi.org/10.1007/s10714-020-02754-3, arXiv:2006.03168 [gr-qc]
620. Dietrich T, Hinderer T, Samajdar A, Interpreting binary neutron star mergers: describing the binary neutron star dynamics, modelling gravitational waveforms, and analyzing detections. arXiv:2004.02527 [gr-qc]
621. Most ER, Weih LR, Rezzolla L, Schaffner-Bielich J (2018) New constraints on radii and tidal deformabilities of neutron stars from GW170817. Phys Rev Lett Ser 120:261103. arXiv:1803.00549 [gr-qc]
622. Abbott BP et al (LIGO Scientific Collaboration and Virgo Collaboration) (2018) GW170817: measurements of neutron star radii and equation of state. Phys Rev Lett 121:161101. arXiv:1805.11581 [gr-qc]
623. MacDuffee C (1926) Euclidean invariants of second degree curves. Am Math Mon 33(5):243–252
624. Rowe DE (2002) Einstein's gravitational field equations and the Bianchi identities. Math Intell 24(4):57–66
625. Vermeil H (1917) Notiz über das mittlere krümmungsmaß einer n-fach ausgedehnten riemann'schen mannigfaltigkeit. Nachrichten von der Gesellschaft der Wissenschaften zu Göttingen, Mathematisch-Physikalische Klasse 1917(3):334–344
626. Renn J, Schemmel M (2007) Gravitation in the twilight of classical physics: an introduction. In: The genesis of general relativity. Springer, pp 926–944
627. Weyl H (1922) Space, time, matter. Dutton
628. Cartan E (1922) Sur les équations de la gravitation d'Einstein. J Mathématiques pures et appliquées 1:141–204
629. Lovelock D (1970) Divergence-free tensorial concomitants. Aequationes Math 4(1):127–138
630. Lovelock D (1971) The Einstein tensor and its generalizations. J Math Phys 12(3):498–501

631. Lovelock D (1972) The four-dimensionality of space and the Einstein tensor. J Math Phys 13(6):874–876
632. Navarro A, Navarro J (2011) Lovelock's theorem revisited. J Geom Phys 61(10):1950–1956. arXiv:1005.2386 [math-ph]
633. Synge JL (1960) Relativity: the general theory. Interscience Publishers, New York
634. Martelli B (2016) An introduction to geometric topology. CreateSpace Independent Publishing Platform. arXiv:1610.02592
635. Stillwell J (2010) Mathematics and its history. Springer, New York
636. Sesiano J (1999) Une introduction à l'histoire de l'algèbre: Résolution des équations des Mésopotamiens à la Renaissance (transl. by A. Pierrehumbert). Presses Polytechniques et Universitaires Romandes
637. Descartes R (1637) La Géométrie. https://www.gutenberg.org/ebooks/26400
638. Mazur J (2014) Enlightening symbols. Princeton Univeristy Press, Princeton
639. Lagrange J-L (1770, 1771) Réflexions sur la résolution algébrique des équations. http://gallica.bnf.fr/ark:/12148/bpt6k229222d/f206
640. Ruffini P (1799) Teoria generale delle equazioni in cui si dimostra impossibile la soluzione algebraica delle equazioni generale di grade superiore al quarto. https://archive.org/details/bub_gb_lNU2AAAAMAAJ/page/n13/mode/2up
641. Abel NH (1824) Mémoire sur les équations algébriques, où l'on démontre l'impossibilité de la résolution de l'équation générale du cinquième degré. https://www.abelprisen.no/nedlastning/verker/oeuvres_1881_del1/oeuvres_completes_de_abel_nouv_ed_1_kap03_opt.pdf
642. Abel NH (1826) Démonstration de l'impossibilité de la résolution algébrique des équations générales qui passent le quatrième degré. https://www.abelprisen.no/nedlastning/verker/oeuvres_1839/oeuvres_completes_de_abel_1_kap02_opt.pdf
643. Galois É (1830) Analyse d'un mémoire sur la résolution algébrique des équations. Bull des Sci Math 271. https://ia802707.us.archive.org/31/items/uvresmathmatiqu00frangoog/uvresmathmatiqu00frangoog.pdf
644. Online youtube video. Katz' personal website
645. Spearman BK, Williams KS (1994) Characterization of solvable quintics $x^5 + ax + b$. Am Math Mon Ser 101(10):986–992
646. Goldmakher L, Arnold's elementary proof of the unsolvability of the quintic. https://web.williams.edu/Mathematics/lg5/394/ArnoldQuintic.pdf
647. Akalin F, Why is the quintic unsolvable? https://www.akalin.com/quintic-unsolvability
648. Zoladek H, The topological proof of Abel-Ruffini's theorem. https://www.tmna.ncu.pl/static/files/v16n2-02.pdf
649. Alekseev V (2004) Abel's theorem in problems and solutions. Kluwer Academic Publishers